I0042102

Classical Mechanics

Classical Mechanics is a textbook for undergraduate students majoring in Physics (or Mathematics and Physics). The book introduces the main ideas and concepts of Newtonian, Lagrangian, and Hamiltonian mechanics, including the basics of rigid body motion and relativistic dynamics, at an intermediate to advanced level. The physical prerequisites are minimal, with a short primer included in the first chapter. As to the mathematical prerequisites, only a working knowledge of linear algebra, basic multivariate calculus, and the rudiments of ordinary differential equations is expected.

Features

- Numerous exercises and examples
- A focus on mathematical rigor that will appeal to Physics students wanting to specialize in theoretical physics, or Mathematics students interested in mathematical physics
- Sufficient material to service either a one- or two-semester course.

Artemio González-López (Madrid, 1959) is a theoretical physicist specializing in mathematical physics, particularly classical and quantum integrability. He earned degrees in Physics (1982) and Mathematics (1995) at Universidad Complutense de Madrid (UCM), and a Ph.D. in Physics (1984) at UCM. A Fulbright Scholar at Princeton University (1985–1987), he has taught at the Universities of Minnesota and Waterloo and at UCM, where he has been a Full Professor since 2009. Author of over 90 scientific papers, he has supervised 4 Ph.D. theses and led 10 research projects. He is also an editor of several scientific journals and heads UCM's Mathematical Physics group.

Classical Mechanics

Artemio González-López
Universidad Complutense de Madrid

CRC Press
Taylor & Francis Group
Boca Raton London New York

CRC Press is an imprint of the
Taylor & Francis Group, an **informa** business
A CHAPMAN & HALL BOOK

First edition published 2026
by CRC Press
2385 NW Executive Center Drive, Suite 320, Boca Raton FL 33431

and by CRC Press
4 Park Square, Milton Park, Abingdon, Oxon, OX14 4RN

CRC Press is an imprint of Taylor & Francis Group, LLC

ISBN: 978-1-032-98796-5 (hbk)
ISBN: 978-1-032-98797-2 (pbk)
ISBN: 978-1-003-60063-3 (ebk)

DOI: 10.1201/9781003600633

Typeset in TeXGyreTermesX font
by KnowledgeWorks Global Ltd.

Publisher's note: This book has been prepared from camera-ready copy provided by the authors.

Dedicated to the loving memory of my parents,
Artemio and Matilde,
to whom I owe everything I am,
and to my wife Debbie, the light of my life.

Contents

Preface

This book originated from the lecture notes for a course on Classical Mechanics that I taught over several years at the Faculty of Physics of Madrid's Complutense University. The course was intended for second-year Physics students, many of whom were also pursuing a double major in Mathematics and Physics. From the outset, the notes were very well received by both students and colleagues teaching parallel sections, and have since been included in the recommended bibliography of the official course guide published annually by the Faculty.

The aim of this book is to provide undergraduate students majoring in Physics, or pursuing a double major in Mathematics and Physics, with a solid understanding of the main ideas and concepts of Newtonian, Lagrangian, and Hamiltonian mechanics, including the fundamentals of rigid body motion and relativistic dynamics, at an intermediate to advanced level. More specifically, the book aims to bridge the gap between elementary textbooks such as Thornton and Marion [57], and advanced undergraduate (or even graduate-level) texts like Goldstein [24], Goldstein, Poole and Safko [25], Landau [37], José and Saletan [33], Gantmacher [22], Arnold [2], and others.

The physical prerequisites are minimal—essentially the material usually covered in a freshman general physics course—and are reviewed and further developed in Chapter 2. As for the mathematical background, only a working knowledge of linear algebra, basic multivariable calculus, and the rudiments of ordinary differential equations is assumed; in fact, some of this material is briefly summarized in Chapter 1.

As should be clear from the preceding remarks, the intended audience includes not only Physics majors but also Mathematics majors with an interest in Physics. In fact, I believe the book will be especially appealing to Physics students planning to specialize in theoretical physics, as well as to Mathematics students interested in mathematical physics. With this audience in mind, I have sought to go beyond most elementary textbooks both in the scope of the material (stressing connections with other areas of physics and mathematics, and introducing more advanced applications whenever possible) and the consistency and rigor of the mathematical developments, making full use of the mathematics that second-year Physics students are expected to know. Topics treated in detail, which are either missing or not covered in depth in comparable textbooks, include:

- General treatment of orthogonal curvilinear coordinates (Chapter 2).
- Derivation of the equations of general Galilean transformations (Chapter 2).

- Introduction of Jacobian elliptic functions through the motion of a particle in a suitable one-dimensional potential (Chapter 2).
- Detailed calculation of the precession of Mercury's perihelion in general relativity (Chapter 3).
- Derivation of the scattering cross section of an attractive central potential (Chapter 3).
- Proof of Noether's theorem for general space-time symmetries, with numerous illustrative examples (Chapter 4).
- Deduction of the *eikonal* equation in geometric optics from Fermat's principle (Chapter 4).
- General treatment of small oscillations in Lagrangian mechanics (Chapter 5).
- Rotations and the rotation group, as the foundation for the description of motion in non-inertial frames (Chapter 6) and the development of rigid body dynamics (Chapter 7).
- Explicit approximate solution of the equations of motion of a particle near Earth's surface, and its application to the solution of many standard problems (Chapter 6).
- Elementary yet mathematically rigorous proof of the existence of body axes in an arbitrary rigid body (Chapter 7).
- Exact solution of Euler's equations for the inertial motion of an asymmetric top derived in an elementary fashion (Chapter 7)
- Derivation of the Lorentz force and the transformation properties of the electromagnetic field in special relativity (Chapter 8).
- Derivation of the Liénard–Wiechert potentials of a moving charge from the principles of special relativity (Chapter 8).

The book is organized into eight chapters. The first three chapters review and further develop the fundamental concepts of Newtonian mechanics. In particular, Chapter 3 includes a detailed study of the motion of a particle in a central potential, with applications to planetary motion and an introduction to the basic ideas of scattering theory. Chapter 4 develops in detail the Lagrangian formulation of mechanics and provides a brief introduction to Hamiltonian mechanics. These ideas are applied in Chapter 5 to the analysis of small oscillations about a stable equilibrium in a general mechanical system. Chapter 6 is devoted to the description of motion in a non-inertial reference frame, making extensive use of the fundamental properties of the three-dimensional rotation group. As a standard application, the approximate motion of a particle near Earth's surface is studied in detail, including the solution of Foucault's pendulum. Rigid body motion is extensively analyzed in Chapter 7, using both the Newtonian and Lagrangian formalisms. Core topics include the inertia tensor, Euler's equations, the inertial motion of a symmetric top, and Lagrange's top. The chapter also presents more advanced material, such as the explicit solution of the equations of motion of an asymmetric top, the definition of the Euler angles, and their use in reconstructing the motion in the inertial frame from the solution of Euler's equations. The final Chapter 8 provides a broad introduction to the special theory of relativity and the fundamentals of relativistic dynamics. Key topics covered include Lorentz transformations and their physical implications, the Lorentz group and Minkowski spacetime, four-momentum conservation and relativistic collisions, the Doppler and Compton effects, hyperbolic motion, and the relativistic Lagrangian. Special emphasis is placed on the deep connection between special relativity and electrodynamics, which is often neglected in elementary treatments. Topics discussed in this context include the relativistic nature of the Lorentz force and the definition and covariance properties of the electromagnetic field tensor. The book concludes with an elementary derivation of the Liénard–Wiechert potentials of a moving charge using the general principles of special relativity.

To support and illustrate the theoretical material, the book includes 99 fully worked-out exercises, 35 in-depth examples, and 73 figures. The exercises are distributed throughout the text and are designed not merely as end-of-section problems, but as integral parts of the exposition. By embedding them alongside the main discussion, the book aims to reinforce key concepts and facilitate the student's understanding through direct application of the ideas presented.

As mentioned above, the book's primary audience are second-year students majoring in Physics and/or Mathematics. If covered in its entirety, the book is appropriate for a full-year (two-semester) course. It can also be readily adapted for a one-semester course by omitting, for instance, Section 3.4, parts of Section 4.5, Chapter 5, and Sections 7.6–7.7, 8.7, and 8.9–8.10. This condensed format is, in fact, the one currently followed at Madrid's Complutense University.

Acknowledgments

I would like to thank the many students at Madrid's Complutense University who, over the years, used the lecture notes on which this book is based. Their continued feedback and warm reception of the notes in their successive versions have been a constant source of motivation. I am also indebted to my colleagues in the Department of Theoretical Physics who have taught the Classical Mechanics course with me, especially José Ramón Peláez and Diego Rubiera, for their input and valuable suggestions. A special mention goes to Luis Manuel ("Manolo") González-Romero, who encouraged me to publish the lecture notes and with whom I have shared countless conversations that significantly improved the original and helped shape this book. I am deeply grateful as well to my colleagues and long-time friends Gabriel Álvarez, Federico Finkel, Miguel Ángel Rodríguez, and Piergiulio Tempesta, with whom I have shared many coffees and wide-ranging conversations about academic (and occasionally non-academic!) life, including the teaching of classical mechanics. It goes without saying that I owe a particular debt of gratitude to my professors during my formative years at Madrid's Complutense University, many of whom later became colleagues and even close friends. A special place among them is held by my dear friend, mentor, and fellow

Real Madrid fan Luis Martínez-Alonso, who has been a constant source of inspiration ever since he taught me a memorable course on functions of a complex variable in 1978–79.

Throughout the years I spent elaborating the lecture notes and writing this book, my immediate family—my beloved parents Artemio and Matilde (now deceased), and my brothers Javier, Alfonso, and Miguel Ángel—has been a steady source of support, helping me navigate the many ups and downs in this long process. But I would like to close this preface with a very special mention of my dearest wife, Debbie, who is—there is no better way of putting it—the light of my life. Thank you for your unwavering love and support, and for filling my life with warmth, beauty, tenderness, and fun!

Madrid, June 2025

CHAPTER 1

Preliminaries

A working knowledge of elementary linear algebra, multivariable calculus, and ordinary differential equations is essential to follow a course in classical mechanics, even at a basic level. In this chapter we briefly review several fundamental results and identities, usually covered in introductory courses in Linear Algebra and Calculus, which shall be extensively used throughout this text.

1.1 VECTORS IN $\mathbb{R}^3$

We shall mostly work with vectors in $\mathbb{R}^3$—the most notable exception being the last chapter on relativistic mechanics, which shall require the use of four-vectors. Throughout this book vectors will be typeset in *roman boldface,* like **a**, **b**, etc., instead of the notation $\vec{a}$, $\vec{b}$, etc., more commonly used when writing at the blackboard. Unless otherwise stated, the components of a vector **a** shall be denoted by a_i $(i = 1, 2, 3)$. We shall normally use the notation $|\mathbf{a}|$ for the length (or magnitude) $\sqrt{a_1^2 + a_2^2 + a_3^2}$ of the vector **a**, and $\mathbf{a} \cdot \mathbf{b}$ or simply **ab** to denote the scalar product $a_1b_1 + a_2b_2 + a_3b_3$ of the vectors **a** and **b**.

Vector product of two vectors $\mathbf{a}, \mathbf{b} \in \mathbb{R}^3$:

$$\mathbf{a} \times \mathbf{b} := \begin{vmatrix} \mathbf{i} & \mathbf{j} & \mathbf{k} \\ a_1 & a_2 & a_3 \\ b_1 & b_2 & b_3 \end{vmatrix}, \tag{1.1}$$

where

$$\mathbf{i} := \mathbf{e}_1 = (1,0,0), \quad \mathbf{j} := \mathbf{e}_2 = (0,1,0), \quad \mathbf{k} := \mathbf{e}_3 = (0,0,1)$$

are the unit vectors of the standard orthonormal frame in $\mathbb{R}^3$. In particular, note that

$$\mathbf{i} \times \mathbf{j} = \mathbf{k}, \qquad \mathbf{j} \times \mathbf{k} = \mathbf{i}, \qquad \mathbf{k} \times \mathbf{i} = \mathbf{j}. \tag{1.2}$$

DOI: 10.1201/9781003600633-1

The vector product is obviously *antisymmetric* ($\mathbf{a} \times \mathbf{b} = -\mathbf{b} \times \mathbf{a}$), so that $\mathbf{a} \times \mathbf{a} = 0$, and *distributive* in both of its arguments:

$$\mathbf{a} \times (\mathbf{b} + \mathbf{c}) = \mathbf{a} \times \mathbf{b} + \mathbf{a} \times \mathbf{c}, \qquad (\mathbf{a} + \mathbf{b}) \times \mathbf{c} = \mathbf{a} \times \mathbf{c} + \mathbf{b} \times \mathbf{c}$$

(note that the second distributive law actually follows from the first by antisymmetry).

Useful identities:

$$\mathbf{a} \times (\mathbf{b} \times \mathbf{c}) = (\mathbf{a} \cdot \mathbf{c})\mathbf{b} - (\mathbf{a} \cdot \mathbf{b})\mathbf{c} \tag{1.3}$$

$$(\mathbf{a} \times \mathbf{b})^2 = \mathbf{a}^2\mathbf{b}^2 - (\mathbf{a} \cdot \mathbf{b})^2 . \tag{1.4}$$

From the identity

$$\mathbf{a} \cdot \mathbf{b} = |\mathbf{a}||\mathbf{b}| \cos \theta_{\mathbf{ab}},$$

where $\theta_{\mathbf{ab}} \in [0, \pi]$ is the *angle* between the vectors $\mathbf{a}$ and $\mathbf{b}$, and Eq. (1.4), it follows that

$$|\mathbf{a} \times \mathbf{b}| = |\mathbf{a}||\mathbf{b}| \sin \theta_{\mathbf{ab}} .$$

Triple product of three vectors $\mathbf{a}, \mathbf{b}, \mathbf{c} \in \mathbb{R}^3$:

$$(\mathbf{a} \times \mathbf{b}) \cdot \mathbf{c} = \begin{vmatrix} a_1 & a_2 & a_3 \\ b_1 & b_2 & b_3 \\ c_1 & c_2 & c_3 \end{vmatrix} =: \det(\mathbf{a}, \mathbf{b}, \mathbf{c}). \tag{1.5}$$

In particular, from the properties of determinants it follows that $\mathbf{a} \times \mathbf{b}$ is *perpendicular* to both $\mathbf{a}$ and $\mathbf{b}$. The triple product is *invariant under cyclic permutations* of its arguments (again by the elementary properties of determinants):

$$(\mathbf{a} \times \mathbf{b}) \cdot \mathbf{c} = (\mathbf{b} \times \mathbf{c}) \cdot \mathbf{a} = (\mathbf{c} \times \mathbf{a}) \cdot \mathbf{b} .$$

Geometrically, $(\mathbf{a} \times \mathbf{b}) \cdot \mathbf{c}$ is equal to the volume of the parallelepiped spanned by the three vectors $\mathbf{a}, \mathbf{b}, \mathbf{c}$ if the frame $\{\mathbf{a}, \mathbf{b}, \mathbf{c}\}$ is *positively oriented,* and minus this volume otherwise.

The components of the vector product of two vectors can be concisely expressed in terms of the components of its factors by means of *Levi-Civita's completely antisymmetric tensor* ε_{ijk} with indices $i, j, k \in \{1, 2, 3\}$, defined by

$$\varepsilon_{ijk} = \begin{cases} 1\,, & (i, j, k) \text{ even permutation of } (1, 2, 3) \\ -1\,, & (i, j, k) \text{ odd permutation of } (1, 2, 3) \\ 0\,, & \text{otherwise.} \end{cases} \tag{1.6}$$

In other words, if i, j, k are all different then ε_{ijk} is equal to the *sign* $\operatorname{sgn}(i, j, k)$ of the permutation (i, j, k), and is zero otherwise. In particular, ε_{ijk} vanishes if two indices are equal, and is clearly antisymmetric under permutation of the indices i, j, k. It is then straightforward to check that

$$\mathbf{e}_i \times \mathbf{e}_j = \sum_{k=1}^{3} \varepsilon_{ijk} \mathbf{e}_k . \tag{1.7}$$

Indeed, if $i = j$ both sides of the previous equation vanish. On the other hand, if $i \neq j$, then by the definition of ε_{ijk} the only nonzero term in the sum (1.7) is the one with k different from both i and j, and in that case $\varepsilon_{ijk} = \operatorname{sgn}(i, j, k)$. In other words, Eq. (1.7) is equivalent to

$$\mathbf{e}_i \times \mathbf{e}_j = \operatorname{sgn}(i, j, k)\mathbf{e}_k, \tag{1.8}$$

with (i, j, k) a permutation of $(1, 2, 3)$, which is easily seen to hold on account of Eq. (1.2). From Eq. (1.7) it follows that we can express the vector product of two three-dimensional vectors $\mathbf{a}$ and $\mathbf{b}$ in terms of the Levi–Civita tensor as follows:

$$\begin{aligned}\mathbf{a} \times \mathbf{b} &= \left(\sum_{i=1}^{3} a_i \mathbf{e}_i\right) \times \left(\sum_{j=1}^{3} b_j \mathbf{e}_j\right) = \sum_{i,j=1}^{3} a_i b_j \mathbf{e}_i \times \mathbf{e}_j = \sum_{i,j=1}^{3} a_i b_j \sum_{k=1}^{3} \varepsilon_{ijk} \mathbf{e}_k \\ &\equiv \sum_{k=1}^{3} (\mathbf{a} \times \mathbf{b})_k \mathbf{e}_k\,,\end{aligned}$$

with

$$\boxed{(\mathbf{a} \times \mathbf{b})_k = \sum_{i,j=1}^{3} \varepsilon_{ijk} a_i b_j\,.} \tag{1.9}$$

Since $\varepsilon_{ijk} = \varepsilon_{kij}$, by cyclically permuting the indices i, j, k we can also write the previous formula as

$$\boxed{(\mathbf{a} \times \mathbf{b})_i = \sum_{j,k=1}^{3} \varepsilon_{ijk} a_j b_k\,.} \tag{1.10}$$

For instance,

$$(\mathbf{a} \times \mathbf{b})_1 = \varepsilon_{123} a_2 b_3 + \varepsilon_{132} a_3 b_2 = a_2 b_3 - a_3 b_2,$$

since $\operatorname{sgn}(1, 2, 3) = +1$ and $\operatorname{sgn}(1, 3, 2) = -1$.

Remark 1.1. In general, if three vectors $\{\mathbf{n}_1, \mathbf{n}_2, \mathbf{n}_3\}$ in $\mathbb{R}^3$ form an orthonormal frame—i.e., if $\mathbf{n}_i \cdot \mathbf{n}_j = \delta_{ij}$ for $i, j \in \{1, 2, 3\}$—then $\mathbf{n}_1 \times \mathbf{n}_2 = \pm\mathbf{n}_3$, and therefore

$$(\mathbf{n}_1 \times \mathbf{n}_2) \cdot \mathbf{n}_3 = (\pm\mathbf{n}_3) \cdot \mathbf{n}_3 = \pm 1. \tag{1.11}$$

We say that the orthonormal frame $\{\mathbf{n}_1, \mathbf{n}_2, \mathbf{n}_3\}$ is *positively oriented* if

$$\mathbf{n}_1 \times \mathbf{n}_2 = \mathbf{n}_3,$$

or equivalently, by Eq. (1.11), if

$$(\mathbf{n}_1 \times \mathbf{n}_2) \cdot \mathbf{n}_3 = \det(\mathbf{n}_1, \mathbf{n}_2, \mathbf{n}_3) = 1.$$

If $\{\mathbf{n}_1, \mathbf{n}_2, \mathbf{n}_3\}$ is a positively oriented orthonormal frame, for all $i \neq j$ we have

$$\mathbf{n}_i \times \mathbf{n}_j = \operatorname{sgn}(i, j, k)\mathbf{n}_k = \varepsilon_{ijk} \mathbf{n}_k\,, \tag{1.12}$$

where (i, j, k) is a permutation of $(1,2,3)$. Indeed, if $i \neq j$ then $\mathbf{n}_i \times \mathbf{n}_j = \varepsilon(i,j,k)\mathbf{n}_k$, with i, j, and k different from one another and $\varepsilon(i,j,k) = \pm 1$. To determine the sign $\varepsilon(i,j,k)$, we take the scalar product of $\mathbf{n}_i \times \mathbf{n}_j$ with $\mathbf{n}_k$ and use the elementary properties of determinants to obtain

$$\begin{aligned}\varepsilon(i,j,k) &= (\mathbf{n}_i \times \mathbf{n}_j)\cdot \mathbf{n}_k = \det(\mathbf{n}_i, \mathbf{n}_j, \mathbf{n}_k) = \operatorname{sgn}(i,j,k)\det(\mathbf{n}_1, \mathbf{n}_2, \mathbf{n}_3)\\ &= \operatorname{sgn}(i,j,k),\end{aligned}$$

since the frame $\{\mathbf{n}_1, \mathbf{n}_2, \mathbf{n}_3\}$ is positively oriented by hypothesis. ■

1.2 VECTOR CALCULUS

A *scalar function* (also called a *scalar field*) is a mapping $f : \mathbb{R}^3 \to \mathbb{R}$. Given a scalar function $f(x_1, x_2, x_3)$ of class C^1 (i.e., with continuous partial derivatives), we define its **gradient** ∇f by

$$\nabla f := \frac{\partial f}{\partial x_1}\mathbf{i} + \frac{\partial f}{\partial x_2}\mathbf{j} + \frac{\partial f}{\partial x_3}\mathbf{k}. \tag{1.13}$$

Note that ∇f is a *vector field*, i.e., a mapping from $\mathbb{R}^3$ to $\mathbb{R}^3$. In other words, $(\nabla f)(x_1, x_2, x_3)$ is a *vector* at each point (x_1, x_2, x_3). Given a vector field $\mathbf{A} = (A_1, A_2, A_3) : \mathbb{R}^3 \to \mathbb{R}^3$ of class C^1, we define its **divergence** $\nabla \cdot \mathbf{A}$ as the *scalar function*

$$\nabla \cdot \mathbf{A} := \frac{\partial A_1}{\partial x_1} + \frac{\partial A_2}{\partial x_2} + \frac{\partial A_3}{\partial x_3}. \tag{1.14}$$

The gradient and the divergence can be easily generalized to scalar functions and vector fields in $\mathbb{R}^n$ with $n > 3$. For instance, the gradient of a scalar function $f(x_1, \dots, x_n)$ (that shall be needed in Chapter 4) is defined by

$$\nabla f(x_1, \dots, x_n) = \sum_{i=1}^{n} \frac{\partial f}{\partial x_i}\mathbf{e}_i,$$

where $\mathbf{e}_i = (0, \dots, 0, \underset{\substack{\downarrow \\ i}}{1}, 0, \dots, 0)$ is the i-th vector of the canonical basis of $\mathbb{R}^n$. There is, however, a differential operator which can only be defined in three dimensions, namely the **curl** $\nabla \times \mathbf{A}$ of a vector field (of class C^1) $\mathbf{A}(x_1, x_2, x_3)$ in $\mathbb{R}^3$:

$$\begin{aligned}\nabla \times \mathbf{A} :&= \left(\frac{\partial A_3}{\partial x_2} - \frac{\partial A_2}{\partial x_3}\right)\mathbf{i} + \left(\frac{\partial A_1}{\partial x_3} - \frac{\partial A_3}{\partial x_1}\right)\mathbf{j} + \left(\frac{\partial A_2}{\partial x_1} - \frac{\partial A_1}{\partial x_2}\right)\mathbf{k}\\ &\equiv \begin{vmatrix} \mathbf{i} & \mathbf{j} & \mathbf{k} \\ \dfrac{\partial}{\partial x_1} & \dfrac{\partial}{\partial x_2} & \dfrac{\partial}{\partial x_3} \\ A_1 & A_2 & A_3 \end{vmatrix}.\end{aligned} \tag{1.15}$$

Note that Eq. (1.10) can be applied to obtain the following concise formula for the components of $\nabla \times \mathbf{A}$:

$$\boxed{\begin{aligned}&(\nabla \times \mathbf{A})_i = \sum_{j,k=1}^{3} \varepsilon_{ijk} \frac{\partial A_k}{\partial x_j} = \frac{\partial A_k}{\partial x_j} - \frac{\partial A_j}{\partial x_k}\,,\\ &(i,j,k) = \text{cyclic permutation of } (1,2,3)\,.\end{aligned}} \tag{1.16}$$

Another important differential operator is the **Laplacian** of a scalar function $f : \mathbb{R}^3 \to \mathbb{R}$ of class C^2, defined by

$$\nabla^2 f := \nabla \cdot (\nabla f) = \sum_{i=1}^{3} \frac{\partial^2 f}{\partial x_i^2}\,. \tag{1.17}$$

Note that $\nabla^2 f$ is again a *scalar function*. The Laplacian operator ∇^2 is often denoted as Δ.

Properties

- The vector $\nabla f(x_1, \ldots, x_n)$ is *orthogonal* at each point to the level (hyper)surfaces $f(x_1, \ldots, x_n) = c$ of the scalar function f (where c is a real constant). In other words, ∇f is orthogonal to the *tangent vectors* of curves lying on level surfaces of f.

- If $f(x_1, x_2, x_3)$ is *any* scalar function of class C^2 we have

$$\nabla \times (\nabla f) = 0. \tag{1.18}$$

  Conversely, if $\mathbf{A}(x_1, x_2, x_3)$ is a vector field of class C^1 on *all* of $\mathbb{R}^3$ such that

$$\nabla \times \mathbf{A} = 0,$$

  then $\mathbf{A} = \nabla f$ for some scalar function f of class C^2.

- Similarly, if $\mathbf{A}(x_1, x_2, x_3)$ is *any* vector field of class C^2 then

$$\nabla \cdot (\nabla \times \mathbf{A}) = 0. \tag{1.19}$$

  Conversely, if $\mathbf{B}(x_1, x_2, x_3)$ is a vector field of class C^1 on all of $\mathbb{R}^3$ such that

$$\nabla \cdot \mathbf{B} = 0$$

  then $\mathbf{B} = \nabla \times \mathbf{A}$ for some vector field $\mathbf{A}(x_1, x_2, x_3)$ of class C^2.

Some useful identities:

$$\nabla \cdot (f\mathbf{A}) = (\nabla f) \cdot \mathbf{A} + f \nabla \cdot \mathbf{A}, \tag{1.20}$$

$$\nabla \times (f\mathbf{A}) = (\nabla f) \times \mathbf{A} + f \nabla \times \mathbf{A}, \tag{1.21}$$

$$\nabla \times (\nabla \times \mathbf{A}) = \nabla(\nabla \cdot \mathbf{A}) - \nabla^2 \mathbf{A}. \tag{1.22}$$

In the first two identities f and $\mathbf{A}$ are respectively a scalar function and a vector field of class C^1 in $\mathbb{R}^3$, while in the third one $\mathbf{A}$ is a vector field of class C^2 and $\nabla^2\mathbf{A}$ is the vector field

$$\nabla^2\mathbf{A} := (\nabla^2 A_1)\mathbf{i} + (\nabla^2 A_2)\mathbf{j} + (\nabla^2 A_3)\mathbf{k}.$$

As we shall see in the next chapter, the identity (1.22) plays an important role in the formulation of electromagnetic theory.

Exercise 1.1. If $\mathbf{A}$ and $\mathbf{B}$ are C^1 vector fields in $\mathbb{R}^3$, express $\nabla\cdot(\mathbf{A}\times\mathbf{B})$ in terms of $\nabla\times\mathbf{A}$ and $\nabla\times\mathbf{B}$.

Solution. Using Eqs. (1.9)–(1.10) for the components of the cross product and Eq. (1.16) for the curl, we obtain

$$\begin{aligned}\nabla\cdot(\mathbf{A}\times\mathbf{B}) &= \sum_{i,j,k=1}^{3}\varepsilon_{ijk}\frac{\partial}{\partial x_k}(A_iB_j) = \sum_{i,j,k=1}^{3}\varepsilon_{ijk}\left(\frac{\partial A_i}{\partial x_k}B_j + A_i\frac{\partial B_j}{\partial x_k}\right)\\ &= \sum_{j=1}^{3}B_j\sum_{i,k=1}^{3}\varepsilon_{jki}\frac{\partial A_i}{\partial x_k} - \sum_{i=1}^{3}A_i\sum_{j,k=1}^{3}\varepsilon_{ikj}\frac{\partial B_j}{\partial x_k}\\ &= \sum_{j=1}^{3}(\nabla\times\mathbf{A})_jB_j - \sum_{i=1}^{3}A_i(\nabla\times\mathbf{B})_i = (\nabla\times\mathbf{A})\cdot\mathbf{B} - \mathbf{A}\cdot(\nabla\times\mathbf{B}).\end{aligned}$$

Note that in the third equality we have applied the identities $\varepsilon_{ijk} = \varepsilon_{jki} = -\varepsilon_{ikj}$, which are a consequence of the antisymmetry of the Levi-Civita tensor.

1.3 CHAIN RULE

A working knowledge of the chain rule is essential to perform even the most basic calculations in classical mechanics (or, as a matter of fact, in any branch of physics). From a formal point of view, the chain rule simply asserts that the derivative $D(f\circ g)$ of the *composition* $f\circ g$ of two differentiable functions $g:\mathbb{R}^n\to\mathbb{R}^m$ and $f:\mathbb{R}^m\to\mathbb{R}^p$ is the composition $Df\circ Dg$ of their derivatives. This abstract formulation yields, however, simple and intuitive formulas for the *partial derivatives* of the composition $f\circ g$, which are the ones we shall use in practice throughout the book.

Consider, for instance, a scalar function $f(x_1,\dots,x_n)$ of the n variables $(x_1,\dots,x_n)$, and suppose that each of these variables x_i is in turn a function of m real variables $(y_1,\dots,y_m)$. Then $f(x_1,\dots,x_n)$ is *implicitly* a function of the y_j's through the x_i's. In other words, when we write (for simplicity's sake) $f(x_1,\dots,x_n)$, what we really mean is $f\big(x_1(y_1,\dots,y_m),\dots,x_n(y_1,\dots,y_m)\big)$. The partial derivative of this function with respect to any of its *independent* variables y_k can be computed using the following variant of the chain rule:

$$\frac{\partial}{\partial y_k}f(x_1,\dots,x_n) = \sum_{i=1}^{n}\frac{\partial f}{\partial x_i}\frac{\partial x_i}{\partial y_k}\,. \tag{1.23}$$

This is a straightforward generalization of the well-known elementary formula for the $n = m = 1$ case, namely

$$\frac{\mathrm{d}}{\mathrm{d}y} f(x) = \frac{\mathrm{d}f}{\mathrm{d}x} \frac{\mathrm{d}x}{\mathrm{d}y}.$$

In classical mechanics (x_1, x_2, x_3) are usually the coordinates of the *position vector* $\mathbf{r}$ of a moving particle, so *they depend implicitly on the time* t. In this case, therefore, *any function* $f(x_1, x_2, x_3)$ *depends implicitly on time through the coordinates* x_i. In other words, $f(x_1, x_2, x_3)$ should be normally understood as shorthand for $f\big(x_1(t), x_2(t), x_3(t)\big)$. Applying Eq. (1.23) with $n = 3$, $m = 1$, and $y_k = y_1 \equiv t$, we obtain the important formula

$$\boxed{\frac{\mathrm{d}}{\mathrm{d}t} f(x_1, x_2, x_3) = \sum_{i=1}^{3} \frac{\partial f}{\partial x_i} \dot{x}_i,} \tag{1.24}$$

where (as is standard in classical mechanics) we have used Newton's notation

$$\boxed{\dot{x}_i \equiv \frac{\mathrm{d}x_i}{\mathrm{d}t}}$$

for the time derivative. Since $\mathbf{r} = (x_1, x_2, x_3)$, we can write $f(x_1, x_2, x_3) = f(\mathbf{r})$ and $\dot{\mathbf{r}} = (\dot{x}_1, \dot{x}_2, \dot{x}_3)$. It is also convenient to use the mnemonic notation $\frac{\partial f}{\partial \mathbf{r}}$ for the gradient of the scalar function $f(\mathbf{r})$, i.e.,

$$\frac{\partial f(\mathbf{r})}{\partial \mathbf{r}} \equiv \nabla f(x_1, x_2, x_3).$$

We can then rewrite the previous formula for $\frac{\mathrm{d}f}{\mathrm{d}t} \equiv \dot{f}$ in vector form as

$$\dot{f} = \frac{\partial f}{\partial \mathbf{r}} \cdot \dot{\mathbf{r}}. \tag{1.25}$$

Exercise 1.2. Compute the gradient of a scalar function $g(r)$ that depends on $\mathbf{r} = (x_1, x_2, x_3)$ only through $r := \sqrt{x_1^2 + x_2^2 + x_3^2}$.

Solution. Applying the chain rule we obtain

$$\frac{\partial g(r)}{\partial x_i} = g'(r)\frac{\partial r}{\partial x_i} = g'(r)\,\frac{x_i}{r}.$$

Hence

$$\nabla\big[g(r)\big] \equiv \frac{\partial g(r)}{\partial \mathbf{r}} = \sum_{i=1}^{3} \frac{\partial g(r)}{\partial x_i}\,\mathbf{e}_i = \frac{g'(r)}{r} \sum_{i=1}^{3} x_i \mathbf{e}_i = g'(r)\,\frac{\mathbf{r}}{r} \equiv g'(r)\mathbf{e}_r.$$

1.4 TOTAL AND PARTIAL TIME DERIVATIVES

Understanding the difference between total and partial time derivatives of a scalar function $f(t, x_1, x_2, x_3) \equiv f(t, \mathbf{r})$ is again crucial to follow even the simplest arguments in classical mechanics. To begin with, as usual the notation $\frac{\partial f}{\partial t}$ stands for the *partial derivative* of f with respect to t considering f as a function of the *independent* variables (t, x_1, x_2, x_3). For instance, if $f(t, \mathbf{r}) = c^2t^2 + \mathbf{r}^2$ (where c is a constant with the dimension of velocity), then

$$\frac{\partial f}{\partial t} = 2ct\,.$$

On the other hand, if—as is customary in classical mechanics—we consider each coordinate x_i as an *implicit* function of t (i.e., we write x_i as shorthand for $x_i(t)$), then $f(t, \mathbf{r})$ *becomes a function of* t *only,* through the explicit dependence of f on t *plus* the implicit dependence on t of $\mathbf{r} = (x_1, x_2, x_3)$. The **total time derivative** of f is the derivative with respect to t of $f(t, \mathbf{r})$ *considered as* $f\big(t, \mathbf{r}(t)\big)$. Applying again the chain rule (with $n = 4$, $m = 1$) we obtain

$$\boxed{\frac{\mathrm{d}}{\mathrm{d}t} f(t, \mathbf{r}) \equiv \dot{f} = \frac{\partial f}{\partial t}\frac{\mathrm{d}t}{\mathrm{d}t} + \sum_{i=1}^{3} \frac{\partial f}{\partial x_i}\dot{x}_i = \frac{\partial f}{\partial t} + \frac{\partial f}{\partial \mathbf{r}} \cdot \dot{\mathbf{r}}\,.} \tag{1.26}$$

This is of course a generalization of Eq. (1.25), to which it reduces when f does not depend *explicitly* on time. For instance, if $f(t, \mathbf{r}) = c^2t^2 + \mathbf{r}^2$ then

$$\frac{\mathrm{d}f}{\mathrm{d}t} \equiv \dot{f} = 2ct + \sum_{i=1}^{3} 2x_i\dot{x}_i = 2ct + 2\mathbf{r} \cdot \dot{\mathbf{r}} \neq \frac{\partial f}{\partial t}\,.$$

Note that, in general, $\dot{f}$ in Eq. (1.26) is a scalar function of the variables $(t, \mathbf{r}, \dot{\mathbf{r}})$, i.e., it depends not only on time and the coordinates of the particle, but also on its *velocity* $\dot{\mathbf{r}}$.

CHAPTER 2

Review of Newtonian mechanics

In this chapter we briefly review the fundamentals of elementary Newtonian mechanics. We start by discussing kinematics and describing orthogonal curvilinear coordinate systems, with special emphasis on spherical, cylindrical, and polar coordinates. Next, we introduce the key concept of inertial reference frame, which lies at the heart of Newton's formulation of the laws of motion, and discuss Galileo's principle of relativity. We then review gravitational and electromagnetic forces, and analyze in detail the motion of a particle in a one-dimensional potential field. The chapter concludes with a discussion of the basic conservation laws in a system of particles.

2.1 KINEMATICS

We shall usually denote the **position vector** of a particle moving in ordinary space ($\mathbb{R}^3$) by

$$\boxed{\mathbf{r} = x_1\mathbf{e}_1 + x_2\mathbf{e}_2 + x_3\mathbf{e}_3 \equiv \sum_{i=1}^{3} x_i\mathbf{e}_i\,,}$$

where $\mathbf{e}_i$ is the i-th coordinate unit vector in a *Cartesian* orthogonal coordinate system:

$$\mathbf{e}_1 = (1,0,0) \equiv \mathbf{i}\,, \quad \mathbf{e}_2 = (0,1,0) \equiv \mathbf{j}\,, \quad \mathbf{e}_3 = (0,0,1) \equiv \mathbf{k}\,.$$

The particle's **velocity** and **acceleration** are respectively the first and second derivatives of the position vector with respect to time:

$$\boxed{\mathbf{v} := \dot{\mathbf{r}}\,, \qquad \mathbf{a} := \dot{\mathbf{v}} = \ddot{\mathbf{r}}\,,}$$

DOI: 10.1201/9781003600633-2

where the dot denotes derivative with respect to *time* t. In Cartesian coordinates,

$$\mathbf{v} = \sum_{i=1}^{3} v_i \mathbf{e}_i\,, \qquad \mathbf{a} = \sum_{i=1}^{3} a_i \mathbf{e}_i\,,$$

where (since the coordinate vectors $\mathbf{e}_i$ are *constant*)

$$v_i = \dot{x}_i\,, \qquad a_i = \ddot{x}_i\,.$$

Notation: $\quad r = |\mathbf{r}|\,, \quad v = |\mathbf{v}|\,.$

Exercise 2.1. Show that if $\mathbf{v} \neq 0$ then

$$\dot{v} = \frac{\mathbf{v} \cdot \mathbf{a}}{v}\,. \tag{2.1}$$

In particular, when v is *constant* the velocity and acceleration vectors are *orthogonal.*

Solution. The easiest way of proving this equality is to differentiate the identity

$$v^2 = \mathbf{v} \cdot \mathbf{v},$$

obtaining

$$2v\dot{v} = \dot{\mathbf{v}} \cdot \mathbf{v} + \mathbf{v} \cdot \dot{\mathbf{v}} = 2\mathbf{v} \cdot \dot{\mathbf{v}} \equiv 2\mathbf{v} \cdot \mathbf{a} \quad \Longrightarrow \quad \dot{v} = \frac{\mathbf{v} \cdot \mathbf{a}}{v}\,.$$

Note that Eq. (2.1) can be written as

$$\dot{v} = \mathbf{a} \cdot \mathbf{t}\,,$$

where (if $v \neq 0$, or equivalently $\mathbf{v} \neq 0$)

$$\boxed{\mathbf{t} := \frac{\mathbf{v}}{v}}$$

is the **unit tangent vector** along the trajectory. Thus the **tangential acceleration**

$$\boxed{a_t := \mathbf{a} \cdot \mathbf{t}}$$

is equal to the time derivative of v.

Exercise 2.2. Show that $\mathbf{r}\dot{\mathbf{r}} = r\dot{r}$.

Solution. Proceeding as in the previous exercise we obtain

$$r^2 = \mathbf{r} \cdot \mathbf{r} \quad \Longrightarrow \quad 2r\dot{r} = \dot{\mathbf{r}} \cdot \mathbf{r} + \mathbf{r} \cdot \dot{\mathbf{r}} = 2\mathbf{r} \cdot \dot{\mathbf{r}} \quad \Longrightarrow \quad \mathbf{r}\dot{\mathbf{r}} = r\dot{r}.$$

2.2 CURVILINEAR COORDINATE SYSTEMS

Consider the system of curvilinear coordinates $\mathbf{q} = (q_1, q_2, q_3)$ defined by a bijective transformation $\mathbf{r} = \mathbf{r}(\mathbf{q})$ with a *non-vanishing Jacobian*

$$\det\left(\frac{\partial \mathbf{r}}{\partial \mathbf{q}}\right) \equiv \det\left(\frac{\partial x_i}{\partial q_j}\right)_{1\leqslant i,j\leqslant 3}.$$

For example, in the case of *spherical coordinates* we have $\mathbf{q} = (r, \theta, \varphi)$, with

$$r \geqslant 0, \quad 0 \leqslant \theta \leqslant \pi, \quad 0 \leqslant \varphi < 2\pi,$$

and

$$\boxed{\mathbf{r} = r(\sin\theta\cos\varphi, \sin\theta\sin\varphi, \cos\theta)} \tag{2.2}$$

(cf. Fig. 2.1). The **unit coordinate vectors** $\{\mathbf{e}_{q_i}\}_{i=1}^3$ are the unit vectors tangent to the *coordinate curves*, along which one of the curvilinear coordinates q_i varies while the rest are held constant. We thus have

$$\boxed{\mathbf{e}_{q_i} = \frac{1}{h_i}\frac{\partial \mathbf{r}}{\partial q_i}, \qquad \text{with} \quad h_i := \left|\frac{\partial \mathbf{r}}{\partial q_i}\right|.} \tag{2.3}$$

Note that $h_i(\mathbf{q}) > 0$ for all i, since the vector $\dfrac{\partial \mathbf{r}}{\partial q_i}$ is the i-th column of the Jacobian matrix $\dfrac{\partial \mathbf{r}}{\partial \mathbf{q}}$, whose determinant—the *Jacobian* of the mapping $\mathbf{r}(\mathbf{q})$—is nonvanishing by hypothesis. It is also important to realize that in general the unit coordinate vectors $\mathbf{e}_{q_i}$ are *not* constant, but depend on the curvilinear coordinates $\mathbf{q}$ of the point at which they are evaluated. In other words, each $\mathbf{e}_{q_i}$ is a *vector field* in $\mathbb{R}^3$.

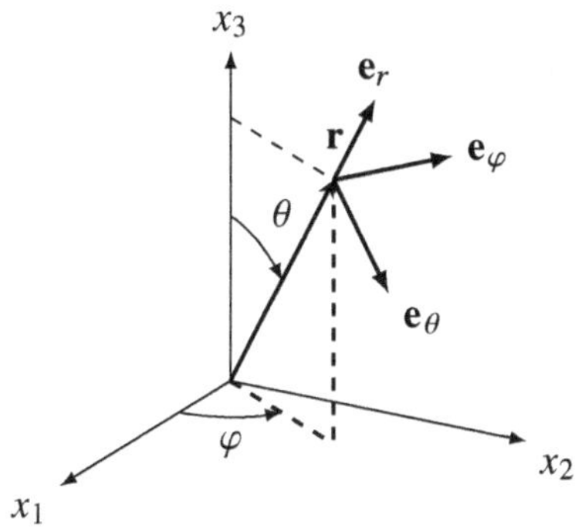

Figure 2.1. Spherical coordinate system.

We shall normally deal with **orthogonal** coordinate systems, whose unit coordinate vectors are mutually orthogonal and thus make up an *orthonormal frame* at each point of $\mathbb{R}^3$. Since $\mathbf{e}_{q_i}$ is a unit vector proportional to $\dfrac{\partial \mathbf{r}}{\partial q_i}$, the necessary and sufficient condition for a coordinate system to be orthogonal is that

$$\boxed{\frac{\partial \mathbf{r}}{\partial q_i} \cdot \frac{\partial \mathbf{r}}{\partial q_j} = 0, \qquad \forall i \neq j\,.}$$

In fact, by Eq. (2.3), in an orthogonal coordinate system we must have

$$\boxed{\frac{\partial \mathbf{r}}{\partial q_i} \cdot \frac{\partial \mathbf{r}}{\partial q_j} = \delta_{ij} h_i^2\,.}$$

We shall also assume that the coordinate system $\mathbf{q}$ is *positively oriented*, by which we mean that

$$\boxed{\mathbf{e}_{q_1} \times \mathbf{e}_{q_2} = \mathbf{e}_{q_3}\,,}$$

or equivalently

$$\det(\mathbf{e}_{q_1}, \mathbf{e}_{q_2}, \mathbf{e}_{q_3}) = 1$$

(cf. Remark 1.1). Since

$$\begin{aligned}\det\left(\frac{\partial \mathbf{r}}{\partial \mathbf{q}}\right) &= \det\left(\frac{\partial \mathbf{r}}{\partial q_1}, \frac{\partial \mathbf{r}}{\partial q_2}, \frac{\partial \mathbf{r}}{\partial q_3}\right) = \det\left(h_1\mathbf{e}_{q_1}, h_2\mathbf{e}_{q_2}, h_3\mathbf{e}_{q_3}\right) \\ &= h_1 h_2 h_3 \det\left(\mathbf{e}_{q_1}, \mathbf{e}_{q_2}, \mathbf{e}_{q_3}\right), \end{aligned} \tag{2.4}$$

with $h_i > 0$ for all i, the necessary and sufficient condition for the orthogonal coordinate system $\mathbf{q}$ to be positively oriented is that the Jacobian of the transformation $\mathbf{r}(\mathbf{q})$ be *positive*:

$$\boxed{\det\left(\frac{\partial \mathbf{r}}{\partial \mathbf{q}}\right) > 0\,.}$$

- The *line element* $\mathrm{d}s^2 := \mathrm{d}\mathbf{r}^2$ can be easily expressed in any orthogonal curvilinear system $\mathbf{q}$ using the chain rule:

$$\mathrm{d}s^2 = \mathrm{d}\mathbf{r}^2 = \left(\sum_{i=1}^{3} \frac{\partial \mathbf{r}}{\partial q_i}\,\mathrm{d}q_i\right)^2 = \left(\sum_{i=1}^{3} h_i \mathbf{e}_{q_i}\,\mathrm{d}q_i\right)^2 = \boxed{\sum_{i=1}^{3} h_i^2\,\mathrm{d}q_i^2\,,}$$

where in the last equality we have used the fact that $\mathbf{e}_{q_i} \cdot \mathbf{e}_{q_j} = \delta_{ij}$. In particular, the line element along the i-th coordinate curve is given by

$$ds = h_i\,\mathrm{d}q_i, \qquad i = 1, 2, 3,$$

a formula that is often used to compute h_i by geometric arguments.

• If $\mathbf{A}$ is a vector field in $\mathbb{R}^3$, we define its components in an orthogonal coordinate system $\mathbf{q}$ by

$$\boxed{A_{q_i} := \mathbf{A} \cdot \mathbf{e}_{q_i},}$$

so that we can write

$$\boxed{\mathbf{A} = \sum_{i=1}^{3} A_{q_i} \mathbf{e}_{q_i} .}$$

The components of the *velocity vector* $\mathbf{v}$ in such a system can also be readily computed applying the chain rule. Indeed,

$$\mathbf{v} = \frac{\mathrm{d}\mathbf{r}}{\mathrm{d}t} = \sum_{i=1}^{3} \frac{\partial \mathbf{r}}{\partial q_i} \dot{q}_i = \boxed{\sum_{i=1}^{3} h_i \dot{q}_i \mathbf{e}_{q_i} ,} \tag{2.5}$$

and hence

$$\boxed{v_{q_i} = h_i \dot{q}_i .} \tag{2.6}$$

2.2.1 Spherical coordinates

In this case

$$\begin{aligned}
\frac{\partial \mathbf{r}}{\partial r} &= (\sin\theta\cos\varphi, \sin\theta\sin\varphi, \cos\theta) , \\
\frac{\partial \mathbf{r}}{\partial \theta} &= r(\cos\theta\cos\varphi, \cos\theta\sin\varphi, -\sin\theta) , \\
\frac{\partial \mathbf{r}}{\partial \varphi} &= r\sin\theta(-\sin\varphi, \cos\varphi, 0),
\end{aligned}$$

and hence

$$\boxed{h_r = \left|\frac{\partial \mathbf{r}}{\partial r}\right| = 1, \qquad h_\theta = \left|\frac{\partial \mathbf{r}}{\partial \theta}\right| = r, \qquad h_\varphi = \left|\frac{\partial \mathbf{r}}{\partial \varphi}\right| = r\sin\theta ,}$$

so that

$$\boxed{\begin{aligned}
\mathbf{e}_r &= \frac{\partial \mathbf{r}}{\partial r} = (\sin\theta\cos\varphi, \sin\theta\sin\varphi, \cos\theta) \equiv \frac{\mathbf{r}}{r} , \\
\mathbf{e}_\theta &= \frac{1}{r}\frac{\partial \mathbf{r}}{\partial \theta} = (\cos\theta\cos\varphi, \cos\theta\sin\varphi, -\sin\theta) , \\
\mathbf{e}_\varphi &= \frac{1}{r\sin\theta}\frac{\partial \mathbf{r}}{\partial \varphi} = (-\sin\varphi, \cos\varphi, 0)
\end{aligned}} \tag{2.7}$$

(cf. Fig. 2.1). Spherical coordinates are *orthogonal*, as from the previous equations it is easily verified that

$$\mathbf{e}_r \cdot \mathbf{e}_\theta = \mathbf{e}_r \cdot \mathbf{e}_\varphi = \mathbf{e}_\theta \cdot \mathbf{e}_\varphi = 0 .$$

Thus the vectors $\{\mathbf{e}_r, \mathbf{e}_\theta, \mathbf{e}_\varphi\}$ form an orthonormal frame at each point. This frame is *positively oriented*, since

$$\mathbf{e}_r \times \mathbf{e}_\theta = \mathbf{e}_\varphi ,$$

or, equivalently,

$$(\mathbf{e}_r \times \mathbf{e}_\theta) \cdot \mathbf{e}_\varphi = 1 .$$

Velocity and acceleration in spherical coordinates.

The components of the velocity vector $\mathbf{v}$ in spherical coordinates can be easily computed using Eq. (2.6). It is also instructive to obtain them directly differentiating the relation $\mathbf{r} = r\mathbf{e}_r$, as we shall next explain. To this end, we first compute the time derivatives of the unit coordinate vectors. Note first of all that, since

$$\mathbf{e}_\alpha \cdot \mathbf{e}_\alpha = 1 \qquad (\alpha = r, \theta, \varphi) ,$$

differentiating with respect to time we obtain

$$\dot{\mathbf{e}}_\alpha \cdot \mathbf{e}_\alpha = 0 .$$

Since the vectors $\mathbf{e}_\alpha$ are mutually orthogonal, $\dot{\mathbf{e}}_r$ must be a linear combination of $\mathbf{e}_\theta$ and $\mathbf{e}_\varphi$, and similarly for the remaining coordinate vectors. More precisely, applying the *chain rule* we arrive at the identities

$$\begin{aligned}
\dot{\mathbf{e}}_r &= \frac{\partial \mathbf{e}_r}{\partial \theta}\,\dot{\theta} + \frac{\partial \mathbf{e}_r}{\partial \varphi}\,\dot{\varphi} = \dot{\theta}\mathbf{e}_\theta + \sin\theta\dot{\varphi}\mathbf{e}_\varphi ,\\
\dot{\mathbf{e}}_\theta &= \frac{\partial \mathbf{e}_\theta}{\partial \theta}\,\dot{\theta} + \frac{\partial \mathbf{e}_\theta}{\partial \varphi}\,\dot{\varphi} = -\dot{\theta}\mathbf{e}_r + \cos\theta\dot{\varphi}\mathbf{e}_\varphi ,\\
\dot{\mathbf{e}}_\varphi &= \frac{\partial \mathbf{e}_\varphi}{\partial \varphi}\,\dot{\varphi} = -(\cos\varphi, \sin\varphi, 0)\,\dot{\varphi} = -\dot{\varphi}(\sin\theta\mathbf{e}_r + \cos\theta\mathbf{e}_\theta) .
\end{aligned} \tag{2.8}$$

From the above relations we easily deduce that

$$\mathbf{v} = \frac{\mathrm{d}\mathbf{r}}{\mathrm{d}t} = \frac{\mathrm{d}}{\mathrm{d}t}(r\mathbf{e}_r) = \dot{r}\mathbf{e}_r + r\dot{\mathbf{e}}_r = \dot{r}\mathbf{e}_r + r\dot{\theta}\mathbf{e}_\theta + r\sin\theta\dot{\varphi}\mathbf{e}_\varphi , \tag{2.9}$$

and hence

$$v_r = \dot{r} , \qquad v_\theta = r\dot{\theta} , \qquad v_\varphi = r\sin\theta\,\dot{\varphi} . \tag{2.10}$$

Note that, since the coordinate vectors are *orthonormal*,

$$v^2 = v_r^2 + v_\theta^2 + v_\varphi^2 = \dot{r}^2 + r^2(\dot{\theta}^2 + \sin\theta^2\dot{\varphi}^2) . \tag{2.11}$$

Similarly, differentiating Eq. (2.9) with respect to t and using Eqs. (2.8) we obtain:

$$\begin{aligned}
\mathbf{a} &= \ddot{r}\mathbf{e}_r + \dot{r}(\dot{\theta}\mathbf{e}_\theta + \sin\theta\dot{\varphi}\mathbf{e}_\varphi) + (r\ddot{\theta} + \dot{r}\dot{\theta})\mathbf{e}_\theta + r\dot{\theta}(-\dot{\theta}\mathbf{e}_r + \cos\theta\dot{\varphi}\mathbf{e}_\varphi)\\
&\quad + (r\sin\theta\ddot{\varphi} + \sin\theta\dot{r}\dot{\varphi} + r\cos\theta\dot{\theta}\dot{\varphi})\mathbf{e}_\varphi - r\sin\theta\dot{\varphi}^2(\sin\theta\mathbf{e}_r + \cos\theta\mathbf{e}_\theta)\\
&= a_r\mathbf{e}_r + a_\theta\mathbf{e}_\theta + a_\varphi\mathbf{e}_\varphi ,
\end{aligned}$$

where the components of the acceleration in spherical coordinates are given by

$$\begin{aligned} a_r &= \ddot{r} - r\dot{\theta}^2 - r\sin^2\theta\,\dot{\varphi}^2 , \\ a_\theta &= r\ddot{\theta} + 2\dot{r}\dot{\theta} - r\sin\theta\cos\theta\,\dot{\varphi}^2 , \\ a_\varphi &= r\sin\theta\,\ddot{\varphi} + 2\sin\theta\dot{r}\dot{\varphi} + 2r\cos\theta\,\dot{\theta}\dot{\varphi} . \end{aligned} \tag{2.12}$$

2.2.2 Cylindrical coordinates

Cylindrical coordinates (ρ, φ, z) are defined by

$$\mathbf{r} = (\rho\cos\varphi, \rho\sin\varphi, z) ,$$

where

$$\rho \geqslant 0 , \qquad 0 \leqslant \varphi < 2\pi , \qquad z \in \mathbb{R}$$

(cf. Fig. 2.2). Now

$$\frac{\partial \mathbf{r}}{\partial \rho} = (\cos\varphi, \sin\varphi, 0), \qquad \frac{\partial \mathbf{r}}{\partial \varphi} = \rho(-\sin\varphi, \cos\varphi, 0), \qquad \frac{\partial \mathbf{r}}{\partial z} = (0, 0, 1) ,$$

and thus

$$h_\rho = 1 , \qquad h_\varphi = \rho , \qquad h_z = 1 .$$

Proceeding as before we obtain:

$$\begin{aligned} \mathbf{e}_\rho &= \frac{\partial \mathbf{r}}{\partial \rho} = (\cos\varphi, \sin\varphi, 0) , \\ \mathbf{e}_\varphi &= \frac{1}{\rho}\frac{\partial \mathbf{r}}{\partial \varphi} = (-\sin\varphi, \cos\varphi, 0) , \\ \mathbf{e}_z &= \frac{\partial \mathbf{r}}{\partial z} = (0, 0, 1) . \end{aligned}$$

Note that, once again,

$$\mathbf{e}_\rho \cdot \mathbf{e}_\varphi = \mathbf{e}_\rho \cdot \mathbf{e}_z = \mathbf{e}_\varphi \cdot \mathbf{e}_z = 0 , \qquad \mathbf{e}_\rho \times \mathbf{e}_\varphi = \mathbf{e}_z ,$$

and hence the cylindrical coordinates (ρ, φ, z) are also orthonormal and positively oriented. Differentiating with respect to t the equations for the coordinate vectors, we immediately obtain the relations

$$\dot{\mathbf{e}}_\rho = \dot{\varphi}\mathbf{e}_\varphi , \qquad \dot{\mathbf{e}}_\varphi = -\dot{\varphi}\mathbf{e}_\rho , \qquad \dot{\mathbf{e}}_z = 0 .$$

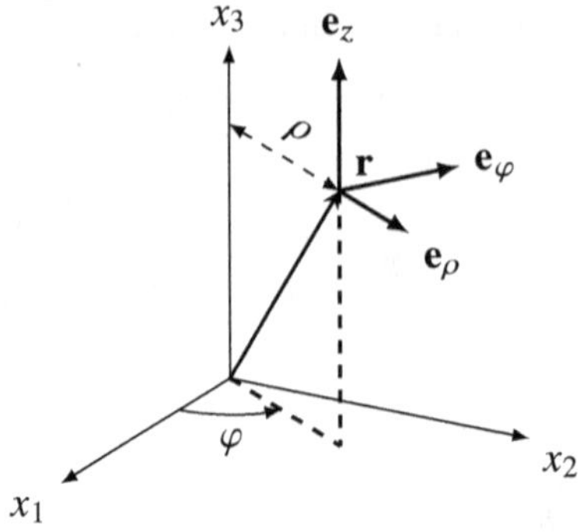

Figure 2.2. Cylindrical coordinates system.

Since now

$$\mathbf{r} = \rho \mathbf{e}_\rho + z\mathbf{e}_z\,, \tag{2.13}$$

(cf. Fig. 2.2), differentiating twice with respect to t and proceeding as before we easily arrive at the following formulas for the components of velocity and acceleration in cylindrical coordinates:

$$\begin{aligned} &v_\rho = \dot\rho\,, \quad v_\varphi = \rho\dot\varphi\,, \quad v_z = \dot z\,; \\ &a_\rho = \ddot\rho - \rho\dot\varphi^2\,, \quad a_\varphi = \rho\ddot\varphi + 2\dot\rho\dot\varphi\,, \quad a_z = \ddot z\,. \end{aligned} \tag{2.14}$$

Note that the equations for the components of the velocity could also have been directly obtained using Eq. (2.6). Again, from the orthonormal character of the coordinate vectors $\{\mathbf{e}_\rho, \mathbf{e}_\varphi, \mathbf{e}_z\}$ it follows that

$$v^2 = v_\rho^2 + v_\varphi^2 + v_z^2 = \dot\rho^2 + \rho^2\dot\varphi^2 + \dot z^2\,.$$

Exercise 2.3. Prove that in any orthogonal curvilinear system of coordinates the components $a_{q_k} := \mathbf{a}\cdot\mathbf{e}_{q_k}$ of the acceleration vector are given by

$$a_{q_k} = h_k\left(\ddot q_k + \sum_{i,j=1}^{3} \Gamma_k^{ij}\dot q_i \dot q_j\right), \qquad \text{with} \quad \Gamma_k^{ij} := \frac{1}{h_k^2}\frac{\partial \mathbf{r}}{\partial q_k}\cdot\frac{\partial^2 \mathbf{r}}{\partial q_i \partial q_j}\,. \tag{2.15}$$

Solution. Indeed, differentiating Eq. (2.5) for the velocity vector with respect to time we obtain

$$\mathbf{a} = \sum_{i=1}^{3} \frac{\partial \mathbf{r}}{\partial q_i} \ddot{q}_i + \sum_{i=1}^{3} \frac{\mathrm{d}}{\mathrm{d}t}\left(\frac{\partial \mathbf{r}}{\partial q_i}\right) \dot{q}_i = \sum_{i=1}^{3} \frac{\partial \mathbf{r}}{\partial q_i} \ddot{q}_i + \sum_{i,j=1}^{3} \frac{\partial^2 \mathbf{r}}{\partial q_i \partial q_j} \dot{q}_i \dot{q}_j$$
$$= \sum_{i=1}^{3} h_i\, \ddot{q}_i \mathbf{e}_i q_i + \sum_{i,j=1}^{3} \frac{\partial^2 \mathbf{r}}{\partial q_i \partial q_j} \dot{q}_i \dot{q}_j.$$

Taking the scalar product of both sides of the last equation with the coordinate vector $\mathbf{e}_{q_k} = \frac{1}{h_k}\frac{\partial \mathbf{r}}{\partial q_h}$, and using the orthogonality relation $\mathbf{e}_{q_i} \cdot \mathbf{e}_{q_k} = \delta_{ik}$, we easily obtain

$$a_{q_k} = \mathbf{a} \cdot \mathbf{e}_{q_k} = h_k\, \ddot{q}_k + \sum_{i,j=1}^{3} \frac{\partial^2 \mathbf{r}}{\partial q_i \partial q_j} \cdot \mathbf{e}_{q_k} \dot{q}_i \dot{q}_j$$
$$= h_k\, \ddot{q}_k + \sum_{i,j=1}^{3} \frac{1}{h_k} \frac{\partial \mathbf{r}}{\partial q_k} \cdot \frac{\partial^2 \mathbf{r}}{\partial q_i \partial q_j} \dot{q}_i \dot{q}_j,$$

which immediately yields Eq. (2.15).

Note. The functions Γ_k^{ij}, which are called the *Christoffel symbols* of the metric $\mathrm{d}s^2 = \sum_{k=1}^{3} h_k^2\, \mathrm{d}q_k^2$, can actually be expressed in terms of the partial derivatives of the metric coefficients h_k. Indeed, differentiating with respect to q_i both sides of the identity

$$\frac{\partial \mathbf{r}}{\partial q_j} \cdot \frac{\partial \mathbf{r}}{\partial q_k} = h_k^2 \delta_{jk}$$

we obtain

$$\frac{\partial}{\partial q_i}\left(h_k^2 \delta_{jk}\right) = \frac{\partial \mathbf{r}}{\partial q_k} \cdot \frac{\partial^2 \mathbf{r}}{\partial q_i q_j} + \frac{\partial \mathbf{r}}{\partial q_j} \cdot \frac{\partial^2 \mathbf{r}}{\partial q_i q_k}.$$

Permuting cyclically the three indices (i, j, k), we arrive at the analogous relations

$$\frac{\partial}{\partial q_j}\left(h_i^2 \delta_{ik}\right) = \frac{\partial \mathbf{r}}{\partial q_i} \cdot \frac{\partial^2 \mathbf{r}}{\partial q_j q_k} + \frac{\partial \mathbf{r}}{\partial q_k} \cdot \frac{\partial^2 \mathbf{r}}{\partial q_i q_j},$$
$$\frac{\partial}{\partial q_k}\left(h_j^2 \delta_{ij}\right) = \frac{\partial \mathbf{r}}{\partial q_j} \cdot \frac{\partial^2 \mathbf{r}}{\partial q_i q_k} + \frac{\partial \mathbf{r}}{\partial q_i} \cdot \frac{\partial^2 \mathbf{r}}{\partial q_j q_k},$$

whence it easily follows that

$$\frac{\partial \mathbf{r}}{\partial q_k} \cdot \frac{\partial^2 \mathbf{r}}{\partial q_i q_j} = \frac{1}{2}\left[\frac{\partial}{\partial q_i}\left(h_k^2 \delta_{jk}\right) + \frac{\partial}{\partial q_j}\left(h_i^2 \delta_{ik}\right) - \frac{\partial}{\partial q_k}\left(h_j^2 \delta_{ij}\right)\right],$$

or equivalently

$$\Gamma^{ij}_{k} = \frac{1}{2h_k^2}\left[\frac{\partial}{\partial q_i}\left(h_k^2\delta_{jk}\right) + \frac{\partial}{\partial q_j}\left(h_i^2\delta_{ik}\right) - \frac{\partial}{\partial q_k}\left(h_j^2\delta_{ij}\right)\right]. \tag{2.16}$$

Exercise 2.4. Consider the system of orthogonal curvilinear coordinates $\mathbf{q} = (q_1, q_2, q_3)$, and let f be a smooth scalar function. Show that

$$\nabla f = \sum_{i=1}^{3} \frac{1}{h_i}\frac{\partial f}{\partial q_i}\,\mathbf{e}_{q_i} \quad \Longrightarrow \quad \boxed{(\nabla f)_{q_i} = \frac{1}{h_i}\frac{\partial f}{\partial q_i}}\,. \tag{2.17}$$

Solution. Although the gradient has been defined in a *Cartesian* (orthogonal) coordinate system (x_1, x_2, x_3), we can find an identity involving ∇f that is independent of any coordinate system. Indeed,

$$\nabla f \cdot \mathrm{d}\mathbf{r} = \sum_{i=1}^{3} \frac{\partial f}{\partial x_i}\,\mathrm{d}x_i = \mathrm{d}f\,,$$

where the *differential* $\mathrm{d}f$ of f can be computed in *any* curvilinear coordinate system $\mathbf{q}$ by the standard formula

$$\mathrm{d}f = \sum_{i=1}^{3} \frac{\partial f}{\partial q_i}\,\mathrm{d}q_i.$$

From the identity

$$\mathrm{d}\mathbf{r} = \sum_{i=1}^{3} \frac{\partial \mathbf{r}}{\partial q_i}\,\mathrm{d}q_i = \sum_{i=1}^{3} h_i\mathbf{e}_{q_i}\,\mathrm{d}q_i$$

we then obtain

$$\mathrm{d}f = \sum_{i=1}^{3} \frac{\partial f}{\partial q_i}\,\mathrm{d}q_i = \nabla f \cdot \mathrm{d}\mathbf{r} = \sum_{i=1}^{3} h_i(\nabla f \cdot \mathbf{e}_{q_i})\,\mathrm{d}q_i = \sum_{i=1}^{3} h_i(\nabla f)_{q_i}\,\mathrm{d}q_i\,.$$

Equating the coefficients of $\mathrm{d}q_i$ in the second and last expression for $\mathrm{d}f$ we obtain Eq. (2.17).

Remark 2.1. Applying Gauss's theorem to the infinitesimal solid whose curvilinear coordinates lie between q_i and $q_i + \mathrm{d}q_i$ $(i = 1, 2, 3)$, it can be shown that the *divergence* of a smooth vector field

$$\mathbf{A} = \sum_{i=1}^{3} A_{q_i}(\mathbf{q})\,\mathbf{e}_{q_i}$$

in a curvilinear *orthogonal* system of coordinates $\mathbf{q}$ can be expressed as

$$\boxed{\nabla\cdot\mathbf{A} = \frac{1}{h_1h_2h_3}\sum_{i=1}^{3}\frac{\partial}{\partial q_i}\left(h_jh_kA_{q_i}\right),}$$

where $\{i,j,k\} = \{1,2,3\}$. From Eq. (2.17) it then follows that the *Laplacian* of a smooth scalar function f is given by

$$\boxed{\nabla^2 f := \nabla\cdot(\nabla f) = \frac{1}{h_1h_2h_3}\sum_{i=1}^{3}\frac{\partial}{\partial q_i}\left(\frac{h_jh_k}{h_i}\frac{\partial f}{\partial q_i}\right).}$$

Similarly, if the orthogonal curvilinear coordinate system $\mathbf{q}$ is *positively oriented*, applying Stokes's theorem to suitable infinitesimal surfaces perpendicular to the unit coordinate vectors $\mathbf{e}_{q_i}$ at an arbitrary point one can show that

$$\boxed{\nabla\times\mathbf{A} = \frac{1}{h_1h_2h_3}\begin{vmatrix} h_1\mathbf{e}_{q_1} & h_2\mathbf{e}_{q_2} & h_3\mathbf{e}_{q_3} \\ \dfrac{\partial}{\partial q_1} & \dfrac{\partial}{\partial q_2} & \dfrac{\partial}{\partial q_3} \\ h_1A_{q_1} & h_2A_{q_2} & h_3A_{q_3}\end{vmatrix}.}$$

For the spherical coordinate system $h_r = 1$, $h_\theta = r$, $h_\varphi = r\sin\theta$, and therefore

$$\boxed{\begin{aligned}
\nabla f &= \frac{\partial f}{\partial r}\mathbf{e}_r + \frac{1}{r}\frac{\partial f}{\partial\theta}\mathbf{e}_\theta + \frac{1}{r\sin\theta}\frac{\partial f}{\partial\varphi}\mathbf{e}_\varphi,\\
\nabla^2 f &= \frac{1}{r^2}\frac{\partial}{\partial r}\left(r^2\frac{\partial f}{\partial r}\right) + \frac{1}{r^2\sin\theta}\frac{\partial}{\partial\theta}\left(\sin\theta\frac{\partial f}{\partial\theta}\right) + \frac{1}{r^2\sin^2\theta}\frac{\partial^2 f}{\partial\varphi^2},\\
\nabla\cdot\mathbf{A} &= \frac{1}{r^2}\frac{\partial}{\partial r}(r^2A_r) + \frac{1}{r\sin\theta}\left[\frac{\partial}{\partial\theta}(\sin\theta A_\theta) + \frac{\partial A_\varphi}{\partial\varphi}\right],\\
\nabla\times\mathbf{A} &= \left[\frac{\partial}{\partial\theta}(\sin\theta\, A_\varphi) - \frac{\partial A_\theta}{\partial\varphi}\right]\frac{\mathbf{e}_r}{r\sin\theta}\\
&\quad + \left[\frac{\partial A_r}{\partial\varphi} - \sin\theta\frac{\partial}{\partial r}(rA_\varphi)\right]\frac{\mathbf{e}_\theta}{r\sin\theta} + \left[\frac{\partial}{\partial r}(rA_\theta) - \frac{\partial A_r}{\partial\theta}\right]\frac{\mathbf{e}_\varphi}{r}.
\end{aligned}} \tag{2.18}$$

Likewise, in cylindrical coordinates $h_\rho = h_z = 1$, $h_\varphi = \rho$, and thus

$$\begin{aligned}
\nabla f &= \frac{\partial f}{\partial \rho}\,\mathbf{e}_\rho + \frac{1}{\rho}\frac{\partial f}{\partial \varphi}\,\mathbf{e}_\varphi + \frac{\partial f}{\partial z}\,\mathbf{e}_z\,,\\
\nabla^2 f &= \frac{1}{\rho}\frac{\partial}{\partial \rho}\left(\rho\frac{\partial f}{\partial \rho}\right) + \frac{1}{\rho^2}\frac{\partial^2 f}{\partial \varphi^2} + \frac{\partial^2 f}{\partial z^2}\,,\\
\nabla\cdot\mathbf{A} &= \frac{1}{\rho}\frac{\partial}{\partial \rho}(\rho A_\rho) + \frac{1}{\rho}\frac{\partial A_\varphi}{\partial \varphi} + \frac{\partial A_z}{\partial z}\,,\\
\nabla\times\mathbf{A} &= \left(\frac{\partial A_z}{\partial \varphi} - \rho\frac{\partial A_\varphi}{\partial z}\right)\frac{\mathbf{e}_\rho}{\rho} + \left(\frac{\partial A_\rho}{\partial z} - \frac{\partial A_z}{\partial \rho}\right)\mathbf{e}_\varphi\\
&\quad + \left[\frac{\partial}{\partial \rho}(\rho A_\varphi) - \frac{\partial A_\rho}{\partial \varphi}\right]\frac{\mathbf{e}_z}{\rho}\,.
\end{aligned} \tag{2.19}$$

Exercise 2.5. Show that the volume element in a (positively oriented) orthogonal curvilinear coordinate system $\mathbf{q}$ is given by

$$\mathrm{d}^3 r = h_1 h_2 h_3\, \mathrm{d}q_1\, \mathrm{d}q_2\, \mathrm{d}q_3\,. \tag{2.20}$$

Solution. Indeed, by the change of coordinates formula for volume integrals we know that the volume element in the curvilinear coordinate system $\mathbf{q}$ is given by

$$\mathrm{d}^3 r = \left|\det\left(\frac{\partial \mathbf{r}}{\partial \mathbf{q}}\right)\right| \mathrm{d}q_1\, \mathrm{d}q_2\, \mathrm{d}q_3.$$

On the other hand, by Eq. (2.4) we have

$$\det\left(\frac{\partial \mathbf{r}}{\partial \mathbf{q}}\right) = h_1 h_2 h_3 \det(\mathbf{e}_{q_1}, \mathbf{e}_{q_2}, \mathbf{e}_{q_3}) = h_1 h_2 h_3,$$

since the orthogonal curvilinear coordinate system $\mathbf{q}$ is assumed to be positively oriented. This establishes Eq. (2.20), as $h_i > 0$ for all i by its definition (2.3).

2.2.3 Motion on a plane in polar coordinates

Suppose that a particle moves on a plane, which we shall take as the plane $z = 0$, so that $z(t) = 0$ for all t. In this case $\dot z = \ddot z = v_z = a_z = 0$, $\rho = r$ (distance to the origin) and (r, φ) are *polar coordinates* in the plane of motion (cf. Fig. 2.3). The previous formulas (2.14) then reduce to the following:

$$v_r = \dot r\,, \qquad v_\varphi = r\dot\varphi\,, \qquad a_r = \ddot r - r\dot\varphi^2\,, \qquad a_\varphi = r\ddot\varphi + 2\dot r\dot\varphi\,. \tag{2.21}$$

In particular, if $r(t) = R$ for all t (*circular motion*), we obtain the familiar formulas

$$v_r = 0\,, \qquad v_\varphi = R\dot\varphi\,, \qquad a_r = -R\dot\varphi^2\,, \qquad a_\varphi = R\ddot\varphi\,. \tag{2.22}$$

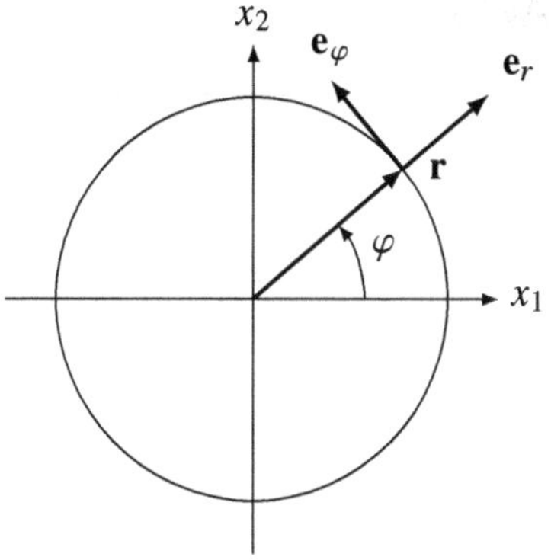

Figure 2.3. Polar coordinates.

Note that, although the radial component of the velocity is identically zero, even when $\ddot{\varphi} = 0$ there is a negative radial acceleration $-R\dot{\varphi}^2 = -v^2/R$ (*centripetal acceleration*).

Example 2.1. *Angular velocity.*

Consider next a particle rotating around a fixed axis. Taking the rotation axis as the z axis and the plane of motion as the plane $z = 0$, the particle's motion is described in polar coordinates by Eqs. (2.22). From these equations it follows that

$$\mathbf{v} = R\dot{\varphi}(t)\,\mathbf{e}_\varphi\,,$$

and since $\mathbf{e}_\varphi = \mathbf{e}_z \times \mathbf{e}_\rho = \mathbf{e}_z \times \mathbf{e}_r$ we have

$$\mathbf{v} = R\dot{\varphi}(t)\mathbf{e}_z \times \mathbf{e}_r = \big(\dot{\varphi}(t)\mathbf{e}_z\big) \times \mathbf{r}\,.$$

Hence in this case we can express the particle's velocity through the intrinsic (i.e., independent of the coordinate system used) formula

$$\boxed{\mathbf{v} = \boldsymbol{\omega} \times \mathbf{r}\,,}$$

where

$$\boldsymbol{\omega}(t) = \dot{\varphi}(t)\mathbf{e}_z$$

is called the **angular velocity**. The angular velocity is therefore a *vector* directed along the axis of rotation, whose magnitude is the *absolute value* $|\dot{\varphi}(t)|$ of the angular velocity of rotation. Rotation around the z axis is called "left-handed" if $\dot{\varphi}(t) > 0$ (i.e., if $\boldsymbol{\omega}$ and $\mathbf{e}_z$ have the same direction) and "right-handed" if $\dot{\varphi}(t) < 0$ (i.e., if $\boldsymbol{\omega}$ and $\mathbf{e}_z$ have opposite directions).

2.3 NEWTON'S LAWS, INERTIAL FRAMES, AND GALILEO'S RELATIVITY PRINCIPLE

2.3.1 Newton's laws

In (non-relativistic) classical mechanics, the **linear momentum** (or momentum, for short) of a particle is defined by

$$\boxed{\mathbf{p} = m\mathbf{v} = m\dot{\mathbf{r}}\,,} \tag{2.23}$$

where m is the particle's **mass**. In classical mechanics the mass is a positive *constant* (in particular, velocity independent) parameter characteristic of each particle. In modern notation and terminology, the first two of **Newton's laws** can be stated as follows:

I. *In the absence of external forces, the momentum (and, hence, the velocity) of a particle remains constant.*

II. *If an external force* **F** *acts on a particle, the rate of variation of its momentum is given by*

$$\frac{\mathrm{d}\mathbf{p}}{\mathrm{d}t} = \mathbf{F}\,. \tag{2.24}$$

Since the particle's mass is constant, the last equation is equivalent to

$$\boxed{\mathbf{F} = m\mathbf{a} = m\ddot{\mathbf{r}}\,.} \tag{2.25}$$

This is the particle's *equation of motion*.

Remarks 2.2.

- As stated above, Newton's first law is a particular case of the second one. Indeed, if $\mathbf{F} = 0$ then $\frac{\mathrm{d}\mathbf{p}}{\mathrm{d}t} = 0$ implies that **p**, and therefore **v**, must be constant.
- Newton's first two laws—or, from what we have just remarked, Eqs. (2.24)–(2.25)—are the foundation of classical mechanics. These laws are valid with very high accuracy for motions involving *small velocities* compared to the speed of light, and at *macroscopic scales*[1]. In particular, they do *not* hold for interactions at the atomic and subatomic scales (between elementary particles, atoms, atomic nuclei, molecules, etc.), which are governed by *quantum mechanics*. Nor are they valid for motion in intense gravitational fields, which is governed by Einstein's theory of *general relativity*. Actually, both quantum mechanics (or even *quantum field*

[1]More precisely, if the typical *action* of the system under study, defined as the product of its typical length and momentum (or energy and time), is much larger than Planck's constant $h = 6.626\,070\,15 \times 10^{-34}$ J s.

theory, which combines quantum mechanics with the special theory of relativity) and general relativity are not universally valid, but rather apply to different physical situations. In fact, at present there is no consistent theory applicable to *all* physical phenomena which unifies quantum mechanics with the general theory of relativity.

- Newton's second law provides an operational definition of mass. Indeed, if we apply the same force **F** to two different particles (denoted by 1 and 2), according to Eq. (2.25) their accelerations have the same direction, and the quotient of their magnitudes is given by

$$\frac{|\mathbf{a}_1|}{|\mathbf{a}_2|} = \frac{m_2}{m_1}.$$

  In this way, the quotient m_2/m_1 can be measured in principle for any pair of particles. From the previous discussion it should also be clear that a particle's mass is a quantitative measure of its *inertia*, i.e., its resistance to being accelerated by an applied force.

- Practically all forces appearing in classical mechanics depend at most on time, position, and velocity, and are therefore *independent of acceleration* (and of derivatives of the position vector of order higher than two)[2]. Newton's second law (2.25) can therefore be written in the form

$$\boxed{\ddot{\mathbf{r}} = \frac{1}{m}\,\mathbf{F}(t, \mathbf{r}, \dot{\mathbf{r}})\,,} \tag{2.26}$$

  where $\mathbf{F}(t, \mathbf{r}, \dot{\mathbf{r}})$ is the force acting on the particle. This vector equation is actually equivalent to the *system of three second-order ordinary differential equations*

$$\ddot{x}_i = \frac{1}{m}\,F_i(t, x_1, x_2, x_3, \dot{x}_1, \dot{x}_2, \dot{x}_3)\,, \qquad i = 1, 2, 3\,, \tag{2.27}$$

  for the three particle coordinates $x_i(t)$. If the function $\mathbf{F}(t, \mathbf{r}, \dot{\mathbf{r}})$ is of class C^1, the equations (2.27) (or (2.26)) with arbitrary *initial conditions*

$$\mathbf{r}(t_0) = \mathbf{r}_0\,, \qquad \dot{\mathbf{r}}(t_0) = \mathbf{v}_0 \tag{2.28}$$

  have (locally) a *unique solution*. In other words, *the position and velocity of the particle at a certain instant* t_0 *determine its trajectory* $\mathbf{r}(t)$ *at any other (past or future) time* t. Classical mechanics is thus an essentially *deterministic* theory.

- *Newton's third law* (or **law of action and reaction**) states that if particle 2 exerts on particle 1 a force $\mathbf{F}_{12}$ then particle 1 exerts on particle 2 a force $\mathbf{F}_{21}$ of equal magnitude and opposite direction:

$$\mathbf{F}_{21} = -\mathbf{F}_{12}\,. \tag{2.29}$$

---

[2]The only exception worth mentioning is the force exerted on an accelerated charge by its own electromagnetic field, the so-called *Abraham–Lorentz–Dirac force*, which is proportional to $\dot{\mathbf{a}}$.

A stronger version of Newton's third law requires that, in addition, the force $\mathbf{F}_{12}$ (and, hence, $\mathbf{F}_{21}$) be parallel to the vector $\mathbf{r}_1 - \mathbf{r}_2$, that is to say, to the straight line joining both particles:

$$\mathbf{F}_{12} = -\mathbf{F}_{21} \parallel \mathbf{r}_1 - \mathbf{r}_2 . \tag{2.30}$$

It is important to bear in mind that Newton's third law—in either of its two versions (2.29) and (2.30)—does *not* have a fundamental character, since (for example) it does *not* hold in general for the electromagnetic force between two charges in relative motion. It is however verified—in fact, in its most restrictive version (2.30)—by the gravitational and electrostatic forces (see below), as well as by most macroscopic forces that occur in ordinary mechanical problems, as for example the tension of a string. ■

2.3.2 Inertial frames

It is obvious that Newton's first law *cannot* be valid in *all* reference frames. Indeed, let S and S' be two reference frames with parallel axes, and denote by $\mathbf{R}(t)$ the coordinates of the origin of S' with respect to the reference frame S at time t. Let us denote by $\mathbf{r}(t)$ the position vector of a particle with respect to the reference frame S at each instant t, so that the particle's velocity (with respect to S) is $\mathbf{v}(t) = \dot{\mathbf{r}}(t)$. *In Newtonian mechanics it is assumed that* time *has a universal character*[3], so that (once the unit of time has been set) the relation between the times t and t' of the same event measured in the frames S and S' is simply

$$t' = t - t_0 ,$$

where t_0 is a constant. From the point of view of S', therefore, the particle's coordinates at the time t' will be the components of the vector

$$\mathbf{r}'(t') = \mathbf{r}(t) - \mathbf{R}(t) = \mathbf{r}(t' + t_0) - \mathbf{R}(t' + t_0) .$$

The particle's velocity with respect to S' is thus

$$\mathbf{v}'(t') = \frac{\mathrm{d}\mathbf{r}'(t')}{\mathrm{d}t'} = \dot{\mathbf{r}}(t' + t_0) - \dot{\mathbf{R}}(t' + t_0) = \mathbf{v}(t' + t_0) - \dot{\mathbf{R}}(t' + t_0) ,$$

where as usual the dot denotes differentiation with respect to t. Suppose now that the particle is **free**, i.e., not subject to any force[4]. If Newton's first law holds in S, then $\mathbf{v}(t) = \mathbf{v}_0$ for all t. By the previous equation, the particle's velocity relative to S' is

$$\mathbf{v}'(t') = \mathbf{v}_0 - \dot{\mathbf{R}}(t' + t_0) ,$$

which is *not* constant unless $\dot{\mathbf{R}}$ is. Note that $\dot{\mathbf{R}}$ is constant if and only if $\ddot{\mathbf{R}} = 0$. We conclude that Newton's first law will hold in the reference frame S'—assuming that it holds in S, and that the axes of S and S' are parallel—if and only if its origin moves *without acceleration* with respect to S.

---

[3]We shall see in Chapter 8 that this postulate no longer holds in the special theory of relativity.

[4]Since at the classical level the magnitude of all known forces between two particles tends to zero as their distance tends to infinity, it is assumed that a particle is free if it is very far away from all other particles.

Definition 2.1. A reference frame in which Newton's first law holds is said to be **inertial**.

In view of the above considerations, Newton's first two laws can be formulated in a more accurate and logically satisfactory fashion as follows:

I. *There is a class of reference frames (called inertial) with respect to which free particles always move with constant velocity.*
II. *In an* inertial *frame, the force* $\mathbf{F}$ *exerted on a particle is equal to the rate of variation of its momentum* $\dfrac{d\mathbf{p}}{dt}$.

It should therefore be clear that:

1) Newton's first two laws are logically independent (in particular, the first law *defines* the class of reference frames in which the second law holds).
2) Both laws are not more or less arbitrary *axioms*, but rather *experimentally verifiable* (and, indeed, *verified*) facts—as remarked above, valid only *approximately*, in a certain range of speeds and forces.
3) The relation (2.25) between force and acceleration is (in general) ***only valid in an inertial reference frame.***

• What reference frames are known to be inertial? Galileo and Newton observed that a reference frame in which distant galaxies are at rest is (to a great approximation) inertial. More recently, it has been found that a reference frame with respect to which the cosmic microwave background radiation (a relic of the *big bang*) appears isotropic is inertial.

2.3.3 Galilean transformations

Let S be an inertial frame, and consider another frame S' whose origin has coordinates $\mathbf{R}(t)$ with respect to S at every instant t (where t denotes the time measured in S). We shall suppose that, at any time t, the axes of S are related to those of S' by a linear (invertible) transformation $A(t)$, i.e.,

$$\mathbf{e}_i = A(t)\mathbf{e}'_i\,, \qquad i = 1, 2, 3\,.$$

We shall also assume from now on that both S and S' are positively oriented orthogonal reference frames (that is, at all times the axes of both S and S' form a positively oriented orthonormal basis of $\mathbb{R}^3$). The matrix A must then be *orthogonal*, i.e., it must verify

$$A^{\mathsf{T}} = A^{-1}\,.$$

The determinant of an orthogonal matrix is equal to ± 1, since[5]

$$A^{\mathsf{T}}A = \mathbf{1} \quad \Longrightarrow \quad (\det A)^2 = 1\,.$$

[5]In what follows, **1** (boldface 1) will denote the identity matrix.

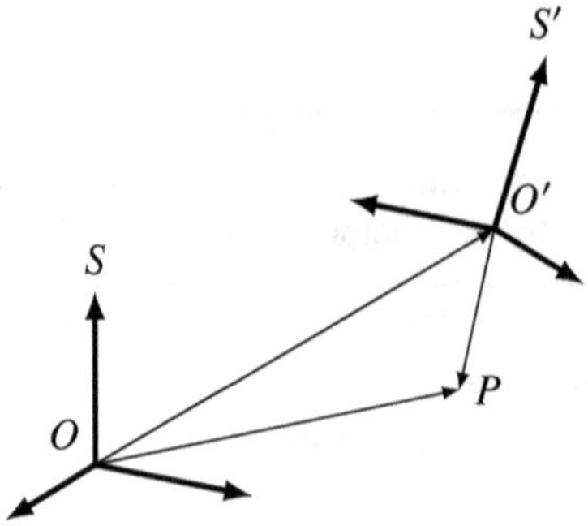

Figure 2.4. Reference frames S and S'. The coordinates of the vectors $\overrightarrow{OO'}$ and $\overrightarrow{OP}$ with respect to S are denoted in the text respectively by $\mathbf{R}$ and $\mathbf{r}$, and thus the coordinates of the vector $\overrightarrow{O'P} = \overrightarrow{OP} - \overrightarrow{OO'}$ with respect to S' are given by $\mathbf{r}' = A \cdot (\mathbf{r} - \mathbf{R})$.

In fact, we must have $\det A = 1$, since[6]

$$(\mathbf{e}_1 \times \mathbf{e}_2) \cdot \mathbf{e}_3 = (\det A)\, (\mathbf{e}'_1 \times \mathbf{e}'_2) \cdot \mathbf{e}'_3 .$$

We ask ourselves how must $A(t)$ and $\mathbf{R}(t)$ be so that the frame S' is inertial (assuming that S is an inertial frame). To answer this question, note that if we denote by $\mathbf{r}(t) = \mathbf{r}_0 + \mathbf{v}_0 t$ the coordinates with respect to S of a free particle at time t, its coordinates with respect to S' at the corresponding time $t' = t - t_0$ are given by[7]

$$\mathbf{r}'(t') = A(t) \cdot \big(\mathbf{r}(t) - \mathbf{R}(t)\big) = A(t) \cdot \big(\mathbf{r}_0 + \mathbf{v}_0 t - \mathbf{R}(t)\big), \qquad t = t' + t_0$$

(cf. Fig. 2.4). Differentiating twice with respect to t' we obtain

$$\frac{\mathrm{d}^2\mathbf{r}'(t')}{\mathrm{d}t'^2} = \ddot{A}(t)\mathbf{r}_0 + \big[t\ddot{A}(t) + 2\dot{A}(t)\big]\mathbf{v}_0 - \big[\ddot{A}(t)\mathbf{R}(t) + 2\dot{A}(t)\dot{\mathbf{R}}(t) + A(t)\ddot{\mathbf{R}}(t)\big] .$$

[6]Indeed, since $\mathbf{e}_i = \sum_{k=1}^{3} A_{ki}\mathbf{e}'_k$ (with $A = (A_{ij})_{i,j=1}^{3}$), we have

$$\begin{aligned}(\mathbf{e}_1 \times \mathbf{e}_2) \cdot \mathbf{e}_3 &= \sum_{i,j,k=1}^{3} A_{i1}A_{j2}A_{k3}(\mathbf{e}'_i \times \mathbf{e}'_j) \cdot \mathbf{e}'_k = \sum_{i,j,k=1}^{3} A_{i1}A_{j2}A_{k3}\varepsilon_{ijk}(\mathbf{e}'_1 \times \mathbf{e}'_2) \cdot \mathbf{e}'_3 \\ &= (\mathbf{e}'_1 \times \mathbf{e}'_2) \cdot \mathbf{e}'_3 \sum_{i,j,k=1}^{3} A_{i1}A_{j2}A_{k3}\varepsilon_{ijk} = \det A\, (\mathbf{e}'_1 \times \mathbf{e}'_2) \cdot \mathbf{e}'_3 .\end{aligned}$$

[7]Indeed, let $\mathbf{c} = (c_1, c_2, c_3)$ be the coordinates of a point in space with respect to the axes $\{\mathbf{e}_1, \mathbf{e}_2, \mathbf{e}_3\}$ of S at a certain time t, and denote by $\mathbf{c}' = (c'_1, c'_2, c'_3)$ the coordinates of the *same* point with respect to the axes $\{\mathbf{e}'_1, \mathbf{e}'_2, \mathbf{e}'_3\}$ of S' at that instant. We then have

$$\sum_{i=1}^{3} c_i\mathbf{e}_i = \sum_{i=1}^{3} c_i A(t) \cdot \mathbf{e}'_i = \sum_{i,k=1}^{3} c_i A_{ki}(t)\,\mathbf{e}'_k = \sum_{k=1}^{3}\left(\sum_{i=1}^{3} A_{ki}(t)c_i\right)\mathbf{e}'_k \quad\Longrightarrow\quad c'_k = \sum_{i=1}^{3} A_{ki}(t)c_i ,$$

or in matrix notation $\mathbf{c}' = A(t)\mathbf{c}$.

If the reference frame S' is also inertial, the right-hand side (RHS) of this equality must vanish identically for all $t \in \mathbb{R}$ and all $\mathbf{r}_0, \mathbf{v}_0 \in \mathbb{R}^3$. We thus have

$$\ddot{A}(t) = t\ddot{A}(t) + 2\dot{A}(t) = 0\,, \qquad \ddot{A}(t)\mathbf{R}(t) + 2\dot{A}(t)\dot{\mathbf{R}}(t) + A(t)\ddot{\mathbf{R}}(t) = 0\,,$$

or, equivalently,

$$\boxed{\dot{A}(t) = 0\,, \qquad \ddot{\mathbf{R}}(t) = 0\,.}$$

In other words:

The necessary and sufficient conditions in order for S' to be an inertial frame are that the rotation matrix $A(t)$ relating the axes of S and S' be *constant*, and that the origin of S' move with *constant velocity* with respect to S, i.e.,

$$\mathbf{R}(t) = \mathbf{R}_0 + \mathbf{V}_0 t\,,$$

with $\mathbf{R}_0$ and $\mathbf{V}_0$ constant vectors.

Moreover, the transformation relating the space-time coordinates $(t, \mathbf{r})$ and $(t', \mathbf{r}')$ of an event in the inertial reference frames S and S' is given by

$$\boxed{\begin{aligned} &t' = t - t_0\,, \qquad \mathbf{r}' = A \cdot \left(\mathbf{r} - \mathbf{R}_0 - \mathbf{V}_0 t\right); \\ &t_0 \in \mathbb{R}\,, \quad \mathbf{R}_0, \mathbf{V}_0 \in \mathbb{R}^3\,, \quad A \in \mathrm{SO}(3, \mathbb{R})\,, \end{aligned}} \tag{2.31}$$

where $\mathrm{SO}(3, \mathbb{R})$ denotes the group[8] of 3×3 real orthogonal matrices with unit determinant.

Definition 2.2. The change of coordinates $(t, \mathbf{r}) \mapsto (t', \mathbf{r}')$ defined by Eq. (2.31) is called a **Galilean transformation**.

Note. A **Galilean boost** is a transformation (2.31) with $t_0 = 0$, $\mathbf{R}_0 = 0$, $A = \mathbb{1}$.

From the previous discussion it then follows that:

The space-time coordinates $(t, \mathbf{r})$ and $(t', \mathbf{r}')$ of the same event in two inertial frames S and S' are related by an appropriate Galilean transformation (2.31).

• It is easy to verify that the composition of two Galilean transformation and the inverse of a Galilean transformation are Galilean transformations (see the exercises at the end of this section). From the mathematical point of view, this means that the set of all Galilean transformations forms a *group*, the so-called **Galilean group**.

• From what we have just seen it follows that, given an inertial frame S, any other inertial frame S' is obtained from S by translating its origin with constant velocity and applying a *constant* (i.e., time-independent) rotation to its axes.

[8]Recall that a *group* is a set G endowed with an associative product (an application $\cdot : G \times G \to G$ such that $g_1(g_2 g_3) = (g_1 g_2) g_3$ for all $g_1, g_2, g_3 \in G$), possessing a unit element and such that every element of G has an inverse.

2.3.4 Galileo's relativity principle

By Eq. (2.31), the acceleration in the reference frame S' is given by

$$\frac{\mathrm{d}^2\mathbf{r}'}{\mathrm{d}t'^2}(t') = A\cdot\ddot{\mathbf{r}}(t)\,,$$

and from (2.25) it then follows that

$$\boxed{m\,\frac{\mathrm{d}^2\mathbf{r}'}{\mathrm{d}t'^2} = \mathbf{F}'\left(t',\mathbf{r}',\frac{\mathrm{d}\mathbf{r}'}{\mathrm{d}t'}\right),} \tag{2.32}$$

with

$$\mathbf{F}'(t',\mathbf{r}',\dot{\mathbf{r}}') = A\cdot\mathbf{F}(t,\mathbf{r},\dot{\mathbf{r}})\,,\qquad \dot{\mathbf{r}}' := \frac{\mathrm{d}\mathbf{r}'}{\mathrm{d}t'}\,. \tag{2.33}$$

Thus, if $\mathbf{F}(t,\mathbf{r},\dot{\mathbf{r}})$ is the force acting at time t on a particle located at a point $\mathbf{r}$ moving with velocity $\dot{\mathbf{r}}$ as measured in the inertial frame S, the corresponding force measured in the second inertial frame S' is given by Eq. (2.33). This equation simply states that the force behaves as a *vector* under a Galilean transformation (2.31). Equivalently, $\mathbf{F}$ and $\mathbf{F}'$ represent the *same* vector in two different frames. In other words, the observers at S and S' measure the *same* force, although of course they assign it different components because their axes do not coincide. Note also that the transformation law (2.33) depends only on the relation between the two inertial frames S and S', and is therefore *independent of the properties of the particle considered* (i.e., its mass, electric charge, etc.).

From Eqs. (2.32)–(2.33) it also follows that Newton's second law—which, as we have seen, is the fundamental law of mechanics—has the same *form* in the inertial frame S' as in the original frame S. In other words:

> The laws of mechanics have the *same form* in *all* inertial frames (**Galileo's relativity principle**).

• What happens to Newton's second law in a *non-inertial* frame? We shall see in Chapter 6 that the force measured by a non-inertial observer differs from that measured by an inertial one by several terms *proportional to the mass of the particle considered*, called **fictitious** or **inertial forces**[9]. In other words, *the laws of physics assume their simplest form* (that is, *without* fictitious forces) *only in* inertial *frames.*

Exercise 2.6. Find the parameters of the Galilean transformation obtained by composing (2.31) with a second Galilean transformation

$$t'' = t' - t'_0\,,\qquad \mathbf{r}'' = A'\cdot(\mathbf{r}' - \mathbf{R}'_0 - \mathbf{V}'_0 t'),$$

with

$$t'_0 \in \mathbb{R}\,,\quad \mathbf{R}'_0, \mathbf{V}'_0 \in \mathbb{R}^3\,,\quad A' \in \mathrm{SO}(3,\mathbb{R})\,.$$

[9] An example of such a force is the *centrifugal force* that appears in a frame whose axes are rotating as seen from an inertial frame.

Solution. Substituting Eq. (2.31) into the previous equations we obtain $t'' = t - t_0 - t_0'$ and

$$\begin{aligned}\mathbf{r}'' &= A' \cdot \left[A(\mathbf{r} - \mathbf{R}_0 - \mathbf{V}_0 t) - \mathbf{R}_0' - \mathbf{V}_0'(t - t_0)\right] \\ &= A'A \cdot \left[\mathbf{r} - \mathbf{R}_0 - \mathbf{V}_0 t - A^{-1}\mathbf{R}_0' - A^{-1}\mathbf{V}_0'(t - t_0)\right] \\ &= A'A \cdot \left[\mathbf{r} - (\mathbf{R}_0 + A^{-1}\mathbf{R}_0' - t_0 A^{-1}\mathbf{V}_0') - (\mathbf{V}_0 + A^{-1}\mathbf{V}_0')t\right].\end{aligned}$$

These are the equations of a Galilean transformation, with parameters

$$t_0'' = t_0 + t_0', \qquad \mathbf{R}_0'' = \mathbf{R}_0 + A^{-1}\mathbf{R}_0' - t_0 A^{-1}\mathbf{V}_0',$$
$$\mathbf{V}_0'' = \mathbf{V}_0 + A^{-1}\mathbf{V}_0', \qquad A'' = A'A.$$

From the mathematical point of view, the previous equations define the *multiplication law* of the Galilean group.

Exercise 2.7. Show that the inverse of (2.31) is a Galilean transformation, and find its parameters.

Solution. From the previous exercise it follows that the inverse of (2.31) is the Galilean transformation with parameters $(t_0', A', \mathbf{R}_0', \mathbf{V}_0')$ satisfying

$$t_0 + t_0' = 0, \qquad A'A = \mathbf{1}, \qquad \mathbf{R}_0 + A^{-1}\mathbf{R}_0' - t_0 A^{-1}\mathbf{V}_0' = \mathbf{V}_0 + A^{-1}\mathbf{V}_0' = 0.$$

Solving for $(t_0', A', \mathbf{R}_0', \mathbf{V}_0')$ we easily obtain

$$t_0' = -t_0, \qquad A' = A^{-1}, \qquad \mathbf{R}_0' = -A(\mathbf{R}_0 + \mathbf{V}_0 t_0), \qquad \mathbf{V}_0' = -A\mathbf{V}_0.$$

Note that the previous equations could also have been obtained by solving for $(t, \mathbf{r})$ in terms of $(t', \mathbf{r}')$ in Eq. (2.31) (exercise).

2.4 CONSERVATION LAWS AND CONSERVATIVE FORCES

2.4.1 Conservation laws

A **conserved quantity** (also called **constant of motion, integral of motion** or **first integral**) is any function of $(t, \mathbf{r}, \dot{\mathbf{r}})$ that remains constant as the particle moves. Knowing a conserved quantity is usually very advantageous, since it provides important information on the nature of the motion. For instance, Newton's first law (2.23) immediately yields a **law of conservation of linear momentum:** in the absence of forces, the linear momentum $\mathbf{p}$ of a particle is *conserved.* Let us next define the particle's **angular momentum** with respect to the origin of coordinates by

$$\boxed{\mathbf{L} = \mathbf{r} \times \mathbf{p} = m\mathbf{r} \times \dot{\mathbf{r}},} \tag{2.34}$$

and the **torque** of the force $\mathbf{F}$ (also with respect to the origin) by

$$\boxed{\mathbf{N} = \mathbf{r} \times \mathbf{F}\,.} \tag{2.35}$$

Differentiating with respect to t the definition of angular momentum and applying Newton's second law we easily obtain the important identity

$$\boxed{\dot{\mathbf{L}} = \mathbf{N}\,.}$$

From this equation it immediately follows the **law of conservation of angular momentum:** if the torque of the force acting on a particle vanishes, its angular momentum is conserved. Note that in this case, since $\mathbf{r}$ is perpendicular to the constant vector $\mathbf{L}$, *the motion takes place in the normal plane to* $\mathbf{L}$ *passing through the origin.*

• From Eq. (2.35) it follows that $\mathbf{N} = 0$ if either $\mathbf{r} = 0$ or the applied force $\mathbf{F}$ is parallel to the particle's position vector $\mathbf{r}$, i.e., (assuming that $\mathbf{F}$ depends only on t, $\mathbf{r}$, and $\dot{\mathbf{r}}$):

$$\mathbf{F} = f(t, \mathbf{r}, \dot{\mathbf{r}})\mathbf{e}_r\,, \tag{2.36}$$

where f is an arbitrary scalar function. This type of force is called **central.**

Consider next the particle's **kinetic energy**, defined by

$$\boxed{T = \frac{1}{2} m\,\dot{\mathbf{r}}^2\,.} \tag{2.37}$$

Taking the scalar product of Newton's second law with the velocity vector $\dot{\mathbf{r}}$ we obtain

$$\boxed{\frac{\mathrm{d}T}{\mathrm{d}t} = m\dot{\mathbf{r}}\ddot{\mathbf{r}} = \mathbf{F}(t, \mathbf{r}, \dot{\mathbf{r}})\dot{\mathbf{r}}\,.} \tag{2.38}$$

In particular, *kinetic energy is conserved if the force* $\mathbf{F}$ *is perpendicular to the velocity vector* $\dot{\mathbf{r}}$ *at all times.* This is what happens, for instance, with the *magnetic force* acting on a charged particle (cf. Eq. (2.51) below).

Definition 2.3. We shall say that a force $\mathbf{F}(\mathbf{r})$ is **conservative** if it can be expressed in terms of a scalar **potential** $V(\mathbf{r})$ through the formula

$$\mathbf{F}(\mathbf{r}) = -\frac{\partial V(\mathbf{r})}{\partial \mathbf{r}} \equiv -\nabla V(\mathbf{r}) = -\sum_{i=1}^{3} \frac{\partial V(\mathbf{r})}{\partial x_i}\,\mathbf{e}_i\,. \tag{2.39}$$

Note, in particular, that by its very definition *a conservative force can depend only on the particle's position vector* $\mathbf{r}$ (i.e., it must be independent of t and $\dot{\mathbf{r}}$). If $\mathbf{F}(\mathbf{r}) = -\dfrac{\partial V(\mathbf{r})}{\partial \mathbf{r}}$ is conservative, we have

$$\mathbf{F}(\mathbf{r})\dot{\mathbf{r}} = -\frac{\partial V(\mathbf{r})}{\partial \mathbf{r}}\,\dot{\mathbf{r}} = -\frac{\mathrm{d}}{\mathrm{d}t}V(\mathbf{r})\,,$$

and Eq. (2.38) becomes

$$\frac{\mathrm{d}}{\mathrm{d}t}(T+V)=0\,.$$

The previous equation is the **law of conservation of energy:** *if the force acting on a particle is conservative, with potential* $V(\mathbf{r})$*, then the* **total energy**

$$\boxed{E := T + V = \frac{1}{2}\,m\,\dot{\mathbf{r}}^2 + V(\mathbf{r})} \tag{2.40}$$

is conserved.

• More generally, we shall say that a *time-dependent* force $\mathbf{F}(t,\mathbf{r})$ is **irrotational** provided that $\nabla \times \mathbf{F}(t,\mathbf{r}) = 0$ for all $(t,\mathbf{r})$, where $\nabla\times\mathbf{F}$ is the **curl** of $\mathbf{F}$ (cf. Eq. (1.16)). It can be shown (assuming, e.g., that $\mathbf{F}$ is of class C^1 on $\mathbb{R}^4$) that $\mathbf{F}$ is irrotational if and only if there is a *time-dependent* function $V(t,\mathbf{r})$ such that

$$\mathbf{F}(t,\mathbf{r}) = -\frac{\partial V(t,\mathbf{r})}{\partial \mathbf{r}}\,.$$

If the force $\mathbf{F}$ is irrotational, differentiating the definition (2.40) of energy we obtain

$$\frac{\mathrm{d}E}{\mathrm{d}t} = \frac{\mathrm{d}T}{\mathrm{d}t} + \frac{\mathrm{d}V}{\mathrm{d}t} = m\dot{\mathbf{r}}\ddot{\mathbf{r}} + \frac{\partial V}{\partial t} + \frac{\partial V}{\partial \mathbf{r}}\,\dot{\mathbf{r}} = \frac{\partial V}{\partial t}\,.$$

Thus if the force is irrotational, but depends explicitly on time, energy is *not* conserved.

2.4.2 Conservative forces

As we saw in the previous subsection, a force $\mathbf{F}(\mathbf{r})$ is conservative if it is the gradient of a function $-V(\mathbf{r})$. Note that (in a *connected* open subset) the potential $V(\mathbf{r})$ is determined by the force $\mathbf{F}(\mathbf{r})$ up to an arbitrary constant, since

$$\mathbf{F}(\mathbf{r}) = \nabla V_1 = \nabla V_2 \quad \Longleftrightarrow \quad \nabla(V_1 - V_2) = 0 \quad \Longrightarrow \quad V_1 - V_2 = \text{const}\,.$$

It can be shown that the conservative character of a force (independent of time and velocity) $\mathbf{F}(\mathbf{r})$ is *equivalent* to any of the following three conditions:

I. The force $\mathbf{F}$ is **irrotational:**

$$\nabla\times\mathbf{F} = 0\,.$$

II. The **work** done by the force $\mathbf{F}$ along *any* closed curve C vanishes:

$$\int_C \mathbf{F}(\mathbf{r})\cdot \mathrm{d}\mathbf{r} = 0\,.$$

III. The work done by the force $\mathbf{F}$ along *any* curve C with fixed endpoints $\mathbf{r}_1$ and $\mathbf{r}_2$ is *independent of the curve.* In other words,

$$\int_{C_1} \mathbf{F}(\mathbf{r})\cdot \mathrm{d}\mathbf{r} = \int_{C_2} \mathbf{F}(\mathbf{r})\cdot \mathrm{d}\mathbf{r}\,,$$

for any two curves C_1 and C_2 with the same endpoints $\mathbf{r}_1$ and $\mathbf{r}_2$.

The *necessity* of conditions I)–III) above (i.e., that if $\mathbf{F}(\mathbf{r})$ is conservative then I)–III) hold) is straightforward. Indeed, condition I) is a direct consequence of the identity

$$\nabla \times \nabla V(\mathbf{r}) = 0 .$$

Likewise, the work done by a *conservative* force $\mathbf{F} = -\nabla V$ along any curve C with endpoints $\mathbf{r}_1$ and $\mathbf{r}_2$ is given by

$$\boxed{\int_C \mathbf{F}(\mathbf{r}) \cdot \mathrm{d}\mathbf{r} = -\int_C \frac{\partial V(\mathbf{r})}{\partial \mathbf{r}} \cdot \mathrm{d}\mathbf{r} = -\int_C \mathrm{d}V = V(\mathbf{r}_1) - V(\mathbf{r}_2) ,} \tag{2.41}$$

and is thus independent of the curve considered (condition III). In particular, if C is closed then we can take $\mathbf{r}_1 = \mathbf{r}_2$, and hence $\mathbf{F}$ does no work (condition II). It is shown in advanced calculus courses that the *converse* (i.e., that if any of the conditions I)–III) above hold, then $\mathbf{F}(\mathbf{r})$ is conservative) is also true provided that $\mathbf{F}(\mathbf{r})$ is of class C^1 in a *simply connected* open subset[10] of $\mathbb{R}^3$ (in particular, on all of $\mathbb{R}^3$).

• From Eq. (2.41) it follows that the work done by a conservative force is equal to the *decrease* in the potential energy as the particle moves from the initial point $\mathbf{r}_1$ to the final one $\mathbf{r}_2$. By the law of conservation of total energy, this coincides with the *increase* in the particle's kinetic energy as it moves from $\mathbf{r}_1$ to $\mathbf{r}_2$.

• More generally, if $\mathbf{F}(t, \mathbf{r}, \dot{\mathbf{r}})$ is an arbitrary (not necessarily conservative) force the work W done by $\mathbf{F}$ along a trajectory $C = \{\mathbf{r} = \mathbf{r}(t) | t \in [t_1, t_2]\}$ starting at a point $\mathbf{r}_1 = \mathbf{r}(t_1)$ with velocity $\dot{\mathbf{r}}_1 = \dot{\mathbf{r}}(t_1)$ and ending at a point $\mathbf{r}_2 = \mathbf{r}(t_2)$ with velocity $\dot{\mathbf{r}}_2 = \dot{\mathbf{r}}(t_2)$ is equal to the increase in the particle's kinetic energy. Indeed,

$$\begin{aligned} W &= \int_C \mathbf{F}(t, \mathbf{r}, \dot{\mathbf{r}}) \, \mathrm{d}t = \int_{t_1}^{t_2} \mathbf{F}\big(t, \mathbf{r}(t), \dot{\mathbf{r}}(t)\big) \cdot \dot{\mathbf{r}}(t) \, \mathrm{d}t = \int_{t_1}^{t_2} m\ddot{\mathbf{r}}(t) \cdot \dot{\mathbf{r}}(t) \, \mathrm{d}t \\ &= \int_{t_1}^{t_2} \frac{\mathrm{d}T}{\mathrm{d}t} \, \mathrm{d}t = T(\dot{\mathbf{r}}_2) - T(\dot{\mathbf{r}}_1), \end{aligned}$$

since $T = \frac{1}{2} m\dot{\mathbf{r}}^2$ depends only on $\dot{\mathbf{r}}$. Note, however, that the work along two trajectories with the same endpoints $\mathbf{r}_1$ and $\mathbf{r}_2$ need *not* be the same, since the initial and/or final velocities $\dot{\mathbf{r}}_{1,2}$ will in general be different for both trajectories.

• A particular case of conservative force of great practical interest is that of a *central force* of the form

$$\boxed{\mathbf{F}(\mathbf{r}) = f(r) \, \frac{\mathbf{r}}{r} .} \tag{2.42}$$

[10]By definition, a connected open subset $U \subset \mathbb{R}^3$ is simply connected if any continuous closed curve contained in U can be continuously contracted to a point within U. For example, $\mathbb{R}^3$, $\mathbb{R}^3$ minus one point, the interior of a sphere, a parallelepiped, a cylinder, etc., are simply connected sets, while $\mathbb{R}^3$ minus a line is not.

Indeed, taking into account that

$$\frac{\partial V(r)}{\partial \mathbf{r}} = V'(r)\,\frac{\partial r}{\partial \mathbf{r}} = V'(r)\,\frac{\mathbf{r}}{r}\,,$$

it is obvious that the force (2.42) is generated by the potential

$$\boxed{V(r) = -\int f(r)\,\mathrm{d}r\,,} \tag{2.43}$$

which depends only on the magnitude of the position vector **r**. Thus *if the force is central and conservative both energy and angular momentum are conserved.*

Exercise 2.8. Show that the central force (2.36) is conservative if and only if the function $f(t, \mathbf{r}, \dot{\mathbf{r}})$ depends only on r.

Solution. To begin with, f can only depend on **r** from the definition of conservative force. If $F_\theta = F_\varphi = 0$ and $F_r = f(\mathbf{r})$ Eq. (2.18) for the curl of **F** in spherical coordinates yields

$$\nabla \times \mathbf{F} = \frac{1}{r\sin\theta}\,\frac{\partial f}{\partial \varphi}\,\mathbf{e}_\theta - \frac{1}{r}\,\frac{\partial f}{\partial \theta}\,\mathbf{e}_\varphi = 0 \iff \frac{\partial f}{\partial \theta} = \frac{\partial f}{\partial \varphi} = 0\,,$$

so that f is a function of r only. Alternatively, using Eq. (2.18) for the gradient of $V(\mathbf{r})$ in spherical coordinates we obtain

$$\frac{\partial V}{\partial \mathbf{r}} = \frac{\partial V}{\partial r}\,\mathbf{e}_r + \frac{1}{r}\,\frac{\partial V}{\partial \theta}\,\mathbf{e}_\theta + \frac{1}{r\sin\theta}\,\frac{\partial V}{\partial \varphi}\,\mathbf{e}_\varphi = -\mathbf{F} = -f\mathbf{e}_r \quad \Longrightarrow \quad \frac{\partial V}{\partial \theta} = \frac{\partial V}{\partial \varphi} = 0\,.$$

Hence V is a function of r only, and so is $f = -V'(r)$.

2.4.3 Gravitational and electrostatic forces

According to Newton's **law of universal gravitation**, the gravitational force exerted by a particle of mass M *fixed* at the origin of coordinates on another particle of mass m located at a point **r** is of the form (2.42) with f inversely proportional to the square of the distance to the origin:

$$\boxed{f(r) = -\frac{GMm}{r^2}\,,} \tag{2.44}$$

where[11]

$$G = 6.674\,30(15)\cdot 10^{-11}\ \mathrm{m^3\,Kg^{-1}\,s^{-2}}$$

[11]In this book we use the CODATA internationally recommended 2022 values of the fundamental physical constants, available at the site `https://physics.nist.gov/cuu/Constants/`.

is the so-called *gravitational constant*. By Eq. (2.43), the potential $V(r)$ generating the gravitational force (2.44) is given (up to an arbitrary constant) by

$$\boxed{V(r) = -\frac{GMm}{r}\,.} \tag{2.45}$$

Note that, since $GMm > 0$, the gravitational force is always *attractive*.

- The acceleration caused by the gravitational force (2.44) on a particle of mass m is

$$\boxed{\mathbf{a} = \frac{\mathbf{F}}{m} = -\frac{GM}{r^3}\,\mathbf{r}\,,}$$

independent of m. This nontrivial fact, first observed by Galileo Galilei, is due to the fact that the mass appearing in Newton's law of universal gravitation (the **gravitational mass**) actually coincides[12] with the mass appearing in Newton's second law (the **inertial mass**). The equality between the gravitational and inertial masses—the so-called **equivalence principle**, on which Einstein's general theory of relativity is based—has been verified with great accuracy (less than one part in 10^{12}) in different experiments.

Similarly, the electric force exerted by a charge Q fixed at the origin on a point charge q located at a point $\mathbf{r}$ is also of the form (2.42), where now

$$\boxed{f(r) = k\,\frac{qQ}{r^2}\,.} \tag{2.46}$$

In the SI system of units, the constant k is given by

$$k = \frac{1}{4\pi\varepsilon_0} \simeq 8.98755 \cdot 10^9\ \mathrm{m\,F^{-1}}\,,$$

where

$$\varepsilon_0 = 8.854\,187\,8188(14) \cdot 10^{-12}\ \mathrm{F\,m^{-1}}$$

is the *vacuum permittivity*. From Eq. (2.46) it follows that the electric force is attractive if the charges q and Q are of opposite signs, and repulsive if they have the same sign. Again, the electric force is obviously conservative, with potential (up to an additive constant)

$$\boxed{V(r) = k\,\frac{qQ}{r}}$$

inversely proportional to the distance between the charges.

[12]Obviously, it is only necessary that inertial and gravitational mass differ by a universal (i.e., the same for all particles) proportionality constant.

More generally, the gravitational force exerted on a particle of mass m located at the point $\mathbf{r}$ by a continuous mass distribution occupying an open subset $U \subset \mathbb{R}^3$ is given by

$$\mathbf{F}(\mathbf{r}) = -Gm \int_U \rho(\mathbf{r}') \frac{\mathbf{r} - \mathbf{r}'}{|\mathbf{r} - \mathbf{r}'|^3} \, \mathrm{d}^3 r' =: m\mathbf{g}(\mathbf{r}) = -m \frac{\partial \Phi(\mathbf{r})}{\partial \mathbf{r}}, \tag{2.47}$$

where $\rho(\mathbf{r}')$ is the mass density at $\mathbf{r}' \in U$, $\mathbf{g}(\mathbf{r})$ is the **gravitational field** created by the mass distribution at the point $\mathbf{r}$, and

$$\Phi(\mathbf{r}) = -G \int_U \frac{\rho(\mathbf{r}')}{|\mathbf{r} - \mathbf{r}'|} \, \mathrm{d}^3 r' \tag{2.48}$$

is the **gravitational potential**. Thus the gravitational force is still conservative in this more general situation, with potential $V(\mathbf{r}) = m\Phi(\mathbf{r})$. Again, the particle's acceleration

$$\mathbf{a} = \frac{\mathbf{F}(\mathbf{r})}{m} = \mathbf{g}(\mathbf{r})$$

is independent of its mass m. Note, however, that in general (unless the mass distribution is spherically symmetric about the origin) the gravitational force (2.47) is *not* central.

Similarly, the force exerted on a point charge q located at a point $\mathbf{r}$ by a *static* charge distribution filling up an open set U is

$$\mathbf{F}(\mathbf{r}) = kq \int_U \rho(\mathbf{r}') \frac{\mathbf{r} - \mathbf{r}'}{|\mathbf{r} - \mathbf{r}'|^3} \, \mathrm{d}^3 r' =: q\mathbf{E}(\mathbf{r}) = -q \frac{\partial \Phi(\mathbf{r})}{\partial \mathbf{r}}, \tag{2.49}$$

where now $\rho(\mathbf{r}')$ is the charge density at a point $\mathbf{r}'$, $\mathbf{E}(\mathbf{r})$ is the **electric field** created by the charge distribution at the point $\mathbf{r}$ and

$$\Phi(\mathbf{r}) = k \int_U \frac{\rho(\mathbf{r}')}{|\mathbf{r} - \mathbf{r}'|} \, \mathrm{d}^3 r' \tag{2.50}$$

is the **electrostatic potential**. Again, the electrostatic force (2.49) is conservative (with potential $V(\mathbf{r}) = q\Phi(\mathbf{r})$), but *not* central unless the charge distribution is spherically symmetric about the origin.

Exercise 2.9. Applying Gauss's theorem to a sphere centered at the origin prove the identity

$$\Delta\left(\frac{1}{r}\right) = -4\pi\delta(\mathbf{r}),$$

where $\Delta \equiv \nabla^2$ and $\delta(\mathbf{r})$ is **Dirac's delta function**. Deduce from Eq. (2.48) that both the gravitational and the electrostatic potential verify **Poisson's equation**

$$\Delta\Phi = 4\pi\alpha\,\rho,$$

where $\alpha = G$ for the gravitational potential and $\alpha = -k$ for the electrostatic one. In particular, the gravitational (resp. electrostatic) potential verifies **Laplace's equation** $\Delta\Phi = 0$ in any region of space where there are no masses (resp. charges).

Note. Dirac's delta function $\delta(\mathbf{r})$ is informally defined by the requirements $\delta(\mathbf{r}) = 0$ for all $\mathbf{r} \neq 0$ and $\int_{\mathbb{R}^3} \delta(\mathbf{r})\, \mathrm{d}^3 r = 1$. It can thus be intuitively viewed as the mass density of a point mass located at the origin. In fact, no ordinary function can simultaneously verify the above two requirements, since for an ordinary function the condition $\delta(\mathbf{r}) = 0$ for all $\mathbf{r} \neq 0$ implies that $\int_{\mathbb{R}^3} \delta(\mathbf{r})\, \mathrm{d}^3 r = 0$. We can think of $\delta(\mathbf{r})$ as the "limit" as $\varepsilon \to 0+$ of any family of functions $\delta_\varepsilon(\mathbf{r})$ satisfying $\int_{\mathbb{R}^3} \delta_\varepsilon(\mathbf{r})\, \mathrm{d}^3 r = 1$ for all $\varepsilon > 0$, and such that $\delta_\varepsilon(\mathbf{r})$ is concentrated inside a ball centered at the origin whose radius tends to zero as $\varepsilon \to 0+$. (One such family is, for instance, $\delta_\varepsilon(\mathbf{r}) = (\pi\varepsilon)^{-3/2} \mathrm{e}^{-r^2/\varepsilon}$.) From this heuristic definition follows the important property $\int_{\mathbb{R}^3} \delta(\mathbf{r}) f(\mathbf{r})\, \mathrm{d}^3 r = f(0)$, for any sufficiently smooth function $f(\mathbf{r})$. In fact, the previous identity can be taken as a working definition of $\delta(\mathbf{r})$. A rigorous mathematical treatment of Dirac's delta function requires the use of the theory of *distributions* (linear functionals defined on spaces of smooth functions vanishing fast enough at infinity).

Solution. To begin with, let us check that $\Delta(1/r) = 0$ for $\mathbf{r} \neq 0$. Indeed, if $\mathbf{r} \neq 0$ (and hence $r \neq 0$) we have

$$\Delta\left(\frac{1}{r}\right) = \nabla \cdot \left[\nabla\left(\frac{1}{r}\right)\right] = \nabla \cdot \left(-\frac{\mathbf{r}}{r^3}\right) = -\frac{\nabla \cdot \mathbf{r}}{r^3} - \mathbf{r} \cdot \nabla\left(\frac{1}{r^3}\right)$$
$$= -\frac{3}{r^3} - \mathbf{r} \cdot \left(-\frac{3\mathbf{e}_r}{r^4}\right) = -\frac{3}{r^3} + \frac{3}{r^3} = 0.$$

To show that $\Delta(1/r) = -4\pi\delta(\mathbf{r})$, we only have to prove the equality

$$\int_{\mathbb{R}^3} \Delta\left(\frac{1}{r}\right) \mathrm{d}^3 r = -4\pi\,.$$

As we have just seen that $\Delta(1/r) = 0$ away from the origin, we can integrate over a ball of arbitrary radius R centered at the origin. We thus have

$$\int_{\mathbb{R}^3} \Delta\left(\frac{1}{r}\right) \mathrm{d}^3 r = \int_{|\mathbf{r}|\leqslant R} \Delta\left(\frac{1}{r}\right) \mathrm{d}^3 r = -\int_{|\mathbf{r}|\leqslant R} \nabla \cdot \left(\frac{\mathbf{e}_r}{r^2}\right) \mathrm{d}^3 r$$
$$= -\int_{|\mathbf{r}|=R} \frac{\mathbf{e}_r}{r^2} \cdot \mathbf{n}\, \mathrm{d}S = -\int_{|\mathbf{r}|=R} \frac{\mathbf{e}_r}{R^2} \cdot \mathbf{e}_r\, \mathrm{d}S$$
$$= -\frac{1}{R^2} \int_{|\mathbf{r}|=R} \mathrm{d}S = -\frac{1}{R^2} \cdot 4\pi R^2 = -4\pi\,,$$

where we have applied Gauss's theorem to obtain the third equality. Taking the Laplacian (with respect to the $\mathbf{r}$ coordinate) of the equation for the gravitational/electrostatic potential,

$$\Phi(\mathbf{r}) = -\alpha \int_U \frac{\rho(\mathbf{r}')}{|\mathbf{r} - \mathbf{r}'|}\, \mathrm{d}^3 r'\,,$$

we then obtain

$$\Delta\Phi(\mathbf{r}) = -\alpha \int_U \Delta_{\mathrm{r}} \left(\frac{\rho(\mathbf{r}')}{|\mathbf{r}-\mathbf{r}'|} \right) \mathrm{d}^3 r' = -\alpha \int_U \rho(\mathbf{r}') \Delta_{\mathrm{r}} \left(\frac{1}{|\mathbf{r}-\mathbf{r}'|} \right) \mathrm{d}^3 r'$$
$$= 4\pi\alpha \int_U \rho(\mathbf{r}') \delta(\mathbf{r}-\mathbf{r}') \, \mathrm{d}^3 r' = 4\pi\alpha\rho(\mathbf{r}).$$

2.4.4 Electromagnetic force

The **electromagnetic force** (also called **Lorentz force**) acting on a point charge q which moves subject to an electric field $\mathbf{E}(t, \mathbf{r})$ and a magnetic field $\mathbf{B}(t, \mathbf{r})$ is given by

$$\mathbf{F}(t, \mathbf{r}, \dot{\mathbf{r}}) = q\big(\mathbf{E}(t, \mathbf{r}) + \dot{\mathbf{r}} \times \mathbf{B}(t, \mathbf{r})\big). \tag{2.51}$$

As is well known, the fields $\mathbf{E}$ and $\mathbf{B}$ verify **Maxwell's equations**

$$\nabla \cdot \mathbf{E} = \frac{\rho}{\varepsilon_0}, \qquad \nabla \times \mathbf{E} = -\frac{\partial \mathbf{B}}{\partial t},$$
$$\nabla \cdot \mathbf{B} = 0, \qquad \nabla \times \mathbf{B} = \mu_0 \mathbf{J} + \frac{1}{c^2} \frac{\partial \mathbf{E}}{\partial t},$$

where $\mathbf{J}$ is the current density,

$$c = 2.997\,924\,58 \times 10^8 \ \mathrm{m\,s^{-1}}$$

is the speed of light *in vacuo*, and

$$\mu_0 := (c^2\varepsilon_0)^{-1} = 1.256\,637\,061\,27(20) \ \mathrm{N\,A^{-2}}$$

is the *vacuum permeability*. From the second and third Maxwell equations it follows that $\mathbf{E}$ and $\mathbf{B}$ can be expressed through a **scalar potential** $\Phi(t, \mathbf{r})$ and a **vector potential** $\mathbf{A}(t, \mathbf{r})$ through the equations

$$\mathbf{E} = -\frac{\partial \Phi}{\partial \mathbf{r}} - \frac{\partial \mathbf{A}}{\partial t}, \qquad \mathbf{B} = \nabla \times \mathbf{A}. \tag{2.52}$$

Remark 2.3. The fields $\mathbf{E}$ and $\mathbf{B}$ do *not* uniquely determine the electromagnetic potentials Φ and $\mathbf{A}$. Indeed, it is easily verified (exercise) that the potentials

$$\widehat{\Phi} = \Phi - \frac{\partial f}{\partial t}, \qquad \widehat{\mathbf{A}} = \mathbf{A} + \frac{\partial f}{\partial \mathbf{r}}, \tag{2.53}$$

where $f(t, \mathbf{r})$ is an arbitrary scalar function[13], generate exactly the same electromagnetic field as Φ and $\mathbf{A}$. It can be shown that it is always possible to choose the function f so

[13]Equation (2.53) is called a *gauge transformation* of the electromagnetic potentials. It can be shown (see Exercise 2.12) that, if the electromagnetic potentials $(\Phi, \mathbf{A})$ and $(\widehat{\Phi}, \widehat{\mathbf{A}})$ generate the same electromagnetic field, then they are related by a gauge transformation (assuming, for simplicity, that the fields are of class C^2 on $\mathbb{R}^4$).

that the new potentials $\widehat{\Phi}$ and $\widehat{\mathbf{A}}$ verify the condition

$$\nabla \cdot \widehat{\mathbf{A}} + \frac{1}{c^2}\frac{\partial \widehat{\Phi}}{\partial t} = 0\,, \tag{2.54}$$

called the *Lorenz gauge*[14]. If the electromagnetic potentials satisfy the Lorenz gauge, it is immediate to check that Maxwell's equations are equivalent to the following two *uncoupled* equations for Φ and $\mathbf{A}$:

$$\frac{1}{c^2}\frac{\partial^2 \Phi}{\partial t^2} - \Delta\Phi = \frac{\rho}{\varepsilon_0}\,, \qquad \frac{1}{c^2}\frac{\partial^2 \mathbf{A}}{\partial t^2} - \Delta\mathbf{A} = \mu_0 \mathbf{J}\,.$$

In particular, *in vacuo* (that is, in any region of space not containing electrical charges or currents), the scalar potential Φ and each component A_i of the vector potential verify the **wave equation**

$$\boxed{\frac{1}{c^2}\frac{\partial^2 u}{\partial t^2} - \Delta u = 0,} \tag{2.55}$$

where c is the velocity of the waves. ■

Exercise 2.10. Using the identity (1.22), show that the components of the fields **E** and **B** also verify the wave equation *in vacuo*.

Solution. Indeed,

$$\nabla \times (\nabla \times \mathbf{E}) = \nabla(\nabla \cdot \mathbf{E}) - \Delta\mathbf{E} = \frac{1}{\varepsilon_0}\nabla\rho - \Delta\mathbf{E} = -\nabla \times \left(\frac{\partial \mathbf{B}}{\partial t}\right)$$
$$= -\frac{\partial}{\partial t}\nabla \times \mathbf{B} = -\mu_0\frac{\partial \mathbf{J}}{\partial t} - \frac{1}{c^2}\frac{\partial^2 \mathbf{E}}{\partial t^2}$$
$$\Longrightarrow \boxed{\frac{1}{c^2}\frac{\partial^2 \mathbf{E}}{\partial t^2} - \Delta\mathbf{E} = -\frac{1}{\varepsilon_0}\nabla\rho - \mu_0\frac{\partial \mathbf{J}}{\partial t},}$$
$$\nabla \times (\nabla \times \mathbf{B}) = -\Delta\mathbf{B} = \nabla \times \left(\mu_0\mathbf{J} + \frac{1}{c^2}\frac{\partial \mathbf{E}}{\partial t}\right) = \mu_0\nabla \times \mathbf{J} + \frac{1}{c^2}\frac{\partial}{\partial t}\nabla \times \mathbf{E}$$
$$= \mu_0\nabla \times \mathbf{J} - \frac{1}{c^2}\frac{\partial^2 \mathbf{B}}{\partial t^2} \quad \Longrightarrow \quad \boxed{\frac{1}{c^2}\frac{\partial^2 \mathbf{B}}{\partial t^2} - \Delta\mathbf{B} = \mu_0\nabla \times \mathbf{J}.}$$

[14] Indeed, it suffices that the function f be a solution of the partial differential equation

$$\frac{1}{c^2}\frac{\partial^2 f}{\partial t^2} - \Delta f = \nabla \cdot \mathbf{A} + \frac{1}{c^2}\frac{\partial \Phi}{\partial t}\,.$$

It is shown in differential equations courses that if the potentials **A** and Φ are analytic functions the previous equation has (locally) a solution dependent on two arbitrary functions of the variable **r** (*Cauchy–Kovalevskaya theorem*).

When ρ and $\mathbf{J}$ vanish both framed equations reduce to the wave equation with wave velocity c.

If both the electric and the magnetic fields are *static*, i.e., if

$$\mathbf{E} = \mathbf{E}(\mathbf{r}), \qquad \mathbf{B} = \mathbf{B}(\mathbf{r}),$$

Maxwell's second equation reduces to $\nabla \times \mathbf{E}(\mathbf{r}) = 0$, and hence

$$\mathbf{E} = -\frac{\partial \Phi(\mathbf{r})}{\partial \mathbf{r}}.$$

From this equation and the expression (2.51) for the Lorentz force it then follows that

$$\frac{\mathrm{d}T}{\mathrm{d}t} = \mathbf{F} \cdot \dot{\mathbf{r}} = q\mathbf{E}(\mathbf{r}) \cdot \dot{\mathbf{r}} = -q\,\frac{\partial \Phi(\mathbf{r})}{\partial \mathbf{r}} \cdot \dot{\mathbf{r}} = -q\frac{\mathrm{d}\Phi}{\mathrm{d}t} \quad \Longrightarrow \quad \frac{\mathrm{d}}{\mathrm{d}t}(T + q\Phi) = 0\,.$$

Thus in this case the function

$$\boxed{T + q\Phi(\mathbf{r}),}$$

which can be regarded as the particle's *electromechanical* energy, is conserved, although the Lorentz force is *not* conservative unless $\mathbf{B} = 0$ (cf. the exercise below). To interpret physically this result it suffices to note that the magnetic force does *no* work, since it is perpendicular to the velocity, and hence to the infinitesimal displacement $\mathrm{d}\mathbf{r}$, and therefore does not contribute to the particle's energy.

Exercise 2.11. Show that the Lorentz force (2.51) is conservative if and only if $\mathbf{B} = \dfrac{\partial \mathbf{E}}{\partial t} = 0$ (that is, if the electromagnetic field is purely electrostatic).

Solution. For a force to be conservative it must be independent of time and of the particle's velocity. The latter condition implies that $\mathbf{B} = 0$, and the former that the electric field $\mathbf{E}$ is static (i.e., time-independent). If these conditions are satisfied then Maxwell's equation $\nabla \times \mathbf{E}(\mathbf{r}) = 0$ implies that $\mathbf{E}(\mathbf{r}) = -\nabla\Phi(\mathbf{r})$, and hence the Lorentz force $\mathbf{F} = q\mathbf{E}(\mathbf{r}) = -\nabla\big(q\Phi(\mathbf{r})\big)$ is indeed conservative.

Exercise 2.12. Show that if the potentials $(\Phi, \mathbf{A})$ and $(\widetilde{\Phi}, \widetilde{\mathbf{A}})$ generate the same electromagnetic field $(\mathbf{E}, \mathbf{B})$ then (2.53) holds for some scalar function $f(t, \mathbf{r})$.

Solution. If $(\Phi, \mathbf{A})$ and $(\widetilde{\Phi}, \widetilde{\mathbf{A}})$ generate the same electromagnetic field then

$$\begin{aligned}
&\mathbf{E} = -\frac{\partial \Phi}{\partial \mathbf{r}} - \frac{\partial \mathbf{A}}{\partial t} = -\frac{\partial \widetilde{\Phi}}{\partial \mathbf{r}} - \frac{\partial \widetilde{\mathbf{A}}}{\partial t} &&\Longrightarrow \quad \frac{\partial}{\partial \mathbf{r}}(\widetilde{\Phi} - \Phi) + \frac{\partial}{\partial t}(\widetilde{\mathbf{A}} - \mathbf{A}) = 0,\\
&\mathbf{B} = \nabla \times \mathbf{A} = \nabla \times \widetilde{\mathbf{A}} &&\Longrightarrow \quad \nabla \times (\widetilde{\mathbf{A}} - \mathbf{A}) = 0.
\end{aligned}$$

From the second equation we deduce that

$$\widetilde{\mathbf{A}} = \mathbf{A} + \frac{\partial g}{\partial \mathbf{r}}$$

for some scalar function $g(t,\mathbf{r})$, and substituting into the first equation we then obtain

$$\frac{\partial}{\partial\mathbf{r}}(\widetilde{\Phi}-\Phi)+\frac{\partial}{\partial t}\frac{\partial g}{\partial\mathbf{r}}=\frac{\partial}{\partial\mathbf{r}}\left(\widetilde{\Phi}-\Phi+\frac{\partial g}{\partial t}\right)=0 \quad\Longrightarrow\quad \widetilde{\Phi}=\Phi-\frac{\partial g}{\partial t}-h(t)$$

for some scalar function of time $h(t)$. Thus Eq. (2.53) holds with

$$f(t,\mathbf{r})=g(t,\mathbf{r})+\int h(t)\,\mathrm{d}t\,.$$

Example 2.2. An electron of mass m and charge $-e<0$ moves in a uniform electromagnetic field $\mathbf{E}=E\mathbf{e}_2$, $\mathbf{B}=B\mathbf{e}_3$ (with $E>0$, $B>0$). Let us compute the electron's trajectory if initially $\mathbf{r}(0)=0$ and $\mathbf{v}(0)=v_0\mathbf{e}_1$, with $v_0>0$.

Taking into account Eq. (2.51), the electron's equations of motion are

$$\boxed{m\ddot{x}_1=-eB\dot{x}_2\,,\qquad m\ddot{x}_2=-eE+eB\dot{x}_1\,,\qquad m\ddot{x}_3=0\,.} \tag{2.56}$$

From the last equation and the initial condition $x_3(0)=\dot{x}_3(0)=0$ it immediately follows that $x_3(t)=0$ for all t. Hence the motion takes place in the horizontal plane $x_3=0$. The equations for the coordinates x_1 and x_2 can be simplified using the dimensionless variables[a]

$$\tau=\frac{eB}{m}\,t\,,\qquad x=\frac{eB^2}{mE}\,x_1\,,\qquad y=\frac{eB^2}{mE}\,x_2\,,$$

in terms of which

$$x''=-y'\,,\qquad y''=x'-1\,, \tag{2.57}$$

where the prime denotes derivative with respect to τ. The initial conditions for the new variables are

$$x(0)=y(0)=0\,,\qquad x'(0)=\frac{eB^2}{mE}\,\dot{x}_1(0)\,\frac{\mathrm{d}t}{\mathrm{d}\tau}=\frac{Bv_0}{E}=:1+a\,,\qquad y'(0)=0\,.$$

The equations of motion for the (x,y) variables can be easily solved through standard techniques, since they are a linear system of second-order differential equations with constant coefficients. In this case, however, the simplest course of action is to introduce the complex variable $z=x+\mathrm{i}y$, in terms of which Eqs. (2.57) reduce to the ordinary differential equation

$$z''=x''+\mathrm{i}y''=-y'+\mathrm{i}x'-\mathrm{i}=\mathrm{i}(z'-1)\,,$$

or equivalently

$$w''=\mathrm{i}\,w'\,,\qquad w:=z-\tau\,.$$

This is just a linear first-order differential equation in w', with initial condition

$$w'(0) = z'(0) - 1 = x'(0) + iy'(0) - 1 = a,$$

whose solution is

$$w' = a e^{i\tau} .$$

Integrating with respect to τ and taking into account the initial condition

$$w(0) = z(0) = 0$$

we obtain

$$w = ia(1 - e^{i\tau}) \quad \Longrightarrow \quad z = \tau + ia(1 - e^{i\tau}) .$$

Taking the real and imaginary parts of z we finally arrive at the equations

$$x = \operatorname{Re} z = \tau + a \sin\tau , \qquad y = \operatorname{Im} z = a(1 - \cos\tau) , \tag{2.58}$$

or, in terms of the original variables,

$$\boxed{x_1 = \frac{Et}{B} + \frac{1}{\omega}\left(v_0 - \frac{E}{B}\right)\sin(\omega t) , \quad x_2 = \frac{1}{\omega}\left(v_0 - \frac{E}{B}\right)\left[1 - \cos(\omega t)\right] ; \quad \omega := \frac{eB}{m} .}$$

Equations (2.58) are the parametric equations of the electron's trajectory. Note that

$$(x(\tau + 2n\pi), y(\tau + 2n\pi)) = (2n\pi + x(\tau), y(\tau)), \tag{2.59}$$

if $n \in \mathbb{Z}$, so that the whole trajectory can be obtained by translating the arc with $0 \leqslant \tau < 2\pi$ by an integer multiple of 2π along the x direction. The qualitative properties of the trajectory depend on the dimensionless parameter

$$a = \frac{Bv_0}{E} - 1 ;$$

note that $a > -1$, since E, B, and v_0 are all positive by hypothesis.

i) If $|a| < 1$, i.e.,

$$0 < v_0 < \frac{2E}{B} ,$$

we have

$$x' = 1 + a\cos\tau > 0 ,$$

and therefore x is an increasing function of τ. In particular, if $a = 0$, or equivalently

$$v_0 = \frac{E}{B} ,$$

the trajectory is the x axis traversed with constant velocity ($x(\tau) = \tau$ or, in the original variables, $x_1(t) = Et/B = v_0 t$). By Eq. (2.59), it suffices to study the arc of the trajectory with $0 \leqslant \tau \leqslant 2\pi$. In general, $0 \leqslant y \leqslant 2a$ if $a > 0$, or $2a \leqslant y \leqslant 0$ if $a < 0$, for all τ. Moreover, in the interval $0 \leqslant \tau \leqslant 2\pi$ we have

$$y = 0 \iff \tau = 0, 2\pi \quad \Longrightarrow \quad x = 0, 2\pi\,,$$

while

$$y = 2a \iff \tau = \pi \quad \Longrightarrow \quad x = \pi\,.$$

At points where y attains its extreme values 0 and $2a$ the electron's velocity is directed along the x axis, since for $\tau = k\pi$ (with $k = 0, 1, 2$) we have

$$x'(k\pi) = 1 + (-1)^k a > 0\,, \qquad y'(k\pi) = a\sin(k\pi) = 0\,.$$

In particular, at such points

$$\frac{\mathrm{d}y}{\mathrm{d}x} = \frac{y'}{x'} = 0\,.$$

The electron's trajectory has thus the qualitative shape shown in Fig. 2.5 (middle curve).

ii) On the other hand, if $a = 1$, i.e.,

$$v_0 = \frac{2E}{B}\,,$$

then $x'(\tau) = 1 + \cos\tau \geqslant 0$, with (for $0 \leqslant \tau \leqslant 2\pi$)

$$x'(\tau) = 0 \iff \tau = \pi \quad \Longrightarrow \quad x = \pi\,,\ y = 2\,.$$

In fact, when $\tau = \pi$ both velocity components x' and $y' = \sin\tau$ vanish simultaneously. It is easily checked that the trajectory has a cusp at the point $(\pi, 2)$ corresponding to $\tau = \pi$, since the slope to its tangent at this point satisfies

$$\frac{\mathrm{d}y}{\mathrm{d}x} = \frac{y'}{x'} = \frac{\sin\tau}{1+\cos\tau} = \frac{2\sin(\tau/2)\cos(\tau/2)}{2\cos^2(\tau/2)} = \tan(\tau/2) \underset{\tau\to\pi\mp}{\longrightarrow} \pm\infty\,.$$

The trajectory is in this case a *cycloid* (cf. Fig. 2.5, top curve).

iii) Finally, if $a > 1$, i.e.,

$$v_0 > \frac{2E}{B}\,,$$

x is no longer a monotonic function of τ in the interval $[0, 2\pi]$. More precisely,

$$x'(\tau) \geqslant 0 \iff \tau \in \big[0, \arccos(-1/a)\big) \cup \big(2\pi - \arccos(-1/a), 2\pi\big]\,,$$

while

$$x'(\tau) < 0 \iff \tau \in \big(\arccos(-1/a), 2\pi - \arccos(-1/a)\big)\,.$$

Moreover, we have

$$y(2\pi - \tau) = y(\tau), \qquad x(2\pi - \tau) + x(\tau) = 2\pi,$$

so that the points $\mathbf{r}(\tau)$ and $\mathbf{r}(2\pi - \tau)$ are symmetric with respect to the vertical line $x = \pi$. Thus the trajectory is symmetric about his line. This can be shown to imply that the trajectory intersects itself at a point on the line $x = \pi$ (cf. Fig. (2.5), bottom curve).

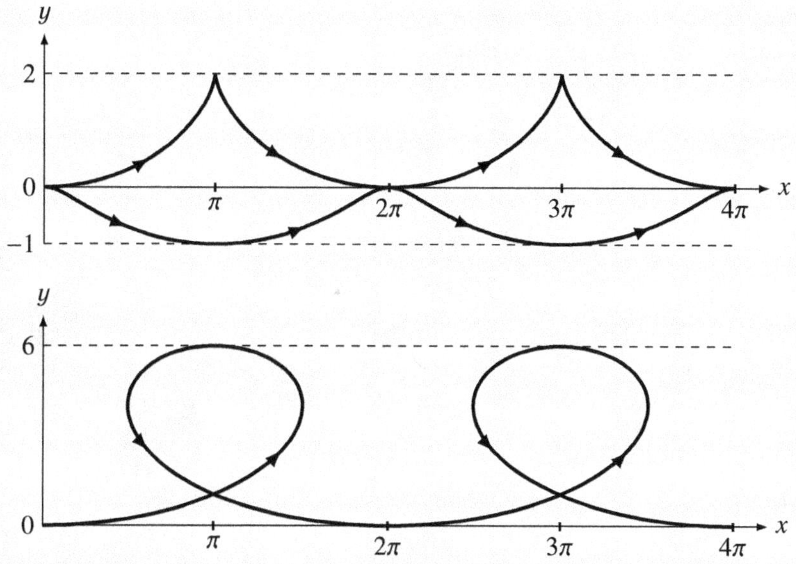

Figure 2.5. Electron's trajectory in Example 2.2 for $v_0 = E/(2B)$ (middle curve), $v_0 = 2E/B$ (top curve), and $v_0 = 4E/B$ (bottom curve).

[a] From the Lorentz force law (2.51) it follows that evB/m and E/B have dimensions of acceleration and velocity, respectively. Hence eB/m has dimensions of $a/v = t^{-1}$, and $(E/B)(m/eB) = mE/(eB^2)$ has dimensions of $vt = l$.

Exercise 2.13. Redo the previous problem assuming that the electric field vanishes ($E = 0$). Show that in this case the particle moves along a circle with constant frequency $\omega = eB/m$, called the *cyclotron frequency*.

Solution. If $E = 0$ the general solution of the equations of motion found in the previous exercise reduces to

$$x_1 = \frac{v_0}{\omega} \sin(\omega t), \qquad x_2 = \frac{v_0}{\omega} \big[1 - \cos(\omega t)\big].$$

This is indeed the equation of a circle of radius v_0/ω centered at the point $(0, v_0/\omega)$ in the $x_3 = 0$ plane, traversed in an anti-clockwise direction (since $e > 0$) with frequency ω.

2.5 MOTION OF A PARTICLE IN A ONE-DIMENSIONAL POTENTIAL

In this section we shall study the motion of a particle in one dimension, subject to a (smooth) force $F(x)$ independent of time and velocity. Such a force is *always conservative*, since $F(x) = -V'(x)$ with

$$V(x) = -\int F(x)\,\mathrm{d}x\,.$$

In this case the law of conservation of energy (2.40) reduces to

$$\boxed{\frac{1}{2}m\dot{x}^2 + V(x) = E\,,} \tag{2.60}$$

where the constant $E \in \mathbb{R}$ is the particle's total energy (which depends on the initial conditions). Conversely, differentiating (2.60) with respect to t we obtain

$$\dot{x}\big(m\ddot{x} - F(x)\big) = 0\,.$$

Hence *if* $\dot{x} \neq 0$ *Eq.* (2.60) *is* equivalent *to the equation of motion* $m\ddot{x} = F(x)$.

2.5.1 Equilibria and turning points

Definition 2.4. The **equilibrium positions** (or **equilibria**) of the potential $V(x)$ are the points $x_0 \in \mathbb{R}$ for which the equation of motion has the constant solution $x(t) = x_0$ for all t.

• If $x(t) = x_0$ for all t is a solution of the equation of motion, from this equation we obtain

$$F\big(x(t)\big) = F(x_0) = -V'(x_0) = 0\,.$$

Thus *the equilibria are the points at which the force acting on the particle vanishes.* From the mathematical point of view, *the equilibria are the* critical points *of the potential* $V(x)$, that is, the roots of the equation

$$\boxed{V'(x) = 0\,.}$$

Note that, by the *existence and uniqueness theorem* for solutions of ordinary differential equations, if x_0 is an equilibrium the *only* solution of the equation of motion satisfying the initial conditions $x(t_0) = x_0$, $\dot{x}(t_0) = 0$ is the constant solution $x(t) = x_0$. In other words:

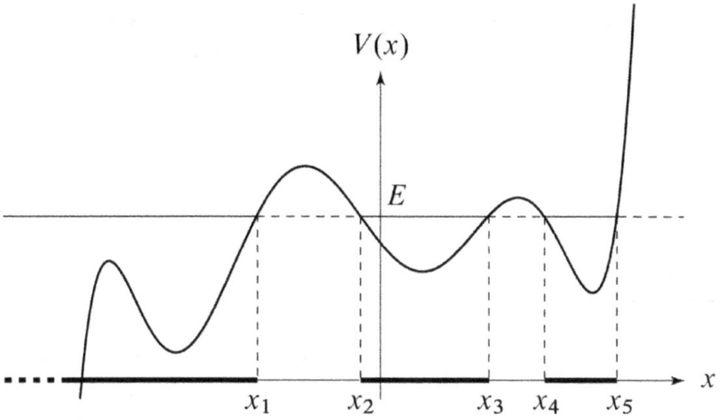

Figure 2.6. One-dimensional potential $V(x)$ with 5 turning points x_i for the energy E shown. The allowed region consists in this case of the three intervals $(-\infty, x_1]$, $[x_2, x_3]$, and $[x_4, x_5]$. Note also that the potential plotted in the figure has exactly 6 equilibria.

If at some instant the particle is at an equilibrium x_0 *with zero velocity*, it will remain at x_0 indefinitely.

From Eq. (2.60) it immediately follows that *for a given energy E the motion can only take place in the region defined by the inequality*

$$V(x) \leqslant E\,,$$

that we shall call the **accessible** (or **allowed**) **region** for the energy E (see Fig. 2.6). In general (i.e., if the potential is sufficiently smooth), the allowed region is a (countable) *disjoint union of closed intervals*, some of which may be infinite to the right or the left (including the limiting case where the allowed region is the whole real line), or even reduce to isolated points (necessarily equilibria).

By continuity, if at some instant the particle lies on one of the disjoint closed intervals making up the allowed region it will always remain inside that interval.

Of particular interest are the endpoints x_i of these intervals, which must satisfy the equation $V(x) = E$. When the particle is at one of these points its velocity vanishes, since

$$x(t) = x_i \quad \Longleftrightarrow \quad V(x_i) = E = \frac{1}{2}m\dot{x}(t)^2 + V(x_i) \quad \Longleftrightarrow \quad \dot{x}(t) = 0 \qquad (2.61)$$

by the law of conservation of energy (2.60). We shall say that such a point x_i is a **turning point** of the trajectory (for the given energy E) if it is not an equilibrium, i.e., if

$$V(x_i) = E\,, \qquad V'(x_i) \neq 0\,.$$

In other words:

For a given energy E, the turning points are the endpoints of the disjoint closed intervals which make up the allowed region, *excluding the equilibria.*

• The reason for this terminology is the fact that *when the particle reaches a turning point its velocity $\dot{x}$ changes sign, and thus the particle "turns."* For example, if $V'(x_i) > 0$ then $V(x) < V(x_i) = E$ on a sufficiently small interval to the left of x_i, and $V(x) > V(x_i) = E$ on a similar interval to the right of x_i, so that the particle cannot reach the region to the *right* of the turning point. Hence the particle must approach the turning point from its *left*, and therefore $\dot{x}$ changes from positive (right before reaching the turning point) to negative (right afterward).

If the particle has energy E, and x_0 is an equilibrium with $V(x_0) = E$, then *the particle's trajectory cannot cross the equilibrium x_0*. Indeed, since $V(x_0) = E$ if the particle is at x_0 at some instant t_0 its velocity $\dot{x}(t_0)$ must vanish. From the remark on equilibria on p. 45 we conclude that $x(t) = x_0$ for all t, and hence the trajectory in this case consists of the single point x_0. Thus, if (for instance) $x(t_0) < x_0$, we must have $x(t) < x_0$ for all t, i.e., the whole trajectory lies at the *left* of x_0. From the previous remarks it then follows that:

The trajectory of a particle with energy E is an *interval* (finite or infinite, which might reduce to a single point) on whose interior $V(x) < E$, limited by *turning points* and/or *equilibria* satisfying $V(x) = E$. Moreover, equilibria with $V(x) = E$ limiting the trajectory *cannot be reached in a finite time*, and in particular *cannot be crossed.*

• The equation of motion $m\ddot{x} = F(x)$ is *invariant* under *time translations* $t \mapsto t + t_0$, for any $t_0 \in \mathbb{R}$, since the time t does not appear explicitly in it. Hence *if $x(t)$ is a solution of the equation of motion so is $x(t + t_0)$, for all $t_0 \in \mathbb{R}$.*

The equation of motion is also invariant under the *time reversal* mapping $t \mapsto -t$. Thus *if $x(t)$ is a solution of* (2.60) *so is $x(-t)$*. Combining this observation with the previous one it follows that $x(t_0 - t)$ is a solution of the equation of motion if $x(t)$ is.

2.5.2 Qualitative description of the motion

The law of conservation of energy (2.60) allows us to easily find the general solution of the equation of motion in implicit form. Indeed, solving for $\dot{x}$ in Eq. (2.60) we obtain

$$\dot{x} = \frac{\mathrm{d}x}{\mathrm{d}t} = \pm\sqrt{\frac{2}{m}\,\big(E - V(x)\big)}\,. \tag{2.62}$$

Each of these *two* equations (corresponding to the two signs before the radical) is a first-order differential equation with *separable variables*, easily solved by separating variables and integrating:

$$t - t_0 = \pm\sqrt{\frac{m}{2}} \int \frac{\mathrm{d}x}{\sqrt{E - V(x)}} \,. \tag{2.63}$$

Here t_0 is an arbitrary integration constant which, without loss of generality, can be taken equal to zero in view of the previous comments. The behavior of the solutions depends crucially on the type of interval inside the allowed region where the motion takes place, as we shall see in more detail below. To simplify the exposition, we shall assume for the time being that *the interval where the motion takes place is limited by turning points* (not by equilibria). Hence on the interior of this interval we must have $V(x) < E$, whereas $V(x) = E$ and $V'(x) \neq 0$ at its endpoints (if any). By the law of conservation of energy (2.60), *the particle's velocity $\dot{x}(t)$ can only change sign at the endpoints of the interval of motion.*

I) Bounded interval $[x_0, x_1]$

Consider first the case in which the particle's motion takes place in a bounded interval $[x_0, x_1]$ limited by two *consecutive turning points* $x_{0,1}$, so that

$$E = V(x_i) \quad \text{and} \quad V'(x_i) \neq 0 \,, \qquad \text{with} \quad i = 0, 1,$$

and $V(x) < E$ for $x_0 < x < x_1$ (cf. Fig. 2.7). Let us suppose, without loss of generality[15], that $x_0 = x(0)$, so that $\dot{x}(0) = 0$. Then $\dot{x} > 0$ for sufficiently small $t > 0$, since otherwise the particle would enter the forbidden region to the left of x_0. We must therefore take the "+" sign in Eq. (2.63), obtaining[16]

$$t = \sqrt{\frac{m}{2}} \int_{x_0}^{x} \frac{\mathrm{d}s}{\sqrt{E - V(s)}} =: \theta(x) \,. \tag{2.64}$$

[15]Indeed, suppose that the particle is at some point $a \in (x_0, x_1)$ at the initial time $t = t_0$. From Eq. (2.63) we then obtain

$$t = t_0 + \operatorname{sgn}(\dot{x}(t_0))\sqrt{\frac{m}{2}} \int_a^x \frac{\mathrm{d}s}{\sqrt{E - V(s)}},$$

and thus the particle will reach the endpoint x_0 at the time

$$t_1 = t_0 - \operatorname{sgn}(\dot{x}(t_0))\sqrt{\frac{m}{2}} \int_{x_0}^{a} \frac{\mathrm{d}s}{\sqrt{E - V(s)}}.$$

(The same is true if $a = x_1$, taking $\operatorname{sgn}(\dot{x}(t_0)) = -1$.) This time is *finite*, since the integral, which is improper at its lower endpoint x_0, is *convergent*. Indeed, since $V'(x_0) \neq 0$ by hypothesis, the integrand behaves as $(s - x_0)^{-1/2}$ in the vicinity of $s = x_0$. Thus $x(t_1) = x_0$ for some finite time t_1, and replacing t by $t - t_1$ we have $x(0) = x_0$.

[16]The integral (2.64), which is improper at its lower limit $s = x_0$, is however convergent (see previous footnote).

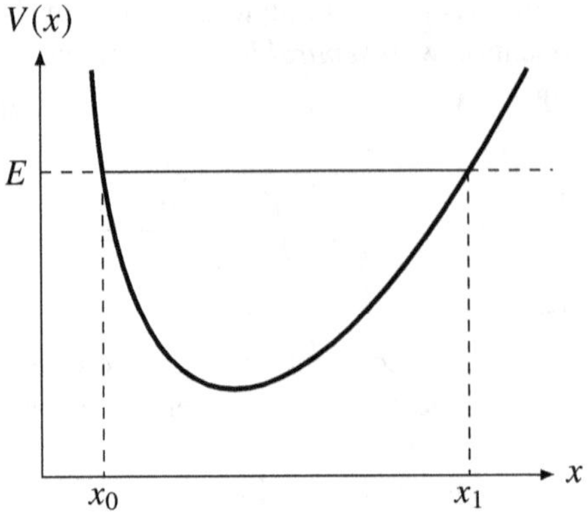

Figure 2.7. One-dimensional potential $V(x)$ with two consecutive turning points x_0, x_1 limiting an allowed interval $[x_0, x_1]$ (for the energy E shown) such that $V(x) < E$ for $x \in (x_0, x_1)$.

Hence the particle will reach the point x_1 at time $t = \tau/2$, with[17]

$$\boxed{\tau = 2\theta(x_1) = \sqrt{2m}\int_{x_0}^{x_1}\frac{ds}{\sqrt{E - V(s)}}\,.} \tag{2.65}$$

Note that, since

$$\theta'(x) = \frac{\sqrt{m/2}}{\sqrt{E - V(x)}} > 0\,, \qquad x_0 < x < x_1\,, \tag{2.66}$$

$\theta(x)$ is monotonically increasing, and hence invertible, in the interval $[x_0, x_1]$ (see Fig. 2.8). Thus for $0 \leqslant t \leqslant \tau/2$ the particle's position as a function of time is given by

$$x = \theta^{-1}(t)\,, \qquad 0 \leqslant t \leqslant \frac{\tau}{2}\,.$$

For $t > \tau/2$ (with $t - (\tau/2)$ small enough) $\dot{x}$ becomes negative, since otherwise the particle would enter the forbidden region to the right of x_1. Using again Eq. (2.63), but this time with the "–" sign, and the initial condition $x(\tau/2) = x_1$, we obtain

$$t = \frac{\tau}{2} - \sqrt{\frac{m}{2}}\int_{x_1}^{x}\frac{ds}{\sqrt{E - V(s)}} = \tau - \sqrt{\frac{m}{2}}\int_{x_0}^{x}\frac{ds}{\sqrt{E - V(s)}} = \tau - \theta(x)\,. \tag{2.67}$$

[17] The integral (2.65) is also improper at its upper limit but again convergent, since $V'(x_1) \neq 0$ implies that the integrand behaves as $(x_1 - s)^{-1/2}$ near $s = x_1$.

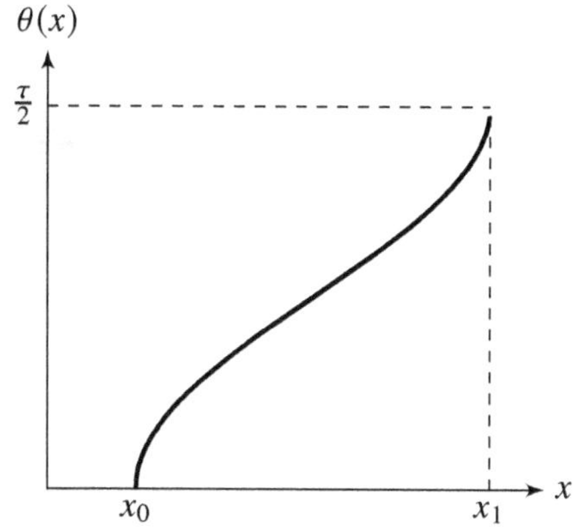

Figure 2.8. Function $\theta(x)$ in Eq. (2.64). Note that $\theta'(x_{0,1}) = +\infty$, by Eq. (2.66).

In particular, the particle will again reach the point x_0 at time $t = \tau$ (since $\theta(x_0) = 0$). Note also that from Eq. (2.67) it follows that[18]

$$x = \theta^{-1}(\tau - t) , \qquad \frac{\tau}{2} \leqslant t \leqslant \tau .$$

Hence the particle's motion for $0 \leqslant t \leqslant \tau$, implicitly given by Eqs. (2.64) and (2.67), can be expressed in terms of the function θ^{-1} by the equations

$$x(t) = \begin{cases} \theta^{-1}(t) , & 0 \leqslant t \leqslant \frac{\tau}{2} ; \\ \theta^{-1}(\tau - t) , & \frac{\tau}{2} \leqslant t \leqslant \tau . \end{cases} \tag{2.68}$$

Note, in particular, that $x(t)$ is symmetric about $t = \tau/2$, since by the previous equation

$$x\Big(\frac{\tau}{2} - s\Big) = \theta^{-1}\Big(\frac{\tau}{2} - s\Big) = x\Big(\frac{\tau}{2} + s\Big) , \qquad 0 \leqslant s \leqslant \frac{\tau}{2} .$$

The solution of the equations of motion valid for all t is just the *periodic extension with period* τ of the function $x(t)$ defined in $[0, \tau]$ by Eq. (2.68) (cf. Fig. 2.9). In other words, if $k\tau \leqslant t \leqslant (k+1)\tau$ with $k \in \mathbb{Z}$ then

$$x(t) = x(t - k\tau) , \tag{2.69}$$

where the RHS is evaluated using Eq. (2.68). Indeed, this function is a solution of the equation of motion due to the invariance of this equation under time translations, satisfies the initial conditions $x(0) = x_0$, $x'(0) = 0$ by construction, and is of class C^2 at the junction points $k\tau$ with $k \in \mathbb{Z}$ (exercise). Summarizing:

[18] This could have also been proved noting that, by the two remarks on p. 46, if $x = \theta^{-1}(t)$ is a solution to the equation of motion for $t \in [0, \tau/2]$, then $x = \theta^{-1}(\tau - t)$ is a solution for $t \in [\tau/2, \tau]$, which satisfies the initial conditions $x(\tau/2) = \theta^{-1}(\tau/2) = x_1$ and $x'(\tau/2) = 0$ (by Eq. (2.60)). By the existence and uniqueness theorem for second-order ordinary differential equations, $x(t) = \theta^{-1}(\tau - t)$ for $\tau/2 \leqslant t \leqslant \tau$.

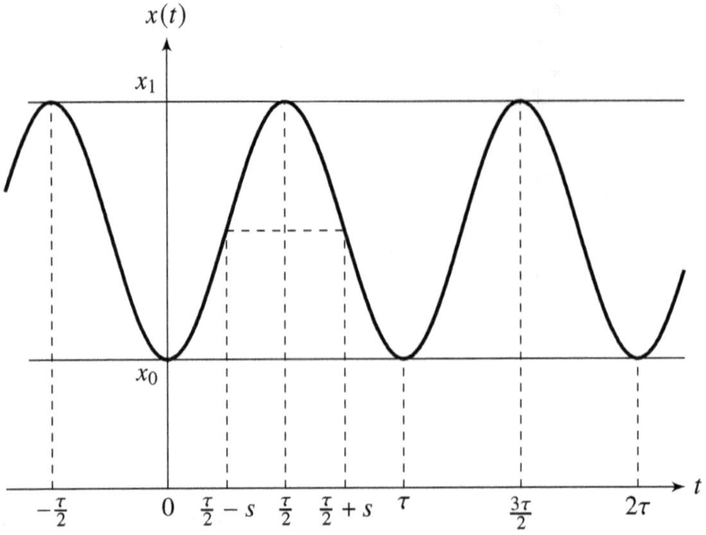

Figure 2.9. Motion of a particle in a one-dimensional potential between two consecutive turning points x_0, x_1.

> The motion of a particle between two consecutive turning points $x_{0,1}$ of a one-dimensional potential is *periodic*, with period τ given by Eq. (2.65).

Exercise 2.14. Show that the function $x(t)$ defined by Eqs. (2.68)–(2.69) is invariant under time reversal, i.e., that $x(t) = x(-t)$.

Solution. The function $f(t) := x(-t)$ is a solution of the equation of motion, due to the invariance of this equation under time reversal. At $t = 0$, the solution $f(t)$ satisfies the *same* initial conditions as $x(t)$, since

$$f(0) = x(0) = x_0\,, \qquad f'(0) = -x'(0) = 0\,.$$

By the existence and uniqueness theorem for second-order ordinary differential equations, $f(t) = x(-t) = x(t)$ for all t.

II) Semi-infinite interval $[x_0, \infty)$

Consider next the case in which the particle moves inside a semi-infinite interval[19] $[x_0, \infty)$ limited by a turning point x_0, so that $V(x_0) = E$, $V'(x_0) \neq 0$, and $V(x) < E$ for $x > x_0$ (cf. Fig. 2.10). If the particle is at the point x_0 for $t = 0$ then $\dot{x}(t) > 0$ for $t > 0$, and the relation between the time t and the position x is given by equation (2.64)

[19]The case in which the motion takes place in a semi-infinite interval $(-\infty, x_0]$ limited by a turning point is dealt with analogously.

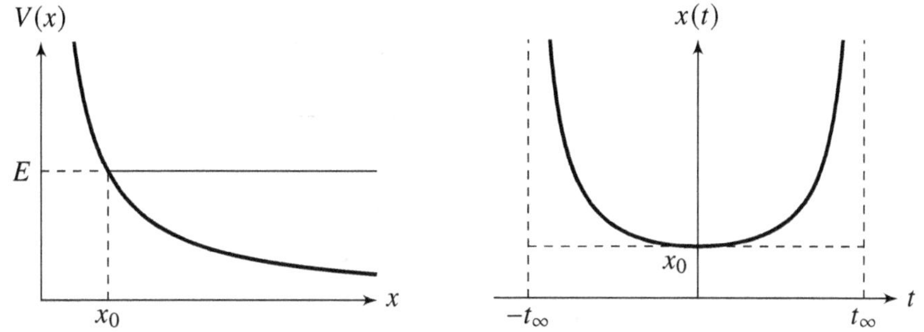

Figure 2.10. Left: one-dimensional potential $V(x)$ with a a turning point x_0 limiting a semi-infinite allowed interval $[x_0, \infty)$ (for the energy E shown) such that $V(x) < E$ for $x > x_0$. Right: corresponding law of motion $x(t)$ (in the case of finite t_∞).

for all $t > 0$. In particular, the particle reaches (positive) infinity at time

$$\boxed{t_\infty = \theta(\infty) = \sqrt{\frac{m}{2}} \int_{x_0}^{\infty} \frac{\mathrm{d}s}{\sqrt{E - V(s)}}\,,}$$

which is finite or infinite depending on whether the integral in the RHS is convergent or divergent at $+\infty$. For instance, if $V(x) \sim -x^a$ with $a \geqslant 0$ for $x \to \infty$ then t_∞ is finite if $a > 2$, and infinite if $0 \leqslant a \leqslant 2$. Taking into account the definition (2.64) of the function $\theta(x)$, the particle's motion for $0 \leqslant t < t_\infty$ is given by the equation

$$x = \theta^{-1}(t)\,, \qquad 0 \leqslant t < t_\infty\,,$$

with $x(t_\infty) = \infty$. On the other hand, for $-t_\infty < t \leqslant 0$ we have

$$x = \theta^{-1}(-t)\,, \qquad -t_\infty < t \leqslant 0\,,$$

where again $x(-t_\infty) = \infty$. Indeed, this function is a solution to the equation of motion (due to the invariance of this equation under time reversal $t \mapsto -t$), and satisfies the correct initial conditions at $t = 0$:

$$x(0) = \theta^{-1}(0) = x_0\,, \qquad \dot{x}(0) = 0$$

(the last equation is actually a consequence of the first, since x_0 is a turning point). An alternative way of reaching the same conclusion is to observe that if $t < 0$ then $\dot{x}(t) < 0$, since the particle is at the point x_0 for $t = 0$. Therefore we must take the "–" sign in Eq. (2.62), which yields the equation

$$t = -\sqrt{\frac{m}{2}} \int_{x_0}^{x} \frac{\mathrm{d}s}{\sqrt{E - V(s)}} = -\theta(x) \quad \Longleftrightarrow \quad x = \theta^{-1}(-t),$$

on account of the initial condition $x(0) = x_0$. In other words, in this case the law of motion is

$$\boxed{x = \theta^{-1}(|t|)\,, \qquad -t_\infty < t < t_\infty\,.}$$

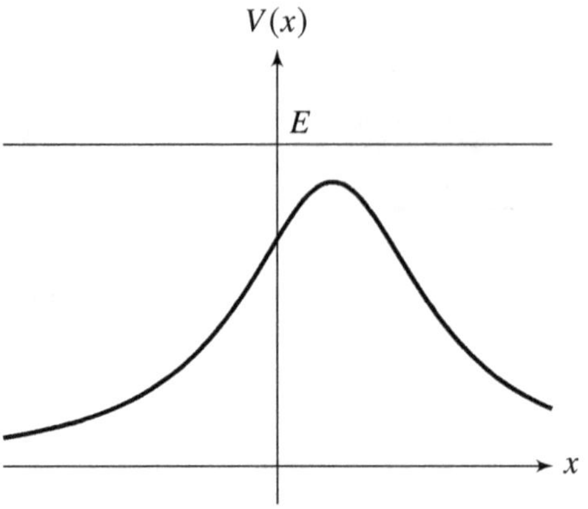

Figure 2.11. One-dimensional potential with $E > V(x)$ for all x (for the energy E shown).

Note, in particular, that (as in Case I) above) $x(t) = x(-t)$.

III) Whole real line $(-\infty, \infty)$
Consider, finally, the case in which for a certain energy E the trajectory is the whole real line, so that $V(x) \leqslant E$ for all x. We must actually have $V(x) < E$ for all $x \in \mathbb{R}$ (cf. Fig. 2.11), since otherwise a point x_0 with $V(x_0) = E$ would be an equilibrium (absolute maximum of V), which cannot be crossed by the trajectory. Let $x(0) = x_0$; then $\dot{x}^2(0)$ is fixed by conservation of energy, namely

$$\dot{x}^2(0) = \frac{2}{m}\left(E - V(x_0)\right) > 0,$$

but the sign of $\dot{x}(0)$ is of course undetermined. If (for example) $\dot{x}(0) > 0$, then $\dot{x}(t) > 0$ for all t, since the velocity cannot vanish in this case by the law of conservation of energy. We must therefore take the "+" sign in (2.62) for all t, which yields the relation

$$t = \sqrt{\frac{m}{2}} \int_{x_0}^{x} \frac{ds}{\sqrt{E - V(s)}} = \theta(x)\,.$$

In particular, the particle reaches $\pm\infty$ at time

$$\boxed{t_{\pm\infty} = \theta(\pm\infty) = \int_{x_0}^{\pm\infty} \frac{ds}{\sqrt{E - V(s)}}}$$

(which may again be finite or infinite, according to whether the integral is convergent of divergent at $\pm\infty$), and the particle's motion is governed by the equation

$$\boxed{x = \theta^{-1}(t)\,, \qquad t_{-\infty} < t < t_{\infty}\,.}$$

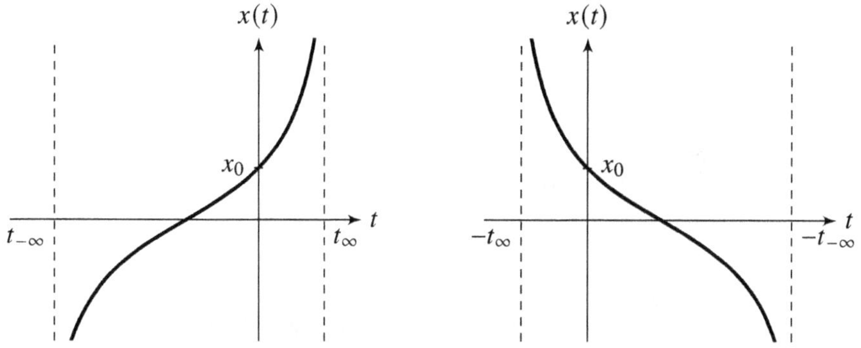

Figure 2.12. Law of motion $x(t)$ for the potential $V(x)$ and the energy E shown in Fig. 2.11 in the cases $\dot{x}(0) > 0$ (left) or $\dot{x}(0) < 0$ (right), assuming that $t_{\pm\infty}$ are finite.

Likewise, if $\dot{x}(0) < 0$ then

$$\boxed{x = \theta^{-1}(-t)\,, \qquad -t_\infty < t < -t_{-\infty}\,,}$$

where now $x(-t_{\pm\infty}) = \pm\infty$ (cf. Fig. 2.12).

Example 2.3. Consider the potential

$$V(x) = k\left(\frac{x^2}{2} - \frac{x^4}{4a^2}\right), \qquad k, a > 0\,,$$

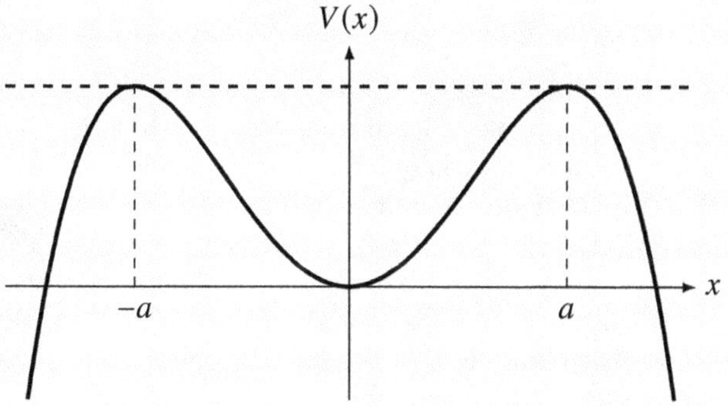

Figure 2.13. Potential in Example 2.3 (solid line) and energy $E = ka^2/4$ (dashed line).

plotted in Fig. 2.13. Differentiating with respect to x we obtain

$$V'(x) = kx\left(1 - \frac{x^2}{a^2}\right) = 0 \iff x = 0,\ \pm a\,.$$

Therefore the equilibria are in this case the points $x = 0$ (relative minimum of V) and $x = \pm a$ (global maxima). The allowed region, and therefore the type of trajectory, depends on the value of the energy E as follows:

i) $E < 0$
The allowed region is the union of the two semi-infinite intervals $(-\infty, -c]$ and $[c, \infty)$, c being the only positive root of the equation $V(x) = E$. Thus in this case the trajectory is *unbounded* (to the right if $x(0) > c$, to the left if $x(0) < -c$).

ii) $E = 0$
The allowed region is the union of the semi-infinite intervals $\big(-\infty, -\sqrt{2}\,a\big]$ and $\big[\sqrt{2}\,a, \infty\big)$ along with the origin, which as we know is an equilibrium. In particular, if $x(0) = 0$ then $x(t) = 0$ for all t (equilibrium solution), while if $|x(0)| \geqslant \sqrt{2}\,a$ the trajectory is *unbounded.*

iii) $0 < E < ka^2/4$
Since $ka^2/4 = V(\pm a)$ is the potential's maximum value, the allowed region is the union of the three intervals $(-\infty, -c_2]$, $[-c_1, c_1]$, and $[c_2, \infty)$, where $c_1 < c_2$ are the two positive roots of the equation $V(x) = E$. Therefore in this case the trajectory is *unbounded* (to the left or right) if $|x(0)| \geqslant c_2$, while if $|x(0)| \leqslant c_1$ the motion is *periodic*, with amplitude c_1.

iv) $E > ka^2/4$
Since $V(x) < E$ for all x, the allowed region (and the trajectory) is *the whole real line*. Note that in this case the time that it takes the particle to reach $\pm\infty$ is *finite*, since for $|x| \to \infty$ the integral

$$\int^{\pm\infty} \frac{dx}{\sqrt{E - V(x)}} \sim \int^{\pm\infty} \frac{dx}{x^2}$$

converges.

v) $E = ka^2/4$
We have left for the end the most interesting case, in which $E = ka^2/4$. Since $V(x) \leqslant ka^2/4$ for all x, the allowed region is again the whole real line, and it might therefore superficially seem that the trajectory is also the whole real line. However, this conclusion is *wrong*, since the allowed region now contains the two *equilibria* $x = \pm a$. If the particle starts at $t = 0$ from a point $x_0 \neq \pm a$, it *cannot* reach the points $\pm a$ in a *finite* time. Indeed, if $x(t_0) = \pm a$ for a certain time $t_0 \in \mathbb{R}$, from (2.60) with $V(\pm a) = ka^2/4 = E$ we obtain $\dot{x}(t_0) = 0$. Since the points $\pm a$ are equilibria, this implies that $x(t) = \pm a$ for all t. Therefore in this case the possible trajectories of the particle are the *open* intervals $(-\infty, -a)$, $(-a, a)$, and (a, ∞), along with the two equilibria $\pm a$. In particular, if $|x(0)| < a$ the trajectory remains in the interval $(-a, a)$ for all $t \in \mathbb{R}$ and is therefore *bounded*. However, it is *not periodic*, but rather verifies $x(\pm\infty) = \pm a$ if $\dot{x}(0) > 0$ or $x(\pm\infty) = \mp a$ if $\dot{x}(0) < 0$. (Why is $\dot{x}(0) \neq 0$ in this case?)

• In this case it is possible to explicitly integrate the equation of motion when $E = ka^2/4$. Indeed, substituting this value of the energy into Eq. (2.62) we obtain

$$\dot{x} = \pm\sqrt{\frac{k}{2ma^2}}\,(x^2 - a^2)\,.$$

Separating variables and integrating we have

$$\pm\sqrt{\frac{2k}{m}}\,t = \int \frac{2a}{x^2 - a^2}\,\mathrm{d}x = \log\left|\frac{x-a}{x+a}\right|$$

$$\implies \left|\frac{x-a}{x+a}\right| = \mathrm{e}^{\pm 2\omega t}\,, \qquad \omega := \sqrt{\frac{k}{2m}}\,,$$

where without loss of generality we have taken the integration constant equal to zero. If the particle is initially in one of the intervals $(-\infty, -a)$ or (a, ∞) then

$$\left|\frac{x-a}{x+a}\right| = \frac{x-a}{x+a}\,,$$

and therefore

$$\boxed{x = a\,\frac{1 + \mathrm{e}^{\pm 2\omega t}}{1 - \mathrm{e}^{\pm 2\omega t}} = \mp a \coth(\omega t)\,.}$$

The previous expression actually defines *four* different solutions. Indeed, if initially the particle is in the region $x > a$ with positive (resp. negative) velocity then we must take the "−" (resp. "+") sign in the previous expression, and the associated solution is therefore defined for $t < 0$ (resp. $t > 0$). This solution corresponds to a motion reaching positive infinity (resp. arriving from positive infinity) in a finite time and tending to the point $x = a$ for $t \to -\infty$ (resp. $t \to +\infty$). Likewise, if $x(0) < -a$ then the solution $x = -a \coth(\omega t)$ with $t > 0$ corresponds to a motion from $x = -\infty$ (for $t \to 0+$) to $x = -a$ (for $t \to \infty$) with positive velocity, while the solution $x = a \coth(\omega t)$ with $t < 0$ corresponds to a motion from $x = -a$ (for $t \to -\infty$) to $x = -\infty$ (for $t \to 0-$) with negative velocity (cf. Fig. 2.14).

Similarly, if the particle lies initially in the interval $(-a, a)$ then

$$\left|\frac{x-a}{x+a}\right| = \frac{a-x}{a+x}\,,$$

and thus

$$\boxed{x = a\,\frac{1 - \mathrm{e}^{\pm 2\omega t}}{1 + \mathrm{e}^{\pm 2\omega t}} = \mp a \tanh(\omega t)}$$

(cf. Fig. 2.14). The solution corresponding to the "+" (resp. "−") sign has always positive (resp. negative) velocity, and tends to $\pm a$ for $t \to \pm\infty$ (resp. $t \to \mp\infty$).

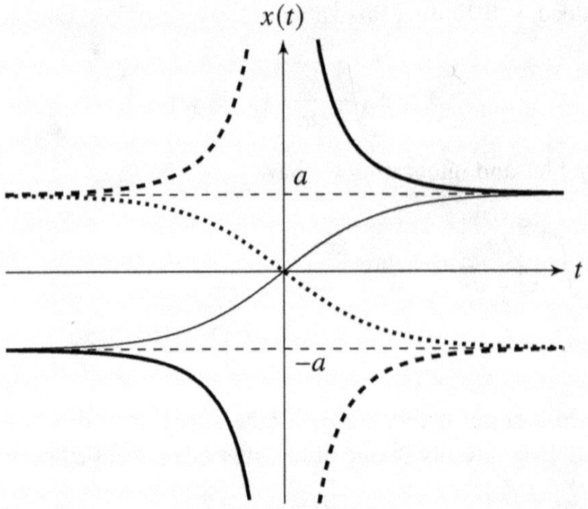

Figure 2.14. Plot of the solutions $x = a\coth(\omega t)$ (solid curve), $x = -a\coth(\omega t)$ (dashed curve), $x = a\tanh(\omega t)$ (thin curve), and $x = -a\tanh(\omega t)$ (dotted curve) in Example 2.3, and their asymptotes $x = \pm a$.

To visualize the different trajectories followed by the particle and qualitatively understand their properties, it is useful to plot the momentum $p = m\dot{x}$ as a function of the position x for different values of the energy E. This plot is usually known as the **phase map** of the system. From the law of conservation of energy it follows that the equation of the trajectories in the phase map is

$$\frac{p^2}{2m} + V(x) = E\,,$$

which in this case reduces to

$$\frac{p^2}{2m} + k\left(\frac{x^2}{2} - \frac{x^4}{4a^2}\right) = E\,.$$

The corresponding trajectories (obviously symmetric with respect to both axes) are represented in Fig. 2.15. Note that the equation of the trajectories with energy equal to the critical energy $E = ka^2/4$ is

$$p = \pm\sqrt{\frac{mk}{2a^2}}\sqrt{x^4 - 2a^2x^2 + a^4} = \pm\sqrt{\frac{mk}{2a^2}}\left(x^2 - a^2\right). \tag{2.70}$$

This is the equation of two parabolas whose axis is the vertical line $x = 0$, intersecting at the equilibria $(\pm a, 0)$. These trajectories divide the phase map into 5 disjoint connected regions, in each of which the trajectories have different qualitative properties (they are bounded or unbounded, reach $x = \pm\infty$ or not, etc.). For this reason, the trajectories (2.70) are called *separatrices*.

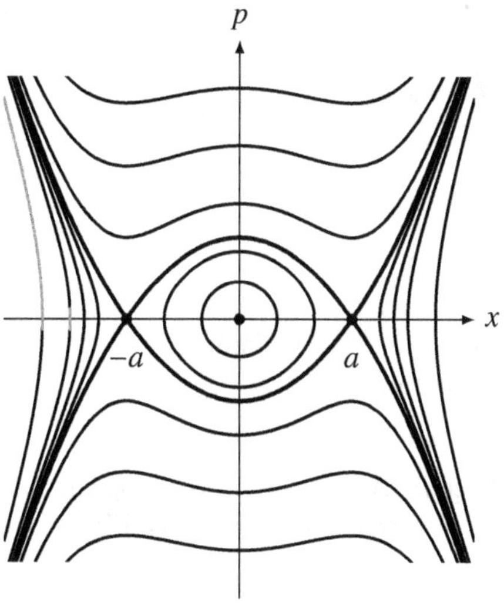

Figure 2.15. Phase map for the potential in Example 2.3. The thick line is the separatrix consisting of the trajectories with energy $E = ka^2/4$ (including the two equilibria $(\pm a, 0)$).

2.5.3 Stability of equilibria and period of the small oscillations

Intuitively speaking, an equilibrium x_0 is **stable** if sufficiently small perturbations of the initial conditions $x(0) = x_0$, $\dot{x}(0) = 0$ lead to solutions $x(t)$ of the equation of motion which remain arbitrarily close to x_0 (and with velocity arbitrarily close to 0) at all times $t > 0$. In the previous example, it is clear that the equilibrium $x = 0$ is *stable*, while $x = \pm a$ are both *unstable* equilibria. Indeed, if we slightly disturb the initial condition $x(0) = \dot{x}(0) = 0$ corresponding to the first of these equilibria, that is, we consider particle motions with $|x(0)|$ and $|\dot{x}(0)|$ small enough, the energy will be slightly positive but much smaller than the critical value $ka^2/4$, and therefore the motion will be periodic and with amplitude close to zero. On the contrary, a perturbation of the initial data $x(0) = \pm a$, $\dot{x}(0) = 0$ such that (for instance) $x(0) = \pm a$ and $|\dot{x}(0)| = \varepsilon > 0$ results in a motion with energy greater than the critical energy $ka^2/4$ no matter how small ε is, and therefore $x(t) \to \pm\infty$ for $t \to \infty$.

In general, an equilibrium is *stable* if and only if it is a *relative minimum* of the potential.

To heuristically justify this statement, suppose that x_0 is a critical point of the potential V, i.e., that $V'(x_0) = 0$. If x_0 is a *relative minimum* of V, then in a sufficiently small interval centered at x_0 we have $V'(x) < 0$ for $x < x_0$ and $V'(x) > 0$ for $x > x_0$. Hence $F(x) = -V'(x)$ and $x - x_0$ have *opposite* signs for x sufficiently close to x_0, i.e., *the force acting on the particle always points* toward *the equilibrium* x_0 in its vicinity. Thus in this case the equilibrium is *stable*. Likewise, if x_0 is a *relative maximum*

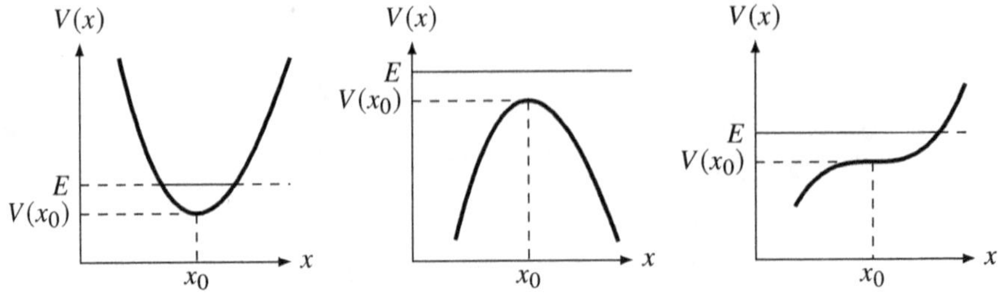

Figure 2.16. Potential $V(x)$ with a local minimum (left), a local maximum (center) or an inflection point (right) at x_0.

of V then near x_0 the force $F(x)$ points *away from* x_0, and hence the equilibrium is *unstable*. Finally, if x_0 is an *inflection point* of V then $V'(x)$ has constant sign (positive or negative) for $x \neq x_0$ in the vicinity of x_0. If (for instance) $V'(x) > 0$, then in a neighborhood of x_0 the force $F(x)$ points *away from* x_0 for $x < x_0$, and the equilibrium is again *unstable*.

An alternative proof of the previous result is based solely on energy considerations. Indeed, suppose to begin with that x_0 is a relative minimum of the potential $V(x)$. A solution of the equation of motion with initial conditions $(x(0), \dot{x}(0))$ close to $(x_0, 0)$ will have an energy

$$E = \frac{1}{2}\, m\dot{x}(0)^2 + V(x(0))$$

slightly greater than $V(x_0)$, as $|\dot{x}(0)| \gtrsim 0$ and (for $x(0)$ close enough to x_0) $V(x(0)) \gtrsim V(x_0)$, since by hypothesis x_0 is a local minimum of V. Hence the motion will consist of oscillations about x_0 with amplitude decreasing as E approaches $V(x_0)$—see Fig. 2.16 (left). On the contrary, if x_0 is a relative maximum of V then a solution with (for instance) $x(0) = x_0$ and $|\dot{x}(0)|$ small but non-vanishing will have energy

$$E = \frac{1}{2}\, m\dot{x}(0)^2 + V(x_0)$$

slightly larger than $V(x_0)$. Hence the particle will move away from x_0 by a finite amount (to the left or right) no matter how small $|\dot{x}(0)|$ is—see Fig. 2.16 (center). Finally, suppose that x_0 is an inflection point of V, with (for instance) $V'(x)$ strictly increasing near x_0. In this case a solution with, e.g., initial conditions $x(0) = x_0$ and $|\dot{x}(0)|$ small but non-vanishing will again have energy $E \gtrsim V(x_0)$, and thus the particle will move away from x_0 by a finite amount (to the *left*) no matter how small $|\dot{x}(0)|$ is—see Fig. 2.16 (right). (In fact, it is apparent from Fig. 2.16 that when x_0 is either a local maximum or an inflection point any solution with initial conditions close to equilibrium will have energy close to $V(x_0)$, and will move away by a finite amount from equilibrium either to the left or to the right.)

Suppose that the potential $V(x)$ is smooth (say, of class C^2) at an equilibrium x_0, and that furthermore $V''(x_0) \neq 0$ (which is the generic case). If the equilibrium is *stable* we must then have

$$\boxed{V'(x_0) = 0, \qquad V''(x_0) > 0.}$$

For initial conditions $(x(0), \dot{x}(0))$ sufficiently close to $(x_0, 0)$ the motion is periodic, as $E \gtrsim V(x_0)$ and the particle oscillates between two consecutive turning points close to x_0; see Fig. 2.16 (left). To find an approximation to the period of these small-amplitude oscillations, taking into account that $|x - x_0| \ll 1$ we Taylor expand the force $F(x)$ about x_0 to first order in $x - x_0$:

$$F(x) = -V'(x) = -V''(x_0)(x - x_0) + \mathrm{O}\left((x - x_0)^2\right).$$

Hence the particle's equation of motion is approximately

$$\ddot{x} = \frac{F(x)}{m} \simeq -\frac{V''(x_0)}{m}(x - x_0),$$

which can be written as

$$\ddot{\xi} + \omega^2 \xi = 0$$

with

$$\xi := x - x_0, \qquad \omega := \sqrt{\frac{V''(x_0)}{m}}.$$

As is well known, the general solution of this equation can be written as

$$\xi = A\cos(\omega t + \alpha),$$

where $\alpha \in [0, 2\pi)$ and $A \geqslant 0$ are arbitrary constants (with $A \ll 1$ to be consistent with the hypothesis $|x - x_0| = |\xi| \ll 1$). Hence the particle's motion near the equilibrium x_0 is approximately described by the equation

$$\boxed{x(t) \simeq x_0 + A\cos(\omega t + \alpha).}$$

In other words, the period τ of the *small oscillations* about the equilibrium x_0 is approximately given by

$$\boxed{\tau \simeq \frac{2\pi}{\omega} = 2\pi\sqrt{\frac{m}{V''(x_0)}}.} \tag{2.71}$$

In particular, from the above formula it follows that the period of the *small* oscillations about a stable equilibrium is approximately *independent of the amplitude* (or, equivalently, the energy). Obviously, this is exactly true for *any* amplitude for the harmonic potential $V(x) = k(x - x_0)^2/2$. For a general potential, the period of the oscillations of *arbitrary* amplitude is given *exactly* by Eq. (2.65), and is thus in general *dependent on the amplitude*. More precisely,

$$\boxed{\tau = \sqrt{2m}\int_{x_1}^{x_2}\frac{\mathrm{d}x}{\sqrt{E - V(x)}} = \sqrt{2m}\int_{x_1}^{x_2}\frac{\mathrm{d}x}{\sqrt{V(x_{1,2}) - V(x)}},} \tag{2.72}$$

where $x_1 < x_2 = x_1 + 2A$ are the two turning points (roots of the equation $V(x) = E$) closest to the equilibrium x_0 and A is the amplitude.

Example 2.4. The period of the small oscillations about the origin for the potential in Example 2.3 is *approximately*

$$\tau \simeq 2\pi\sqrt{\frac{m}{k}}\,. \tag{2.73}$$

Note, however, that this approximation is only correct for *small* amplitudes $A \ll a$. For *arbitrary* amplitude $0 < A < a$, the period is *exactly* given by the formula

$$\tau = \sqrt{2m}\int_{-A}^{A}\frac{\mathrm{d}x}{\sqrt{E - V(x)}} = 2\sqrt{2m}\int_{0}^{A}\frac{\mathrm{d}x}{\sqrt{E - V(x)}}\,,$$

with

$$E = V(A) = \frac{kA^2}{4a^2}\,(2a^2 - A^2)\,.$$

Substituting this value of E into the previous formula for τ, setting $x = As$ and operating we obtain

$$\tau = 4\sqrt{\frac{m}{k}}\int_0^1\frac{\mathrm{d}s}{\sqrt{(1-s^2)\left(1-\frac{\varepsilon^2}{2}(1+s^2)\right)}}\,,\qquad \varepsilon := \frac{A}{a} \in (0,1)\,. \tag{2.74}$$

Note that when ε tends to 1, that is A tends to a, the integral tends to infinity, which is consistent with the fact that for $A = a$ the particle takes an infinite time to reach the equilibria $x = \pm a$. If ε is small, taking into account that

$$\left(1-\frac{\varepsilon^2}{2}(1+s^2)\right)^{-1/2} = 1 + \frac{\varepsilon^2}{4}(1+s^2) + \mathrm{O}(\varepsilon^4)$$

and using the above formula for the period we obtain the more accurate expansion

$$\begin{aligned}
\tau &= 4\sqrt{\frac{m}{k}}\left(\int_0^1\frac{\mathrm{d}s}{\sqrt{1-s^2}} + \frac{\varepsilon^2}{4}\int_0^1\frac{1+s^2}{\sqrt{1-s^2}}\,\mathrm{d}s + \mathrm{O}(\varepsilon^4)\right)\\
&= \sqrt{\frac{m}{k}}\left[2\pi + \varepsilon^2\int_0^{\pi/2}(1+\cos^2\theta)\,\mathrm{d}\theta + \mathrm{O}(\varepsilon^4)\right]\\
&= 2\pi\sqrt{\frac{m}{k}}\left(1 + \frac{3}{8}\,\varepsilon^2 + \mathrm{O}(\varepsilon^4)\right)
\end{aligned} \tag{2.75}$$

(cf. Fig. 2.17). (In the previous formulas we have used the standard notation $O(s)$ to denote any function $f(s)$ such that $|f(s)| \leqslant M|s|$ for some positive constant M as $s \to 0$.)

Note. In this case, the *exact* value of the period can be expressed in terms of the *complete elliptic integral of the first kind*

$$K(\alpha) := \int_0^{\pi/2} \frac{ds}{\sqrt{1-\alpha^2 \sin^2 s}} = \int_0^1 \frac{dx}{\sqrt{(1-x^2)(1-\alpha^2 x^2)}}, \tag{2.76}$$

with $0 \leqslant \alpha < 1$, namely

$$\tau = 4\sqrt{\frac{m}{k}}\left(1-\frac{\varepsilon^2}{2}\right)^{-1/2} K\left(\frac{\varepsilon}{\sqrt{2-\varepsilon^2}}\right).$$

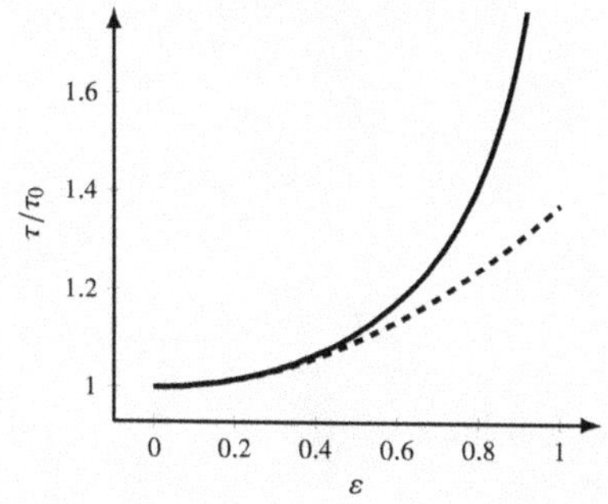

Figure 2.17. Period of the oscillations about the origin for the potential in Example 2.3 (in units of the approximate period of the small oscillations $\tau_0 = 2\pi\sqrt{m/k}$) as a function of the parameter $\varepsilon = A/a$ (solid line) compared to its approximation (2.75) (dashed line).

Exercise 2.15. Find the relation between the energy E and the amplitude A of the small oscillations about a stable equilibrium x_0 with $V''(x_0) > 0$.

Solution. The turning points (points on the trajectory at a maximum and minimum distance from the equilibrium x_0) are the two roots $x_{1,2}$ of the equation $E = V(x)$ to the left and right of x_0. Using the approximation

$$V(x) \simeq V(x_0) + \frac{1}{2} V''(x_0)(x-x_0)^2$$

we obtain

$$E = V(x_{1,2}) \simeq V(x_0) + \frac{1}{2} V''(x_0)(x_{1,2}-x_0)^2$$

$$\Longrightarrow \quad A := \frac{1}{2}(x_2 - x_1) \simeq |x_{1,2} - x_0| \simeq \sqrt{\frac{2(E-V(x_0))}{V''(x_0)}}.$$

Exercise 2.16. Find the dependence on the amplitude and energy of the period of the oscillations about $x = 0$ of a particle of mass m moving subject to the potential $V(x) = k|x|^n$, with $k > 0$ and $n \in \mathbb{N}$.

Solution. The period of the oscillations about the origin for the given potential is

$$\tau = \sqrt{2m}\int_{-A}^{A}\frac{dx}{\sqrt{E - k|x|^n}} = 2\sqrt{2m}\int_0^A \frac{dx}{\sqrt{E - kx^n}} = 2\sqrt{\frac{2m}{k}}\int_0^A \frac{dx}{\sqrt{A^n - x^n}},$$

where the amplitude A is the positive root of the equation $V(x) = E$, namely $A = (E/k)^{1/n}$, and we have taken into account that the potential is an even function of x. This integral cannot be expressed in terms of elementary functions unless $n = 2$. However, performing the change of variables $x = As$ we obtain

$$\tau = 2A\sqrt{\frac{2m}{k}}\,A^{-n/2}\int_0^1 \frac{ds}{\sqrt{1 - s^n}} = \sqrt{\frac{m}{k}}\,A^{1-\frac{n}{2}}\,I_n\,,$$

where

$$I_n := 2\sqrt{2}\int_0^1 \frac{ds}{\sqrt{1 - s^n}}$$

is a pure number. In terms of the energy,

$$\tau = \sqrt{m}\,k^{-\frac{1}{n}}\,E^{\frac{1}{n}-\frac{1}{2}}\,I_n\,.$$

Note that only when $n = 2$ the period is independent of the amplitude. Note also that if $0 < n < 2$ the period increases with increasing amplitude, while if $n > 2$ it decreases.

Note.

- The integral I_n can actually be expressed in terms of *Euler's gamma function*

$$\Gamma(x) = \int_0^\infty t^{x-1}e^{-t}\,dt \qquad (x > 0)$$

  by the formula

$$I_n = 2\sqrt{2\pi}\,\frac{\Gamma(1 + \frac{1}{n})}{\Gamma(\frac{1}{2} + \frac{1}{n})}\,.$$

- If a one-dimensional potential $V(x)$ behaves near a point x_0 as

$$V(x) = V(x_0) + \frac{V^{(n)}(x_0)}{n!}(x - x_0)^n + O\big(|x - x_0|^{n+1}\big)$$

—with n even and $V^{(n)}(x_0) > 0$, so that x_0 is a stable equilibrium—the period of the small oscillations about x_0 of amplitude A is given by the previous formula with $k = V^{(n)}(x_0)/n!$, namely

$$\tau \simeq I_n \sqrt{\frac{n!m}{V^{(n)}(x_0)}}\, A^{1-\frac{n}{2}} = 2\sqrt{\frac{2\pi n!m}{V^{(n)}(x_0)}}\, \frac{\Gamma\left(1+\frac{1}{n}\right)}{\Gamma\left(\frac{1}{2}+\frac{1}{n}\right)}\, A^{1-\frac{n}{2}}.$$

Exercise 2.17. Redo the discussion in Example 2.3 for the potential

$$V(x) = k\left(\frac{x^4}{4a^2} - \frac{x^2}{2}\right), \qquad k,\ a > 0\,.$$

In particular, determine the stable equilibria and compute the period of small oscillations about them. Show that the period of the oscillations of amplitude $A \gg a$ about the origin is approximately proportional to a/A, and find the proportionality constant.

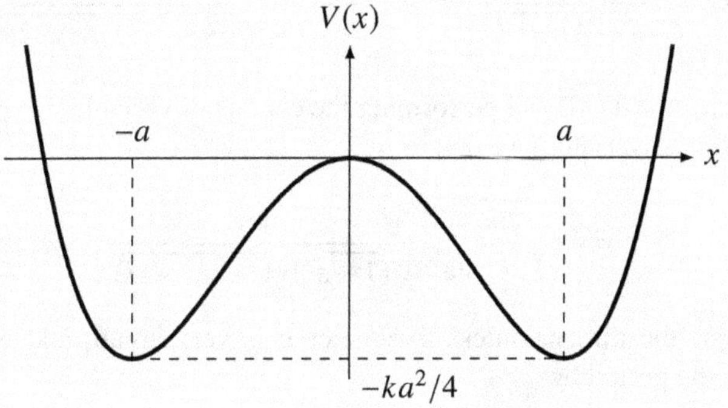

Figure 2.18. Potential $V(x) = k\left(\frac{x^4}{4a^2} - \frac{x^2}{2}\right)$ with $k > 0$.

Solution. The potential in this exercise differs only by a sign from that of Example 2.3, and thus its graph is as shown in Fig. 2.18. In particular, the equilibria are again 0 (unstable) and $\pm a$ (both stable). In this case E must be greater than the absolute minimum $V(\pm a) = -ka^2/4$ of V, and therefore we can have the following types of trajectories:

i) $E = -ka^2/4$

In this case either $x(t) = -a$ or $x(t) = a$ for all t (equilibrium solutions).

ii) $-ka^2/4 < E < 0$

The particle oscillates about the equilibrium $x = -a$ if $x(0) < 0$, or about $x = a$ if $x(0) > 0$, so that the motion is periodic and bounded. The period of the small oscillations about either equilibrium is approximately given by

$$\tau \simeq 2\pi\sqrt{\frac{m}{V''(\pm a)}} = 2\pi\sqrt{\frac{m}{2k}}\,.$$

iii) $E = 0$
If $x(0) > 0$ then the motion is bounded ($0 < x \leqslant \sqrt{2}\,a$, where $\sqrt{2}\,a$ is the positive root of $V(x) = 0$) but *not periodic*, since the unstable equilibrium at the origin cannot be reached in a finite time. Similarly, if $x(0) < 0$ the motion is again bounded (with $-\sqrt{2}\,a \leqslant x < 0$) but not periodic. Finally, if $x(0) = 0$, then $x(t) = 0$ for all t (equilibrium solution).

iv) $E > 0$
In this case the motion is bounded and periodic, since the two roots $\pm A$ of the equation $V(x) = 0$ are turning points. Note that $A > \sqrt{2}\,a$ is not infinitesimally small, so that Eq. (2.71) does not apply in this case. The period of the oscillations as a function of their amplitude A is given, however, by the *exact* formula

$$\tau = 2\sqrt{2m}\int_0^A \frac{\mathrm{d}x}{\sqrt{V(A) - V(x)}} = 4a\sqrt{\frac{2m}{k}}\int_0^A \frac{\mathrm{d}x}{\sqrt{(A^2 - x^2)(A^2 - 2a^2 + x^2)}}\,.$$

Calling $\varepsilon := a/A < 1/\sqrt{2}$ and performing the change of variable $x = As$ (so that s is a dimensionless variable) we obtain

$$\tau = 4\varepsilon\sqrt{\frac{2m}{k}}\int_0^1 \frac{\mathrm{d}s}{\sqrt{(1 - s^2)(1 - 2\varepsilon^2 + s^2)}}\,.$$

When $A \gg a$ the dimensionless parameter ε is very small, and we can thus approximate the period by

$$\tau \simeq 4C\sqrt{\frac{2m}{k}}\,\varepsilon = 4C\sqrt{\frac{2m}{k}}\,\frac{a}{A}\,,$$

where the constant C is given by

$$C := \int_0^1 \frac{\mathrm{d}s}{\sqrt{1 - s^4}} \overset{s=\cos\theta}{=} \int_0^{\pi/2} \frac{\mathrm{d}\theta}{\sqrt{2 - \sin^2\theta}} = \frac{1}{\sqrt{2}}K(1/\sqrt{2}) = 1.31103\ldots,$$

where $K(\alpha)$ is the complete elliptic integral of the first kind defined by Eq. (2.76). (The constant C can also be expressed in terms of Euler's gamma function as $C = \dfrac{\sqrt{\pi}}{4}\dfrac{\Gamma(1/4)}{\Gamma(3/4)}$.) Thus the period of the oscillations with *large* amplitude $A \gg a$ is (approximately) *inversely proportional* to the amplitude.

Note. The exact value of the period of the oscillations of amplitude $A > \sqrt{2}\,a$ can also be determined in terms of a complete elliptic integral of the first kind. Indeed, performing the change of variable $s = \cos\theta$ in the integral for τ we obtain

$$\tau = 4\varepsilon\sqrt{\frac{2m}{k}}\int_0^{\pi/2}\frac{d\theta}{\sqrt{2(1-\varepsilon^2)-\sin^2\theta}} = 4\varepsilon\sqrt{\frac{m}{k(1-\varepsilon^2)}}\,K\left(\frac{1}{\sqrt{2(1-\varepsilon^2)}}\right).$$

Exercise 2.18. A particle of mass $m = 2$ and energy $E = 0$ starts from the origin at $t = 0$ moving toward the right subject to the one-dimensional potential

$$V(x) = -(1-x^2)(1-k^2x^2)$$

(in appropriate units), where $k \in [0, 1]$ is a parameter (cf. Fig. 2.19).

i) Prove that for $k \in [0, 1)$ the motion is periodic, and find its amplitude and period. What happens if $k = 1$?

ii) The solution $x(t)$ of the equation of motion in part i) is called *Jacobi's elliptic sine*, and is usually denoted by $\mathrm{sn}(t; k)$. Its quarter period (for $0 \leqslant k < 1$) is the *complete elliptic integral of the first kind* $K(k)$ defined in Eq. (2.76). Express $\mathrm{sn}(t; 0)$ and $\mathrm{sn}(t; 1)$ in terms of elementary functions, and compute $K(0)$. What is the value of $K(1)$?

iii) For $0 \leqslant k < 1$, write down the solution of the equation of motion with $E = 0$ and $x(0) = 1/k$ in terms of sn. How long does the particle take to reach infinity?

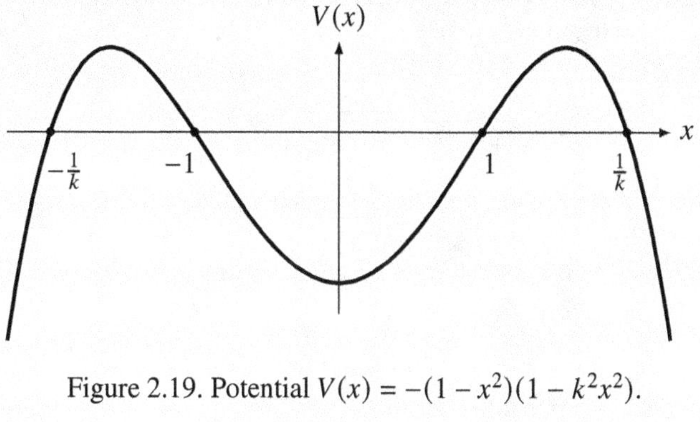

Figure 2.19. Potential $V(x) = -(1-x^2)(1-k^2x^2)$.

Solution.

i) The roots of the equation $V(x) = E = 0$ are $x = \pm 1, \pm 1/k$, so that $V(x) \leqslant E = 0$ for $|x| \leqslant 1$ or $|x| \geqslant 1/k$. If $0 \leqslant k < 1$, the accessible region is the union of the three disjoint intervals $(-\infty, -1/k]$, $[-1, 1]$, $[1/k, \infty)$, the second of which contains the origin. Thus if $x(0) = 0$ the motion is confined to the interval $[-1, 1]$. On the other hand,

$$V'(x) = 2x(1 - k^2x^2) + 2k^2x(1 - x^2) = 2x\big(1 + k^2 - 2k^2x^2\big)$$

does not vanish for $x = \pm 1$ unless $k = 1$. Hence if $0 \leqslant k < 1$ the points $x = \pm 1$ are turning points, and the motion is therefore periodic with amplitude 1. The period of the motion is given by

$$\tau(k) = 2\int_{-1}^{1} \frac{dx}{\sqrt{-V(x)}} = 4\int_{0}^{1} \frac{dx}{\sqrt{(1 - x^2)(1 - k^2x^2)}},$$

where in the last equality we have taken into account that $V(x) = V(-x)$. The situation is very different for $k = 1$, since in this case the points $x = \pm 1$ are equilibria, and the motion is thus bounded but not periodic (the particle can only reach the points $x = \pm 1$ in an infinite time). In this case it is actually straightforward to integrate the energy equation:

$$\dot{x}^2 = (1 - x^2)^2 \iff \dot{x} = \pm(1 - x^2),$$

where we must take the "+" sign since initially the particle is moving to the right ($\dot{x}(0) > 0$). Taking into account that $x(0) = 0$ and $-1 < x < 1$ we obtain

$$t = \int_0^x \frac{ds}{1 - s^2} = \frac{1}{2}\log\left|\frac{1 + x}{1 - x}\right| = \frac{1}{2}\log\left(\frac{1 + x}{1 - x}\right)$$

$$\implies e^{2t}(1 - x) = 1 + x \implies x = \frac{e^{2t} - 1}{e^{2t} + 1} = \tanh t\,.$$

ii) The solution sought is obtained integrating the energy equation with the initial conditions $x(0) = 0$ and $\dot{x}(0) > 0$:

$$t = \int_0^x \frac{ds}{\sqrt{(1 - s^2)(1 - k^2s^2)}} \iff x = \operatorname{sn}(t; k),$$

where we have used the definition of sn. In other words, defining

$$F_k(x) \equiv \int_0^x \frac{ds}{\sqrt{(1 - s^2)(1 - k^2s^2)}},$$

we then have $\operatorname{sn}(t; k) = F_k^{-1}(t)$. The period $\tau(k)$ of the motion was computed in the previous part, whence it follows that

$$\frac{\tau(k)}{4} = \int_0^1 \frac{dx}{\sqrt{(1 - x^2)(1 - k^2x^2)}} = K(k)\,.$$

For $k = 1$ we have already seen that $\mathrm{sn}(t;1) = \tanh t$. This function is *not* periodic, and indeed in this case

$$K(1) = \int_0^1 \frac{\mathrm{d}x}{1 - x^2} = +\infty$$

(the improper integral diverges logarithmically at $x = 1$). On the other hand, for $k = 0$ we obtain

$$F_0(x) = \int_0^x \frac{\mathrm{d}x}{\sqrt{1 - x^2}} = \arcsin x \iff \mathrm{sn}(t;0) = F_0^{-1}(t) = \sin t\,.$$

In this case

$$K(0) = \int_0^1 \frac{\mathrm{d}x}{\sqrt{1 - x^2}} = \int_0^{\pi/2} \mathrm{d}t = \frac{\pi}{2}$$

(and, indeed, the period of the sin function is $2\pi = 4K(0)$).

iii) Integrating the energy equation (taking into account that the particle must initially move to the right, since it would otherwise leave the accessible region $V(x) \leqslant 0$) we have

$$t = \int_{1/k}^x \frac{\mathrm{d}s}{\sqrt{(s^2 - 1)(k^2 s^2 - 1)}}\,.$$

Performing the change of variable $s = 1/(kz)$ we obtain

$$t = \int_{1/(kx)}^1 \frac{\mathrm{d}z}{kz^2\sqrt{\left(\frac{1}{k^2z^2} - 1\right)\left(\frac{1}{z^2} - 1\right)}} = \int_{1/(kx)}^1 \frac{\mathrm{d}z}{\sqrt{(1 - z^2)(1 - k^2z^2)}}$$
$$= K(k) - F_k\big(1/(kx)\big)\,,$$

whence it follows that

$$x = \frac{1}{k\,\mathrm{sn}\big(K(k) - t; k\big)}\,.$$

In particular, the particle reaches infinity when the denominator vanishes for the first time, i.e., for $t = K(k)$.

Exercise 2.19. Show that the solution of the simple pendulum's equation of motion

$$\ddot{\theta} + \frac{g}{l}\sin\theta = 0$$

with the initial conditions $\theta(0) = \theta_0 \in (0, \pi)$, $\dot{\theta}(0) = 0$ is

$$\sin(\theta/2) = k\,\mathrm{sn}\big(K(k) - \sqrt{g/l}\,t; k\big)$$

where $k = \sin(\theta_0/2)$, and that its period is given by $4K(k)\sqrt{l/g}$. [*Hint:* perform the change of variable $\sin(\theta/2) = kx$ in the energy equation.]

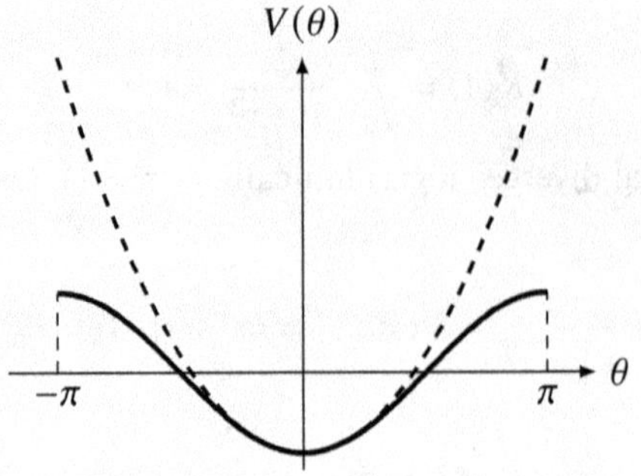

Figure 2.20. Simple pendulum potential $V(\theta) = -\frac{g}{l}\cos\theta$ (solid line) compared to its harmonic approximation $V(\theta) = -\frac{g}{l}\left(1 - \frac{\theta^2}{2}\right)$ (dashed line).

Solution. The equation of motion is that of a particle with unit mass moving in one dimension subject to the effective potential

$$V(\theta) = -\int F(\theta)\,d\theta = \frac{g}{l}\int \sin\theta\,d\theta = -\frac{g}{l}\cos\theta$$

(cf. Fig. 2.20). The energy equation is therefore

$$\frac{1}{2}\dot{\theta}^2 - \frac{g}{l}\cos\theta = E\,.$$

Using the initial conditions we obtain $E = -(g/l)\cos\theta_0$, and hence

$$\dot{\theta}^2 = \frac{2g}{l}\left(\cos\theta - \cos\theta_0\right) = \frac{4g}{l}\left(\sin^2(\theta_0/2) - \sin^2(\theta/2)\right) \equiv \frac{4g}{l}\left(k^2 - \sin^2(\theta/2)\right),$$

whence it also follows that $-\theta_0 \leqslant \theta \leqslant \theta_0$ (i.e., the pendulum's motion is oscillatory and θ_0 is its angular amplitude). Integrating the previous equation we obtain

$$\sqrt{\frac{g}{l}}\,t = -\frac{1}{2}\int_{\theta_0}^{\theta} \frac{d\alpha}{\sqrt{k^2 - \sin^2(\alpha/2)}}$$

(why the "–" sign?). Performing the change of variable $\sin(\alpha/2) = kx$ we arrive at the equation

$$\sqrt{\frac{g}{l}}\,t = -k\int_{1}^{\frac{1}{k}\sin(\theta/2)} \frac{dx}{\cos(\alpha/2)\sqrt{k^2 - k^2x^2}} = \int_{\frac{1}{k}\sin(\theta/2)}^{1} \frac{dx}{\sqrt{(1-x^2)(1-k^2x^2)}}\,,$$

or equivalently

$$\sqrt{\frac{g}{l}}\,t = K(k) - F_k\big(\sin(\theta/2)/k\big) \iff F_k\big(\sin(\theta/2)/k\big) = K(k) - \sqrt{\frac{g}{l}}\,t$$

$$\iff \frac{1}{k}\,\sin(\theta/2) = \operatorname{sn}\left(K(k) - \sqrt{\frac{g}{l}}\,t; k\right),$$

which is the solution sought Finally, since the period of sn is $4K(k)$ the pendulum's period is $4K(k)\sqrt{l/g}$. See Fig. 2.21 for a plot of the exact solution $\theta(t)$ just computed compared to its approximation $\theta = \theta_0 \cos(\sqrt{g/l}\,t)$, obtained by solving the linearized pendulum equation $\ddot{\theta} + (g/l)\theta = 0$ with the initial conditions $\theta(0) = \theta_0$ and $\dot{\theta}(0) = 0$.

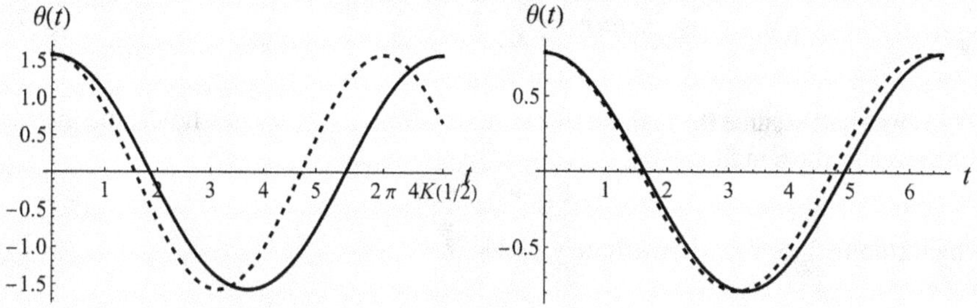

Figure 2.21. Exact solution of the simple pendulum's equation with the initial conditions $\theta(0) = \theta_0$, $\dot{\theta}(0) = 0$ (solid line) compared to its approximation $\theta = \theta_0 \cos(\sqrt{g/l}\,t)$ (dotted line) for $\theta_0 = \pi/2$ (left) and $\theta_0 = \pi/4$ (right). In both plots we have taken $\sqrt{l/g}$ as the unit of time.

2.6 DYNAMICS OF A SYSTEM OF PARTICLES

2.6.1 Dynamics of a system of particles

We shall study in this section the motion of a system of N particles of mass m_i ($i = 1, \dots, N$). Let us denote by $\mathbf{r}_i$ the position vector of the i-th particle in a certain inertial system, and define $\mathbf{F}_{ij}$ as the force exerted by particle j on particle i (in particular, $\mathbf{F}_{ii} = 0$) and $\mathbf{F}_i^{(e)}$ as the external force acting on the i-th particle. Newton's second law of motion applied to the i-th particle of the system then states that

$$\boxed{m_i \ddot{\mathbf{r}}_i = \sum_{j=1}^{N} \mathbf{F}_{ij} + \mathbf{F}_i^{(e)}, \qquad i = 1, \dots, N,} \tag{2.77}$$

where the sum over j on the RHS represents the internal force exerted on particle i by the remaining particles of the system. These N *vector* equations are actually a system of $3N$ *scalar* second-order ordinary differential equations in the unknowns

$\mathbf{r}_1, \ldots, \mathbf{r}_N$. By the existence and uniqueness theorem for such systems, if the RHS of Eq. (2.77) is a function of class C^1 in the variables $(t, \mathbf{r}_1, \ldots, \mathbf{r}_N, \dot{\mathbf{r}}_1, \ldots, \dot{\mathbf{r}}_N)$ in a certain open subset $A \subset \mathbb{R}^{6N+1}$ then the system's equations of motion (2.77) have (locally) a *unique* solution verifying any initial condition of the form

$$\mathbf{r}_1(t_0) = \mathbf{r}_{10}, \ \ldots, \mathbf{r}_N(t_0) = \mathbf{r}_{N0}; \quad \dot{\mathbf{r}}_1(t_0) = \mathbf{v}_{10}, \ \ldots, \dot{\mathbf{r}}_N(t_0) = \mathbf{v}_{N0}$$

with $(t_0, \mathbf{r}_{10}, \ \ldots, \mathbf{r}_{N0}, \mathbf{v}_{10}, \ \ldots, \mathbf{v}_{N0}) \in A$. In other words, *the trajectories of all the particles in the system are determined by their positions and velocities at any instant.* In this sense, Newtonian mechanics is a *completely deterministic* theory.

Summing over i in Eq. (2.77) we obtain

$$\sum_{i=1}^{N} m_i \ddot{\mathbf{r}}_i = \sum_{i,j=1}^{N} \mathbf{F}_{ij} + \sum_{i=1}^{N} \mathbf{F}_i^{(e)} . \tag{2.78}$$

If—as we shall assume throughout this section—*Newton's third law* holds, the internal forces verify the condition

$$\mathbf{F}_{ij} + \mathbf{F}_{ji} = 0 ,$$

which summed over i, j immediately yields

$$0 = \sum_{i,j=1}^{N} \mathbf{F}_{ij} + \sum_{i,j=1}^{N} \mathbf{F}_{ji} = 2 \sum_{i,j=1}^{N} \mathbf{F}_{ij} ,$$

where in the last step we have used the fact that the summation indices (i, j) are dummy (i.e., $\sum_{i,j=1}^{N} \mathbf{F}_{ji} = \sum_{i,j=1}^{N} \mathbf{F}_{ij}$). Denoting by

$$\boxed{\mathbf{F}^{(e)} := \sum_{i=1}^{N} \mathbf{F}_i^{(e)}}$$

the **total external force** acting on the system, Eq. (2.78) can be written more concisely as

$$\boxed{\sum_{i=1}^{N} m_i \ddot{\mathbf{r}}_i = \mathbf{F}^{(e)} .} \tag{2.79}$$

Let us next define the system's **center of mass** as the point with coordinates

$$\boxed{\mathbf{R} := \frac{1}{M} \sum_{i=1}^{N} m_i \mathbf{r}_i ,} \tag{2.80}$$

where

$$M = \sum_{i=1}^{N} m_i$$

is the total mass of the system. In other words, the center of mass (which is generally abbreviated by CM) is the average of the particles' coordinates weighted by their masses. In terms of the CM, Eq. (2.79) adopts the simple form

$$\boxed{M\ddot{\mathbf{R}} = \mathbf{F}^{(e)}\,.} \tag{2.81}$$

Hence:

The center of mass moves as a single particle of mass M on which the total *external* force acting on the system is exerted.

As a consequence, *the motion of the center of mass is not affected by the internal forces acting on the system.* In particular, if the total external force vanishes then $\ddot{\mathbf{R}} = 0$. Hence:

In the absence of external forces, the center of mass moves with constant velocity.

2.6.2 Conservation laws

Consider first the system's total **momentum P**, defined as the sum of the momenta of its constituent particles:

$$\boxed{\mathbf{P} := \sum_{i=1}^{N} m_i \dot{\mathbf{r}}_i = M\dot{\mathbf{R}}\,.} \tag{2.82}$$

Thus *the total momentum of the system coincides with the momentum of its center of mass regarded as a single particle of mass M*. It also follows from the previous equation that Eq. (2.81) is equivalent to

$$\boxed{\dot{\mathbf{P}} = \mathbf{F}^{(e)}\,.}$$

In particular, *in the absence of* external *forces the system's total momentum is* conserved.

Consider next the system's **angular momentum** with respect to the origin of coordinates, defined as the sum of the angular momenta of its N constituent particles:

$$\boxed{\mathbf{L} := \sum_{i=1}^{N} m_i \mathbf{r}_i \times \dot{\mathbf{r}}_i\,.} \tag{2.83}$$

Let us denote by $\mathbf{r}'_i$ the position vector of the i-th particle with respect to the CM, so that

$$\mathbf{r}_i = \mathbf{R} + \mathbf{r}'_i \,. \tag{2.84}$$

Substituting this expression for $\mathbf{r}_i$ in Eq. (2.83) we obtain

$$\mathbf{L} = \sum_{i=1}^{N} m_i(\mathbf{R} + \mathbf{r}'_i) \times (\dot{\mathbf{R}} + \dot{\mathbf{r}}'_i) = M\mathbf{R} \times \dot{\mathbf{R}} + \mathbf{R} \times \sum_{i=1}^{N} m_i \dot{\mathbf{r}}'_i + \left(\sum_{i=1}^{N} m_i \mathbf{r}'_i\right) \times \dot{\mathbf{R}} + \sum_{i=1}^{N} m_i \mathbf{r}'_i \times \dot{\mathbf{r}}'_i \,. \tag{2.85}$$

On the other hand, from Eq. (2.84) and the definition (2.80) of the CM it easily follows that

$$\sum_{i=1}^{N} m_i \mathbf{r}_i = M\mathbf{R} = \sum_{i=1}^{N} m_i \mathbf{R} + \sum_{i=1}^{N} m_i \mathbf{r}'_i = M\mathbf{R} + \sum_{i=1}^{N} m_i \mathbf{r}'_i$$

$$\Longrightarrow \boxed{\sum_{i=1}^{N} m_i \mathbf{r}'_i = 0 \,,} \tag{2.86}$$

and therefore the second and third terms in the RHS of Eq. (2.85) vanish identically. We thus have

$$\boxed{\mathbf{L} = M\mathbf{R} \times \dot{\mathbf{R}} + \sum_{i=1}^{N} m_i \mathbf{r}'_i \times \dot{\mathbf{r}}'_i \,.} \tag{2.87}$$

In other words, *the angular momentum of the system is the sum of the angular momentum of its center of mass and the* **internal angular momentum** (last term in the RHS of Eq. (2.87)) due to the motion of the particles about the CM.

Example 2.5. Suppose that the system moves as a whole with velocity (not necessarily uniform) $\mathbf{v}(t)$, i.e., that

$$\dot{\mathbf{r}}_i = \mathbf{v}(t) \,, \qquad i = 1, \ldots, N \,.$$

In this case

$$\dot{\mathbf{R}} = \frac{1}{M} \sum_{i=1}^{N} m_i \dot{\mathbf{r}}_i = \mathbf{v}(t) \,,$$

and thus the internal angular momentum vanishes:

$$\dot{\mathbf{r}}'_i = \dot{\mathbf{r}}_i - \dot{\mathbf{R}} = 0 \,, \quad i = 1, \ldots, N \quad \Longrightarrow \quad \mathbf{L} = M\mathbf{R} \times \dot{\mathbf{R}} = M\mathbf{R} \times \mathbf{v} \,.$$

By Eq. (2.83), the time derivative of the angular momentum is given by

$$\dot{\mathbf{L}} = \sum_{i=1}^{N} m_i \mathbf{r}_i \times \ddot{\mathbf{r}}_i = \sum_{i=1}^{N} \mathbf{r}_i \times \mathbf{F}_i^{(e)} + \sum_{i,j=1}^{N} \mathbf{r}_i \times \mathbf{F}_{ij} .$$

Again, it is easy to check that the last term vanishes if Newton's third law holds in its *stronger version*, that is if

$$\mathbf{F}_{ji} = -\mathbf{F}_{ij} \parallel \mathbf{r}_i - \mathbf{r}_j , \qquad i \neq j . \tag{2.88}$$

Indeed,

$$\begin{aligned} 0 &= \sum_{i,j=1}^{N} (\mathbf{r}_i - \mathbf{r}_j) \times \mathbf{F}_{ij} = \sum_{i,j=1}^{N} \mathbf{r}_i \times \mathbf{F}_{ij} - \sum_{i,j=1}^{N} \mathbf{r}_j \times \mathbf{F}_{ij} \\ &= \sum_{i,j=1}^{N} \mathbf{r}_i \times \mathbf{F}_{ij} + \sum_{i,j=1}^{N} \mathbf{r}_j \times \mathbf{F}_{ji} = 2 \sum_{i,j=1}^{N} \mathbf{r}_i \times \mathbf{F}_{ij} . \end{aligned}$$

Thus in this case we have

$$\boxed{\dot{\mathbf{L}} = \sum_{i=1}^{N} \mathbf{r}_i \times \mathbf{F}_i^{(e)} =: \mathbf{N}^{(e)} ,} \tag{2.89}$$

where by definition $\mathbf{N}^{(e)}$ is the **total torque of the external forces** acting on the system. In other words:

If Newton's third law holds in its stronger version (2.88) then the time derivative of the system's angular momentum is equal to the total torque of the *external* forces acting on it. In particular, if the total torque of the *external* forces acting on the system vanishes its angular momentum is conserved.

Exercise 2.20. Show that in general the total torque of the external forces is different from the torque of the total external force with respect to the CM.

Solution. By definition , the torque of the total external force with respect to the CM is the vector $\mathbf{R} \times \mathbf{F}^{(e)}$. Taking into account the definitions of $\mathbf{R}$ and $\mathbf{F}^{e}$ we easily obtain

$$\mathbf{R} \times \mathbf{F}^{(e)} = \frac{1}{M} \left(\sum_{i=1}^{N} m_i \mathbf{r}_i \right) \times \left(\sum_{j=1}^{N} F_j^{(e)} \right) = \sum_{i,j=1}^{N} \frac{m_i}{M} \mathbf{r}_i \times \mathbf{F}_j^{(e)} ,$$

which is different in general than $\sum_{i=1}^{N} \mathbf{r}_i \times \mathbf{F}_i^{(e)}$. For example, for two particles we have

$$\begin{aligned}\mathbf{R}\times\mathbf{F}^{(e)} - \sum_{i=1}^{2} \mathbf{r}_i \times \mathbf{F}_i^{(e)} &= \sum_{i=1}^{2} \mathbf{r}_i \times \left(\frac{m_i}{M} \sum_{j=1}^{2} \mathbf{F}_j^{(e)} - \mathbf{F}_i^{(e)} \right) \\ &= \frac{\mathbf{r}_1}{M} \times \left(m_1 \mathbf{F}_2^{(e)} - m_2 \mathbf{F}_1^{(e)} \right) + \frac{\mathbf{r}_2}{M} \times \left(m_2 \mathbf{F}_1^{(e)} - m_1 \mathbf{F}_2^{(e)} \right) \\ &= \frac{\mathbf{r}_1 - \mathbf{r}_2}{M} \times \left(m_1 \mathbf{F}_2^{(e)} - m_2 \mathbf{F}_1^{(e)} \right),\end{aligned}$$

which does not vanish unless

$$m_1 \mathbf{F}_2^{(e)} - m_2 \mathbf{F}_1^{(e)} \parallel \mathbf{r}_1 - \mathbf{r}_2 \,.$$

Note. A common situation, however, in which $\mathbf{R} \times \mathbf{F}^{(e)}$ is equal to the total torque of the external forces arises when all the particles have the same mass and the total external force $\mathbf{F}_i^{(e)}$ is independent of i. For example, this will be the case if the particles move in a constant gravitational field, or if they move in a constant electric field and have the same charge. Indeed, if $m_i = m$ and $\mathbf{F}_i^{(e)} = \mathbf{F}$ for all i we have $M = Nm$, $\mathbf{F}^{(e)} = N\mathbf{F}$, and therefore

$$\mathbf{R}\times\mathbf{F}^{(e)} = \sum_{i=1}^{N} \frac{m}{Nm} \mathbf{r}_i \times (N\mathbf{F}) = \sum_{i=1}^{N} \mathbf{r}_i \times \mathbf{F} = \sum_{i=1}^{N} \mathbf{r}_i \times \mathbf{F}_i^{(e)}.$$

Let us next study the **kinetic energy** of the system, defined as

$$T = \frac{1}{2} \sum_{i=1}^{N} m_i \dot{\mathbf{r}}_i^2 \,. \tag{2.90}$$

Using again the decomposition (2.84) and the identity (2.86) we easily obtain:

$$T = \frac{1}{2} M \dot{\mathbf{R}}^2 + \frac{1}{2} \sum_{i=1}^{N} m_i \dot{\mathbf{r}}_i'^2 \,, \tag{2.91}$$

where the last term is the **internal kinetic energy** due to the motion of the particles with respect to the CM. Hence *the kinetic energy of the system is the sum of the kinetic energy of its CM and the internal kinetic energy.*

We shall say that the forces acting on the system are **conservative**—or, equivalently, that the system itself is conservative—if there is a *single* scalar function $V(\mathbf{r}_1, \ldots, \mathbf{r}_N)$

such that

$$\mathbf{F}_i := \mathbf{F}_i^{(e)} + \sum_{j=1}^{N} \mathbf{F}_{ij} = -\frac{\partial V}{\partial \mathbf{r}_i}, \qquad i = 1, \dots, N. \tag{2.92}$$

As in the case of a single particle addressed in Section 2.4.2, *if the forces acting on the system are conservative the system's* ***total energy***

$$E = T + V(\mathbf{r}_1, \dots, \mathbf{r}_N)$$

is conserved. Indeed, if the forces acting on the system are conservative we have

$$\frac{dE}{dt} = \sum_{i=1}^{N} m_i \dot{\mathbf{r}}_i \cdot \ddot{\mathbf{r}}_i + \frac{dV}{dt} = \sum_{i=1}^{N} \dot{\mathbf{r}}_i \cdot \mathbf{F}_i + \sum_{i=1}^{N} \frac{\partial V}{\partial \mathbf{r}_i} \dot{\mathbf{r}}_i = \sum_{i=1}^{N} \left(\mathbf{F}_i + \frac{\partial V}{\partial \mathbf{r}_i} \right) \dot{\mathbf{r}}_i = 0 .$$

Let us assume that there exist certain functions $V_i(\mathbf{r}_i)$, $V_{ij}(\mathbf{r}_i, \mathbf{r}_j)$ (with $i \neq j$, $1 \leqslant i, j \leqslant N$) such that $V_{ij}(\mathbf{r}_i, \mathbf{r}_j) = V_{ji}(\mathbf{r}_j, \mathbf{r}_i)$ (i.e., $V_{ij}(\mathbf{r}_i, \mathbf{r}_j)$ is symmetric under exchange of the particles i and j) and

$$\mathbf{F}_i^{(e)} = -\frac{\partial V_i(\mathbf{r}_i)}{\partial \mathbf{r}_i}, \qquad \mathbf{F}_{ij} = -\frac{\partial V_{ij}(\mathbf{r}_i, \mathbf{r}_j)}{\partial \mathbf{r}_i}, \qquad 1 \leqslant i \neq j \leqslant N . \tag{2.93a}$$

Then the system is conservative, with potential (up to an additive constant)

$$V = \sum_{i=1}^{N} V_i(\mathbf{r}_i) + \sum_{1 \leqslant i < j \leqslant N} V_{ij}(\mathbf{r}_i, \mathbf{r}_j) . \tag{2.93b}$$

Indeed, it is easily verified that Eqs. (2.93) imply the more general relation (2.92):

$$\begin{aligned}
\mathbf{F}_i + \frac{\partial V}{\partial \mathbf{r}_i} &= \sum_{j=1}^{N} \mathbf{F}_{ij} + \frac{\partial}{\partial \mathbf{r}_i} \sum_{1 \leqslant j < k \leqslant N} V_{jk}(\mathbf{r}_j, \mathbf{r}_k) \\
&= \sum_{j=1}^{N} \mathbf{F}_{ij} + \sum_{k=i+1}^{N} \frac{\partial}{\partial \mathbf{r}_i} V_{ik}(\mathbf{r}_i, \mathbf{r}_k) + \sum_{j=1}^{i-1} \frac{\partial}{\partial \mathbf{r}_i} V_{ji}(\mathbf{r}_j, \mathbf{r}_i) \\
&= \sum_{j=1}^{N} \mathbf{F}_{ij} + \sum_{j=i+1}^{N} \frac{\partial}{\partial \mathbf{r}_i} V_{ij}(\mathbf{r}_i, \mathbf{r}_j) + \sum_{j=1}^{i-1} \frac{\partial}{\partial \mathbf{r}_i} V_{ij}(\mathbf{r}_i, \mathbf{r}_j) \\
&= \sum_{\substack{j=1 \\ j \neq i}}^{N} \left(\mathbf{F}_{ij} + \frac{\partial V_{ij}}{\partial \mathbf{r}_i} \right) = 0 ,
\end{aligned}$$

where we have used the identities $\mathbf{F}_{ii} = 0$ and $V_{ji}(\mathbf{r}_j, \mathbf{r}_i) = V_{ij}(\mathbf{r}_i, \mathbf{r}_j)$.

- By Newton's third law we must have

$$\mathbf{F}_{ji} = -\frac{\partial V_{ji}(\mathbf{r}_j, \mathbf{r}_i)}{\partial \mathbf{r}_j} = -\frac{\partial V_{ij}(\mathbf{r}_i, \mathbf{r}_j)}{\partial \mathbf{r}_j} = -\mathbf{F}_{ij} = \frac{\partial V_{ij}(\mathbf{r}_i, \mathbf{r}_j)}{\partial \mathbf{r}_i},$$

so that the function $V_{ij}(\mathbf{r}_i, \mathbf{r}_j)$ should verify the system of partial differential equations

$$\frac{\partial V_{ij}}{\partial \mathbf{r}_i} + \frac{\partial V_{ij}}{\partial \mathbf{r}_j} = 0.$$

It can be shown (exercise) that the general solution of this system is an arbitrary function of the difference $\mathbf{r}_i - \mathbf{r}_j$. In other words,

$$\boxed{V_{ij}(\mathbf{r}_i, \mathbf{r}_j) = U_{ij}(\mathbf{r}_i - \mathbf{r}_j), \qquad \text{with } U_{ji}(\mathbf{r}) = U_{ij}(-\mathbf{r}).} \tag{2.94}$$

Substituting into (2.93b) we obtain the following more explicit formula for the potential V:

$$\boxed{V = \sum_{i=1}^{N} V_i(\mathbf{r}_i) + \sum_{1 \leqslant i < j \leqslant N} U_{ij}(\mathbf{r}_i - \mathbf{r}_j).} \tag{2.95}$$

In fact, in most conservative physical systems of interest the potential is of the form (2.95).

Exercise 2.21. Study under what conditions on the functions U_{ij} Newton's third law holds in its stronger version (2.88).

Solution. If $\mathbf{r} := \mathbf{r}_i - \mathbf{r}_j$, Eq. (2.88) will hold if and only if

$$0 = \mathbf{r} \times \mathbf{F}_{ij} = -\mathbf{r} \times \frac{\partial U_{ij}(\mathbf{r}_i - \mathbf{r}_j)}{\partial \mathbf{r}_i} = -\mathbf{r} \times \frac{\partial U_{ij}(\mathbf{r})}{\partial \mathbf{r}},$$

i.e., if the gradient of $U_{ij}(\mathbf{r})$ has only a radial component. By the formula for the gradient in spherical coordinates

$$\frac{\partial U_{ij}}{\partial \mathbf{r}} = \frac{\partial U_{ij}}{\partial r}\mathbf{e}_r + \frac{1}{r}\frac{\partial U_{ij}}{\partial \theta}\mathbf{e}_\theta + \frac{1}{r\sin\theta}\frac{\partial U_{ij}}{\partial \varphi}\mathbf{e}_\varphi,$$

this will be the case if and only if U_{ij} depends only on $r = |\mathbf{r}_i - \mathbf{r}_j|$, i.e., if there is a function of one variable u_{ij} such that $U_{ij} = u_{ij}(|\mathbf{r}_i - \mathbf{r}_j|)$ (with $u_{ij} = u_{ji}$, in view of Eq. (2.94)). This condition is in fact satisfied by most conservative forces, like the gravitational or the electrostatic ones (see next exercise). When it holds, Eq. (2.95) reads

$$\boxed{V = \sum_{i=1}^{N} V_i(\mathbf{r}_i) + \sum_{1 \leqslant i < j \leqslant N} u_{ij}(|\mathbf{r}_i - \mathbf{r}_j|).} \tag{2.96}$$

Exercise 2.22. Write down the potential for a system of charged particles of mass m_i and charge q_i ($i = 1, \dots, N$) moving in an external electric field generated by an electrostatic potential $\Phi(\mathbf{r})$.

Solution. The external force acting on the i-th particle is due to its interaction with the electric field $\mathbf{E}(\mathbf{r}) = -\dfrac{\partial \Phi(\mathbf{r})}{\partial \mathbf{r}}$ generated by the electrostatic potential $\Phi(\mathbf{r})$, namely

$$\mathbf{F}_i^{(e)} = q_i \mathbf{E}(\mathbf{r}_i) = -q_i \frac{\partial \Phi}{\partial \mathbf{r}}(\mathbf{r}_i) = -q_i \frac{\partial \Phi(\mathbf{r}_i)}{\partial \mathbf{r}_i} = -\frac{\partial V_i(\mathbf{r}_i)}{\partial \mathbf{r}_i},$$

with

$$V_i(\mathbf{r}_i) = q_i \Phi(\mathbf{r}_i).$$

On the other hand, the force exerted by particle j on particle i is the sum of the electric and gravitational forces between both particles, given by

$$\mathbf{F}_{ij} = (kq_iq_j - Gm_im_j) \frac{\mathbf{r}_i - \mathbf{r}_j}{|\mathbf{r}_i - \mathbf{r}_j|^3} = -\frac{\partial}{\partial \mathbf{r}_i} \frac{kq_iq_j - Gm_im_j}{|\mathbf{r}_i - \mathbf{r}_j|} = -\frac{\partial u_{ij}(|\mathbf{r}_i - \mathbf{r}_j|)}{\partial \mathbf{r}_i},$$

with $k = 1/(4\pi\varepsilon_0)$ and

$$u_{ij}(r) = \frac{kq_iq_j - Gm_im_j}{r} = u_{ji}(r).$$

The system is thus conservative, with potential

$$V = \sum_{i=1}^{N} q_i \Phi(\mathbf{r}_i) + \sum_{1 \leqslant i < j \leqslant N} \frac{kq_iq_j - Gm_im_j}{|\mathbf{r}_i - \mathbf{r}_j|}.$$

In practice, the electrostatic coupling constant kq_iq_j is usually much greater than the gravitational one Gm_im_j. For instance, for protons

$$q_i = q_j = 1.602\,176\,634 \cdot 10^{-19}\,\mathrm{C} \ \text{(exact)},$$
$$m_i = m_j = 1.672\,621\,925\,95(52) \cdot 10^{-27}\,\mathrm{kg},$$

so that

$$\frac{Gm_im_j}{kq_iq_j} \simeq 8.09355 \cdot 10^{-37}.$$

For this reason, the gravitational interaction between the charges is usually neglected, and the potential reduces accordingly to

$$V = \sum_{i=1}^{N} q_i \Phi(\mathbf{r}_i) + \frac{1}{4\pi\varepsilon_0} \sum_{1 \leqslant i < j \leqslant N} \frac{q_iq_j}{|\mathbf{r}_i - \mathbf{r}_j|}.$$

CHAPTER 3

Motion in a central potential

THIS chapter is devoted to analyzing the motion of a point particle subject to a conservative central force, generated by a central—i.e., spherically symmetric—potential. We begin by reducing the two-body problem to an equivalent one-body problem for the motion of the relative coordinate of the two particles. We deduce the main conserved quantities of the motion (energy and angular momentum), and use them to derive the law of motion. We then study in detail the motion of a particle under the Kepler $(1/r)$ potential, with particular emphasis on the important case of planetary motion. Finally, we discuss the scattering of a particle by a central potential, and derive Rutherford's formula for the scattering cross section of the Kepler potential.

3.1 REDUCTION OF THE TWO-BODY PROBLEM

We shall study in this section the motion of two point masses m_1 and m_2 not subject to any external forces. If $\mathbf{F}_{12}$ denotes the force exerted by the second particle on the first, by Newton's third law the first particle exerts a force $-\mathbf{F}_{12}$ on the second one, and the system's equations of motion are thus

$$\boxed{\begin{aligned} m_1\ddot{\mathbf{r}}_1 &= \mathbf{F}_{12}(t, \mathbf{r}_1, \mathbf{r}_2, \dot{\mathbf{r}}_1, \dot{\mathbf{r}}_2) \\ m_2\ddot{\mathbf{r}}_2 &= -\mathbf{F}_{12}(t, \mathbf{r}_1, \mathbf{r}_2, \dot{\mathbf{r}}_1, \dot{\mathbf{r}}_2) . \end{aligned}} \tag{3.1}$$

It is convenient to rewrite these equations in terms of the variables

$$\boxed{\mathbf{R} = \frac{1}{M}(m_1\mathbf{r}_1 + m_2\mathbf{r}_2) , \qquad \mathbf{r} = \mathbf{r}_1 - \mathbf{r}_2} \tag{3.2}$$

DOI: 10.1201/9781003600633-3

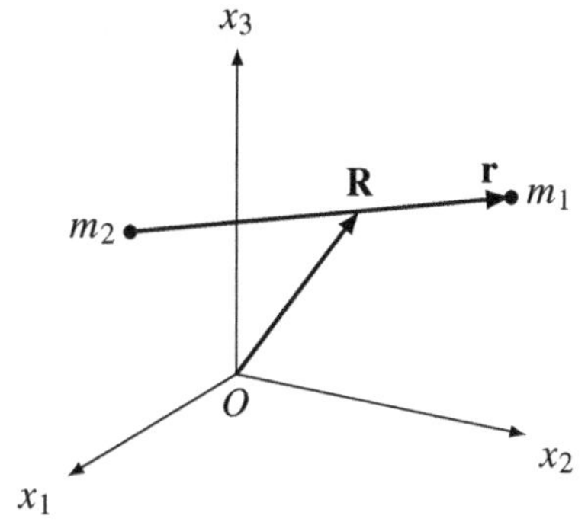

Figure 3.1. Coordinates **R** and **r** in the two-body problem.

(cf. Fig. 3.1), where $M := m_1 + m_2$ denotes the system's total mass. Solving for $\mathbf{r}_1$ and $\mathbf{r}_2$ in the previous equations we easily obtain the inverse relations

$$\boxed{\mathbf{r}_1 = \mathbf{R} + \frac{m_2}{M}\,\mathbf{r}\,, \qquad \mathbf{r}_2 = \mathbf{R} - \frac{m_1}{M}\,\mathbf{r}\,.} \tag{3.3}$$

As we saw in Section 2.6.1, since there are no external forces the center of mass **R** moves without acceleration, i.e., $\ddot{\mathbf{R}} = 0$. As to the **relative coordinate r**, using Eqs. (3.1) and (3.3) we immediately obtain

$$\mu\ddot{\mathbf{r}} = \mathbf{F}_{12}\left(t, \mathbf{R} + \frac{m_2}{M}\,\mathbf{r}\,, \mathbf{R} - \frac{m_1}{M}\,\mathbf{r}\,, \dot{\mathbf{R}} + \frac{m_2}{M}\,\dot{\mathbf{r}}\,, \dot{\mathbf{R}} - \frac{m_1}{M}\,\dot{\mathbf{r}}\right), \tag{3.4}$$

where

$$\boxed{\mu := \frac{m_1 m_2}{m_1 + m_2}} \tag{3.5}$$

is the so-called **reduced mass** of the system. Note that this is a second-order differential equation in the single (vector) variable **r**, since the motion of the center of mass is known ($\mathbf{R} = \mathbf{R}_0 + \mathbf{V}_0 t$, with $\mathbf{R}_0$ and $\mathbf{V}_0$ constant vectors determined by the initial conditions). Hence:

> The two-body problem (3.1) is always equivalent to the one-body problem (3.4).

If (as is usually the case) the internal force satisfies the condition[1]

$$\mathbf{F}_{12} = \mathbf{F}(t, \mathbf{r}_1 - \mathbf{r}_2, \dot{\mathbf{r}}_1 - \dot{\mathbf{r}}_2)\,, \tag{3.6}$$

[1]The dependence of **F** only on the *relative* coordinates and velocities is very natural, since it is clearly related to the *homogeneity* of space. Note that the homogeneity of time would also require that **F** be time-independent, which is almost always the case.

that is, if it depends only on the particles' *relative* coordinates and velocities, the equation of motion of the relative coordinate $\mathbf{r}$ reduces to

$$\boxed{\mu\ddot{\mathbf{r}} = \mathbf{F}(t, \mathbf{r}, \dot{\mathbf{r}}).} \tag{3.7}$$

In other words, if the force $\mathbf{F}_{12}$ is of the form (3.6) the relative coordinate $\mathbf{r}$ moves as the position vector of a particle of mass μ under the force $\mathbf{F}(t, \mathbf{r}, \dot{\mathbf{r}})$, and the two-body problem (3.1) reduces to the one-body problem (3.7).

Once Eq. (3.7) is solved, the motion of the coordinates $\mathbf{r}_1$ and $\mathbf{r}_2$ is easily found using Eqs. (3.3). Since $\ddot{\mathbf{R}} = 0$, if we move the origin to the center of mass the resulting reference frame, called **center of mass frame**, remains inertial. In the CM frame, Eqs. (3.3) simplify to

$$\mathbf{r}_1 = \frac{m_2}{M}\mathbf{r}\,, \qquad \mathbf{r}_2 = -\frac{m_1}{M}\mathbf{r}\,.$$

In many applications, the mass m_2 is much larger than m_1. In this case $m_1/M \simeq 0$, $m_2/M \simeq 1$, and thus (in the CM frame)

$$\mathbf{r}_1 \simeq \mathbf{r}\,, \qquad \mathbf{r}_2 \simeq 0\,.$$

In other words, in this case the heavy particle is approximately fixed at the origin (i.e., the CM), and the relative coordinate $\mathbf{r}$ is approximately equal to the radius vector of the light particle.

3.2 SOLUTION OF THE EQUATIONS OF MOTION

3.2.1 Constants of motion

The most important example of force satisfying condition (3.6) is that of a *central force* of the form

$$\boxed{\mathbf{F}_{12} = f(|\mathbf{r}_1 - \mathbf{r}_2|)\,\frac{\mathbf{r}_1 - \mathbf{r}_2}{|\mathbf{r}_1 - \mathbf{r}_2|}\,.}$$

In this case Eq. (3.7) reduces to

$$\boxed{\mu\ddot{\mathbf{r}} = f(r)\,\frac{\mathbf{r}}{r},} \tag{3.8}$$

which is the equation of motion of a particle of mass μ subject to the central force

$$\mathbf{F}(\mathbf{r}) = f(r)\,\frac{\mathbf{r}}{r}\,. \tag{3.9}$$

We shall study in this section how to find the general solution of Eq. (3.8), and analyze the qualitative behavior of its trajectories.

As we saw in Section 2.4.2, the force (3.9) is *conservative*, since

$$\mathbf{F}(\mathbf{r}) = -\frac{\partial V(r)}{\partial \mathbf{r}}, \qquad \text{with} \qquad \boxed{V(r) = -\int f(r)\, dr\,.}$$

Thus the total energy is conserved, i.e.,

$$\boxed{\frac{1}{2}\mu\dot{\mathbf{r}}^2 + V(r) = E,}$$

remains constant throughout the motion. Moreover, since the force (3.9) is *central*, the angular momentum $\mathbf{L} = \mu\mathbf{r} \times \dot{\mathbf{r}}$ is also conserved. If, as we shall assume from now on,

$$\mathbf{L} \neq 0,$$

the motion takes place in the plane perpendicular to $\mathbf{L}$ *passing through the origin* (i.e., the center of force)[2]. We shall choose the z axis in the direction of the constant vector $\mathbf{L}$, so that

$$\boxed{\mathbf{L} = L\mathbf{e}_z, \qquad \text{with} \quad L = |\mathbf{L}| > 0.} \tag{3.10}$$

In particular, with this choice of axes *the motion takes place in the* $z = 0$ *plane.* Let us introduce polar coordinates (r, φ) in this plane through the usual equations

$$\mathbf{r} = r(\cos\varphi, \sin\varphi, 0) \qquad (r > 0\,, \quad 0 \leqslant \varphi < 2\pi)\,.$$

The unit coordinate vectors are

$$\mathbf{e}_r = (\cos\varphi, \sin\varphi, 0) = \frac{\mathbf{r}}{r}\,, \qquad \mathbf{e}_\varphi = (-\sin\varphi, \cos\varphi, 0)\,,$$

and therefore

$$\dot{\mathbf{e}}_r = \dot{\varphi}\mathbf{e}_\varphi\,, \qquad \dot{\mathbf{e}}_\varphi = -\dot{\varphi}\mathbf{e}_r\,,$$

whence we easily obtain the formulas (cf. Section 2.2.3)

$$v_r = \dot{r}\,, \quad v_\varphi = r\dot{\varphi}\,; \qquad a_r = \ddot{r} - r\dot{\varphi}^2\,, \quad a_\varphi = r\ddot{\varphi} + 2\dot{r}\dot{\varphi}\,.$$

[2]If $\mathbf{L} = 0$ the trajectory is a straight line passing through the origin, which is actually a particular (degenerate) case of motion on a plane through the origin. Indeed, if $\mathbf{r}$ is not identically zero (degenerate case of motion along a line through the origin) the velocity vector must be parallel to $\mathbf{r}$ at all times, and consequently $v_\theta = r\dot{\theta}$ and $v_\varphi = r\sin\theta\,\dot{\varphi}$ vanish identically. Therefore the angles θ and φ are constant, and the trajectory lies on the straight line passing through the origin in the direction of the *constant* vector $\mathbf{e}_r$ (including the degenerate case in which $\dot{\mathbf{r}}$ is identically zero and the particle is at rest at a point on this line). Taking the line of the motion as the x axis we have $r = |x|$ and $\mathbf{e}_r = \operatorname{sgn} x\,\mathbf{i}$, so that $\mathbf{F}(\mathbf{r}) = \operatorname{sgn} x f(|x|)\,\mathbf{i}$ and $V(r) = V(|x|)$. Thus when $\mathbf{L} = 0$ the motion takes place in one dimension under the conservative force $F(x) = \operatorname{sgn} x f(|x|)$—or, equivalently, the potential $V(|x|)$—a problem studied in the previous chapter.

The equations of motion in polar coordinates are therefore

$$\boxed{\begin{aligned} \ddot{r} - r\dot{\varphi}^2 &= \frac{f(r)}{\mu} \\ r\ddot{\varphi} + 2\dot{r}\dot{\varphi} &= 0 . \end{aligned}} \tag{3.11}$$

3.2.2 Law of motion and equation of the trajectory

In order to determine the **trajectory** described by the particle (i.e., r as a function of φ or vice versa) and the **law of motion** (i.e., r and φ as functions of t), it is easier to use the laws of conservation of energy and angular momentum, as we shall next see. Indeed, the angular momentum of the particle is given by

$$\mathbf{L} = L\mathbf{e}_z = \mu\mathbf{r} \times \dot{\mathbf{r}} = \mu r\mathbf{e}_r \times (\dot{r}\mathbf{e}_r + r\dot{\varphi}\mathbf{e}_\varphi) = \mu r^2\dot{\varphi}\mathbf{e}_z ,$$

so that

$$\boxed{\mu r^2\dot{\varphi} = L > 0 .} \tag{3.12}$$

From the previous equation (or, more precisely, our choice of the z axis in the direction of $\mathbf{L}$) it follows that

$$\boxed{\dot{\varphi}(t) > 0 , \qquad \forall t ;} \tag{3.13}$$

in particular, if the trajectory surrounds the origin it must be traversed in an *anticlockwise* direction. Note also that the second equation of motion (3.11) is just the time derivative of Eq. (3.12) (divided by μr).

An immediate consequence of the conservation of angular momentum is the so-called *law of areas*, first formulated by Johannes Kepler in the early 17th century. Indeed, note that the area $A(\varphi)$ swept by the particle's position vector when moving between two points on its trajectory with polar coordinates $(r(\varphi_0), \varphi_0)$ and $(r(\varphi), \varphi)$ is given by

$$A(\varphi) = \frac{1}{2}\int_{\varphi_0}^{\varphi} r^2(\alpha)\,\mathrm{d}\alpha$$

(cf. Fig. 3.2). Since

$$\dot{A} = \frac{\mathrm{d}A}{\mathrm{d}\varphi}\,\dot{\varphi} = \frac{1}{2}\,r^2\dot{\varphi} = \frac{L}{2\mu} \tag{3.14}$$

is constant, the area ΔA swept over a time Δt is simply

$$\boxed{\Delta A = \frac{L}{2\mu}\,\Delta t .}$$

In other words, *the particle sweeps out equal areas in equal times* as it moves along its trajectory (**law of areas**). Note that this property is valid for *any* central force

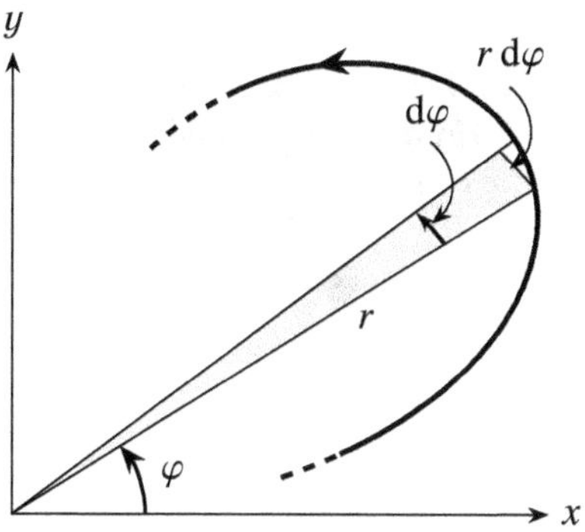

Figure 3.2. Infinitesimal area swept by the position vector **r**.

$f(t, \mathbf{r}, \dot{\mathbf{r}})\, \mathbf{e}_r$, more general than (3.9), since it only requires the conservation of angular momentum.

We can use the conservation law of angular momentum to express $\dot{\varphi}$ in terms of r and the angular momentum L as

$$\boxed{\dot{\varphi} = \frac{L}{\mu r^2}\,.} \tag{3.15}$$

Substituting this expression into the conservation of energy equation we immediately obtain

$$E = \frac{1}{2}\mu \dot{\mathbf{r}}^2 + V(r) = \frac{1}{2}\mu \dot{r}^2 + \frac{1}{2}\mu r^2 \dot{\varphi}^2 + V(r) = \frac{1}{2}\mu \dot{r}^2 + \frac{L^2}{2\mu r^2} + V(r)\,, \tag{3.16}$$

or equivalently,

$$\boxed{\frac{1}{2}\mu \dot{r}^2 + U_L(r) = E\,,} \tag{3.17}$$

where the **effective potential** $U_L(r)$ is defined by

$$\boxed{U_L(r) := V(r) + \frac{L^2}{2\mu r^2}\,.} \tag{3.18}$$

Although U_L *depends on the (constant) value of the angular momentum*, we shall from now on adhere to the customary practice of dropping the subindex L and simply write U instead of U_L. Note also that the effective force generated by the last term in $U(r)$, namely

$$-\frac{\partial}{\partial \mathbf{r}}\left(\frac{L^2}{2\mu r^2}\right) = -\frac{\partial}{\partial r}\left(\frac{L^2}{2\mu r^2}\right)\mathbf{e}_r = \frac{L^2}{\mu r^3}\,\mathbf{e}_r = \mu r \dot{\varphi}^2\,\mathbf{e}_r = \frac{\mu v_\varphi^2}{r}\,\mathbf{e}_r\,,$$

can be interpreted as a *centrifugal force*. Again, it is easy to see that the first equation of motion (3.11) is the time derivative of Eq. (3.16) (divided by $\mu\dot{r}$). In other words, *the laws of conservation of energy and angular momentum are obtained by integrating once with respect to time t the equations of motion* (3.11).

Equations (3.12) and (3.17) easily yield the law of motion and the equation of the trajectory *in implicit form*. Indeed, the law of motion is directly obtained by integrating Eq. (3.17) (after separating variables) and using the conservation of angular momentum:

$$\boxed{t = \pm\sqrt{\frac{\mu}{2}}\int\frac{\mathrm{d}r}{\sqrt{E-U(r)}}\,, \qquad \varphi = \frac{L}{\mu}\int\frac{\mathrm{d}t}{r^2(t)}\,,} \tag{3.19}$$

where it is understood that in the second equation we must substitute the value of $r(t)$ obtained from the first one. (We shall see later, however, that these equations are seldom the easiest way of finding the law of motion.)

As to the equation of the trajectory, from Eq. (3.17) it follows that

$$\frac{\mu}{2}\,\dot{\varphi}^2 r'^2(\varphi) + U(r) = E\,,$$

where (as we shall do throughout this section) we have denoted by a prime the derivative with respect to the angle φ. Using Eq. (3.15) to express $\dot{\varphi}$ as a function of r we obtain

$$\frac{L^2}{2\mu}\left(\frac{r'}{r^2}\right)^2 + U(r) = E\,.$$

Introducing the dependent variable

$$\boxed{u = \frac{1}{r}\,,}$$

we can rewrite the previous equation as

$$u'^2 = \frac{2\mu}{L^2}\left(E - U(1/u)\right), \tag{3.20}$$

whose integration yields the equation of the trajectory:

$$\boxed{\varphi = \pm\frac{L}{\sqrt{2\mu}}\int^{1/r}\frac{\mathrm{d}u}{\sqrt{E-U(1/u)}}\,.} \tag{3.21}$$

In practice, to find the equation of the trajectory it is sometimes convenient to differentiate Eq. (3.20) with respect to φ, thus obtaining a second-order equation that is often easier to integrate than (3.20). Indeed, proceeding in this way we obtain

$$2u'u'' = \frac{2\mu}{L^2u^2}\frac{\mathrm{d}U}{\mathrm{d}r}\,u'\,, \tag{3.22}$$

and therefore, taking into account the definition (3.18) of U,

$$u'' = \frac{\mu}{L^2u^2}\frac{dU}{dr} = \frac{\mu}{L^2u^2}\left(\frac{dV}{dr} - \frac{L^2}{\mu}u^3\right),$$

or equivalently,

$$\boxed{u'' + u = -\frac{\mu}{L^2u^2}f(1/u).} \tag{3.23}$$

The previous equation is known as **Binet's equation**. Binet's equation, written as

$$\boxed{f(r) = -\frac{L^2}{\mu r^2}(u'' + u),}$$

is often used to compute the central force law $f(r)$ if the equation of the trajectory $r = r(\varphi)$ is known.

• In general, *if the equation of the trajectory is known it is possible to find the law of motion* implicitly. Indeed, suppose that the equation of the trajectory is $r = r(\varphi)$. From the conservation of angular momentum we obtain

$$\boxed{t = \frac{\mu}{L}\int r^2(\varphi)\,d\varphi,} \tag{3.24}$$

which yields t as a function of φ. Inverting this relation we obtain $\varphi(t)$, while the motion of the radial coordinate can of course be found by substituting $\varphi(t)$ in the equation of the trajectory:

$$r = r\big(\varphi(t)\big).$$

Remark 3.1. Binet's equation actually holds for a *general* central force $f(r,\varphi)\mathbf{e}_r$ in two dimensions, even if energy is not conserved unless f is independent of the polar angle φ. To derive Binet's equation in this more general setting, we start from the equation of motion in the radial direction

$$\ddot{r} - r\dot{\varphi}^2 = \frac{f(r,\varphi)}{\mu}.$$

Since angular momentum is still conserved, we can replace $r\dot{\varphi}^2$ in the previous equation by $L^2/(\mu^2 r^3)$. Using this fact and the identity

$$\frac{d}{dt} = \dot{\varphi}\frac{d}{d\varphi} = \frac{L}{\mu r^2}\frac{d}{d\varphi}$$

we obtain

$$\frac{L}{\mu r^2}\frac{d}{d\varphi}\left(\frac{L}{\mu r^2}r'\right) - \frac{L^2}{\mu^2 r^3} = -\frac{L^2u^2}{\mu^2}u'' - \frac{L^2u^3}{\mu^2} = \frac{f(r,\varphi)}{\mu},$$

and therefore

$$u'' + u = -\frac{\mu}{L^2u^2}f(1/u,\varphi).$$

■

Example 3.1. Let us find the equation of the trajectories of a particle of mass μ subject to the central force

$$\boxed{\mathbf{F} = \frac{k}{r^3}\,\mathbf{e}_r\,.}$$

In this case $f(1/u) = ku^3$, and thus Binet's equation reduces to

$$\boxed{u'' + Cu = 0\,, \qquad C := 1 + \frac{k\mu}{L^2}\,.}$$

The solutions of this equation depend on the sign of the dimensionless constant C. More precisely:

I) $C < 0$

In this case—which can only happen if $k < 0$, i.e., if the force is *attractive*—the general solution of Binet's equation is

$$u = a\,\mathrm{e}^{\gamma\varphi} + b\,\mathrm{e}^{-\gamma\varphi} \iff r = \left(a\,\mathrm{e}^{\gamma\varphi} + b\,\mathrm{e}^{-\gamma\varphi}\right)^{-1} \qquad (a, b \in \mathbb{R})\,,$$

with

$$\gamma := \sqrt{|C|} > 0\,.$$

It is easy to see that if a and b are both positive the solution can be expressed as

$$\boxed{r = A\,\mathrm{sech}\big(\gamma(\varphi - \varphi_0)\big) \qquad (A > 0)\,,} \tag{3.25}$$

if either a or b vanish then

$$\boxed{r = \mathrm{e}^{\pm\gamma(\varphi - \varphi_0)}\,,} \tag{3.26}$$

whereas if a and b have opposite signs we have[a]

$$\boxed{r = A\,\mathrm{csch}\big(\gamma(\varphi - \varphi_0)\big) \qquad (A \neq 0)\,.} \tag{3.27}$$

The trajectories (3.25) are *bounded* ($r \leqslant A$), while (3.26) and (3.27) are not (in the former case $r \to \infty$ for $\varphi \to \pm\infty$, while in the latter $r \to \infty$ for $\varphi \to \varphi_0$). It is also easy to check that all of these trajectories are of *spiral* type, since the angle φ can take arbitrarily large (positive or negative) values and r tends to 0 as $\varphi \to \pm\infty$ (cf. Fig. 3.3).

More precisely, for (3.25) $\varphi \in \mathbb{R}$ and

$$\lim_{\varphi\to\pm\infty} r(\varphi) = 0\,.$$

Let $t_{\pm\infty}$ denote the times corresponding to $\varphi = \pm\infty$ (i.e., such that $\varphi(t_{\pm\infty}) = \pm\infty$). Then the trajectory (3.25) spirals away from the origin as t increases from $t_{-\infty}$, reaches its point of maximum distance to the origin ($r = A$), and spirals back into the origin as $t \longrightarrow t_\infty$. Likewise, for the trajectory (3.26) $\varphi \in \mathbb{R}$ and $r \longrightarrow 0$ as $\varphi \longrightarrow \infty$ (for the "−" sign in the exponential) or $\varphi \longrightarrow -\infty$ (for the "+" sign). Hence in the first case the trajectory spirals into the origin as $t \longrightarrow t_\infty$, while in the second one it spirals away from the origin as t increases from $t_{-\infty}$. Finally, for the trajectory (3.27) we have $\varphi > \varphi_0$ for $A > 0$ and $\varphi < \varphi_0$ for $A < 0$, and $r \longrightarrow 0$ as $\varphi \longrightarrow \infty$ or $\varphi \longrightarrow -\infty$, respectively. Thus for $A > 0$ the trajectory spirals into the origin for $t \longrightarrow t_\infty$, while for $A < 0$ it spirals away from the origin as t increases from $t_{-\infty}$. Moreover, in both cases $r \longrightarrow \infty$ as $\varphi \longrightarrow \varphi_0$, which is a necessary condition for the trajectory to have an asymptote making an angle φ_0 with the positive x axis. In fact, it can be shown that in this case the trajectory does have such an asymptote (see next exercise).

II) $C = 0$

In this case $k = -L^2/\mu < 0$, and the general solution of Binet's equation is

$$u = a + b\varphi \quad \Longleftrightarrow \quad \boxed{r = \frac{1}{a + b\varphi}} \qquad (a, b \in \mathbb{R}), \tag{3.28}$$

so that $\varphi > -a/b$ for $b > 0$ and $\varphi < -a/b$ for $b < 0$. If $b = 0$ the trajectory is simply a *circle* centered at the origin. On the other hand, for $b \neq 0$ the trajectories are *unbounded* ($r \longrightarrow \infty$ for $\varphi \longrightarrow -a/b$) and spiraling toward the origin (when $b > 0$) or away from it (when $b < 0$), as $r \longrightarrow 0$ when φ tends to ∞ or $-\infty$. Moreover, it can be shown that in this case the trajectory does have an asymptote making an angle $-a/b$ with the positive real axis (see next exercise and Fig. 3.3).

III) $C > 0$

In this case (which takes place, in particular, if $k > 0$) the equation of the trajectories is

$$\begin{cases} u = \dfrac{1}{A}\cos\bigl(\gamma(\varphi - \varphi_0)\bigr) \quad \Longleftrightarrow \quad \boxed{r = A\sec\bigl(\gamma(\varphi - \varphi_0)\bigr)} \\ (A > 0,\ \ 0 \leqslant \varphi_0 < 2\pi), \end{cases} \tag{3.29}$$

so that (since $r > 0$) the angle φ can be taken (for instance) in the range $\bigl(\varphi_0 - \frac{\pi}{2\gamma}, \varphi_0 + \frac{\pi}{2\gamma}\bigr)$. The trajectories are *not spiraling,* since $r \geqslant A > 0$, but are *unbounded* ($r \longrightarrow \infty$ as $\varphi \longrightarrow \varphi_0 \pm \frac{\pi}{2\gamma}$). In fact, it can be shown that in this case the trajectory has two asymptotes making an angle $\varphi_0 \pm \frac{\pi}{2\gamma}$ with the positive real axis (cf. Fig. 3.3).

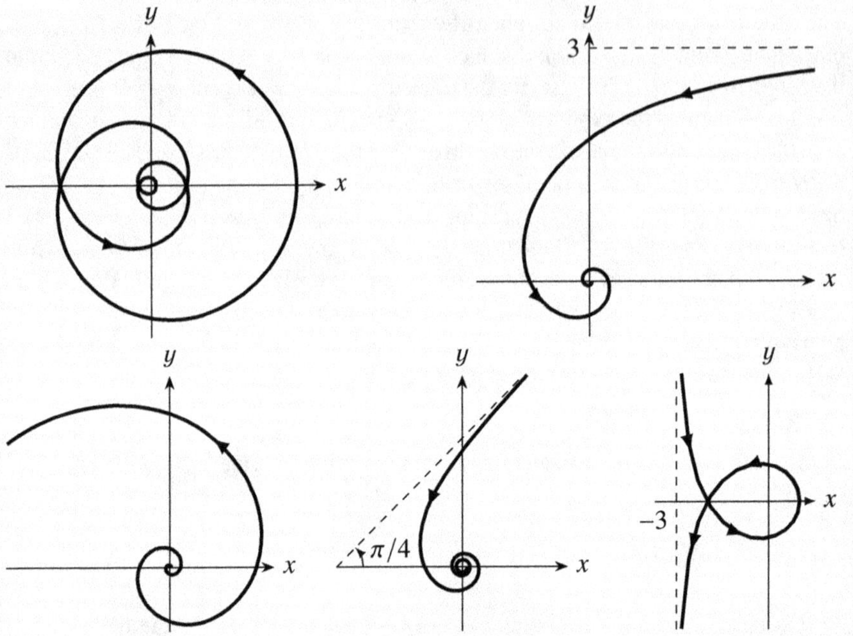

Figure 3.3. Trajectories of a particle in the central force field $\mathbf{F} = k\mathbf{e}_r/r^3$. From left to right and top to bottom, $r = \operatorname{sech}(\varphi/3)$, $r = \operatorname{csch}(\varphi/3)$, $r = e^{\varphi/3}$, $r = \left(\varphi - \frac{\pi}{4}\right)^{-1}$, and $r = \sec(\varphi/3)$. (The orientation of the trajectories agrees with the convention used in this chapter, according to which $L_z > 0$ and hence $\dot{\varphi} > 0$.)

Note. If $C \leqslant 0$ (and, in particular, $k < 0$), the trajectories in the previous example are generically called *Cotes spirals*.

[a] More precisely, in cases (3.25) and (3.27) the parameters A and φ_0 are given by

$$\varphi_0 = \frac{\log|b/a|}{2\gamma}, \qquad A = \frac{\operatorname{sgn} a}{2\sqrt{|ab|}}.$$

Exercise 3.1. Show that a plane curve with polar equation $r = f(\varphi)$ has an asymptote making an angle α with the positive x axis provided that

$$\lim_{\varphi\to\alpha} f(\varphi) = \infty, \qquad c := \lim_{\varphi\to\alpha} f(\varphi)\sin(\varphi - \alpha) < \infty, \tag{3.30}$$

and in that case the Cartesian equation of the asymptote is $\cos\alpha\, y - \sin\alpha\, x = c$. Using this result, prove that:

i) The trajectory (3.27) has an asymptote of equation

$$\cos\varphi_0\, y - \sin\varphi_0\, x = \frac{A}{\gamma}\,. \tag{3.31}$$

ii) The trajectory (3.28) with $b \neq 0$ has an asymptote of equation

$$\cos\varphi_0\, y - \sin\varphi_0\, x = \frac{1}{b}\,, \qquad \varphi_0 := -a/b\,. \tag{3.32}$$

iii) The trajectory (3.29) has two asymptotes of equations

$$\cos\big(\varphi_0 \pm \tfrac{\pi}{2\gamma}\big)\, y - \sin\big(\varphi_0 \pm \tfrac{\pi}{2\gamma}\big)\, x = \mp\frac{A}{\gamma}\,. \tag{3.33}$$

Solution. Suppose that a plane curve has an asymptote as $x \to x_0$, where x_0 could be finite (for a vertical asymptote) or $\pm\infty$. Without loss of generality, we can take the equation of the asymptote as

$$\cos\alpha\, y - \sin\alpha\, x = c,$$

where α is the angle between the asymptote and the positive x axis. Let us assume, for definiteness, that $x_0 = \pm\infty$ (i.e., that the asymptote is not vertical; the case of a vertical asymptote is dealt with similarly, exchanging the roles of x and y). Then $\cos\alpha \neq 0$, and we can take the curve as the graph of a function $y(x)$. By hypothesis, we have

$$\lim_{x\to\pm\infty}\left[y(x) - \left(\tan\alpha\, x + \frac{c}{\cos\alpha}\right)\right] = 0. \tag{3.34}$$

In particular, as $x \to \pm\infty$ the points on the curve $y = y(x)$ (or, equivalently, $r = f(\varphi)$) satisfy

$$f(\varphi) = r = \sqrt{x^2 + y(x)^2} \to \infty$$

and, by Eq. (3.34),

$$\lim_{x\to\pm\infty}\frac{y(x)}{x} = \lim_{x\to\pm\infty}\tan\varphi = \tan\alpha \quad\Longrightarrow\quad \varphi \to \alpha.$$

This proves the first equation (3.30). On the other hand, since $y = r\sin\varphi = f(\varphi)\sin\varphi$ and $x = r\cos\varphi = f(\varphi)\cos\varphi$ along the curve, from Eq. (3.34) multiplied by $\cos\alpha \neq 0$ we deduce that

$$\begin{aligned} 0 &= \lim_{\varphi\to\alpha}\left[\cos\alpha \cdot f(\varphi)\sin\varphi - \sin\alpha \cdot f(\varphi)\cos\varphi - c\right] \\ &= \lim_{\varphi\to\alpha}\left[f(\varphi)\sin(\varphi - \alpha) - c\right], \end{aligned}$$

which is equivalent to the second equation (3.30).

For the trajectory (3.27), $r \longrightarrow \infty$ as φ tends to φ_0. Hence $\alpha = \varphi_0$, and

$$c = \lim_{\varphi\to\varphi_0} [f(\varphi)\sin(\varphi-\varphi_0)] = A \lim_{\varphi\to\varphi_0} \frac{\sin(\varphi-\varphi_0)}{\sinh\big(\gamma(\varphi-\varphi_0)\big)}$$
$$= A \lim_{\varphi\to\varphi_0} \frac{\varphi-\varphi_0+o(\varphi-\varphi_0)}{\gamma(\varphi-\varphi_0)+o(\varphi-\varphi_0)} = A \lim_{\varphi\to\varphi_0} \frac{1+o(1)}{\gamma+o(1)} = \frac{A}{\gamma}.$$

(Recall that $h(s) = o\big(g(s)\big)$ as $s \longrightarrow s_0$ if g does not vanish for $s \neq s_0$ in a sufficiently small neighborhood of s_0, and the quotient $h(s)/g(s)$ tends to 0 as $s \longrightarrow s_0$.) This proves Eq. (3.31). Similarly, for the trajectory (3.28) (with $b \neq 0$) $r \longrightarrow \infty$ as $\varphi \longrightarrow -a/b$. Thus $\alpha = -a/b \equiv \varphi_0$, $a + b\varphi = b(\varphi - \varphi_0)$, and therefore

$$c = \lim_{\varphi\to\varphi_0} \frac{\sin(\varphi-\varphi_0)}{b(\varphi-\varphi_0)} = \frac{1}{b},$$

which establishes Eq. (3.32). Finally, for the trajectory (3.29) $r \longrightarrow \infty$ as $\varphi \longrightarrow \varphi_0 \pm \frac{\pi}{2\gamma}$, and

$$c = A \lim_{\varphi\to\varphi_0\pm\frac{\pi}{2\gamma}} \frac{\sin\big(\varphi-\varphi_0\mp\frac{\pi}{2\gamma}\big)}{\cos\big(\gamma(\varphi-\varphi_0)\big)} = A \lim_{\beta\to\pm\frac{\pi}{2\gamma}} \frac{\sin\big(\beta\mp\frac{\pi}{2\gamma}\big)}{\cos\big(\gamma\beta\big)}$$
$$= A \lim_{\beta\to\pm\frac{\pi}{2\gamma}} \frac{\cos\big(\beta\mp\frac{\pi}{2\gamma}\big)}{-\gamma\sin\big(\gamma\beta\big)} = \mp\frac{A}{\gamma},$$

which proves Eq. (3.33).

The energy of a trajectory $r = r(\varphi)$ can be computed from Eq. (3.20) and the definition (3.18) of U, namely

$$E = \frac{L^2}{2\mu}\,(u'^2 + u^2) + V(1/u)\,. \tag{3.35}$$

Since $E = \mu v^2/2 + V(1/u)$, the particle's speed as a function of its distance r to the origin is given by

$$v = \frac{L}{\mu}\,\sqrt{u'^2 + u^2}\,. \tag{3.36}$$

Both formulas can also be obtained directly, taking into account that

$$v^2 = \dot r^2 + r^2\dot\varphi^2 = \dot\varphi^2(r'^2 + r^2) = \frac{L^2}{\mu^2 r^4}\left(\frac{u'^2}{u^4} + \frac{1}{u^2}\right) = \frac{L^2}{\mu^2}\,(u'^2 + u^2)\,.$$

In Example 3.1 we can take

$$V(r) = -\int f(r)\,\mathrm{d}r = -k\int \frac{\mathrm{d}r}{r^3} = \frac{k}{2r^2},$$

and thus

$$E = \frac{L^2}{2\mu}\left(u'^2 + u^2\right) + \frac{k}{2}\,u^2 = \frac{L^2}{2\mu}\left(u'^2 + Cu^2\right).$$

For instance, for the trajectories (3.25) it is easily checked that

$$E = -\frac{L^2|C|}{2\mu A^2} < 0\,,$$

while for (3.27) and (3.29) we have

$$E = \frac{L^2|C|}{2\mu A^2} > 0\,.$$

It is also immediate to verify that the energy of the trajectories (3.28) is

$$E = \frac{L^2 b^2}{2\mu} \geqslant 0\,,$$

while the trajectories (3.26) have energy $E = 0$. These results agree with the fact that the trajectories (3.25) are *bounded*, while (3.27), (3.28) (if $b \neq 0$), and (3.29) are *unbounded*. Indeed, note that, as in this case

$$\lim_{r\to\infty} V(r) = 0\,,$$

if the particle reaches infinity we must necessarily have

$$E = \frac{1}{2}\,\mu v_\infty^2 \geqslant 0\,.$$

Example 3.2. Let us find the central force causing a particle to describe the spiral $r = a\varphi$. To this end, it suffices to substitute $u = 1/(a\varphi)$ into Binet's equation, which yields

$$f(r) = -\frac{L^2}{\mu r^2}\left(u'' + u\right) = -\frac{L^2}{\mu a r^2}\left(\frac{2}{\varphi^3} + \frac{1}{\varphi}\right) = -\frac{L^2}{\mu a r^2}\left(\frac{2a^3}{r^3} + \frac{a}{r}\right)$$
$$= \boxed{-\frac{L^2}{\mu a^3}\left(\frac{2a^5}{r^5} + \frac{a^3}{r^3}\right)}\,.$$

The motion of the angular coordinate φ is easily determined from Eq. (3.24):

$$t = \frac{\mu}{L}\int a^2\varphi^2\,\mathrm{d}\varphi = \frac{\mu a^2}{3L}\left(\varphi^3 - \varphi_0^3\right) \quad\Longrightarrow\quad \boxed{\varphi = \left(\frac{3L}{\mu a^2}\,t + \varphi_0^3\right)^{1/3}}\,,$$

with $\varphi_0 = \varphi(0)$, whence it follows that

$$\boxed{r = a\varphi = a\left(\frac{3L}{\mu a^2}\,t + \varphi_0^3\right)^{1/3}}\,.$$

The energy of this trajectory is computed without difficulty using Eq. (3.35). To this end, we first need to find the potential, which is given by

$$V(r) = \frac{L^2}{\mu a^3} \int \left(\frac{2a^5}{r^5} + \frac{a^3}{r^3} \right) dr = -\frac{L^2}{2\mu a^2} \left(\frac{a^4}{r^4} + \frac{a^2}{r^2} \right),$$

up to an arbitrary constant that has been taken equal to zero so that

$$\lim_{r \to \infty} V(r) = 0 .$$

Substituting into Eq. (3.35) we obtain

$$E = \frac{L^2}{2\mu} (u'^2 + u^2) - \frac{L^2}{2\mu} (a^2 u^4 + u^2) = \frac{L^2}{2\mu} (u'^2 - a^2 u^4) = 0 .$$

By the law of conservation of energy, the particle reaches infinity with zero speed. The speed of the particle at any point on the trajectory can be computed using Eq. (3.36), but since we know the potential V and the energy E it can be more directly obtained from the energy equation:

$$E = 0 = \frac{1}{2} \mu v^2 + V(r) = \frac{1}{2} \mu v^2 - \frac{L^2}{2\mu a^2} \left(\frac{a^4}{r^4} + \frac{a^2}{r^2} \right) \quad \Longrightarrow \quad \boxed{v = \frac{L}{\mu r^2} \sqrt{r^2 + a^2}} .$$

From this equation it also follows that, as we already knew, $v \to 0$ as $r \to \infty$.

Exercise 3.2. If we apply Binet's equation to a circle of radius a centered at the origin we apparently obtain a force inversely proportional to the square of the distance from origin:

$$f(r) = -\frac{L^2}{a\mu} \frac{1}{r^2} . \tag{3.37}$$

Is this result correct?

Solution. The result is clearly *false* as stated, since *any* potential V whose corresponding effective potential U has a critical point r_0 admits a circular orbit $r = r_0$ with energy $U(r_0)$ (cf. Eq. (3.17)). Hence Eq. (3.37) is *not* correct. This fact is not totally surprising, since to obtain Binet's equation from Eq. (3.22) it is necessary to divide by u', which is not allowed if $u = 1/r$ is constant. Equation (3.37) is however *true* if properly interpreted. Indeed, since $r = a$ along the circular orbit, what this equation actually states is that

$$f(a) = -\frac{L^2}{\mu a^3} = -\mu a \dot{\varphi}^2 = -\frac{\mu v^2}{a} , \tag{3.38}$$

i.e., that the central force $f(a)$ along the orbit generates the centripetal acceleration $-\mu v^2/a$. Equivalently, (3.38) is the necessary and sufficient condition for the potential $U(r)$ to have a critical point at $r = a$. Thus in the case of a circular orbit centered at the origin Binet's equation is simply the condition for the existence of such an orbit, and does not provide any information about the force law at distances from the origin different from the orbit's radius.

3.2.3 Advance of the periapsis

As we have just shown in the previous section, the motion of the radial coordinate r is determined by the law of conservation of energy (3.17)–(3.18). Formally, this is the equation of motion of a particle of mass μ in the one-dimensional potential $U(r)$. For this reason, most of the results from Section 2.5 are also valid in the present context. For instance, from Eq. (3.17) it follows that the motion can only take place in the region defined by the inequality

$$U(r) = V(r) + \frac{L^2}{2\mu r^2} \leqslant E\,. \tag{3.39}$$

It is however important to note that, unlike the variable x in Section 2.5, the coordinate *r can only take non-negative values.*

Example 3.3. For the **Kepler potential**

$$V(r) = -k/r\,, \qquad \text{with } k > 0\,,$$

the effective potential

$$U(r) = -\frac{k}{r} + \frac{L^2}{2\mu r^2}$$

behaves as shown in Fig. 3.4. Indeed, $U(r)$ diverges as $L^2/(2\mu r^2)$ for $r \to 0$ and tends to zero as $-k/r$ for $r \to \infty$. Moreover,

$$U'(r) = \frac{k}{r^2} - \frac{L^2}{\mu r^3} = \frac{k}{r^3}\left(r - \frac{L^2}{k\mu}\right),$$

so that U decreases for $0 < r < a := L^2/(k\mu)$ and increases for $r > a$, attaining its minimum value at $r = a$. The particle's energy must be greater than or equal to this minimum value, given by

$$U(a) = -\frac{k}{2a} = -\frac{k^2\mu}{2L^2} =: E_{\min}\,.$$

From Fig. 3.4 it follows that the trajectories with energy $E \geqslant 0$ are *unbounded*, since in this case the inequality (3.39) implies that $r \in [r_0, \infty)$, where $r_0 > 0$ is the only root of the equation $U(r) = E$. Therefore in this case the particle "comes" from infinity, reaches a minimum distance from the origin equal to r_0 (which is a turning point of U), and goes back to infinity. On the contrary, if $E_{\min} < E < 0$ then $r_1 \leqslant r \leqslant r_2$, where $r_1 < r_2$ are the two roots of the equation $U(r) = E$ (cf. Fig. 3.4), which are again turning points. Therefore in this case the trajectory is *bounded* and stays away from the origin. Finally, if $E = E_{\min}$ the trajectory is the *circle* $r = a$ (cf. Fig. 3.4).

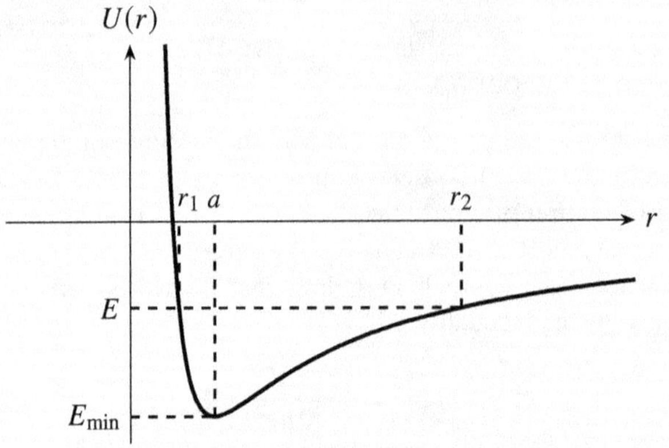

Figure 3.4. Effective potential $U(r)$ for the Kepler potential $V(r) = -k/r$ (with $k > 0$).

Exercise 3.3. Repeat the previous discussion for the potential $V(r) = k/(2r^2)$ in Example 3.1.

Solution. The effective potential is in this case given by

$$U(r) = \frac{k}{2r^2} + \frac{L^2}{2\mu r^2} = \frac{L^2}{2\mu}\frac{C}{r^2}, \tag{3.40}$$

with $C = 1 + (k\mu/L^2)$. Since $L^2/(2\mu) > 0$, the behavior of U depends on the sign of C (cf. Fig. 3.5).

i) For $C < 0$, the trajectories with energy $E \geqslant 0$ are unbounded and can reach the origin, since the allowed region for such energies is the whole semiaxis $r \geqslant 0$. On the other hand, for $E < 0$ the allowed region is an interval of the form $[0, r_0]$, where $r_0 > 0$ is a turning point. Thus in this case the trajectories with negative energy are bounded but fall into the origin. Note that this is consistent with our previous analysis, since for $C < 0$ the trajectories with positive energy $L^2|C|/(2\mu A^2)$ are given by Eq. (3.27) and those with zero energy by Eq. (3.26), whereas the trajectories with negative energy $-L^2|C|/(2\mu A^2)$ obey Eq. (3.25).

ii) For $C = 0$ the effective potential vanishes identically. Thus in this case the trajectories with positive energy are unbounded and fall into the origin, since the allowed region is again the semiaxis $r \geqslant 0$. These trajectories are the curves (3.28) with $b \neq 0$, whose energy is indeed $L^2b^2/(2\mu) > 0$. On the other hand, for $E = 0$ *all* the points on the semiaxis $r \geqslant 0$ are equilibrium solutions of $U(r)$ (since $U'(r) = 0$

and $U(r) = E = 0$ for all r). Hence in this case the possible trajectories are the circles $r = r_0$ with arbitrary $r_0 \geqslant 0$ (i.e., the curves (3.28) with $b = 0$).

iii) Finally, for $C > 0$ we must have $E > 0$, and all the trajectories reach infinity but stay away from the origin (indeed, the allowed region is an interval of the form $[r_0, \infty)$, where r_0 is a turning point). As we saw above, these trajectories have equation (3.29) and their energy is indeed $L^2C/(2\mu A^2) > 0$.

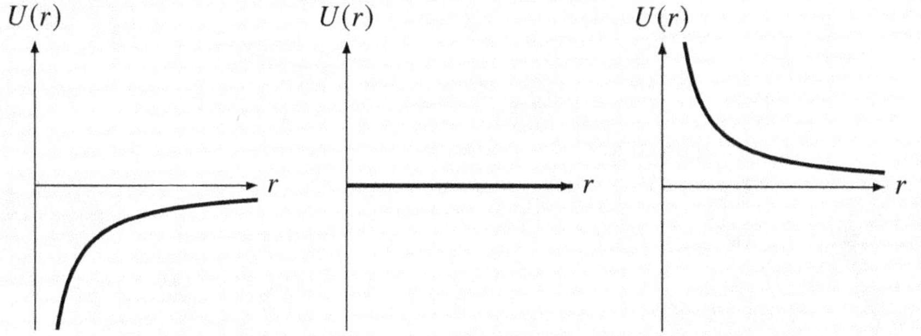

Figure 3.5. Plot of the effective potential (3.40) for $C < 0$ (left), $C = 0$ (middle), and $C > 0$ (right).

Of particular interest are *bounded orbits*, in which the radial coordinate moves between two consecutive turning points $0 < r_1 < r_2$ of the effective potential $U(r)$ (cf. Fig. 3.6). In this case, the points on the trajectory at the minimum distance r_1 from the origin are called **periapsides** or **pericenters** (**perigees**, **perihelia**, or **periastra**[3] if the center of force is respectively Earth, the Sun or a star), while those at the maximum distance r_2 are called **apoapsides** or **apocenters** (**apogees**, **aphelia**, or **apoastra**[4]). Both of these types of points are jointly referred to as **apsides**[5] (or *apsidal points*).

As we saw in Section 2.5, the motion of the radial coordinate is in this case *periodic* in time, with period (depending in general on the energy and the angular momentum)

$$\tau_r = \sqrt{2\mu} \int_{r_1}^{r_2} \frac{dr}{\sqrt{E - U(r)}} . \tag{3.41}$$

However, *this does not mean that the particle's motion is periodic*. Indeed, the necessary and sufficient condition for the motion to be periodic is that when the radial coordinate r completes a certain integer number of its periods the angle φ increases by an integer multiple of 2π, so that the particle returns to its starting point. Obviously, this is equivalent to requiring that the bounded orbit be *closed*. Let us next examine under what conditions this will be the case. To begin with, when the radial coordinate r increases from r_1 to r_2 the variable $u = 1/r$ decreases, and hence $\frac{d\varphi}{du} < 0$ (since

[3]Singular *periapsis*, *perigee*, *perihelion*, and *periastron*.
[4]Singular *apoapsis*, *apogee*, *aphelion*, and *apoastron*.
[5]Singular *apsis*.

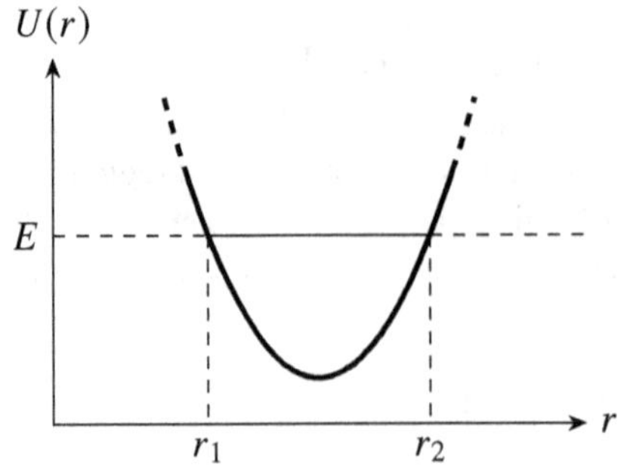

Figure 3.6. Effective potential $U(r)$ admitting a bounded orbit of energy E with $r_1 \leqslant r \leqslant r_2$, where $r_1 < r_2$ are two consecutive turning points of U for the energy E.

$\dot{\varphi} = \frac{d\varphi}{du}\,\dot{u} > 0$), so that we must take the minus sign in Eq. (3.21). Setting (without loss of generality) $\varphi(r_1) = 0$ we obtain

$$\varphi = \frac{L}{\sqrt{2\mu}} \int_{1/r}^{1/r_1} \frac{du}{\sqrt{E - U(1/u)}} =: \varphi_1(r)\,. \tag{3.42}$$

In particular, when r reaches its maximum value r_2 the angle φ increases by

$$\Delta\varphi_{12} = \varphi_1(r_2) = \frac{L}{\sqrt{2\mu}} \int_{1/r_2}^{1/r_1} \frac{du}{\sqrt{E - U(1/u)}}\,. \tag{3.43}$$

On the other hand, when r decreases from r_2 to r_1 the variable u increases, so that we should take the "+" in Eq. (3.21). In other words, we have

$$\varphi = \Delta\varphi_{12} + \frac{L}{\sqrt{2\mu}} \int_{1/r_2}^{1/r} \frac{du}{\sqrt{E - U(1/u)}} =: \varphi_2(r) \tag{3.44}$$

Thus when r takes once again its minimum value r_1 the angle φ has increased by

$$\Delta\varphi = \varphi_2(r_1) = 2\Delta\varphi_{12}\,.$$

The **advance** (or **precession**) **of the periapsis** $\Delta\varphi$, defined as the increase in the azimuthal angle φ between two consecutive periapsides, is thus given by

$$\Delta\varphi = \sqrt{\frac{2L^2}{\mu}} \int_{1/r_2}^{1/r_1} \frac{du}{\sqrt{E - U(1/u)}} = \sqrt{\frac{2L^2}{\mu}} \int_{r_1}^{r_2} \frac{dr}{r^2\sqrt{E - U(r)}}\,. \tag{3.45}$$

In general, $\Delta\varphi$ is *not* an integer multiple of 2π, and thus the particles does *not* return to its initial position (i.e., the point with polar coordinates $r = r_1$, $\varphi = 0$) when the coordinate r takes the value r_1 for the second time (cf. Fig. (3.7)). The necessary and

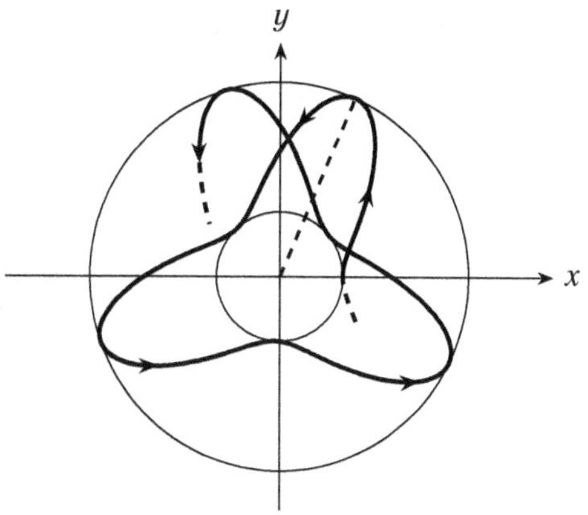

Figure 3.7. Precession of the periapsis in a bounded orbit. The circles $r = r_1$ and $r = r_2$ (minimum and maximum values of the radial coordinate) have been drawn with a thin line, while the dashed black segment represents the line $\varphi = \Delta\varphi_{12}$ joining the origin with an apoapsis.

sufficient condition for the motion to be *periodic*—or, equivalently, the orbit to be *closed*—is that after n periods of the r coordinate the increase of the angle φ, which is obviously equal to $n\Delta\varphi$, be an integer multiple $2m\pi$ of 2π. We have thus proved the following:

The motion on a bounded orbit in which r varies between two consecutive turning points $r_1 < r_2$ of the effective potential U is *periodic* if and only if the orbit is *closed*. The necessary and sufficient condition for this to happen is that the advance of the periapsis (3.45) be a *rational multiple* of 2π.

Remark 3.2. According to *Bertrand's theorem*, the only central potentials for which *all* bounded orbits are closed (and, hence, periodic) are the harmonic potential ($V(r) = \frac{1}{2}kr^2$ with $k > 0$) and the Kepler potential ($V(r) = -k/r$ with $k > 0$). ■

Exercise 3.4. Prove that the orbits in a central force field are symmetric about the line joining the origin with an apsis.

Solution. The function $u(\varphi)$ is a solution to Binet's equation, which is invariant under the transformations $\varphi \mapsto -\varphi$ and $\varphi \mapsto \varphi + \varphi_0$ with φ_0 an arbitrary constant. Hence if $u(\varphi)$ is a solution so are $u(-\varphi)$, $u(\varphi+\varphi_0)$, and $u(\varphi_0-\varphi)$, for every $\varphi_0 \in \mathbb{R}$. Since the angle φ does not appear explicitly in Binet's equation, we can assume without loss of generality that the apsis considered has polar angle $\varphi = 0$. The orbit will then be symmetric about this apsis provided that $u(\varphi) = u(-\varphi)$. Since $u(-\varphi)$ is also a solution of Binet's equation, in order to prove the previous equality it suffices to show that $u(\varphi)$ and $g(\varphi) = u(-\varphi)$ satisfy the same initial conditions at $\varphi = 0$, i.e., that $u(0) = g(0)$ and $u'(0) = g'(0)$. The first of these equalities is obvious, while the second one easily follows from the fact that $u'(0) = 0$, since by definition of apsis u has a maximum or a minimum at $\varphi = 0$.

Exercise 3.5. Show that the *advance of the apoapsis* (increase in the angle φ between two consecutive apoapsides) is also given by Eq. (3.45).

Solution. Since $u(-\varphi) = u(\varphi)$ by the previous exercise, if the polar angle of a periapsis is $\varphi = 0$ there are two consecutive apoapsides of the orbit at angles $\varphi = \pm\Delta\varphi_{12}$, and thus their angular displacement is again $2\Delta\varphi_{12} = \Delta\varphi$.

Remark 3.3. From the previous considerations it also follows that $u(\varphi + \Delta\varphi) = u(\varphi)$ (indeed, by definition of $\Delta\varphi$ both $u(\varphi)$ and $u(\varphi + \Delta\varphi)$ are solutions of Binet's equation with the same initial conditions $u(0) = 1/r_1$, $u'(0) = 0$ at $\varphi = 0$.) Thus $u(\varphi)$ is an even periodic function of period $\Delta\varphi$. ■

Exercise 3.6. Find the period of a circular orbit $r = a$ in a central potential.

Solution. The conservation of angular momentum implies that

$$\varphi(t) = \varphi(0) + \frac{Lt}{\mu a^2}.$$

Hence the motion is periodic, with period $\tau = 2\pi\mu a^2/L$. In general, the period of a closed orbit around the origin in a central potential can be found from the law of areas through the formula $\tau = 2\mu A/L$, where A is the area enclosed by the curve. (We are actually assuming that, as is usually the case, $r(\varphi)$ is a one-valued function of φ).

Example 3.4. For the harmonic potential $V(r) = \frac{1}{2}kr^2$, with $k > 0$, the effective potential $U(r)$ is as shown Fig. 3.8. Therefore in this case all orbits are bounded. The periapsis advance of any of these orbits is given by

$$\Delta\varphi = \sqrt{\frac{2L^2}{\mu}} \int_{1/r_2}^{1/r_1} \frac{du}{\sqrt{E - \frac{L^2}{2\mu}u^2 - \frac{k}{2u^2}}} = \sqrt{\frac{2L^2}{\mu}} \int_{u_2}^{u_1} \frac{u\,du}{\sqrt{-\frac{L^2}{2\mu}u^4 + Eu^2 - \frac{k}{2}}},$$

where $u_1 > u_2$ are the two roots of the equation $-\frac{L^2}{2\mu}u^4 + Eu^2 - \frac{k}{2} = 0$. Performing the change of variable $s = u^2$ the last integral becomes

$$\Delta\varphi = \int_{s_2}^{s_1} \frac{ds}{\sqrt{p(s)}},$$

where

$$p(s) = -s^2 + \frac{2\mu E}{L^2}\,s - \frac{k\mu}{L^2} = -\left(s - \frac{\mu E}{L^2}\right)^2 + \frac{\mu^2 E^2}{L^4}\left(1 - \frac{kL^2}{\mu E^2}\right) \tag{3.46}$$

and $s_1 > s_2$ are the two roots of the equation $p(s) = 0$ (obviously, $s_i = u_i^2$). Performing then the change of variable

$$s = \frac{\mu E}{L^2} + \frac{\mu E}{L^2}\left(1 - \frac{kL^2}{\mu E^2}\right)^{1/2} \sin\theta\,, \qquad -\frac{\pi}{2} \leqslant \theta \leqslant \frac{\pi}{2}\,, \tag{3.47}$$

we finally obtain

$$\Delta\varphi = \int_{-\pi/2}^{\pi/2} \mathrm{d}\theta = \pi\,.$$

Since $\Delta\varphi$ is a rational multiple of 2π, *all* the orbits are periodic (in agreement with Bertrand's theorem).

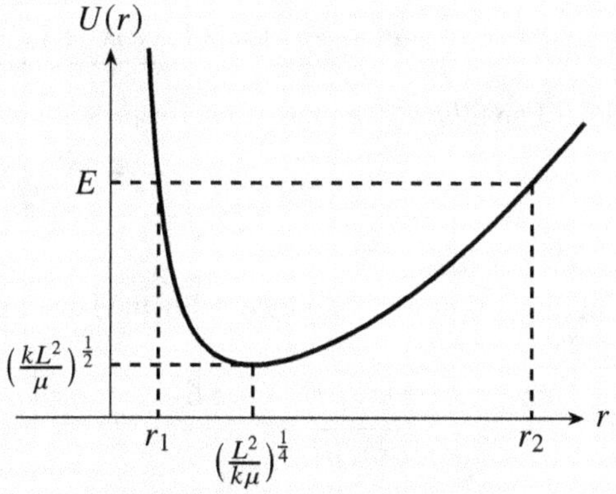

Figure 3.8. Effective potential $U(r)$ for the harmonic potential $V(r) = kr^2/2$ (with $k > 0$) for $L > 0$.

Exercise 3.7. Show that all the orbits of the harmonic potential $V(r) = kr^2/2$ (with $k > 0$) are ellipses centered at the origin, and compute the period of the motion.

Solution. The equation of the orbits is in this case

$$\varphi = \sqrt{\frac{L^2}{2\mu}} \int^{1/r} \frac{\mathrm{d}u}{\sqrt{E - \frac{L^2}{2\mu}u^2 - \frac{k}{2u^2}}} = \int^{1/r} \frac{u\,\mathrm{d}u}{\sqrt{-u^4 + \frac{2\mu E}{L^2}u^2 - \frac{k\mu}{L^2}}} = \frac{1}{2}\int^{1/r^2} \frac{\mathrm{d}s}{\sqrt{p(s)}}\,,$$

with $p(s)$ defined by Eq. (3.46). Performing the change of variable (3.47) (note that, as seen in Example 3.4, in this case $E^2 \geqslant kL^2/\mu$) we obtain

$$\varphi = \varphi_0 + \frac{1}{2}\arcsin\left(\frac{\frac{L^2}{\mu E r^2} - 1}{\sqrt{1 - \frac{L^2 k}{\mu E^2}}}\right)$$

$$\Longrightarrow \qquad \frac{L^2}{\mu E r^2} = 1 + \sqrt{1 - \frac{L^2 k}{\mu E^2}}\, \sin\big(2(\varphi - \varphi_0)\big)\,.$$

Taking, without loss of generality, $\varphi_0 = \pi/4$, the above equation can be written as

$$r^2 = \frac{\alpha}{1 - e\cos 2\varphi}\,, \qquad \text{with} \quad \alpha = \frac{L^2}{\mu E}\,, \qquad e = \sqrt{1 - \frac{L^2 k}{\mu E^2}} \leqslant 1$$

(and $e = 1$ if and only if $L = 0$). In Cartesian coordinates,

$$r^2 - er^2\cos 2\varphi = x^2 + y^2 - e(x^2 - y^2) = \boxed{(1-e)x^2 + (1+e)y^2 = \alpha}\,,$$

which is the equation of an *ellipse centered at the origin* with semiaxes

$$a = \sqrt{\frac{\alpha}{1-e}}\,, \qquad b = \sqrt{\frac{\alpha}{1+e}}\,.$$

The period of the motion τ is easily computed using the law of areas:

$$\frac{L\tau}{2\mu} = \pi ab = \frac{\pi\alpha}{\sqrt{1-e^2}} = \frac{\pi L^2/\mu E}{\sqrt{L^2 k/\mu E^2}} = \frac{\pi L}{\sqrt{k\mu}} \qquad \Longrightarrow \qquad \boxed{\tau = 2\pi\sqrt{\frac{\mu}{k}}}\,.$$

Note, in particular, that in this case the period is *independent* of E and L, and is therefore the same for all orbits.

Note. In this case, the equation of the orbits can be obtained more easily by solving the equations of motion in Cartesian coordinates, namely

$$\ddot{x} + \frac{k}{\mu}x = 0\,, \qquad \ddot{y} + \frac{k}{\mu}y = 0\,.$$

Indeed, calling $\omega = \sqrt{k/\mu}$ and setting, without loss of generality, $\dot{x}(0) = 0$, $y(0) = 0$ (i.e., taking the x axis in the direction of an apsis[a]) we obtain

$$x = a\,\cos(\omega t)\,, \qquad y = b\,\sin(\omega t)\,,$$

with a and b real constants, which are the parametric equations of an ellipse centered at the origin with semiaxes $|a|$ and $|b|$. (In fact, choosing the x axis so that $x(0) > 0$

it follows that $a > 0$, and the condition $\dot{\varphi} > 0$ implies that $\dot{y}(0) = b\omega > 0$, i.e., that $b > 0$.) From the above equations it also follows that the period of the motion is

$$\tau = \frac{2\pi}{\omega} = 2\pi\sqrt{\frac{\mu}{k}}\,.$$

[a] Indeed, at an apsis we have $\dot{r} = v_r = 0$, and hence $\mathbf{r}$ is perpendicular to $\dot{\mathbf{r}}$.

Exercise 3.8. Find the periapsis advance for the central potential

$$V(r) = -\frac{k}{r} + \frac{h}{2r^2}, \qquad \text{with} \quad k, h > 0.$$

Solution. The effective potential is

$$U(r) = -\frac{k}{r} + \frac{\gamma^2}{r^2}\,, \qquad \text{with} \quad \gamma \equiv \sqrt{\frac{L^2}{2\mu} + \frac{h}{2}}\,,$$

and thus behaves qualitatively like Kepler's effective potential (cf. Fig. 3.4). Since the minimum value of U is now $-k^2/(4\gamma^2)$, if $-k^2/(4\gamma^2) < E < 0$ the orbits are bounded and the periapsis advance is given by

$$\Delta\varphi = \sqrt{\frac{2L^2}{\mu}} \int_{u_1}^{u_2} \frac{du}{\sqrt{E + ku - \gamma^2 u^2}}\,,$$

where $u_1 < u_2$ are the two roots of the equation $E + ku - \gamma^2 u^2 = 0$. Since

$$E + ku - \gamma^2 u^2 = E + \frac{k^2}{4\gamma^2} - \gamma^2\left(u - \frac{k}{2\gamma^2}\right)^2$$

with $E + k^2/(4\gamma^2) > 0$, the change of variable

$$u = \frac{k}{2\gamma^2} + \frac{1}{\gamma}\sqrt{E + \frac{k^2}{4\gamma^2}}\,\sin\theta$$

yields

$$\Delta\varphi = \sqrt{\frac{2L^2}{\mu\gamma^2}} \int_{-\pi/2}^{\pi/2} d\theta = \pi\sqrt{\frac{2L^2}{\mu\gamma^2}} = \frac{2\pi}{\sqrt{1 + \frac{\mu h}{L^2}}}\,.$$

In particular, for Kepler's potential ($h = 0$) we have $\Delta\varphi = 2\pi$, so that all bounded orbits are closed (again in agreement with Bertrand's theorem). Note, finally, that in the generic case $h \neq 0$ the closed (and hence periodic) bounded orbits are those satisfying the condition

$$1 + \frac{\mu h}{L^2} = q^2, \qquad \text{with} \quad q \in \mathbb{Q}.$$

Exercise 3.9. Find the equation of the orbits for the *repulsive* harmonic potential $V(r) = -kr^2/2$ (with $k > 0$).

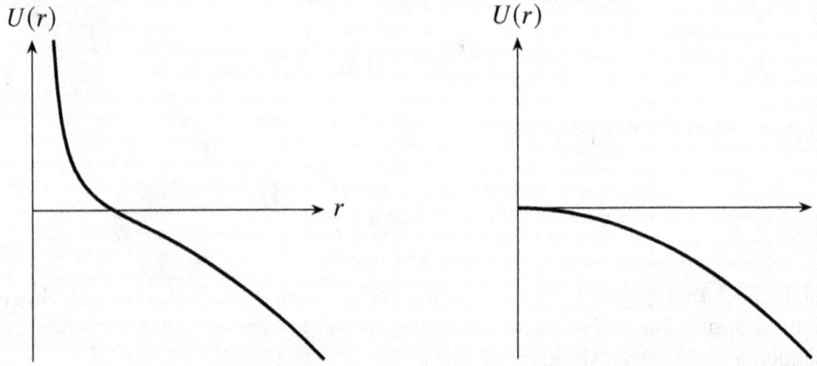

Figure 3.9. Effective potential for the repulsive harmonic potential $V(r) = -\frac{1}{2}kr^2$ (with $k > 0$) for $L > 0$ (left) and $L = 0$ (right).

Solution. The effective potential $U(r) = \dfrac{L^2}{2\mu r^2} - \dfrac{1}{2}kr^2$ is monotonically decreasing, since

$$U'(r) = -\frac{L^2}{\mu r^3} - kr < 0, \qquad \forall r > 0.$$

For $L > 0$, $U(r)$ tends to ∞ for $r \to 0+$ and to $-\infty$ for $r \to \infty$ (cf. Fig. 3.9 left). Hence in this case the trajectories are unbounded and bounded away from the origin for any energy $E \in \mathbb{R}$. On the other hand, for $L = 0$ the effective potential $U(r)$ reduces to $-\frac{1}{2}kr^2$ (cf. Fig. 3.9 right). Hence the trajectories with positive energy are unbounded but reach the origin, the trajectory with zero energy is the equilibrium $\mathbf{r} = 0$, and the trajectories with negative energy are again are unbounded and bounded away from the origin. As in the attractive case, it is easier to solve the equation of motion in Cartesian coordinates,

$$\ddot{x} - \frac{k}{\mu}x = 0, \qquad \ddot{y} - \frac{k}{\mu}y = 0,$$

with the initial conditions $\dot{x}(0) = y(0) = 0$. In this way we obtain

$$x = a\cosh(\omega t), \qquad y = b\sinh(\omega t), \tag{3.48}$$

with a, b arbitrary real constants and $\omega = (k/\mu)^{1/2}$ as before. Note that the z component of the angular momentum of the solution (3.48) is

$$L_z = x\dot{y} - y\dot{x} = \omega ab\left(\cosh^2(\omega t) - \sinh^2(\omega t)\right) = \omega ab;$$

hence $L = L_z$—i.e., $\dot{\varphi} > 0$, in agreement with our convention (3.13)—if and only if a and b have the same sign. When a and b are both nonzero (i.e., for $L > 0$), eliminating the time t we obtain the equation

$$\frac{x^2}{a^2} - \frac{y^2}{b^2} = 1.$$

Hence in this case the trajectories are *rectangular hyperbolas* with center at the origin and axes parallel to the coordinate axes. Therefore in this case the trajectories are unbounded and bounded away from the origin, in agreement with the previous qualitative discussion. When $a = 0$ the trajectory is the y axis, while for $b = 0$ the trajectory is the half-line $[a, \infty)$ (for $a > 0$) or $(-\infty, a]$ (for $a < 0$) on the x axis. Finally, for $a = b = 0$ the trajectory is the equilibrium $\mathbf{r} = 0$. This is again consistent with the qualitative discussion above, since the energy of the trajectories (3.48) is

$$\begin{aligned} E &= \frac{1}{2}\,m(\dot{x}^2 + \dot{y}^2) - \frac{1}{2}\,k(x^2 + y^2) \\ &= \frac{1}{2}\,m\omega^2\big(a^2\sinh^2(\omega t) + b^2\cosh^2(\omega t)\big) - \frac{1}{2}\,k\big(a^2\cosh^2(\omega t) + b^2\sinh^2(\omega t)\big) \\ &= \frac{k}{2}(b^2 - a^2). \end{aligned}$$

Example 3.5. According to Einstein's theory of general relativity, the potential felt by a particle of mass m at a distance r from a mass $M \gg m$ fixed at the origin is effectively given by

$$V(r) = -\frac{GMm}{r} - \frac{GML^2}{mc^2r^3},$$

where $L > 0$ is the particle's angular momentum. In planetary motion the general relativity correction is much smaller than the Kepler potential term. Indeed, the quotient of the two terms in $V(r)$ is given by $L^2/(m^2c^2r^2)$, which for nearly circular orbits can be estimated by taking $L = mrv$:

$$\frac{(mrv)^2}{m^2c^2r^2} = \frac{v^2}{c^2}.$$

The velocity v of a planet in the solar system is at most 59 Km/s (Mercury's maximum velocity), so that $v^2/c^2 = O(10^{-8})$. In spite of this, the general relativity correction causes a precession of the periapsis of the planetary orbits slightly different from 2π, so that these orbits are in general *not* closed as is the case for the

Kepler potential. The periapsis advance given by (see Note a) below)

$$\Delta\varphi = \sqrt{\frac{2L^2}{m}} \int_{u_2}^{u_1} \frac{du}{\sqrt{E + GMmu - \frac{L^2}{2m} u^2 \left(1 - \frac{2GM}{c^2} u\right)}}, \qquad (3.49)$$

where $u_2 < u_1$ are the two positive roots of the cubic polynomial under the square root, cannot be expressed in terms of elementary functions. However, since the general relativity correction is very small it is possible to compute it approximately to order c^{-2} or, more accurately, to order 2 in the small *dimensionless* parameter (see Note b) below) $GMm/(Lc)$, as follows. We start by writing

$$\begin{aligned}
\Delta\varphi &= 2 \int_{u_2}^{u_1} \left(1 - \frac{2GM}{c^2} u\right)^{-1/2} \left[\frac{2m}{L^2} \frac{E + GMmu}{1 - \frac{2GM}{c^2} u} - u^2\right]^{-1/2} du \\
&\simeq 2 \int_{u_2}^{u_1} \left(1 + \frac{GM}{c^2} u\right) \left[\frac{2m}{L^2} (E + GMmu) \left(1 + \frac{2GM}{c^2} u\right) - u^2\right]^{-1/2} du \\
&= 2 \left(1 - \frac{4G^2M^2m^2}{L^2c^2}\right)^{-1/2} \int_{u_2}^{u_1} \left(1 + \frac{GM}{c^2} u\right) P(u)^{-1/2} du,
\end{aligned}$$

where $P(u)$ is the cubic polynomial

$$P(u) = -u^2 + \frac{2m}{L^2} \left(1 - \frac{4G^2M^2m^2}{L^2c^2}\right)^{-1} \left[E + GMm \left(1 + \frac{2E}{mc^2}\right) u\right]$$

and we have taken into account that $GMu/c^2 \sim GM/(c^2 a) = O(v^2/c^2)$ (see Note b) below) to approximate the term $\left(1 - \frac{2GM}{c^2} u\right)^{-1/2}$. Likewise,

$$\begin{aligned}
\left(1 - \frac{4G^2M^2m^2}{L^2c^2}\right)^{-1/2} \left(1 + \frac{GM}{c^2} u\right) &\simeq \left(1 + \frac{2G^2M^2m^2}{L^2c^2}\right) \left(1 + \frac{GM}{c^2} u\right) \\
&\simeq 1 + \frac{2G^2M^2m^2}{L^2c^2} + \frac{GM}{c^2} u,
\end{aligned}$$

and hence

$$\Delta\varphi \simeq 2 \int_{u_2}^{u_1} \left(1 + \frac{2G^2M^2m^2}{L^2c^2} + \frac{GM}{c^2} u\right) P(u)^{-1/2} du.$$

The *quadratic* polynomial $P(u)$ differs from the cubic polynomial appearing under the square root in Eq. (3.49) by terms of order c^{-2}, so the positive roots $u_2 < u_1$ of the latter polynomial are approximately equal to the two roots $u_2^* < u_1^*$ of $P(u)$.

Since $P(u) = (u_1^* - u)(u - u_2^*)$ we have

$$\Delta\varphi \simeq 2\int_{u_2^*}^{u_1^*}\left(1+\frac{2G^2M^2m^2}{L^2c^2}+\frac{GM}{c^2}u\right)[(u_1^*-u)(u-u_2^*)]^{-1/2}\,\mathrm{d}u$$
$$=\left(1+\frac{2G^2M^2m^2}{L^2c^2}\right)I_0+\frac{GM}{c^2}I_1,$$

with

$$I_k := 2\int_{u_2^*}^{u_1^*} u^k[(u_1^*-u)(u-u_2^*)]^{-1/2}\,\mathrm{d}u\,.$$

These integrals are easily computed by the standard change of variable

$$u=\frac{1}{2}(u_1^*+u_2^*)+\frac{1}{2}(u_1^*-u_2^*)\sin\theta\,,\qquad -\frac{\pi}{2}\leqslant\theta\leqslant\frac{\pi}{2}\,,$$

with the result

$$I_0 = 2\int_{-\pi/2}^{\pi/2}\mathrm{d}\theta = 2\pi\,,\quad I_1=\int_{-\pi/2}^{\pi/2}\left[u_1^*+u_2^*+(u_1^*-u_2^*)\sin\theta\right]\mathrm{d}\theta=\pi(u_1^*+u_2^*)\,.$$

Inserting these values into the last formula for $\Delta\varphi$ we obtain

$$\Delta\varphi-2\pi\simeq\frac{4\pi G^2M^2m^2}{L^2c^2}+\frac{\pi GM}{c^2}(u_1^*+u_2^*)\,.$$

Since the constant multiplying $u_1^*+u_2^*$ in the last term is already of order c^{-2}, we can compute the roots u_i^* to order 1, i.e., as the roots of the polynomial

$$\lim_{c\to\infty}P(u)=-u^2+\frac{2GMm^2}{L^2}u+\frac{2m}{L^2}E\,.$$

We thus obtain

$$u_1^*+u_2^*\simeq\frac{2GMm^2}{L^2}\,,$$

which yields the following formula for the *advance* of the periapsis after one period of the radial coordinate r:

$$\Delta\varphi-2\pi\simeq 6\pi\left(\frac{GMm}{Lc}\right)^2\,.$$

Due to the smallness of the dimensionless parameter GMm/LC in the previous formula, we can use the expression we shall derive in the next section relating the angular momentum to the semi-major axis a and the eccentricity e of a Keplerian orbit, namely

$$L^2 = GMm^2a(1-e^2)\,.$$

We thus have

$$\boxed{\Delta\varphi - 2\pi \simeq \frac{6\pi GM}{c^2 a(1-e^2)}\,.}$$

For planetary motion this effect is very small (of order v^2/c^2), but it *accumulates* with each period. The *rate of advance* of the periapsis is given by

$$\frac{\Delta\varphi - 2\pi}{\tau} \simeq \frac{6\pi GM}{c^2 a(1-e^2)\tau}\,,$$

where τ is the period of the radial coordinate r. To order c^{-2}, we can use the formula for τ we shall derive in the next section for the Kepler potential, namely

$$\tau = \frac{2\pi a^{3/2}}{\sqrt{GM}}\,.$$

We thus finally obtain

$$\frac{\Delta\varphi - 2\pi}{\tau} \simeq \frac{3(GM)^{3/2}}{c^2 a^{5/2}(1-e^2)}\,.$$

If we measure lengths in astronomical units (AU) and times in years we have

$$GM = 4\pi^2\ \mathrm{AU}^3/\mathrm{year}^2,$$

and the previous formula reads

$$\frac{\Delta\varphi - 2\pi}{\tau} \simeq \frac{24\pi^3}{c^2 a^{5/2}(1-e^2)}\ \mathrm{rad/year}\,.$$

In the solar system the rate of advance of the perihelion is maximum for Mercury, since its orbit has the smallest semi-major axis ($a = 0.38709893$ AU) and one of the largest eccentricities ($e = 0.20563069$). Taking into account that

$$c = 2.99792458 \cdot 10^8\ \frac{\mathrm{m}}{\mathrm{s}} = 2.99792458 \cdot 10^8\ \frac{3.1558149504 \cdot 10^7}{1.495978707 \cdot 10^{11}}\ \frac{\mathrm{AU}}{\mathrm{year}}$$
$$= 6.3241077 \times 10^4\ \frac{\mathrm{AU}}{\mathrm{year}}\,,$$

we obtain the following value for the rate of advance of Mercury's perihelion:

$$\boxed{\frac{\Delta\varphi - 2\pi}{\tau} \simeq 2.08387 \cdot 10^{-6}\ \text{rad/year} = 42.9829''/\text{century}\,.}$$

Notes.

a) More precisely, in the previous formula the energy E should be replaced by

$$\frac{mc^2}{2}\left[\left(1 + \frac{E}{mc^2}\right)^2 - 1\right] = E + \frac{E^2}{2mc^2}\,.$$

However, the last term is much smaller than the first one, since in planetary motion $E \ll mc^2$ (see next footnote).

b) To determine the order of magnitude of $GMm/(Lc)$ in planetary motion, we can use the value of L we shall obtain from the analysis of the Kepler problem in the next section,

$$L = m\sqrt{GMa(1-e^2)}\,,$$

where a and e are respectively the semi-major axis and the eccentricity of the Keplerian orbit. We thus obtain

$$\frac{GMm}{Lc} \simeq \frac{1}{c}\sqrt{\frac{GM}{a(1-e^2)}}\,.$$

Using the formula for the period τ from the next section,

$$\tau = \frac{2\pi a^{3/2}}{\sqrt{GM}}\,,$$

we have

$$\frac{GMm}{Lc} \simeq \frac{2\pi a}{\tau c\sqrt{1-e^2}} = \frac{v/c}{\sqrt{1-e^2}}\,, \qquad v := \frac{2\pi a}{\tau}\,.$$

In planetary motion the orbital velocity $v = 2\pi a/\tau$ is at most 48.9 Km/s (Mercury's orbital velocity), while the factor of $(1-e^2)^{-1/2}$ is only slightly larger than one even for relatively eccentric orbits (1.02 for Mercury). Hence $GMm/(Lc)$ is typically of order 10^{-4} in planetary motion. Similarly, using the formula in the next section for the energy of a Keplerian orbit as a good estimate for E we obtain

$$\frac{E}{mc^2} \simeq \frac{GM}{2ac^2} = \frac{2\pi^2 a^2}{c^2\tau^2} = \frac{1}{2}\left(\frac{v}{c}\right)^2 = O(10^{-8})\,.$$

3.3 KEPLER'S PROBLEM AND PLANETARY MOTION

3.3.1 Kepler's problem

We shall study in this section **Kepler's problem**, i.e., the motion of two bodies of masses m_1 and m_2 subject only to their mutual gravitational attraction

$$\boxed{\mathbf{F}_{12} = -\mathbf{F}_{21} = -\frac{k(\mathbf{r}_1 - \mathbf{r}_2)}{|\mathbf{r}_1 - \mathbf{r}_2|^3},}$$

where the constant k is given by

$$\boxed{k = Gm_1m_2 = GM\mu > 0\,.}$$

Therefore in this case

$$\mathbf{F}(\mathbf{r}) = -\frac{k}{r^2}\,\mathbf{e}_r \quad \Longrightarrow \quad \boxed{f(r) = -\frac{k}{r^2}\,, \quad V(r) = -\frac{k}{r}\,,}$$

and the associated one-body problem is

$$\mu\ddot{\mathbf{r}} = -k\frac{\mathbf{r}}{r^3}\,,$$

or equivalently

$$\boxed{\ddot{\mathbf{r}} = -GM\frac{\mathbf{r}}{r^3}\,.}$$

The equation of the orbits is easily obtained from Binet's equation, which for this potential is particularly simple:

$$\boxed{u'' + u = \frac{\mu k}{L^2}\,.}$$

The general solution of this equation can be expressed in the form

$$u = \frac{\mu k}{L^2}\big(1 + e\cos(\varphi - \varphi_0)\big)\,,$$

with e and φ_0 integration constants. Note that we can assume without loss of generality that $e \geqslant 0$, since if $e < 0$ it suffices to replace φ_0 with $\pi + \varphi_0$ in the previous equation. Clearly φ_0 is the polar angle of the orbit's periapsis, so that taking the x axis as the line joining the origin to the periapsis we can set $\varphi_0 = 0$. The equation of the orbits reduces then to

$$\boxed{r = \frac{\alpha}{1 + e\cos\varphi}\,, \qquad \text{with} \quad \alpha := \frac{L^2}{\mu k}\,.} \tag{3.50}$$

The parameter e can be related to the energy and angular momentum of the orbit using Eq. (3.35):

$$\begin{aligned} E &= \frac{L^2}{2\mu}(u'^2 + u^2) - ku = \frac{\mu k^2}{2L^2}\left[e^2 \sin^2\varphi + (1 + e\cos\varphi)^2 - 2(1 + e\cos\varphi)\right] \\ &= \frac{\mu k^2}{2L^2}(e^2 - 1)\,. \end{aligned}$$

Since $e \geqslant 0$, we have

$$\boxed{e = \sqrt{1 + \frac{2EL^2}{\mu k^2}}\,.} \tag{3.51}$$

Note that, as we saw in Example 3.3, for the Kepler potential $E \geqslant E_{\text{min}} = -\mu k^2/(2L^2)$, and thus the quantity under the radical is nonnegative.

• The orbits of Kepler's potential are **conic sections**. Indeed, from Eq. (3.50) we obtain

$$r = \alpha - ex \implies x^2 + y^2 = \alpha^2 - 2\alpha ex + e^2x^2 \implies (1 - e^2)x^2 + y^2 + 2\alpha ex = \alpha^2\,,$$

which is a second-degree polynomial equation in (x, y). The type of conic depends on the sign of $1 - e^2$ as follows:

$$\boxed{\begin{aligned} e > 1 \quad &\implies \quad \textbf{hyperbola} \\ e = 1 \quad &\implies \quad \textbf{parabola} \\ 0 < e < 1 \quad &\implies \quad \textbf{ellipse} \\ e = 0 \quad &\implies \quad \textbf{circle}\,. \end{aligned}}$$

In terms of the energy (cf. Eq. (3.51)),

$$\boxed{\begin{aligned} E > 0 \quad &\implies \quad \textbf{hyperbola} \\ E = 0 \quad &\implies \quad \textbf{parabola} \\ -\frac{\mu k^2}{2L^2} < E < 0 \quad &\implies \quad \textbf{ellipse} \\ E = -\frac{\mu k^2}{2L^2} \quad &\implies \quad \textbf{circle}\,, \end{aligned}}$$

where $-\mu k^2/(2L^2)$ is the minimum energy that a particle of mass μ and angular momentum L can have. Note, in particular, that this result is consistent with the qualitative discussion of Example 3.3. Note also that in the Kepler potential *all bounded orbits are closed* (and hence *periodic*), confirming once again Bertrand's theorem.

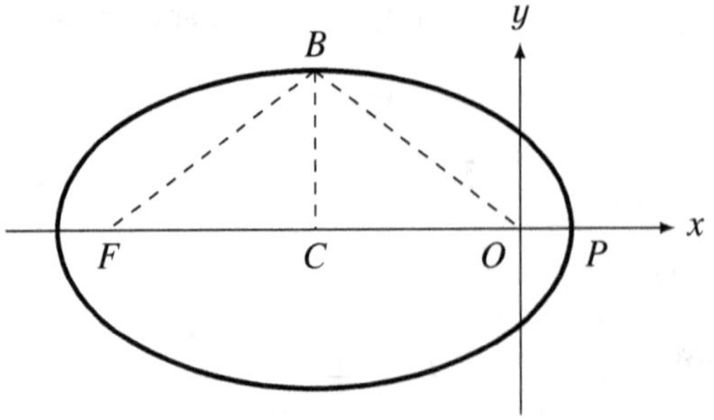

Figure 3.10. Geometry of an elliptical orbit in planetary motion. The point C is the center of the ellipse, F and O (the origin) its foci and P the periapsis. The distances $\overline{CP} = a$ and $\overline{CB} = b$ are respectively the ellipse's semi-major and semi-minor axes, and $\overline{OC} = \overline{FC} = c$ is its focal distance. By the defining property of the ellipse, we have $\overline{BO} + \overline{BF} = 2\sqrt{b^2 + c^2} = \overline{PO} + \overline{PF} = (a - c) + (a + c) = 2a \implies a = \sqrt{b^2 + c^2}$.

3.3.2 Planetary motion

The most interesting case is that of elliptical orbits (including, in particular, circular ones), in which $0 \leqslant e < 1$ or $E < 0$, since it is the relevant case when studying the motion of the planets around the Sun. The Cartesian equation of the orbits can be rewritten as

$$(1 - e^2)\left(x + \frac{\alpha e}{1 - e^2}\right)^2 + y^2 = \alpha^2 + \frac{\alpha^2 e^2}{1 - e^2} = \frac{\alpha^2}{1 - e^2}, \tag{3.52}$$

which is the equation of an ellipse centered at the point

$$\left(-\frac{\alpha e}{1 - e^2}, 0\right) \tag{3.53}$$

with semi-major and semi-minor axes respectively given by

$$\boxed{a = \frac{\alpha}{1 - e^2}, \qquad b = \frac{\alpha}{\sqrt{1 - e^2}}.} \tag{3.54}$$

Recall that the *focal distance* c (defined as the distance of the center of the ellipse to either of its foci) and the *eccentricity* ε of an ellipse with semiaxes $a \geqslant b$ are given by

$$\boxed{c = \sqrt{a^2 - b^2}, \qquad \varepsilon = \frac{c}{a}} \tag{3.55}$$

(cf. Fig. 3.10). Using the previous expressions for a and b we easily obtain

$$c = \frac{\alpha}{1 - e^2}\sqrt{1 - (1 - e^2)} = \frac{\alpha e}{1 - e^2} = ea \quad \implies \quad \boxed{e = \varepsilon.}$$

Therefore the constant e appearing in the equation of the orbits is the *eccentricity* of the ellipse, and Eq. (3.51) relates the particle's energy and angular momentum to the eccentricity of its orbit. The above equations also determine the position of the *foci* of the ellipse, which by definition are the two points on the major axis (in this case, the x axis) at a distance c from the center of the ellipse. Indeed, from Eqs. (3.53)–(3.55) it follows that the center of the ellipse has coordinates $(-c, 0)$, and thus the foci are the points $(-2c, 0)$ and $(0, 0)$. In particular, this shows that one of the foci is the origin of coordinates, that is, the center of gravitational attraction. Hence *the bounded orbits in planetary motion are ellipses, one of whose foci is the Sun* (Kepler's first law).

• From Eq. (3.54) and the expression (3.51) for the eccentricity it follows that the energy of an elliptical orbit is

$$\boxed{E = -\frac{\mu k^2}{2L^2}(1 - e^2) = -\frac{\mu k^2 \alpha}{2aL^2} = -\frac{k}{2a}\,.} \tag{3.56}$$

We see, therefore, that *the energy depends only on the semi-major axis of the orbit* (i.e., it is independent of its eccentricity).

• The period τ of elliptic orbits in planetary motion is easily determined from the law of areas (3.14), taking into account that the area of an ellipse is equal to πab:

$$\frac{L\tau}{2\mu} = \pi ab = \pi\sqrt{\alpha}\, a^{3/2} \quad \Longrightarrow \quad \boxed{\tau = \frac{2\pi\mu}{L}\sqrt{\alpha}\, a^{3/2} = 2\pi\sqrt{\frac{\mu}{k}}\, a^{3/2}\,,} \tag{3.57}$$

or in terms of the energy

$$\boxed{\tau = \pi k\sqrt{\frac{\mu}{2}}\, |E|^{-3/2}\,.}$$

In particular, *the period depends only on the semi-major axis, and is independent of the eccentricity*. In planetary motion the formula for the period is usually expressed in the form

$$\boxed{\tau = \frac{2\pi a^{3/2}}{\sqrt{GM}} \simeq \frac{2\pi a^{3/2}}{\sqrt{GM_\odot}}\,,}$$

$M_\odot$ being the Sun's mass. Thus (with great approximation) *the ratio τ^2/a^3 is the same for all planets* (Kepler's third law).

• Let us denote by p and p' respectively the distance of the perihelion and aphelion of the ellipse to the origin. From the equation of the orbit (3.50) it easily follows that the particle is in the perihelion (resp. in the aphelion) when $\varphi = 0$ (resp. $\varphi = \pi$), and therefore

$$p = \frac{\alpha}{1+e} = a(1-e)\,, \qquad p' = \frac{\alpha}{1-e} = a(1+e)\,.$$

This is also apparent from the geometry of the ellipse (cf. Fig. 3.10), since $a(1-e) = a - c, = a(1+e) = a + c$.

• It is also straightforward to compute the speed at any point in the orbit from the law of conservation of energy:

$$v^2 = \frac{2}{\mu}\,\big(E - V(r)\big) = \frac{k}{\mu}\left(\frac{2}{r} - \frac{1}{a}\right) = \frac{k}{\mu\alpha}\big[2(1 + e\cos\varphi) - (1 - e^2)\big]$$

$$= \boxed{\frac{k^2}{L^2}\,(1 + e^2 + 2e\cos\varphi)\,.}$$

Note that v is maximal at the perihelion ($\varphi = 0$) and minimal at the aphelion ($\varphi = \pi$), with respective values

$$v_p = \frac{k}{L}\,(1 + e)\,, \qquad v_{p'} = \frac{k}{L}\,(1 - e)\,.$$

In particular, the quotient

$$\frac{v_p}{v_{p'}} = \frac{1 + e}{1 - e}$$

depends only on the eccentricity of the orbit. It is sometimes of interest to express the speeds v_p and $v_{p'}$ as a function of p and p', instead of L. To this end, it suffices to note that

$$\alpha = \frac{L^2}{k\mu} = p(1 + e) = p'(1 - e) \quad\Longrightarrow\quad L = \sqrt{k\mu p(1 + e)} = \sqrt{k\mu p'(1 - e)}\,,$$

and therefore

$$v_p = \sqrt{\frac{k}{\mu p}\,(1 + e)}\,, \qquad v_{p'} = \sqrt{\frac{k}{\mu p'}\,(1 - e)}.$$

Example 3.6. The mean value over a period of a planetary orbit of any quantity $f(r)$ is defined by

$$\langle f(r)\rangle := \frac{1}{\tau}\int_0^\tau f(r)\,\mathrm{d}t\,.$$

Taking into account that

$$\mathrm{d}t = \frac{\mathrm{d}\varphi}{\dot\varphi} = \frac{\mu}{L}\,r^2\,\mathrm{d}\varphi\,,$$

the time integral can be transformed into the following integral over the polar angle φ:

$$\langle f(r)\rangle = \frac{\mu}{\tau L}\int_0^{2\pi} r^2 f(r)\,\mathrm{d}\varphi = \frac{\mu\alpha^2}{\tau L}\int_0^{2\pi} f\!\left(\frac{\alpha}{1 + e\cos\varphi}\right)\frac{\mathrm{d}\varphi}{(1 + e\cos\varphi)^2}\,.$$

Using Eqs. (3.50), (3.54), and (3.57) we obtain

$$\frac{\mu\alpha^2}{\tau L} = \frac{\mu\alpha^2}{\tau\sqrt{\alpha k\mu}} = \frac{1}{\tau}\sqrt{\frac{\mu}{k}}\,a^{3/2}(1 - e^2)^{3/2} = \frac{(1 - e^2)^{3/2}}{2\pi}\,,$$

and thus

$$\langle f(r)\rangle = \frac{(1-e^2)^{3/2}}{2\pi}\int_0^{2\pi} f\left(\frac{a(1-e^2)}{1+e\cos\varphi}\right)\frac{\mathrm{d}\varphi}{(1+e\cos\varphi)^2}\,.$$

For instance, the mean distance of a planet to the Sun is given by

$$\langle r\rangle = a(1-e^2)^{5/2} I(e)\,,\qquad I(e) := \frac{1}{2\pi}\int_0^{2\pi}\frac{\mathrm{d}\varphi}{(1+e\cos\varphi)^3}\,.$$

The integral $I(e)$ can be computed using the residue theorem taught in complex analysis courses, with the result

$$I(e) = \frac{e^2+2}{2(1-e^2)^{5/2}}\,.$$

We finally obtain

$$\boxed{\langle r\rangle = \left(1+\frac{e^2}{2}\right)a\,.}$$

Exercise 3.10. Integrate Eq. (3.24) to find the relation between t and φ in planetary motion.

Solution. Using Eq. (3.50) for the Kepler orbits and the first relation (3.54) we obtain

$$t = \frac{L^3}{\mu k^2}\int\frac{\mathrm{d}\varphi}{(1+e\cos\varphi)^2} = \sqrt{\frac{\mu}{k}}\,a^{3/2}(1-e^2)^{3/2}\int\frac{\mathrm{d}\varphi}{(1+e\cos\varphi)^2}\,. \tag{3.58}$$

To compute the integral, we start by making the change of variable $u = \tan(\varphi/2)$, so that

$$\begin{cases} \cos\varphi = 2\cos^2(\varphi/2) - 1 = \dfrac{2}{\sec^2(\varphi/2)} - 1 = \dfrac{2}{1+u^2} - 1 = \dfrac{1-u^2}{1+u^2}\,, \\ \mathrm{d}u = \dfrac{1}{2}\sec^2(\varphi/2)\,\mathrm{d}\varphi = \dfrac{1}{2}(1+u^2)\,\mathrm{d}\varphi, \end{cases}$$

and hence

$$\int\frac{\mathrm{d}\varphi}{(1+e\cos\varphi)^2} = 2\int\frac{\mathrm{d}u}{(1+u^2)\left[1+\frac{e(1-u^2)}{1+u^2}\right]^2} = 2\int\frac{(1+u^2)\,\mathrm{d}u}{\left[1+e+(1-e)u^2\right]^2}\,. \tag{3.59}$$

Setting now

$$u = \sqrt{\frac{1+e}{1-e}}\,v$$

we obtain

$$\int \frac{d\varphi}{(1+e\cos\varphi)^2} = \frac{2}{(1+e)^2}\sqrt{\frac{1+e}{1-e}} \int \frac{1+\frac{1+e}{1-e}v^2}{(1+v^2)^2}\,dv$$
$$= 2(1-e^2)^{-3/2} \int \frac{1-e+(1+e)v^2}{(1+v^2)^2}\,dv$$
$$= 2(1-e^2)^{-3/2} \left[(1+e)\arctan v - 2e\int \frac{dv}{(1+v^2)^2}\right].$$

The last integral is computed integrating by parts in the integral of $(1+v^2)^{-1}$:

$$\arctan v = \int \frac{dv}{1+v^2} = \frac{v}{1+v^2} + \int \frac{2v^2\,dv}{(1+v^2)^2}$$
$$= \frac{v}{1+v^2} + 2\arctan v - 2\int \frac{dv}{(1+v^2)^2}$$
$$\Longrightarrow\ 2\int \frac{dv}{(1+v^2)^2} = \frac{v}{1+v^2} + \arctan v.$$

Putting everything together we obtain:

$$\int \frac{d\varphi}{(1+e\cos\varphi)^2} = 2(1-e^2)^{-3/2}\left(\arctan v - \frac{ev}{1+v^2}\right).$$

Since

$$\frac{v}{1+v^2} = \sqrt{\frac{1-e}{1+e}}\,\frac{u}{\frac{1-e}{1+e}u^2+1} = \frac{\sqrt{1-e^2}\,u}{(1-e)u^2+1+e} = \frac{\sqrt{1-e^2}\tan\left(\frac{\varphi}{2}\right)}{2e+(1-e)\sec^2\left(\frac{\varphi}{2}\right)}$$
$$= \sqrt{1-e^2}\,\frac{\sin\left(\frac{\varphi}{2}\right)\cos\left(\frac{\varphi}{2}\right)}{2e\cos^2\left(\frac{\varphi}{2}\right)+1-e} = \frac{1}{2}\sqrt{1-e^2}\,\frac{\sin\varphi}{1+e\cos\varphi},$$

we finally arrive at the formula

$$\boxed{t = \sqrt{\frac{\mu}{k}}\,a^{3/2}\left[2\arctan\left(\sqrt{\frac{1-e}{1+e}}\tan\left(\frac{\varphi}{2}\right)\right) - e\sqrt{1-e^2}\,\frac{\sin\varphi}{1+e\cos\varphi}\right]},$$

where we have discarded the integration constant so that at $t=0$ the particles is at the periapsis $\varphi = 0$. This expression is too unwieldy in practice, and the time dependence of r (and hence φ) in the Kepler problem is usually computed inverting Kepler's equation introduced in Exercise 3.12.

Exercise 3.11. Redo the previous calculation for parabolic and hyperbolic orbits.

Solution. For parabolic orbits $e = 1$, and Eq. (3.58) yields

$$t = \frac{L^3}{\mu k^2} \int \frac{\mathrm{d}\varphi}{(1+\cos\varphi)^2} = \frac{L^3}{4\mu k^2} \int \mathrm{d}\varphi \, \sec^4\left(\tfrac{\varphi}{2}\right)$$
$$= \frac{L^3}{2\mu k^2} \int \mathrm{d}\left(\tan\left(\tfrac{\varphi}{2}\right)\right) \left(1 + \tan^2\left(\tfrac{\varphi}{2}\right)\right) = \frac{L^3}{2\mu k^2} \left(\tan\left(\tfrac{\varphi}{2}\right) + \frac{1}{3}\tan^3\left(\tfrac{\varphi}{2}\right)\right).$$

For hyperbolic orbits $e > 1$, and Eq. (3.52) can be written as

$$\frac{(x-c)^2}{a^2} - \frac{y^2}{b^2} = 1,$$

with

$$a = \frac{\alpha}{e^2 - 1}, \qquad b = \frac{\alpha}{\sqrt{e^2-1}}, \qquad c = ea.$$

By Eq. (3.50)

$$L^3 = (\alpha k \mu)^{3/2} = (k\mu)^{3/2} a^{3/2} (e^2 - 1)^{3/2},$$

which substituted into the first equality in Eq. (3.58) yields

$$t = \sqrt{\frac{\mu}{k}}\, a^{3/2} (e^2-1)^{3/2} \int \frac{\mathrm{d}\varphi}{(1 + e\cos\varphi)^2}.$$

Performing the change of variable $u = \tan(\varphi/2)$ and using Eq. (3.59) we obtain

$$\int \frac{\mathrm{d}\varphi}{(1+e\cos\varphi)^2} = 2 \int \frac{(1+u^2)\,\mathrm{d}u}{\left[e+1-(e-1)u^2\right]^2}.$$

Setting next

$$u = \sqrt{\frac{e+1}{e-1}}\, v$$

we have

$$(e^2-1)^{3/2} \int \frac{(1+u^2)\,\mathrm{d}u}{\left[e+1-(e-1)u^2\right]^2}$$
$$= \int \frac{e - 1 + (e+1)v^2}{(1-v^2)^2}\,\mathrm{d}v = \left[2e \int \frac{\mathrm{d}v}{(1-v^2)^2} - (e+1)\,\mathrm{arctanh}\, v\right].$$

The last integral is computed as the integral of $(1+v^2)^{-2}$ above, with the result

$$\int \frac{\mathrm{d}v}{(1-v^2)^2} = \frac{1}{2}\left(\mathrm{arctanh}\, v + \frac{v}{1-v^2}\right).$$

We thus obtain

$$(e^2-1)^{3/2}\int\frac{(1+u^2)\,\mathrm{d}u}{\left[e+1-(e-1)u^2\right]^2}=\frac{ev}{1-v^2}-\operatorname{arctanh}v,$$

and taking into account that

$$\frac{v}{1-v^2}=\sqrt{\frac{e-1}{e+1}}\,\frac{u}{1-\frac{e-1}{e+1}u^2}=\frac{\sqrt{e^2-1}\,u}{e+1-(e-1)u^2}=\frac{\sqrt{e^2-1}\,\tan\left(\frac{\varphi}{2}\right)}{e+1-(e-1)\tan^2\left(\frac{\varphi}{2}\right)}$$
$$=\frac{1}{2}\frac{\sqrt{e^2-1}\,\sin\varphi}{1+e\cos\varphi},$$

we finally arrive at the following expression for $t(\varphi)$:

$$t=\sqrt{\frac{\mu}{k}}\,a^{3/2}\left[e\sqrt{e^2-1}\,\frac{\sin\varphi}{1+e\cos\varphi}-2\operatorname{arctanh}\left(\sqrt{\frac{e-1}{e+1}}\,\tan\left(\frac{\varphi}{2}\right)\right)\right].$$

Exercise 3.12. Given an elliptic orbit of eccentricity e and major semiaxis a, define the *eccentric anomaly* $\psi(t)$ by the equation

$$\boxed{\omega t=\psi-e\sin\psi,}\tag{3.60}$$

where $\omega=\dfrac{2\pi}{\tau}=\sqrt{\dfrac{k}{\mu}}\,a^{-3/2}$ is the mean orbital frequency. (Note that

$$\frac{\mathrm{d}}{\mathrm{d}\psi}(\psi-e\sin\psi)=1-e\cos\psi\geqslant 1-e>0,$$

so that (3.60) uniquely determines ψ as a function of t by the inverse function theorem.) Show that the radius vector $r(t)$ can be expressed in terms of the eccentric anomaly through the equation

$$\boxed{r=a(1-e\cos\psi)\,.}\tag{3.61}$$

Solution. From the first Eq. (3.19) with

$$U(r)=\frac{L^2}{2\mu r^2}-\frac{k}{r}=\frac{ak(1-e^2)}{2r^2}-\frac{k}{r}$$

and Eq. (3.56) we easily obtain

$$t=\sqrt{\frac{\mu}{2k}}\int_p^r\frac{s\,\mathrm{d}s}{\sqrt{-\frac{s^2}{2a}+s-\frac{a}{2}\left(1-e^2\right)}},$$

where we have chosen as lower limit in the integral the orbit's perihelion p so that $dt/dr > 0$ till r reaches the next aphelion (i.e., for $0 \leqslant t \leqslant \tau/2$). Since

$$P(s) := -\frac{s^2}{2a} + s - \frac{a}{2}(1 - e^2) = \frac{1}{2a}\left[a^2e^2 - (s-a)^2\right]$$

we perform the natural change of variable $s = a(1 - e\cos\beta)$ in the integral, so that $ds = ae\sin\beta\, d\beta$ and

$$P(s) = \frac{ae^2}{2}\sin^2\beta\,.$$

Taking into account that $p = a(1-e)$ implies that $\beta = 0$ when $s = p$ we finally obtain

$$t = \sqrt{\frac{\mu a^3}{k}}\int_0^{\psi}(1 - e\cos\beta)\, d\beta = \frac{1}{\omega}(\psi - e\sin\psi)$$

with $r = a(1 - e\cos\psi)$, as claimed.

Equation (3.60) is usually called in the literature *Kepler's equation*. The geometric meaning of ψ can be understood from Fig. 3.11. Indeed, since the point P on the elliptical orbit and the point P' on the circle of radius a centered at C lie on the same vertical they must have the same abscissa (measured from the focus F at the origin), namely

$$a\cos\psi - c = a(\cos\psi - e) = r\cos\varphi\,.$$

On the other hand, from the equation of the ellipse in polar coordinates

$$r = \frac{a(1-e^2)}{1 + e\cos\varphi}$$

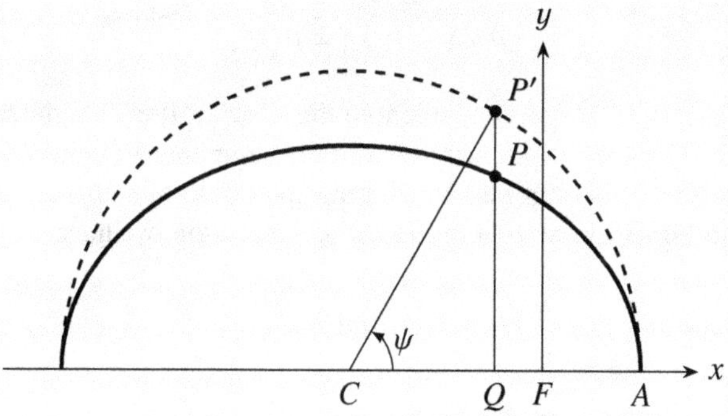

Figure 3.11. Eccentric anomaly ψ of point P on an elliptical orbit (solid curve). Note that $\overline{CF} = c = ea$, $\overline{CA} = a$, and the dashed curve represents the circle of radius a centered at C.

we obtain

$$r = a(1-e^2) - er\cos\varphi = a(1-e^2) - ea(\cos\psi - e) = a(1 - e\cos\psi),$$

namely Eq. (3.61).

Note. Strictly speaking, we have proved (3.60)–(3.61) only for one half-period, i.e., for $0 \leqslant t \leqslant \tau/2$ or equivalently $0 \leqslant \psi \leqslant \pi$. However, using the identities $r(t) = r(-t) = r(t + k\tau)$ (with $k \in \mathbb{Z}$) it is straightforward to show that these equations are in fact valid for *all* values of t and ψ. Indeed, if $\psi \mapsto -\psi$ then r does not change in Kepler's equation and $t \mapsto -t$, which is again consistent with the identity $r(t) = r(-t)$. Thus Kepler's equation can be extended to the interval $-\tau/2 \leqslant t \leqslant \tau/2$, i.e., to a whole period of the motion. Likewise, when ψ changes by $2k\pi$ (with $k \in \mathbb{Z}$) r does not change in Kepler's equation and t changes by $2k\pi/\omega = k\tau$, which is consistent with the identity $r(t + k\tau) = r(\tau)$. This establishes Kepler's equation for an arbitrary time $t \in \mathbb{R}$.

Exercise 3.13. If ψ is the angle ACP' in Fig. 3.11, derive Kepler's equation using the law of areas.

Solution. According to the law of areas, if t is the time taken by the planet to travel from the periapsis A to the point P in Fig. 3.11 we have

$$\frac{PFA}{\pi ab} = \frac{t}{\tau} = \frac{\omega t}{2\pi} \quad \Longrightarrow \quad PFA = \frac{1}{2}\, ab \cdot \omega t\,,$$

where PFA denotes the area swept by the planet's position vector as it travels from A to P along its orbit. On the other hand, from Fig. 3.11 it follows that

$$PFA = PQA - PQF, \tag{3.62}$$

where PQF and PQA respectively denote the areas of the triangle PQF and the sector delimited by the elliptic arc AP and the segments PQ and QA. Since the ellipse in Fig. (3.11) is obtained by dilating the circle of radius a and center C (dashed line in Fig. 3.11) by b/a in the vertical direction, we have

$$\begin{aligned} PQA &= \frac{b}{a}\, P'QA = \frac{b}{a}\,(P'AC - P'QC) = \frac{b}{a}\left(\frac{a^2\psi}{2} - \frac{a^2}{2}\sin\psi\cos\psi\right) \\ &= \frac{1}{2}\, ab(\psi - \sin\psi\cos\psi). \end{aligned} \tag{3.63}$$

Here $P'QA$ is the area of the circular sector delimited by the arc AP' and the segments $P'Q$ and QA, $P'AC$ is the area of the circular sector determined by the points P', A, and C, and $P'QC$ is the area of the triangle $P'QC$. On the other hand,

$$QP = \frac{b}{a}\,QP' = b\sin\psi, \quad QF = CF - CQ = ae - a\cos\psi$$

$$\Longrightarrow \quad PQF = \frac{1}{2}\,b\sin\psi \cdot a(e - \cos\psi). \tag{3.64}$$

Combining Eqs. (3.62)–(3.64) we finally obtain

$$\frac{1}{2}\,ab\cdot\omega t = \frac{1}{2}\,ab(\psi - \sin\psi\cos\psi) - \frac{1}{2}\,ab\sin\psi(e - \cos\psi) = \frac{1}{2}\,ab(\psi - e\sin\psi),$$

which yields Kepler's equation (3.60).

Exercise 3.14. Find the analogue of Kepler's equation for *hyperbolic* orbits of the Kepler potential.

Solution. In this case the energy is positive, and can be expressed as $k/(2a)$ if we define $a = \alpha/(e^2 - 1)$ (cf. Eq. (3.51)). Proceeding as for elliptical orbits and taking into account that

$$L^2 = k\mu\alpha = k\mu a(e^2 - 1)$$

we arrive at the formula

$$t = \sqrt{\frac{\mu}{2k}} \int_p^r \frac{s\,\mathrm{d}s}{\sqrt{P(s)}}, \tag{3.65}$$

where now $p = \alpha/(e + 1) = a(e - 1)$ and

$$P(s) := \frac{s^2}{2a} + s - \frac{a}{2}\,(e^2 - 1) = \frac{1}{2a}\big[(s + a)^2 - a^2e^2\big].$$

This suggests performing the change of variable $s = a(e\cosh\beta - 1)$, so that $\mathrm{d}s = ae\sinh\beta\,\mathrm{d}\beta$, $P(s) = \frac{1}{2}\,a^2e^2\sinh^2\beta$ and

$$t = \sqrt{\frac{\mu}{k}}\,a^{3/2} \int_0^\psi (e\cosh\beta - 1)\,\mathrm{d}\beta = \frac{1}{\omega}\,(e\sinh\psi - \psi)$$

$$\Longrightarrow \quad \boxed{\omega t = e\sinh\psi - \psi,} \tag{3.66}$$

with $\omega = \sqrt{k/\mu}\,a^{-3/2}$ and

$$\boxed{r = a(e\cosh\psi - 1).} \tag{3.67}$$

Equations (3.66)–(3.67) are the analogue of Eqs. (3.60)–(3.61) for hyperbolic orbits. The motion of the radial coordinate r is obtained inverting the relation $\omega t = (e\sinh\psi - \psi)$ for t as a function of ψ, and substituting the result into Eq. (3.67). This is possible, since the RHS of the previous equation has derivative

$e\cosh\psi - 1 \geqslant e - 1 > 0$, and is therefore a monotonically increasing function of ψ. Strictly speaking, we have established Eqs. (3.66)–(3.67) only for $t \geqslant 0$, or equivalently $\psi \geqslant 0$. However, from the identity $r(t) = r(-t)$ and the fact that $t \mapsto -t$ implies $\psi \mapsto -\psi$ and $r \mapsto r$ in Kepler's equation we deduce that Eqs. (3.66) hold for all real values of t and β.

Exercise 3.15. Find the equation of the orbits in the *repulsive* $1/r$ potential $V(r) = k/r$ (with $k > 0$).

Solution. Binet's equation is in this case

$$u'' + u = -\frac{k\mu}{L^2},$$

whose general solution can be taken as

$$u = \frac{k\mu}{L^2}\big(e\cos(\varphi - \varphi_0) - 1\big).$$

Again, we can assume w.l.o.g. that $e > 0$ and $\varphi_0 = 0$ by an appropriate choice of the x axis. Moreover, since $r > 0$ we must have $e > 1$. We can thus write

$$\boxed{r = \frac{\alpha}{e\cos\varphi - 1}, \qquad \alpha := \frac{L^2}{k\mu}.}$$

All the orbits in this case are clearly *unbounded*, since $r \to \infty$ for $\varphi \to \pm\arccos(1/e)$. It is also clear that the polar angle of the periapsis is $\varphi = 0$, and its distance to the origin is equal to $\alpha/(e-1)$. To find the Cartesian equation of the orbit we multiply both sides of the previous equation by $e\cos\varphi - 1$, thus obtaining

$$\begin{aligned} r = ex - \alpha \implies & x^2 + y^2 = e^2x^2 - 2\alpha ex + \alpha^2 \\ \iff & (e^2 - 1)x^2 - 2\alpha ex - y^2 = -\alpha^2, \end{aligned}$$

or equivalently

$$(e^2 - 1)\left(x - \frac{\alpha e}{e^2 - 1}\right)^2 - y^2 = -\alpha^2 + \frac{\alpha^2 e^2}{e^2 - 1} = \frac{\alpha^2}{e^2 - 1}.$$

This is the equation of a *hyperbola* with center $(\alpha e/(e^2 - 1), 0)$ and semiaxes

$$a = \frac{\alpha}{e^2 - 1}, \qquad b = \frac{\alpha}{\sqrt{e^2 - 1}}.$$

In fact, since $ex = \alpha + r > 0$ the orbit is the branch of the hyperbola in the half-plane $x > 0$. The focal distance of the hyperbola is given by

$$c = \sqrt{a^2 + b^2} = \frac{\alpha^2}{e^2 - 1}\sqrt{1 + (e^2 - 1)} = \frac{\alpha e}{e^2 - 1} = ea,$$

and hence its eccentricity c/a is equal to the parameter e. This also implies that the center of the hyperbola is the point $(c, 0)$, and hence the foci are the points $(0, 0)$ and $(0, 2c)$. The energy is now given by

$$E = \frac{L^2}{2\mu}(u'^2 + u^2) + ku = \frac{\mu k^2}{2L^2}\left[e^2 \sin^2\varphi + (e\cos\varphi - 1)^2 + 2(e\cos\varphi - 1)\right]$$
$$= \frac{\mu k^2}{2L^2}(e^2 - 1),$$

or, using the equation for the semi-major axis,

$$E = \frac{\mu k^2 \alpha}{2aL^2} = \frac{k}{2a}.$$

Finally, the eccentricity can be related to the energy and angular momentum of the orbit through the equation

$$e = \sqrt{1 + \frac{2EL^2}{\mu k^2}},$$

which is the same as for the attractive Kepler potential.

Exercise 3.16. Find the analogue of Kepler's equation for the orbits of the repulsive Kepler potential.

Solution. Proceeding as in Exercise 3.12 and using the formulas

$$E = \frac{k}{2a}, \qquad L^2 = k\mu\alpha = k\mu a(e^2 - 1)$$

derived in the previous exercise we again arrive at Eq. (3.65) with $p = a(1 + e)$, where now

$$P(s) := \frac{s^2}{2a} - s - \frac{a}{2}(e^2 - 1) = \frac{1}{2a}\left[(s - a)^2 - a^2 e^2\right].$$

We therefore perform the change of variable $s = a(1 + e\cosh\beta)$, obtaining

$$t = \sqrt{\frac{\mu}{k}}\, a^{3/2} \int_0^{\psi} (e\cosh\beta + 1)\, d\beta = \frac{1}{\omega}(\psi + e\sinh\psi) \implies \boxed{\omega t = \psi + e\sinh\psi,}$$

with

$$\boxed{r = a(1 + e\cosh\psi).}$$

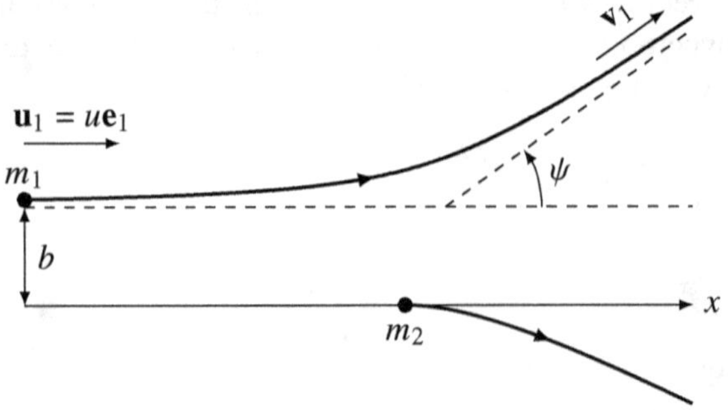

Figure 3.12. Ideal scattering experiment of a particle of mass m_1 by a target of mass m_2.

3.4 SCATTERING BY A CENTRAL POTENTIAL

3.4.1 Differential scattering cross section

In an ideal **scattering** experiment, a particle of mass m_1 (projectile) is launched against a target of mass m_2. It is assumed that the projectile and the target form an *isolated system*, and that the force $\mathbf{F}_{12} = -\mathbf{F}_{21}$ between them is *conservative*[6], i.e.,

$$\mathbf{F}_{12}(\mathbf{r}_1, \mathbf{r}_2) = -\frac{\partial V(\mathbf{r}_1 - \mathbf{r}_2)}{\partial \mathbf{r}_1},$$

and tends to zero if $r \equiv |\mathbf{r}_1 - \mathbf{r}_2| \to \infty$. In other words, the interaction potential V is approximately constant if $r \gg 1$, and we can thus assume without loss of generality that

$$\lim_{r\to\infty} V(\mathbf{r}) = 0\,. \tag{3.68}$$

The projectile is assumed to be initially (i.e., for $t \to -\infty$) very far away from the target, so that $\mathbf{F}_{12} \simeq 0$ and both the target and the projectile move with approximately constant velocity. Initially the target is at rest at the origin in a suitable inertial system, while if we orient the x axis in the direction of the projectile's initial velocity $\mathbf{u}_1$ we have $\mathbf{u}_1 = u\mathbf{e}_1$ with $u > 0$. As t increases the projectile starts interacting with the target and is deflected by it, until for $t \to \infty$ the distance between the projectile and the target tends to infinity, $\mathbf{F}_{12}$ is again approximately zero, and both particles move with approximately constant speed (cf. Fig. 3.12).

Let us denote by $\psi \in [0, \pi]$ the angle between the final velocity $\mathbf{v}_1$ of the projectile and the initial one $\mathbf{u}_1$—i.e., between $\mathbf{v}_1$ and the positive x axis—known as the **deflection angle**. For a certain initial speed u, the final speed $\mathbf{v}_1$ depends solely on the projectile's initial position in a plane P_∞ perpendicular to the x axis and at a very large (ideally infinite) distance from the origin. This position is determined by its *polar coordinates*

[6]As shown in Section 2.6.2, if $\mathbf{F}_{12} = -\mathbf{F}_{21}$ the potential V can only depend on $\mathbf{r}_1$ and $\mathbf{r}_2$ through $\mathbf{r}_1 - \mathbf{r}_2$.

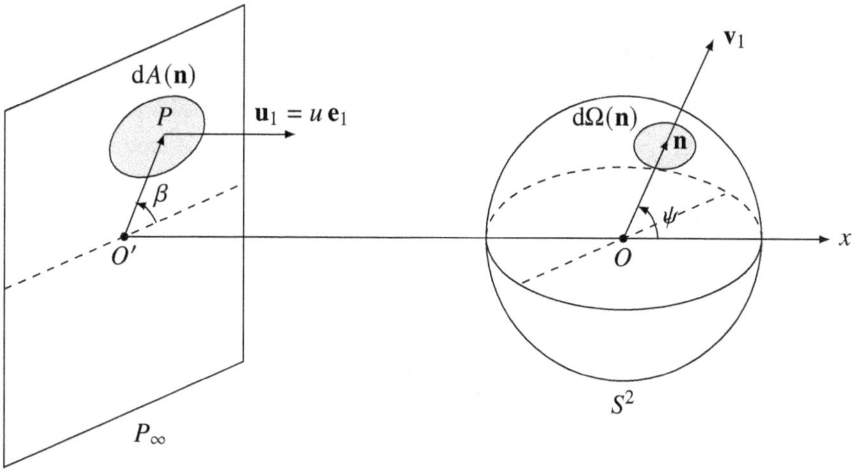

Figure 3.13. Definition of $\sigma(\mathbf{n})$ (the distance $|\overrightarrow{O'P}|$ is the impact parameter b).

on P_∞, i.e., by the initial distance b of the projectile to the x axis, called the **impact parameter**, and by its *azimuthal angle* β about this axis. Note, in particular, that if the interaction potential is *central* the angle ψ can only depend on the impact parameter b. Conversely, from the final speed $\mathbf{v}_1$ we can in principle infer the initial position of the projectile on the plane P_∞, that is, the values of its impact parameter b and its azimuthal angle β. In fact, in a scattering experiment what is actually observed is not the final velocity $\mathbf{v}_1$ but only its *direction* $\mathbf{v}_1/v_1 =: \mathbf{n}$. This direction is a dimensionless *unit* vector, i.e., a point on the *unit sphere* S^2, determined by the angle ψ between the vector $\mathbf{v}_1$ (or, equivalently, $\mathbf{n}$) and the x axis and by the azimuthal angle ϕ of $\mathbf{v}_1$ about this axis. In fact, if the mass of the target is much greater than that of the projectile the target will move with approximately constant velocity for all t. Hence the **laboratory frame**, in which by definition the target is always at rest, is an *inertial frame*. In this frame the law of conservation of energy applied at $t = \pm\infty$ yields

$$\frac{1}{2}m_1 u^2 = \frac{1}{2}m_1 v_1^2 \quad \Longrightarrow \quad \boxed{v_1 = u}\,.$$

Suppose that the direction of the projectile's final velocity lies in a differential surface element (technically called a *differential solid angle*) $\mathrm{d}\Omega(\mathbf{n}) \subset S^2$ centered at a unit vector $\mathbf{n}$. Let $\mathrm{d}A(\mathbf{n})$ be the area of the infinitesimal region of the plane P_∞ corresponding to the projectile's initial positions resulting in a final velocity $\mathbf{v}_1$ whose direction lies in $\mathrm{d}\Omega(\mathbf{n})$ (cf. Fig. 3.13). By definition, the quotient

$$\boxed{\sigma(\mathbf{n}) = \frac{\mathrm{d}A(\mathbf{n})}{\mathrm{d}\Omega(\mathbf{n})}} \tag{3.69}$$

is called the **differential scattering cross section** for the direction $\mathbf{n}$. Note, in particular, that $\sigma(\mathbf{n})$ *has dimensions of area*, since $\mathrm{d}\Omega$ is dimensionless.

In an actual scattering experiment, a **beam** of particles with the same mass m and initial velocity $u\mathbf{e}_1$ is directed at the target and the number of particles scattered in a certain solid angle from the target is measured. If we denote by ν the beam's *density*, i.e., the number of incoming projectiles per unit cross-sectional area perpendicular to the initial velocity $\mathbf{u}_1$, then the number of particles $\mathrm{d}N(\mathbf{n})$ scattered in the solid angle $\mathrm{d}\Omega(\mathbf{n})$ centered at $\mathbf{n}$ is equal to the number of beam particles starting from the corresponding surface $\mathrm{d}A(\mathbf{n})$ in P_∞, that is

$$\mathrm{d}N(\mathbf{n}) = \nu\,\mathrm{d}A(\mathbf{n}) \equiv \nu\,\sigma(\mathbf{n})\,\mathrm{d}\Omega(\mathbf{n})\,, \tag{3.70}$$

and hence

$$\boxed{\sigma(\mathbf{n}) = \frac{1}{\nu}\frac{\mathrm{d}N(\mathbf{n})}{\mathrm{d}\Omega(\mathbf{n})}\,.} \tag{3.71}$$

Note that the same formula is true if (as is usually done) we interpret $N(\mathbf{n})$ as the number of projectiles scattered in the differential solid angle $\mathrm{d}\Omega(\mathbf{n})$ *per unit time* and ν as the number of projectiles that cross P_∞ per unit area per unit time (the so-called projectile *flux*). Since $\mathrm{d}N(\mathbf{n})$ is measurable in a scattering experiment, the previous equation allows one to *experimentally* determine $\sigma(\mathbf{n})$. On the other hand, given a potential $V(\mathbf{r})$ one can in principle evaluate the RHS of Eq. (3.69), thus obtaining a *theoretical* value for the differential scattering cross section $\sigma(\mathbf{n})$. The comparison of this theoretical value with the actual result from a scattering experiment makes it possible to check experimentally whether the interaction potential between the projectiles and the target is equal to $V(\mathbf{r})$.

Since $\mathrm{d}A = b\,\mathrm{d}b\,\mathrm{d}\beta$ and $\mathrm{d}\Omega = \sin\psi\,\mathrm{d}\psi\,\mathrm{d}\phi$, Eq. (3.69) can be written in the form

$$\boxed{b|\,\mathrm{d}b|\,\mathrm{d}\beta = \sigma(\psi,\phi)\sin\psi|\,\mathrm{d}\psi|\,\mathrm{d}\phi\,,}$$

where the absolute value is used to take into account that $\mathrm{d}b/\mathrm{d}\psi$ is *negative* if (as is usually the case) the potential is *repulsive*. Moreover, when the interaction potential is *central*, the azimuthal angles ϕ and β are *equal*. Indeed, the motion takes place in the plane determined by the initial velocity $\mathbf{u}_1$ and the projectile's initial position vector $\mathbf{r}_1 = \overrightarrow{O'P}$ on the plane P_∞ (cf. Fig. 3.13), and hence the azimuthal angle remains constant. Moreover $\mathrm{d}A$, and therefore σ, only depends on ψ (by the symmetry under rotations about the initial velocity $\mathbf{u}_1$). Thus for a central interaction potential the previous equation reduces to

$$b|\,\mathrm{d}b| = \sigma(\psi)\sin\psi\,|\,\mathrm{d}\psi| \quad\Longrightarrow\quad \boxed{\sigma(\psi) = \frac{b}{\sin\psi}\left|\frac{\mathrm{d}b}{\mathrm{d}\psi}\right|\,.} \tag{3.72}$$

Moreover, since σ is independent of ϕ Eq. (3.70) can be written as

$$\mathrm{d}N(\psi,\phi) = \nu\sigma(\psi)\sin\psi\,\mathrm{d}\psi\,\mathrm{d}\phi\,.$$

Integrating over ϕ between 0 and 2π we deduce that *the number of particles scattered with deflection angles in the interval* $(\psi, \psi + d\psi)$ *and at any angle* $\phi \in [0, 2\pi)$ *is given by*

$$\boxed{2\pi \nu \sin\psi \, \sigma(\psi) \, d\psi \, .}$$

In general, the **total scattering cross section** is defined as

$$\boxed{\sigma_{\text{tot}} = \int_{S^2} \sigma(\mathbf{n}) \, d\Omega = \int_0^{\pi} d\psi \, \sin\psi \int_0^{2\pi} d\phi \, \sigma(\psi, \phi) \, ,}$$

where the integral is extended to the surface of the unit sphere S^2 (that is, to all possible directions of the final velocity of the scattered particles). If $\sigma(\mathbf{n})$ does not depend on the azimuthal angle ϕ (as is the case, for example, for a central interaction potential), the previous expression reduces to

$$\boxed{\sigma_{\text{tot}} = 2\pi \int_0^{\pi} \sigma(\psi) \sin\psi \, d\psi} \, .$$

From the definition (3.69) of scattering cross section it follows that

$$\sigma_{\text{tot}} = \int_{S^2} dA(\mathbf{n})$$

is the total area in the plane P_∞ perpendicular to the direction of the initial velocity $\mathbf{u}_1$ crossed by projectiles that are scattered in *any* direction. In other words, the total number of particles scattered by the target is equal to $\nu\sigma_{\text{tot}}$. For interaction potentials with infinite range it is clear that σ_{tot} is infinite, since every particle is scattered to some extent.

Exercise 3.17. Express the total scattering cross section for *backward scattering* σ_b in terms of the differential scattering cross section.

Solution. By definition, a particle is scattered backward if the x component of its final velocity $\mathbf{v}_1$ is negative, i.e., if the deflection angle ψ is between $\pi/2$ and π. Thus σ_b is obtained by integrating $dA(\mathbf{n}) = \sigma(\psi, \phi) \sin\psi \, d\psi \, d\phi$ over the range $\phi \in [0, 2\pi]$ and $\psi \in [\pi/2, \pi]$:

$$\sigma_b = \int_{\pi/2}^{\pi} d\psi \, \sin\psi \int_0^{2\pi} d\phi \, \sigma(\psi, \phi) \, .$$

If the potential is central σ is independent of ϕ, and the previous formula simplifies to

$$\sigma_b = 2\pi \int_{\pi/2}^{\pi} \sin\psi \, \sigma(\psi) \, d\psi \, .$$

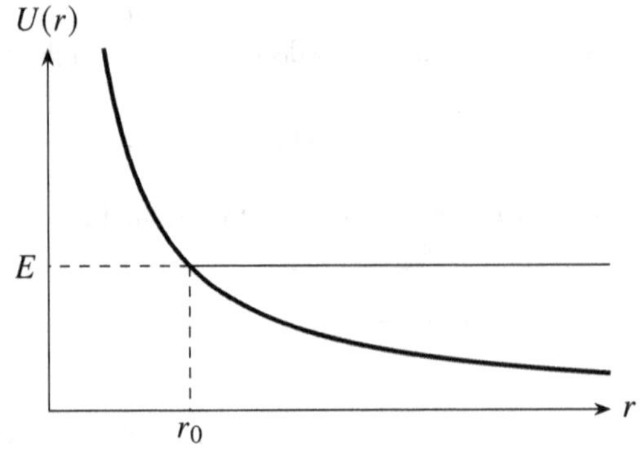

Figure 3.14. Effective potential for a repulsive central force field with $\lim_{r\to\infty} V(r) = 0$.

3.4.2 Scattering by a central potential

We shall study in this subsection the scattering of a particle of mass m by a *repulsive* central force field with potential $V(r)$ (cf. Fig. 3.15). By the previous discussion, this problem is equivalent to the scattering of a particle of mass m by a target of mass $m_2 = \infty$ located at the origin (center of force), the force $\mathbf{F}_{12}$ being equal to

$$\mathbf{F}_{12} = -\frac{\partial V(|\mathbf{r}_1 - \mathbf{r}_2|)}{\partial \mathbf{r}_1}.$$

Note that $V'(r) \leqslant 0$ for all $r \geqslant 0$, since we are assuming that the force is repulsive and thus $f(r) = -V'(r) \geqslant 0$. Thus V is monotonically decreasing, and since $\lim_{r\to\infty} V(r) = 0$ we must have $V(r) > 0$ for all $r \geqslant 0$. It follows that the effective potential $U(r)$ is also positive and monotonically decreasing (since $U'(r) = -L^2/(mr^3) + V'(r) < 0$), with $\lim_{r\to 0} U(r) = \infty$ (since $U(r) \geqslant L^2/(2mr^2)$) and $\lim_{r\to\infty} U(r) = 0$. Hence the energy must be positive, and for all $E > 0$ the effective potential U has a single turning point (i.e., the equation $U(r) = 0$ has a unique positive root; cf. Fig. 3.14).

Let us start by relating the deflection angle ψ to the angle χ between the line $\overrightarrow{OA}$ joining the origin with the periapsis and the outgoing (i.e., for $t \to \infty$) asymptote (cf. Fig. 3.15). To this end, note that the angles χ and χ' in this figure are equal, since the orbit is *symmetric* about the line joining the center of force with the periapsis A. Hence the deflection angle is given by

$$\boxed{\psi = \pi - 2\chi\,.} \tag{3.73}$$

On the other hand, the angle χ is the difference between the angular coordinates of a point in the orbit with $x \to -\infty$ ($\varphi = \pi$) and the periapsis A ($\varphi = \pi - \chi$). If we denote by r_0 the distance of the periapsis A to the origin, from Eq. (3.21) (with the "$-$" sign, since in this case $s = 1/r$ decreases when the angle φ increases from $\pi - \chi$ to π) it

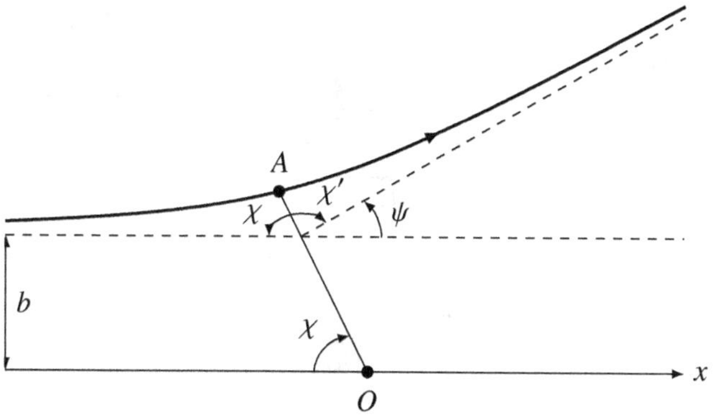

Figure 3.15. Scattering of a particle by a repulsive central potential (O is the origin of coordinates and A is the periapsis).

follows that[7]

$$\chi = \varphi(\infty) - \varphi(r_0) = -\int_{1/r_0}^{0} \frac{\mathrm{d}s}{\sqrt{\frac{2m}{L^2}\big(E - V(1/s)\big) - s^2}}$$
$$= \int_{0}^{1/r_0} \frac{\mathrm{d}s}{\sqrt{\frac{2m}{L^2}\big(E - V(1/s)\big) - s^2}}\,.$$

Note that the periapsis is a turning point, and thus r_0 is the (unique) positive root of the equation

$$\frac{2m}{L^2}\big(E - V(r)\big) - \frac{1}{r^2} = 0\,.$$

On the other hand, from Eq. (3.68) and the law of conservation of energy we obtain

$$\boxed{E = \frac{1}{2}mu^2\,,} \tag{3.74}$$

while the conservation of angular momentum yields

$$\mathbf{L} = (b\mathbf{e}_2) \times (mu\mathbf{e}_1) \quad \Longrightarrow \quad \boxed{L = mub = \sqrt{2mE}\,b\,.} \tag{3.75}$$

[7]Most textbooks adopt the convention that the projectile moves from left to right in the laboratory frame, as we have done in these notes (cf. Fig. (3.15)). It should be noted that this convention is the opposite of the one adopted in Section 3.2.2 (cf. Eq. (3.12)), since now L_z and $\dot\varphi$ are both negative. All the formulas obtained in the previous sections remain valid, however, if we simply change L (which with the previous convention was equal to L_z) by $-L$. As a matter of fact, since the equations we shall use in this section depend only on L^2, no modifications whatsoever will be necessary.

Performing the change of variable $x = bs$ in the equation for χ we obtain the more compact expression

$$\boxed{\chi = \int_0^{x_0} \frac{dx}{\sqrt{1 - x^2 - \frac{V(b/x)}{E}}}}\,, \tag{3.76}$$

where $x_0 = b/r_0$ is the positive root of the equation

$$\boxed{1 - x^2 - \frac{V(b/x)}{E} = 0}\,.$$

In other words, the integral in Eq (3.76) is extended to all values of x such that

$$1 - x^2 - \frac{V(b/x)}{E} \geqslant 0\,.$$

Remark 3.4. For an *attractive* central potential $V(r)$, the deflection angle is given by $\psi = 2\chi - \pi$, where χ is again computed using Eq. (3.76). Indeed, from Fig. 3.16 it follows that in this case

$$\boxed{\psi = \pi - 2\theta = \pi - 2(\pi - \chi) = 2\chi - \pi}\,. \tag{3.77}$$

It is important to note that when the potential is attractive the value of ψ computed in this way can be greater than π, since the orbits can go around the origin one or more times. This is the case, for example, for the potential $-k/(2r^2)$ with $k > 0$ (for certain values of the impact parameter). When this occurs, the computation of the differential scattering cross section becomes considerably more involved.

More precisely, suppose that the particle can be deflected by arbitrarily large angles $\theta \geqslant 0$ (as is the case, for instance, for the attractive $1/r^2$ potential). Given a deflection angle $\psi \in [0, \pi]$, in order to compute $\sigma(\psi)$ we must add up the contributions

$$\sigma_0(\alpha) = \frac{b}{|\sin\alpha|}\left|\frac{db}{d\alpha}\right|$$

to $\sigma(\psi)$ coming not only from $\alpha = \psi$, but also from all angles of the form $\alpha = \psi + 2n\pi$ and $\alpha = 2n\pi - \psi$ with $n \in \mathbb{N}$, since these angles define the same scattering direction as ψ. In other words, in this case we have

$$\sigma(\psi) = \sum_{n=0}^{\infty} \sigma_0(2n\pi + \psi) + \sum_{n=1}^{\infty} \sigma_0(2n\pi - \psi)]\,. \tag{3.78}$$

Note, however, that for the attractive Kepler potential $-k/r$ (with $k > 0$) the deflection angle is always between 0 and π, since the orbits of positive energy are in this case branches of hyperbolas. Hence in this important case we can just use the simple formula (3.72). ■

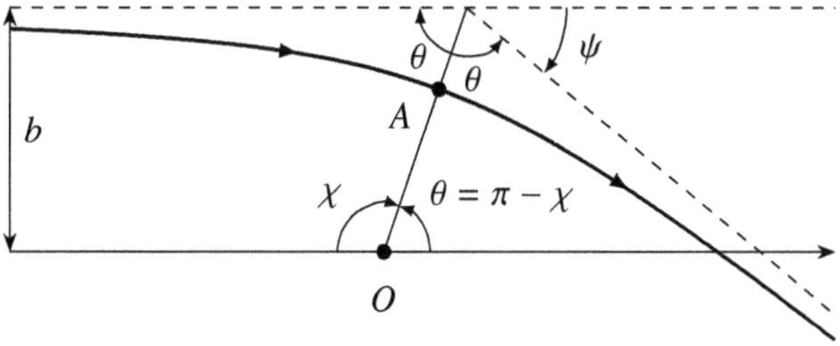

Figure 3.16. Scattering by an attractive central potential (O is the origin of coordinates and A is a periapsis).

Exercise 3.18. Find the minimum distance from the origin that can be reached by particles with a fixed energy E scattered by a *repulsive* central potential $V(r)$.

Solution. The minimum distance from the origin r_0 reached by a particle with a given energy $E = mu^2/2$ and impact parameter b is the coordinate of the only turning point of its orbit, determined by the equation

$$U(r_0) = E \iff V(r_0) + \frac{L^2}{2mr_0^2} = V(r_0) + \frac{mu^2b^2}{2r_0^2} = \frac{1}{2}mu^2 .$$

If the potential is repulsive, it is intuitively clear that the minimum value of r_0 is attained when the impact parameter vanishes (i.e., for a head-on collision). This is also easily checked using the previous formula, since if d is the value of r_0 corresponding to $b = 0$ we have

$$V(d) = \frac{1}{2}mu^2 > V(r_0) \implies d < r_0 ,$$

since V is decreasing (repulsive). Therefore the minimum distance d from the origin for a repulsive central potential $V(r)$ is determined by the equation

$$\boxed{V(d) = \frac{1}{2}mu^2} . \tag{3.79}$$

This distance depends only on u—that is, on the energy of the scattered beam—and, again due to the repulsive nature of the potential, it decreases as this energy increases.

Example 3.7. Let us find the differential scattering cross section by the central force field $f(r) = k/r^3$, with $k > 0$.

In this case $V(r) = k/(2r^2)$, and hence the angle χ is given by

$$\chi = \int_0^{x_0} \frac{dx}{\sqrt{1 - \gamma^2 x^2}} ,$$

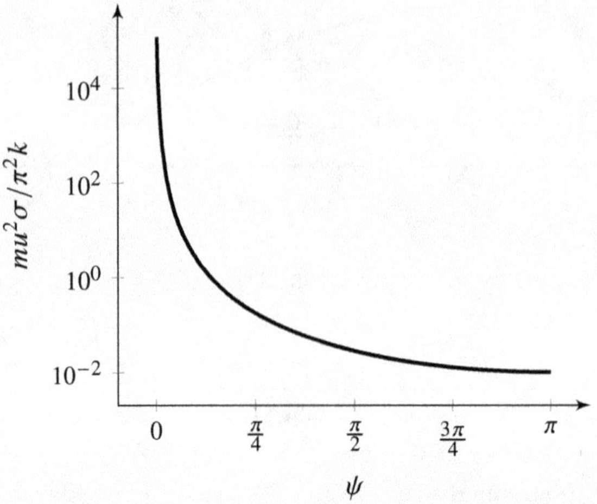

Figure 3.17. Differential scattering cross section by the central potential $V(r) = k/(2r^2)$ with $k > 0$ (note that the vertical axis scale is logarithmic).

with

$$\gamma = \sqrt{1 + \frac{k}{2b^2E}} = \sqrt{1 + \frac{k}{mu^2b^2}}, \qquad x_0 = 1/\gamma.$$

Performing the change of variable $\gamma x = \xi$ we obtain

$$\chi = \frac{1}{\gamma}\int_0^1 \frac{\mathrm{d}\xi}{\sqrt{1-\xi^2}} = \frac{\pi}{2\gamma} \implies \psi = \pi - 2\chi = \pi\left(1 - \frac{1}{\gamma}\right). \tag{3.80}$$

The differential scattering cross section $\sigma(\psi)$ is easily computed from Eq. (3.72). Indeed, from the previous equation for ψ we obtain

$$\gamma^2 = \frac{\pi^2}{(\pi-\psi)^2} = 1 + \frac{k}{mu^2b^2} \implies b^2 = \frac{k}{mu^2}\,\frac{(\pi-\psi)^2}{\psi(2\pi-\psi)}, \tag{3.81}$$

and thus

$$\begin{aligned}\sigma(\psi) &= \frac{1}{2}\csc\psi\left|\frac{\mathrm{d}(b^2)}{\mathrm{d}\psi}\right| = \frac{k}{2mu^2}\csc\psi\left|\frac{-2\psi(2\pi-\psi)(\pi-\psi) - 2(\pi-\psi)^3}{\psi^2(2\pi-\psi)^2}\right| \\ &= \boxed{\frac{\pi^2 k}{mu^2}\,\frac{(\pi-\psi)\csc\psi}{\psi^2(2\pi-\psi)^2}}\end{aligned} \tag{3.82}$$

(cf. Fig. 3.17). In this case the minimum distance to the origin is determined by the equation

$$\frac{1}{2}mu^2 = V(d) = \frac{k}{2d^2} \implies d = \sqrt{\frac{k}{mu^2}},$$

in terms of which

$$\boxed{\sigma(\psi) = (\pi d)^2 \frac{(\pi - \psi)\csc\psi}{\psi^2(2\pi - \psi)^2}}\,.$$

Note that for $\psi \to 0$ the differential scattering cross section $\sigma(\psi)$ diverges as ψ^{-3}, since

$$\sigma(\psi) \underset{\psi\to 0}{\simeq} \frac{\pi d^2}{4\psi^3}\,,$$

while for $\psi \to \pi$ we have

$$\sigma(\psi) \underset{\psi\to\pi}{\longrightarrow} \frac{d^2}{\pi^2}\,.$$

Note. In general, *in a repulsive central potential of infinite range* $\sigma(\psi)$ *should diverge as* $1/\psi^2$ *or faster as* $\psi \to 0$. Indeed, let $\psi(b)$ be the deflection angle corresponding to an impact parameter b. If the potential $V(r)$ is repulsive and of infinite range, $\psi(b)$ is a decreasing function of π satisfying $\psi(0) = \pi$ and $\lim_{b\to\infty}\psi(b) = 0$ (cf. Fig. 3.18). If $\varepsilon \in (0, \pi)$ is any fixed angle, particles starting from P_∞ with an impact parameter greater than $b_\varepsilon = \psi^{-1}(\varepsilon)$ will be scattered with a deflection angle less than ε. Therefore the area of the region of P_∞ from which these projectiles start—i.e., the exterior of the circle centered at the origin with radius b_ε—is infinite. By definition of σ, this area is equal to the integral

$$2\pi \int_0^\varepsilon \sigma(\psi)\sin\psi\; d\psi\,.$$

Thus the previous integral must diverge at its lower limit, and hence ψ must tend to infinity as $1/\psi^\alpha$ with $\alpha \geqslant 2$ as ψ tends to 0 (recall that $\int_0^\varepsilon \psi^{-\alpha}\, d\psi$ diverges for $\alpha \geqslant 1$). ■

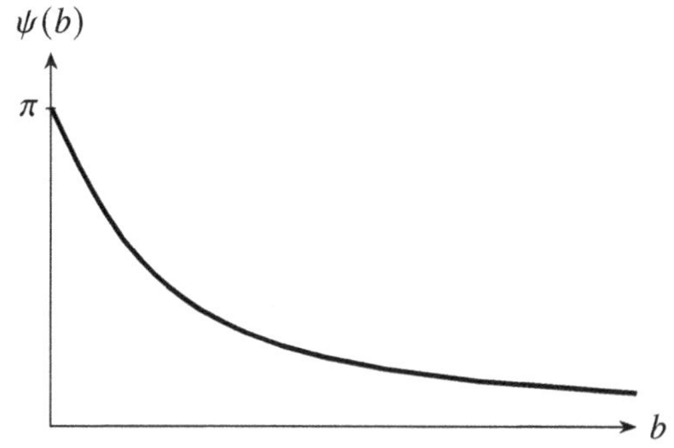

Figure 3.18. Deflection angle ψ as a function of the impact parameter b for a central repulsive potential of infinite range.

Exercise 3.19. Find $\sigma(\psi)$ for the potential in the previous example using the equation of its orbits (cf. Example 3.1).

Solution. If $k > 0$, the equation of the orbits of the potential $V = k/(2r^2)$ is given by (3.29), namely (calling $A = r_0$ and taking $\varphi_0 = 0$ without loss of generality)

$$r = r_0 \sec(\gamma\varphi), \qquad \text{with} \quad \gamma = \sqrt{1 + \frac{km}{L^2}} = \sqrt{1 + \frac{k}{mu^2b^2}} > 1.$$

Since the angle φ ranges between $-\pi/(2\gamma)$ and $\pi/(2\gamma)$, this curve has two asymptotes each of which makes an angle of $\pi/(2\gamma) < \pi/2$ with the positive x axis (cf. Fig. 3.19). On the other hand, from Fig. 3.15 it follows that the angle formed by the asymptotes of the trajectory in a repulsive central potential is 2χ. Therefore in this case $\chi = \pi/(2\gamma)$, which is precisely equation (3.80).

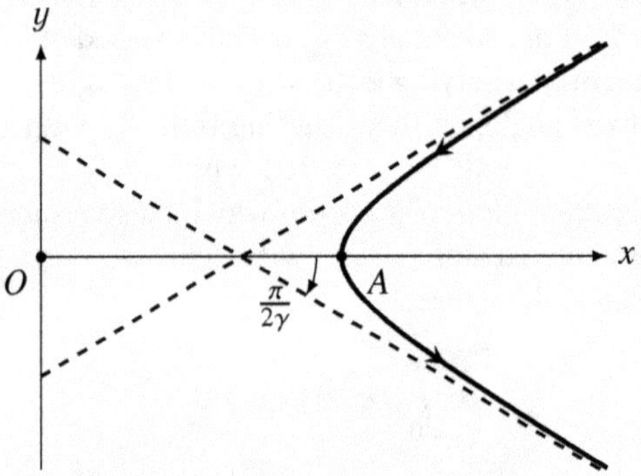

Figure 3.19. Orbit of equation $r = r_0 \sec(\gamma\varphi)$ for the potential $V(r) = k/(2r^2)$ with $k > 0$ and its asymptotes (dashed lines).

Exercise 3.20. Find the scattering cross section $\sigma(\psi)$ for the *attractive* potential $V(r) = -k/(2r^2)$ with $k > 0$.

Solution. From Example 3.1 it follows that the given potential has scattering orbits (i.e., orbits for which $r \to \infty$ for large negative or positive times) only if the constant C is positive. Replacing k by $-k$ in the formula for C in the previous example we obtain

$$C = 1 - \frac{km}{L^2} = 1 - \frac{k}{mu^2b^2} > 0 \iff b > \sqrt{\frac{k}{mu^2}}.$$

Choosing the x axis appropriately, the orbits satisfying this condition have equation

$$r = r_0 \sec(\gamma\varphi),$$

where $r_0 > 0$ and $-\pi/(2\gamma) < \varphi < \pi/(2\gamma)$ with

$$\gamma := \sqrt{C} = \sqrt{1 - \frac{k}{mu^2b^2}} \in (0, 1).$$

Since $\chi = \pi/(2\gamma)$ by the previous example, from Eq. (3.77) we obtain

$$\psi = 2\chi - \pi = \pi\left(\frac{1}{\gamma} - 1\right).$$

In this case, however, γ can be arbitrarily small, and thus ψ arbitrarily large. Hence we must use Eq. (3.78) to evaluate the scattering cross section $\sigma(\psi)$. To this end, we first compute σ_0 proceeding as in Example 3.7, with the result

$$\sigma_0(\theta) = \frac{\pi^2 k}{mu^2}\,|\csc\theta|\,\frac{\theta+\pi}{\theta^2(2\pi+\theta)^2}$$

(exercise). Noting that

$$\frac{\theta+\pi}{\theta^2(2\pi+\theta)^2} = \frac{1}{4\pi}\left[\frac{1}{\theta^2} - \frac{1}{(2\pi+\theta)^2}\right]$$

we easily obtain

$$\begin{aligned}\sum_{n=0}^{\infty}\sigma_0(2n\pi+\psi) &= \frac{\pi k}{4mu^2}\csc\psi\sum_{n=0}^{\infty}\left[\frac{1}{(2n\pi+\psi)^2} - \frac{1}{(2(n+1)\pi+\psi)^2}\right]\\ &= \frac{\pi k}{4mu^2}\,\frac{\csc\psi}{\psi^2}.\end{aligned}$$

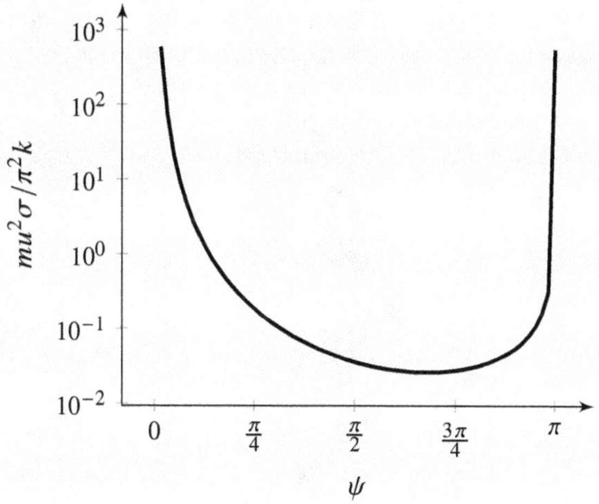

Figure 3.20. Scattering cross section for the attractive potential $V(r) = -k/(2r^2)$ with $k > 0$ (note that the vertical axis scale is logarithmic).

Likewise,

$$\sum_{n=1}^{\infty} \sigma_0(2n\pi - \psi) = \frac{\pi k}{4mu^2} \csc\psi \sum_{n=1}^{\infty} \left[\frac{1}{(2n\pi - \psi)^2} - \frac{1}{(2(n+1)\pi - \psi)^2} \right]$$
$$= \frac{\pi k}{4mu^2} \frac{\csc\psi}{(2\pi - \psi)^2},$$

and therefore

$$\sigma(\psi) = \frac{\pi k}{4mu^2} \csc\psi \left[\frac{1}{\psi^2} + \frac{1}{(2\pi - \psi)^2} \right] = \frac{\pi k}{2mu^2} \csc\psi \, \frac{\psi^2 + 2\pi(\pi - \psi)}{\psi^2(2\pi - \psi)^2},$$

which differs from the cross section of the repulsive $k/(2r^2)$ potential (with $k > 0$) in Eq. (3.82) by the positive quantity

$$\frac{\pi k}{2mu^2} \frac{\csc\psi}{(2\pi - \psi)^2}.$$

Note also that in this case $\sigma(\psi)$ diverges not only as $\psi \to 0$ but also as $\psi \to \pi$ (cf. Fig. 3.20). In fact, when $\psi \to \pi$ we have

$$\sigma(\psi) \simeq \frac{k}{2\pi mu^2} (\pi - \psi)^{-1}.$$

Example 3.8. Let us compute the differential scattering cross section of a particle of mass m by an impenetrable sphere of radius a, whose center is fixed at the origin.

Obviously, if $b > a$ the particles are not scattered, so we shall assume that $b \leqslant a$. In this case the potential is formally given by

$$V(r) = \begin{cases} \infty, & r < a \\ 0, & r > a, \end{cases}$$

and therefore

$$V(b/x) = \begin{cases} \infty, & x > b/a \\ 0, & x < b/a. \end{cases}$$

Since $b \leqslant a$, taking into account that the minimum distance from the origin is obviously $r_0 = a$ and that $V(b/x) = 0$ for $0 \leqslant x < b/a$, Eq. (3.76) yields

$$\chi = \int_0^{b/a} \frac{dx}{\sqrt{1 - x^2 - \frac{V(b/x)}{E}}} = \int_0^{b/a} \frac{dx}{\sqrt{1 - x^2}} = \arcsin(b/a)$$
$$\Longrightarrow \quad b = a \sin\chi = a \sin\left(\frac{\pi}{2} - \frac{\psi}{2}\right) = a\cos(\psi/2).$$

From Eq. (3.72) it then follows that

$$\sigma(\psi) = \frac{a\cos(\psi/2)}{\sin\psi} \cdot \frac{a}{2}\sin(\psi/2) = \frac{a^2}{4},$$

independently of ψ. Thus in this case the total scattering cross section is given by

$$\sigma_{\text{tot}} = 2\pi\,\frac{a^2}{4}\int_0^\pi \sin\psi\; d\psi = \pi a^2,$$

in agreement with the fact that particles traveling at a distance greater than a from the x axis are not scattered.

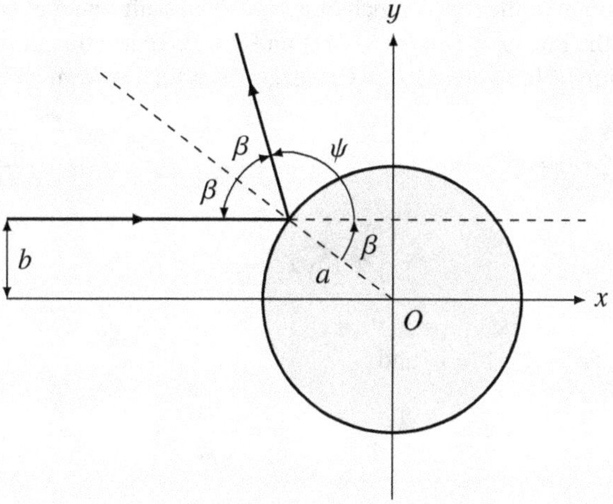

Figure 3.21. Scattering of a particle by an impenetrable sphere of infinite mass.

Alternative method: From Fig. 3.21 it follows that[a]

$$\psi = \pi - 2\beta, \qquad b = a\sin\beta = a\sin\left(\frac{\pi}{2} - \frac{\psi}{2}\right) = a\cos(\psi/2),$$

which is the equation for b as a function of the deflection angle ψ found above.

[a]Indeed, since the potential is central the force exerted by the sphere when the particle collides with it is normal to the surface of the sphere at the impact point. Therefore the only component of the projectile's velocity $\mathbf{v}$ that can change after the collision is the one normal to the sphere. In addition, by the law of conservation of energy $\mathbf{v}^2 = v_\perp^2 + v_\parallel^2$, and hence $v_\perp^2$, is conserved. It follows that $v_\perp$ changes sign after the collision, which is equivalent to the equality of the angles that the projectiles' velocity makes with the normal to the sphere before and after the collision.

3.4.3 Rutherford's formula

We shall next consider the scattering of a particle of mass m by a *repulsive* Kepler potential $V(r) = k/r$, with $k > 0$. This is also the interaction potential between two point charges q_1 and q_2 of the same sign, if $k = q_1 q_2/(4\pi\varepsilon_0)$. The particle's trajectory can be simply obtained changing k by $-k$ in Eq. (3.50), namely (cf. Exercise 3.15)

$$r = \frac{-\alpha}{1 + e\cos(\varphi - \varphi_0)}\,; \qquad \alpha = \frac{L^2}{km}\,, \quad e = \sqrt{1 + \frac{2EL^2}{mk^2}} > 1\,,$$

where the energy and angular momentum are given in terms of m and b by Eqs. (3.74) and (3.75). Taking, without loss of generality, $\varphi_0 = \pi$ we obtain the simpler expression

$$\boxed{r = \frac{\alpha}{e\cos\varphi - 1}\,,} \tag{3.83}$$

which is the equation of the right branch of a *hyperbola* with center at $(e\alpha/(e^2-1), 0)$ whose axes are the lines $x = (e\alpha/(e^2-1))$ and $y = 0$. To see this, it is convenient to recall the equation of the trajectory in Cartesian coordinates from Exercise 3.15:

$$x^2 + y^2 = e^2x^2 - 2\alpha e x + \alpha^2 \implies (e^2-1)\left(x - \frac{\alpha e}{e^2-1}\right)^2 - y^2 = -\alpha^2 + \frac{\alpha^2 e^2}{e^2-1} = \frac{\alpha^2}{e^2-1}\,.$$

This can be written as

$$\frac{X^2}{a^2} - \frac{Y^2}{b^2} = 1\,, \tag{3.84}$$

with $X = x - \alpha/(e^2-1)$, $Y = y$, and

$$a = \frac{\alpha}{e^2-1}\,, \qquad b = \frac{\alpha}{\sqrt{e^2-1}}\,.$$

Equation (3.84) is the equation of a hyperbola having as axes the lines $X = 0$ and $Y = 0$, centered at the point with coordinates $X = Y = 0$ where its axes intersect. In the original (x, y) coordinates the center is thus the point $C = (\alpha e/(e^2-1), 0)$ on the positive x axis. The curve's asymptotes are the lines $Y = \pm(b/a)\,X$, or

$$y = \pm\frac{b}{a}\left(x - \frac{\alpha e}{e^2-1}\right),$$

which meet at the center C. Moreover, the hyperbola's focal distance c is given by

$$c^2 = a^2 + b^2 = \frac{\alpha^2}{(e^2-1)^2}(1 + e^2 - 1) = \frac{\alpha^2 e^2}{(e^2-1)^2} \implies c = \frac{\alpha e}{e^2-1} = ea\,,$$

so that $e = c/a$ is again the eccentricity. The foci are the two points on the x axis at a distance c from the center $(\alpha e/(e^2-1), 0) = (c, 0)$, i.e., the origin and the point $F = (2c, 0)$. Finally, by Eq. (3.83) we have $ex = \alpha + r > 0$, and hence the trajectory coincides with the *right* branch of the hyperbola.

Since $r \geqslant 0$, the angle φ in Eq. (3.83) must range between $-\arccos(1/e)$ and $\arccos(1/e)$. Thus the asymptotes of the hyperbola make an angle $\arccos(1/e)$ with the

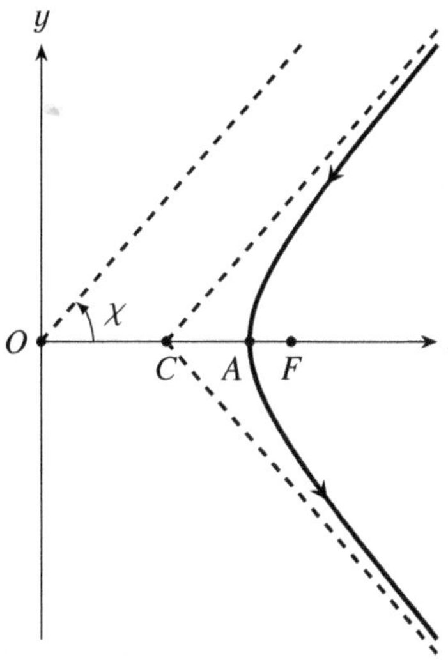

Figure 3.22. Trajectory of a particle in the repulsive Kepler potential $V(r) = k/r$, with $k > 0$. The point $C = (\alpha e/(e^2 - 1), 0)$ is the center of the hyperbola, while O and $F = (2\alpha e/(e^2 - 1), 0)$ are its foci. The focal distance is thus $c = |\overrightarrow{OC}| = |\overrightarrow{CF}| = e\alpha/(e^2 - 1) = ea$, where $a = |\overrightarrow{CA}|$ and e is the eccentricity. Geometrically, the hyperbola is the locus of points P in the (x, y) plane defined by the condition $|\overrightarrow{OP}| - |\overrightarrow{PF}| = 2a$, but the trajectory includes only the hyperbola's right branch shown in the figure.

x axis, which in this case is the line joining the origin with the periapsis $\big(\alpha/(e-1), 0\big)$, whose polar angle is $\varphi = 0$ (cf. Fig. 3.22). Therefore

$$\boxed{\chi = \arccos(1/e)} \quad \Longrightarrow \quad \frac{1}{e} = \cos\chi = \cos\left(\frac{\pi}{2} - \frac{\psi}{2}\right) = \sin(\psi/2)\,,$$

whence it follows that

$$e^2 - 1 = \csc^2(\psi/2) - 1 = \cot^2(\psi/2) = \frac{2EL^2}{mk^2}\,.$$

From Eqs. (3.74) and (3.75) we obtain

$$\frac{2EL^2}{mk^2} = \frac{m^2u^4b^2}{k^2} \quad \Longrightarrow \quad b^2 = \frac{k^2}{m^2u^4}\cot^2(\psi/2)\,.$$

Taking into account that $\cot(\psi/2) \geqslant 0$ (since $0 \leqslant \psi \leqslant \pi$), from the previous expression we finally conclude that

$$\boxed{b = \frac{k}{mu^2}\cot(\psi/2)}\,. \tag{3.85}$$

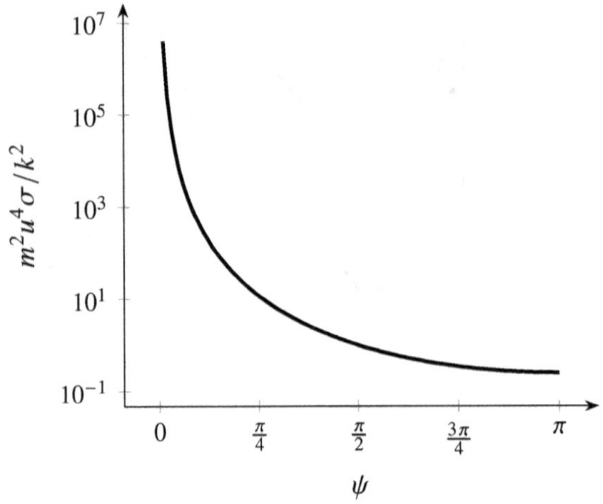

Figure 3.23. Differential scattering cross section by the potential $V(r) = k/r$, with $k > 0$ (note that the vertical axis scale is logarithmic).

Substituting this expression for $b(\psi)$ into Eq. (3.72) we finally obtain

$$\sigma(\psi) = \frac{k^2}{m^2u^4}\,\frac{\cot(\psi/2)}{\sin\psi}\cdot\frac{1}{2}\csc^2(\psi/2) = \boxed{\frac{k^2}{4m^2u^4}\csc^4(\psi/2)\,,} \tag{3.86}$$

which is **Rutherford's** celebrated **formula** (cf. Fig. 3.23). The minimum distance to the origin reached by particles scattered by the repulsive Kepler potential is given by Eq. (3.79), i.e.,

$$\frac{k}{d} = \frac{1}{2}mu^2 \quad\Longrightarrow\quad d = \frac{2k}{mu^2}\,.$$

In terms of this minimum distance, the differential scattering cross section is expressed as

$$\boxed{\sigma(\psi) = \frac{d^2}{16}\csc^4(\psi/2)\,.}$$

The asymptotic behavior of $\sigma(\psi)$ is in this case

$$\sigma(\psi) \underset{\psi\to 0}{\simeq} \frac{d^2}{\psi^4}\,, \qquad \sigma(\psi) \underset{\psi\to\pi}{\longrightarrow} \frac{d^2}{16}\,.$$

Exercise 3.21. Derive Eq. (3.85) using Eqs. (3.73) and (3.76).

Solution. In this case

$$1 - x^2 - \frac{V(b/x)}{E} = 1 - \frac{kx}{bE} - x^2 = 1 - \frac{2kx}{mu^2b} - x^2 = 1 + \frac{k^2}{m^2u^4b^2} - \left(x + \frac{k}{mu^2b}\right)^2,$$

so that it is convenient to perform the change of variable

$$x + \frac{k}{mu^2 b} = \sqrt{1 + \frac{k^2}{m^2 u^4 b^2}}\ \xi$$

in the integral (3.76). We then have

$$1 - x^2 - \frac{V(b/x)}{E} = \left(1 + \frac{k^2}{m^2 u^4 b^2}\right)(1 - \xi^2),$$

and thus

$$\chi = \int_{\xi_0}^{1} \frac{d\xi}{\sqrt{1 - \xi^2}} = \frac{\pi}{2} - \arcsin \xi_0 ,$$

where ξ_0 is the value of ξ for $x = 0$:

$$\xi_0 = \frac{k/(mu^2 b)}{\sqrt{1 + \frac{k^2}{m^2 u^4 b^2}}} .$$

(Note that the upper limit of the integral, corresponding to the value $x = x_0$, is the positive root of the equation $1 - \xi^2 = 0$.) Using now Eq. (3.73) we obtain

$$\psi = 2 \arcsin \xi_0 \implies \xi_0 = \sin(\psi/2) \implies \frac{k^2}{m^2 u^4 b^2} = \left(1 + \frac{k^2}{m^2 u^4 b^2}\right) \sin^2(\psi/2)$$

$$\implies \frac{k^2}{m^2 u^4 b^2} = \tan^2(\psi/2) \implies b = \frac{k}{mu^2} \cot(\psi/2) ,$$

since $\psi \in [0, \pi]$.

Exercise 3.22. Prove that the differential scattering cross section for the *attractive* Kepler potential $V(r) = -k/r$, with $k > 0$, is also given by Rutherford's formula (3.86).

Solution. The positive energy orbits in the attractive Kepler potential are branches of hyperbolas (cf. Section 3.3.1 and Fig. 3.24), and thus they do not go around the origin. Hence we can apply Eqs. (3.72)–(3.76), taking into account that for an attractive potential the deflection angle ψ is equal to $2\chi - \pi$. From the equation of the orbits (3.50) it follows that in this case $\chi = \arccos(-1/e)$, and therefore

$$-\frac{1}{e} = \cos \chi = \cos\left(\frac{\pi}{2} + \frac{\psi}{2}\right) = -\sin(\psi/2) \implies \frac{1}{e} = \sin(\psi/2) ,$$

as in the repulsive case. The rest of the calculation is identical, since the formula for the eccentricity depends on k^2, and thus

$$\cot^2(\psi/2) = e^2 - 1 = \frac{m^2u^4b^2}{k^2} \quad \Longrightarrow \quad b = \frac{k}{mu^2}\cot(\psi/2)\,,$$

which is precisely Eq. (3.85).

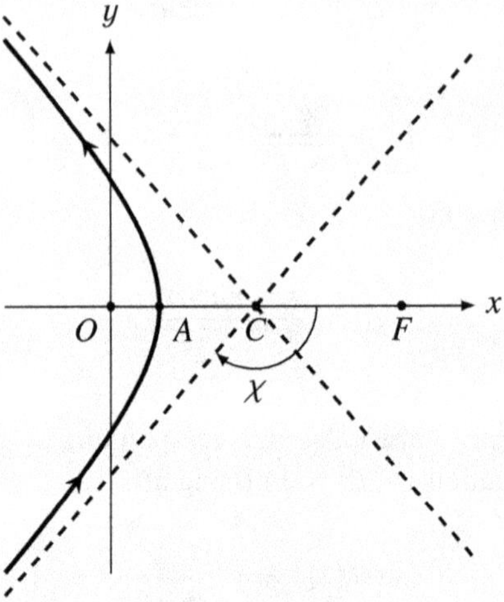

Figure 3.24. Trajectory of a particle in the attractive Kepler potential $V(r) = -k/r$ with $k > 0$ (left branch of a hyperbola with O and F as foci and C as center, where $|\overrightarrow{OC}| = |\overrightarrow{CF}| = c = e\alpha/(e^2 - 1)$).

Example 3.9. Let us compute the total cross section σ_b for *backward* scattering of a particle by the potential $V(r) = k/r$ with $k > 0$.

By definition, σ_b is the area of the region of the plane P_∞ crossed by particles that are scattered backward, i.e., with a deflection angle ψ between $\pi/2$ and π. From Eq. (3.85) it follows that the impact parameter b of particles scattered backward ranges from 0 (corresponding to $\psi = \pi$) to k/mu^2 (corresponding to $\psi = \pi/2$). In other words, the particles that are scattered backward are those crossing the plane P_∞ at a distance at most k/mu^2 from the x axis. Hence the requested cross section is equal to the area of a circle of radius k/mu^2, i.e.,

$$\boxed{\sigma_b = \frac{\pi k^2}{m^2u^4} = \frac{\pi d^2}{4}}\,.$$

Obviously, the same result can be obtained using Rutherford's formula:

$$\sigma_b = 2\pi \int_{\pi/2}^{\pi} \sin\psi \, \sigma(\psi) \, d\psi = \frac{\pi k^2}{m^2 u^4} \int_{\pi/2}^{\pi} \frac{\cos(\psi/2)}{\sin^3(\psi/2)} \, d\psi$$

$$= \frac{2\pi k^2}{m^2 u^4} \int_{\pi/4}^{\pi/2} \frac{\cos\alpha}{\sin^3\alpha} \, d\alpha = -\frac{\pi k^2}{m^2 u^4} \frac{1}{\sin^2\alpha}\bigg|_{\pi/4}^{\pi/2} = \frac{\pi k^2}{m^2 u^4} .$$

The previous result was of great importance in the history of physics. Indeed, at the beginning of the 20th century it was widely believed that the positive charge within an atom was evenly distributed (this is the so-called "plum pudding model" of J. J. Thomson). To test this hypothesis, Ernest Rutherford devised an experiment that consisted of directing a beam of α-particles (He^{++} nuclei) toward a very thin gold foil (to minimize multiple collisions). As the mass of the α-particles is much greater than that of the electrons—which, according to Thomson's model, would move in the uniform background of the positive charge of the atoms—these particles should barely be scattered. In 1909 Hans Geiger and Ernest Marsden performed this experiment under the direction of Ernest Rutherford, and to their surprise found that a small but non-zero number of α-particles were scattered even at angles close to 180°. Rutherford, Geiger and Marsden correctly concluded that this fact was due to individual collisions of the α-particles with the positive charges of the Au atoms, which, contrary to Thomson's hypothesis, should be concentrated in a very small region. In fact, the theoretical value found previously for σ_b makes it possible to find an upper bound of the radius a of the Au nucleus. Indeed, if ν is the density of α-particles in the beam per unit cross-sectional area, the number of particles scattered backward is given by

$$N_b = \nu\sigma_b = \frac{\pi\nu}{4} d^2 \quad \Longrightarrow \quad d = \sqrt{\frac{4N_b}{\pi\nu}} .$$

Since the distance d of maximum approach of the α-particles to the Au nucleus must be greater than or equal to its radius a, from the previous formula we deduce that

$$\boxed{a \leqslant \sqrt{\frac{4N_b}{\pi\nu}}} .$$

Exercise 3.23. Find the total cross section for backward scattering for the potential $V(r) = k/(2r^2)$, with $k > 0$.

Solution. In this potential, the relation between the deflection angle ψ and the impact parameter b is given by Eq. (3.81). Therefore a particle will be scattered at angles equal to or greater than $\pi/2$ if

$$b^2 \leqslant \frac{k}{mu^2} \frac{\pi^2/4}{\pi/2(2\pi - \pi/2)} = \frac{k}{3mu^2} .$$

Hence the total cross section for backward scattering is given by

$$\boxed{\sigma_b = \frac{\pi k}{3mu^2} = \frac{\pi d^2}{3} .}$$

CHAPTER 4

Lagrangian and Hamiltonian mechanics

In this chapter we develop the Lagrangian and Hamiltonian formalisms of classical mechanics. We begin with a brief introduction to the calculus of variations, deducing the Euler–Lagrange equations of a variational problem and studying its basic conservation laws. We next show that Newton's equations of motion of an unconstrained system are nothing but the Euler–Lagrange equations of its action functional. Thus the trajectories of a dynamical system can be viewed as the curves that maximize—or, more rigorously, render stationary—the action. This fundamental idea is the essence of Hamilton's principle, which we extend to systems with holonomic (i.e., velocity independent) constraints through the principle of virtual work. Using the Lagrangian formalism, we explore the connection between symmetries of the action and conservation laws, encapsulated in Noether's celebrated theorem. The chapter ends with a concise overview of the fundamentals of Hamiltonian mechanics. We derive Hamilton's canonical equations of motion from Lagrange's equations, introduce the classical Poisson bracket, and briefly discuss canonical transformations and the Hamilton–Jacobi equation.

4.1 INTRODUCTION TO THE CALCULUS OF VARIATIONS

4.1.1 Fundamental problem of the calculus of variations

The fundamental problem of the **calculus of variations** (in its simplest version) is that of finding the **extrema** (i.e., maxima or minima) of a function of the form

$$\boxed{F[y] = \int_{x_1}^{x_2} f\big(x, y(x), y'(x)\big)\,\mathrm{d}x} \qquad \left({}' = \frac{\mathrm{d}}{\mathrm{d}x}\right), \tag{4.1}$$

where $f : \mathbb{R}^3 \to \mathbb{R}$ is of class C^2 (i.e., twice continuously differentiable). The *domain* of F consists of the functions $y : [x_1, x_2] \to \mathbb{R}$ (also assumed to be of class C^2) that

DOI: 10.1201/9781003600633-4

satisfy the **boundary conditions**

$$\boxed{y(x_1) = y_1\,, \qquad y(x_2) = y_2,} \tag{4.2}$$

with $y_1, y_2 \in \mathbb{R}$ *fixed*. From the mathematical viewpoint,

$$F : C_0^2([x_1, x_2]) \to \mathbb{R}$$

is therefore a function whose domain is the space $C_0^2([x_1, x_2])$ of scalar functions $y : [x_1, x_2] \to \mathbb{R}$ of class C^2 in the interval $[x_1, x_2]$ satisfying the conditions (4.2). In other words, F assigns to each function $y : [x_1, x_2] \to \mathbb{R}$, which can be identified with its **graph**

$$\{(x, y(x)) : x_1 \leqslant x \leqslant x_2\} \subset \mathbb{R}^2\,,$$

the number given by the RHS of Eq. (4.1). A function like F, whose domain is a set of functions, is usually called a **functional**. The function f appearing in Eq. (4.1) is called the **density** of the corresponding functional F.

Many interesting problems in mathematics and physics can be reduced to finding the extrema of an appropriate functional of the form (4.1)–(4.2). We shall list below a few of the most prominent ones.

Example 4.1. What is the shortest curve joining two fixed points on a plane?

Let us denote the two fixed points by (x_i, y_i) $(i = 1, 2)$, with $x_1 \neq x_2$ (this can always be arranged by suitably choosing the axes). If we restrict ourselves, for the sake of simplicity, to plane curves that are graphs of functions $y : [x_1, x_2] \to \mathbb{R}$, the problem considered is equivalent to finding the minimum of the *length functional*

$$\boxed{F[y] = \int_{x_1}^{x_2} \sqrt{1 + y'(x)^2}\, \mathrm{d}x}$$

with the condition (4.2).

Example 4.2. *The brachistochrone problem.* A particle of mass m is forced to move in a vertical plane along a curve with fixed endpoints (x_1, y_1) and (x_2, y_2), with $x_1 < x_2$ and $y_1 > y_2$. For which curve is the time taken by the particle to travel between both endpoints a *minimum*?

If we neglect friction, the reaction force exerted by the curve on the particle is normal to the curve at each point. Hence it does no work, and consequently the particle's energy is conserved:

$$\frac{1}{2} m v^2 + m g y = E\,.$$

If the curve in question is the graph of a function $y(x)$, the differential of time along the curve is given by

$$\mathrm{d}t = \frac{\mathrm{d}s}{v} = \frac{\sqrt{1+y'(x)^2}}{\sqrt{2g\left(\frac{E}{mg} - y(x)\right)}}\,\mathrm{d}x\,.$$

Thus the problem proposed is equivalent to finding the minimum of the functional (proportional to the travel time along the curve)

$$\boxed{F[y] = \int_{x_1}^{x_2} \sqrt{\frac{1+y'(x)^2}{y_0 - y(x)}}\,\mathrm{d}x} \qquad \left(y_0 := \frac{E}{mg}\right) \tag{4.3}$$

with the condition (4.2).

Note that from energy conservation it follows that $y \leqslant y_0$, and that the energy (and hence y_0) depends only on the particle's initial velocity v_0:

$$E = \frac{1}{2}mv_0^2 + mgy_1 \quad \Longrightarrow \quad y_0 = y_1 + \frac{v_0^2}{2g}\,.$$

In particular, $y_0 = y_1$ if the particle is initially at rest.

Example 4.3. *Fermat's principle.* What is the trajectory followed by a light ray traveling from a point (x_1, y_1) to a second point (x_2, y_2) in a flat optical medium with refractive index $n(x, y)$?

According to **Fermat's principle** (in the approximation of geometric optics), the trajectory of the light ray joining the points (x_1, y_1) and (x_2, y_2) is the curve for which the time taken by light to cover the distance between both points is *minimum*. Suppose, again, that $x_1 \neq x_2$, and that the trajectory is the graph of a function $y(x)$. By definition of index of refraction, the speed of light at a point (x, y) in the medium is given by

$$v(x,y) = \frac{c}{n(x,y)}\,,$$

where c is the speed of light *in vacuo*. Since

$$\mathrm{d}t = \frac{\mathrm{d}s}{v(x,y)} = n(x,y)\,\frac{\mathrm{d}s}{c}\,,$$

the problem proposed is equivalent to determining the minimum of the functional (proportional to light's travel time)

$$\boxed{F[y] = \int_{x_1}^{x_2} n\big(x, y(x)\big)\sqrt{1+y'(x)^2}\,\mathrm{d}x\,,} \tag{4.4}$$

again with the condition (4.2).

The functional (4.4), which has dimensions of length, is called *optical length*. Note that if the refractive index is *constant* F is proportional to the length functional of the first example, and thus the path followed by light rays in this case is the shortest curve joining the points (x_1, y_1) and (x_2, y_2). Likewise, if the refractive index is proportional to $(y_0 - y)^{-1/2}$ the path followed by light is the brachistochrone in the previous example.

4.1.2 Euler–Lagrange equations

In order to solve the fundamental problem of the calculus of variations formulated in the previous subsection, we shall proceed in essentially the same way as in real analysis when determining the extrema of an ordinary function $F : \mathbb{R} \to \mathbb{R}$. The key idea is that in both cases the extrema are points (functions, in this case) for which the *variation* of the function F when we infinitesimally increase its argument vanishes at first order.

More precisely, suppose that $y(x)$ is an extremum (maximum or minimum) of the functional (4.1) with the condition (4.2). Let $\eta(x)$ be an arbitrary function (of class C^2) satisfying the conditions

$$\eta(x_1) = \eta(x_2) = 0\,, \tag{4.5}$$

so that for all $\varepsilon \in \mathbb{R}$ the function $y_\varepsilon := y + \varepsilon\eta$ satisfies the boundary conditions (4.2). The functions $y_\varepsilon(x)$ (with $\varepsilon \in \mathbb{R}$) form a **one-parameter family** containing the extremum $y(x)$ for $\varepsilon = 0$. More informally, if ε is small we can think of $y_\varepsilon(x)$ as a small *variation* of the extremum $y(x)$. In any case, if we restrict the functional F to these functions we obtain the *scalar function of one variable*

$$g(\varepsilon) := F[y_\varepsilon] = \int_{x_1}^{x_2} f\big(x, y(x) + \varepsilon\eta(x), y'(x) + \varepsilon\eta'(x)\big)\,\mathrm{d}x,$$

which by construction has an extremum at $\varepsilon = 0$. We know that the *necessary* (although in general not sufficient) condition for this to happen is that $g'(0) = 0$. Since

$$\begin{aligned} g'(\varepsilon) &= \int_{x_1}^{x_2} \frac{\partial}{\partial\varepsilon} f\big(x, y(x) + \varepsilon\eta(x), y'(x) + \varepsilon\eta'(x)\big)\,\mathrm{d}x \\ &= \int_{x_1}^{x_2} \bigg[\frac{\partial f}{\partial y}\big(x, y(x) + \varepsilon\eta(x), y'(x) + \varepsilon\eta(x)\big)\,\eta(x) \\ &\qquad\qquad + \frac{\partial f}{\partial y'}\big(x, y(x) + \varepsilon\eta(x), y'(x) + \varepsilon\eta(x)\big)\eta'(x)\bigg]\,\mathrm{d}x \end{aligned}$$

we have

$$g'(0) = \int_{x_1}^{x_2} \bigg[\frac{\partial f}{\partial y}\big(x, y(x), y'(x)\big)\,\eta(x) + \frac{\partial f}{\partial y'}\big(x, y(x), y'(x)\big)\,\eta'(x)\bigg]\,\mathrm{d}x\,.$$

Hence, if the function $y(x)$ is an extremum of the functional F with the boundary conditions (4.2) it must satisfy

$$\int_{x_1}^{x_2} \left[\frac{\partial f}{\partial y}(x, y(x), y'(x))\, \eta(x) + \frac{\partial f}{\partial y'}(x, y(x), y'(x))\, \eta'(x) \right] \mathrm{d}x = 0 \tag{4.6}$$

for *any* function $\eta(x)$ satisfying (4.5). Equation (4.6) can be simplified by integrating the last term by parts, since

$$\begin{aligned}
&\int_{x_1}^{x_2} \frac{\partial f}{\partial y'}(x, y(x), y'(x))\, \eta'(x)\, \mathrm{d}x \\
&= \frac{\partial f}{\partial y'}(x, y(x), y'(x))\, \eta(x) \Big|_{x_1}^{x_2} - \int_{x_1}^{x_2} \eta(x) \frac{\mathrm{d}}{\mathrm{d}x}\left(\frac{\partial f}{\partial y'}(x, y(x), y'(x)) \right) \mathrm{d}x \\
&= - \int_{x_1}^{x_2} \eta(x) \frac{\mathrm{d}}{\mathrm{d}x}\left(\frac{\partial f}{\partial y'}(x, y(x), y'(x)) \right) \mathrm{d}x\,,
\end{aligned}$$

where we have taken into account condition (4.5). Substituting back into Eq. (4.6) we finally obtain

$$\begin{aligned}
g'(0) &= \frac{\mathrm{d}}{\mathrm{d}\varepsilon}\Big|_{\varepsilon=0} F[y_\varepsilon] \\
&= \int_{x_1}^{x_2} \left[\frac{\partial f}{\partial y}(x, y(x), y'(x)) - \frac{\mathrm{d}}{\mathrm{d}x}\left(\frac{\partial f}{\partial y'}(x, y(x), y'(x)) \right) \right] \eta(x)\, \mathrm{d}x = 0\,.
\end{aligned} \tag{4.7}$$

Since this condition must hold for *any* function η satisfying Eq. (4.5), the term in square brackets must vanish identically in the interval $[x_1, x_2]$. Hence *the extremum* $y(x)$ *must satisfy the* **Euler–Lagrange equation**

$$\boxed{\frac{\mathrm{d}}{\mathrm{d}x}\left(\frac{\partial f}{\partial y'}(x, y(x), y'(x)) \right) - \frac{\partial f}{\partial y}(x, y(x), y'(x)) = 0\,, \quad \forall x \in [x_1, x_2]\,.} \tag{4.8}$$

• Clearly, the argument leading to the Euler–Lagrange equation (4.8) is still valid if $y(x)$ is only a *local* extremum of the functional F.

• It is important to remember that the Euler–Lagrange equation (4.8) is a *necessary*, but in general *not sufficient*, condition for the function $y(x)$ to be an extremum of the functional F. In fact, the solutions of this differential equation can be regarded as the *critical points* of F, in the same way as the points at which the derivative of an ordinary function $F : \mathbb{R} \to \mathbb{R}$ vanishes are its critical points. Indeed, what the previous argument shows is that the functional $F[y]$ is *stationary* (i.e., approximately constant) when $y(x)$ is a solution of the Euler–Lagrange equations. For this reason, the functions $y(x)$ satisfying these equation are usually called *stationary points* of the functional (4.1).

• If (as we are assuming throughout) the function f is of class C^2, Eq. (4.8) can be written in the equivalent form

$$\frac{\partial^2 f}{\partial y'^2}\, y'' + \frac{\partial^2 f}{\partial y \partial y'}\, y' + \frac{\partial^2 f}{\partial x \partial y'} - \frac{\partial f}{\partial y} = 0\,.$$

In particular, if the condition

$$\frac{\partial^2 f}{\partial y'^2} \not\equiv 0$$

is satisfied, the previous equation is a *second-order ordinary differential equation* in the unknown function $y(x)$. To find the stationary points of the functional F, we must supplement this differential equation with the *boundary conditions* (4.2).

• Multiplying the left-hand side (LHS) of the Euler–Lagrange equation (4.8) by y' we obtain

$$y'\frac{\mathrm{d}}{\mathrm{d}x}\frac{\partial f}{\partial y'} - y'\frac{\partial f}{\partial y} = \frac{\mathrm{d}}{\mathrm{d}x}\left(y'\frac{\partial f}{\partial y'}\right) - y'\frac{\partial f}{\partial y} - y''\frac{\partial f}{\partial y'} = \frac{\mathrm{d}}{\mathrm{d}x}\left(y'\frac{\partial f}{\partial y'} - f\right) + \frac{\partial f}{\partial x}\,.$$

Hence if $y' \neq 0$ the Euler–Lagrange equation can be written in the equivalent form

$$\frac{\mathrm{d}}{\mathrm{d}x}\left(y'\frac{\partial f}{\partial y'} - f\right) + \frac{\partial f}{\partial x} = 0\,. \tag{4.9}$$

In particular, if f does not explicitly depend on x (that is, if it is a function of y and y' only), the function in parentheses in the LHS of Eq. (4.9) is conserved:

$$\boxed{\frac{\partial f}{\partial x} = 0 \quad \Longrightarrow \quad h := y'\frac{\partial f}{\partial y'} - f = \text{const.}} \tag{4.10}$$

It is said in this case that the function h is a **first integral** of the Euler–Lagrange equation (4.8), since when h is conserved the *second-order* equation (4.8) is equivalent to the *first-order* equation (4.10). We shall call h the **energy integral**, since in many mechanical problems it is equal to the system's mechanical energy[1]. Likewise, if f does not depend on y it follows from the Euler–Lagrange equation that the partial derivative of f with respect to y' is conserved:

$$\boxed{\frac{\partial f}{\partial y} = 0 \quad \Longrightarrow \quad \frac{\partial f}{\partial y'} = \text{const.}} \tag{4.11}$$

[1]The energy integral is also called *Jacobi integral* by some authors.

Example 4.4. The problems set forth in Examples 4.1–4.2 are easily solved using the Euler–Lagrange equation and its conservation laws (4.10)–(4.11). Indeed, for the length functional Eq. (4.11) yields

$$\frac{\partial f}{\partial y'} = \frac{y'}{\sqrt{1+y'^2}} = \text{const.} \quad \Longrightarrow \quad y' = \text{const.}$$

Therefore $y(x) = ax + b$, where the constants a and b must be chosen so that conditions (4.2) are satisfied. Hence the curve of minimum length is the line segment joining the given points. (*Stricto sensu*, we have only shown that the straight line is a *stationary point* of the length functional.)

For the brachistochrone functional (4.3), Eq. (4.10) reads

$$\left(\frac{y'^2}{\sqrt{1+y'^2}} - \sqrt{1+y'^2}\right)(y_0 - y)^{-1/2} = -(1+y'^2)^{-1/2}(y_0-y)^{-1/2} = \text{const.}$$
$$\Longrightarrow \quad (y_0 - y)(1+y'^2) = 2a\,,$$

with $a > 0$ constant. Hence

$$y' = \pm\sqrt{\frac{2a}{y_0 - y} - 1} = \pm\sqrt{\frac{2a - y_0 + y}{y_0 - y}}\,,$$

and thus

$$x - x_0 = \pm \int \sqrt{\frac{y_0 - y}{2a - y_0 + y}}\, \mathrm{d}y\,,$$

with x_0 constant. Performing the change of variable

$$y_0 - y = 2a\sin^2\theta$$

we obtain

$$x - x_0 = \mp 4a \int \frac{\sin^2\theta\cos\theta}{\cos\theta}\,\mathrm{d}\theta = \mp 4a\int \sin^2\theta\,\mathrm{d}\theta = \mp 2a\int(1-\cos 2\theta)\,\mathrm{d}\theta$$
$$= \mp a\,(2\theta - \sin 2\theta).$$

The parametric equations of the curve sought are therefore

$$\boxed{x = x_0 \mp a(2\theta - \sin 2\theta)\,, \qquad y = y_0 - 2a\sin^2\theta = y_0 - a\,(1-\cos 2\theta)\,.} \tag{4.12}$$

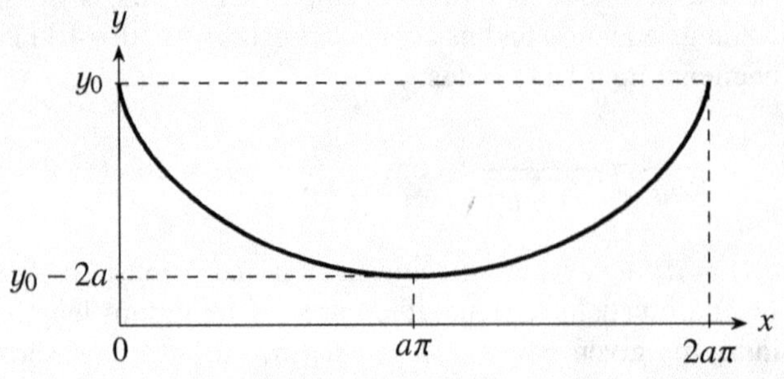

Figure 4.1. Arc of the cycloid (4.12), with $x_0 = 0$ and "+" sign, in a period $0 \leqslant \theta \leqslant \pi$.

These are the equations of an inverted *cycloid*[a] traced out by a circle of radius a (see Fig. 4.1 above), where the constants x_0 and a must again be determined imposing the boundary conditions (4.2).

[a]Note that the double sign can actually be omitted, since the points on the curve corresponding to the "−" sign can be obtained from those with the "+" sign changing θ by $-\theta$.

Example 4.5. The Euler–Lagrange equation for the optical length functional (4.4) reads

$$\frac{\mathrm{d}}{\mathrm{d}x}\left(\frac{n(x,y)\,y'}{\sqrt{1+y'^2}}\right) - \sqrt{1+y'^2}\,\frac{\partial n(x,y)}{\partial y} = 0\,.$$

This equation can be expressed in a more compact form taking into account that, if s is the arc length along the path of the light ray, then

$$\frac{\mathrm{d}}{\mathrm{d}s} = \left(\frac{\mathrm{d}s}{\mathrm{d}x}\right)^{-1}\frac{\mathrm{d}}{\mathrm{d}x} = \left(1+y'^2\right)^{-1/2}\frac{\mathrm{d}}{\mathrm{d}x}\,.$$

In this way we obtain the equation

$$\boxed{\frac{\mathrm{d}}{\mathrm{d}s}\left(n(x,y)\,\frac{\mathrm{d}y}{\mathrm{d}s}\right) = \frac{\partial n(x,y)}{\partial y}}\,.$$

For instance, if the refractive index does not depend on the y coordinate the previous equation yields

$$n(x)\,\frac{\mathrm{d}y}{\mathrm{d}s} = k \quad\Longrightarrow\quad \frac{n^2(x)\,y'^2}{1+y'^2} = k^2 \quad\Longrightarrow\quad y' = \pm\frac{k}{\sqrt{n^2(x)-k^2}}\,,$$

where k is a constant. Thus in this case the equation of the light rays is

$$y = y_0 \pm k \int \frac{dx}{\sqrt{n^2(x) - k^2}} .$$

In particular, if

$$n(x) = \frac{n_0}{x} \qquad (n_0 > 0, \quad x > 0)$$

we have

$$y - y_0 = \pm k \int \frac{x\, dx}{\sqrt{n_0^2 - k^2 x^2}} = \mp \frac{1}{k} \sqrt{n_0^2 - k^2 x^2} \implies x^2 + (y - y_0)^2 = \frac{n_0^2}{k^2} .$$

Therefore the paths of the light rays are in this case arcs of *circles* whose centers lie on the y axis.

We shall next consider a more general version of the fundamental problem of the calculus of variations, in which the functional F depends on n scalar functions $y_1, \dots, y_n$ of one real variable x. Equivalently (and more advantageously in practice), we can regard F as a function of a single *vector-valued* function $\mathbf{y} := (y_1, \dots, y_n) : \mathbb{R} \to \mathbb{R}^n$. More precisely, consider the functional

$$F[\mathbf{y}] = \int_{x_1}^{x_2} f\big(x, \mathbf{y}(x), \mathbf{y}'(x)\big)\, dx , \tag{4.13}$$

whose domain is the space of functions $\mathbf{y} : [x_1, x_2] \to \mathbb{R}^n$ of class C^2 on the interval $[x_1, x_2]$ satisfying boundary conditions similar to (4.2):

$$\mathbf{y}(x_1) = \mathbf{y}_1 , \qquad \mathbf{y}(x_2) = \mathbf{y}_2 , \tag{4.14}$$

for certain fixed vectors $\mathbf{y}_1, \mathbf{y}_2 \in \mathbb{R}^n$.

As before, in order to find the extrema of the functional (4.13) subject to the conditions (4.14) we consider a variation

$$\mathbf{y}_\varepsilon(x) = \mathbf{y}(x) + \varepsilon \boldsymbol{\eta}(x)$$

about a hypothetical extremum $\mathbf{y}(x)$, where the vector-valued function $\boldsymbol{\eta} =: (\eta_1, \dots, \eta_n)$ must satisfy

$$\boldsymbol{\eta}(x_1) = \boldsymbol{\eta}(x_2) = 0 \tag{4.15}$$

so that $\mathbf{y}_\varepsilon$ verifies (4.14) for all ε. Restricting the functional F to the perturbed extremum $\mathbf{y}_\varepsilon$ we obtain, as before, the scalar function of one variable

$$g(\varepsilon) := F[\mathbf{y}_\varepsilon] = \int_{x_1}^{x_2} f\big(x, \mathbf{y}_\varepsilon(x), \mathbf{y}_\varepsilon'(x)\big)\, dx , \tag{4.16}$$

whose derivative at $\varepsilon = 0$ must vanish. Computing this derivative, integrating by parts, and taking into account conditions (4.15) we readily obtain

$$g'(0) = \int_{x_1}^{x_2} \sum_{i=1}^{n} \left[\frac{\partial f}{\partial y_i}(x, \mathbf{y}(x), \mathbf{y}'(x)) - \frac{\mathrm{d}}{\mathrm{d}x} \frac{\partial f}{\partial y_i'}(x, \mathbf{y}(x), \mathbf{y}'(x)) \right] \eta_i(x) . \tag{4.17}$$

Since this expression must vanish identically for all functions η_i satisfying conditions (4.15), we conclude that the extrema of the functional (4.13) must verify the n **Euler–Lagrange equations**

$$\boxed{\frac{\mathrm{d}}{\mathrm{d}x} \frac{\partial f}{\partial y_i'} - \frac{\partial f}{\partial y_i} = 0 , \qquad i = 1, \dots, n .} \tag{4.18}$$

Again, these equations are only *necessary* for the function $\mathbf{y}(x)$ to be an extremum of the functional F. Indeed, the solutions of Eqs. (4.18) are actually the *critical* or *stationary points* of the functional (4.13).

- Expanding Eqs. (4.18) we obtain

$$\sum_{j=1}^{n} \frac{\partial^2 f}{\partial y_i' \partial y_j'} y_j'' + \sum_{j=1}^{n} \frac{\partial^2 f}{\partial y_j \partial y_i'} y_j' + \frac{\partial^2 f}{\partial x \partial y_i'} - \frac{\partial f}{\partial y_i} = 0 , \qquad i = 1, \dots, n .$$

Hence if the *Hessian* of the density f with respect to the variables y_i' does not vanish identically, i.e., if

$$\det \left(\frac{\partial^2 f}{\partial y_i' \partial y_j'} \right)_{1 \leqslant i,j \leqslant n} \not\equiv 0 ,$$

the Euler–Lagrange equations (4.18) are a *system of n second-order ordinary differential equations in the n unknown scalar functions* $y_i(x)$ ($i = 1, \dots, n$), which must be supplemented by the $2n$ boundary conditions (4.14).

- Multiplying the LHS of the Euler–Lagrange (4.18) by y_i' and summing over i we obtain

$$\sum_{i=1}^{n} y_i' \frac{\mathrm{d}}{\mathrm{d}x} \frac{\partial f}{\partial y_i'} - \sum_{i=1}^{n} y_i' \frac{\partial f}{\partial y_i} = \frac{\mathrm{d}}{\mathrm{d}x} \left(\sum_{i=1}^{n} y_i' \frac{\partial f}{\partial y_i'} - f \right) + \frac{\partial f}{\partial x} = 0 .$$

Hence *if f does not depend explicitly on the variable x the function*

$$\boxed{h := \sum_{i=1}^{n} y_i' \frac{\partial f}{\partial y_i'} - f = \mathbf{y}' \frac{\partial f}{\partial \mathbf{y}'} - f}$$

is conserved. As in the scalar case, h is usually called the **energy** (or Jacobi) **integral**. It is also evident that if the density f is independent of the variable y_i, the derivative of f with respect to y_i' is conserved:

$$\boxed{\frac{\partial f}{\partial y_i} = 0 \quad \Longrightarrow \quad \frac{\partial f}{\partial y_i'} = \text{const.}}$$

Example 4.6. Let us find the equation of the paths followed by light rays in an optical three-dimensional medium with refractive index $n(\mathbf{r})$.

According to Fermat's principle, the trajectory $\mathbf{r} = \mathbf{r}(u)$ of the light ray joining two points $\mathbf{r}_1, \mathbf{r}_2 \in \mathbb{R}^3$ (where $u \in [u_1, u_2]$ is any parameter along the path) must minimize the time taken by light to cover the distance between both points. Since

$$\frac{\mathrm{d}s}{\mathrm{d}u} = \sqrt{\mathbf{r}'^2(u)}\,,$$

where the prime denotes derivative with respect to the parameter u, we have

$$\mathrm{d}t = \frac{\mathrm{d}t}{\mathrm{d}s}\,\frac{\mathrm{d}s}{\mathrm{d}u}\,\mathrm{d}u = \frac{1}{v}\,\sqrt{\mathbf{r}'^2(u)}\,\mathrm{d}u = \frac{n\big(\mathbf{r}(u)\big)}{c}\,\sqrt{\mathbf{r}'^2(u)}\,\mathrm{d}u\,.$$

Thus the trajectory sought must minimize the optical length functional (proportional to the travel time)

$$F[\mathbf{r}] = \int_{u_1}^{u_2} n\big(\mathbf{r}(u)\big)\sqrt{\mathbf{r}'^2(u)}\,\mathrm{d}u$$

with the boundary conditions

$$\mathbf{r}(u_1) = \mathbf{r}_1\,, \qquad \mathbf{r}(u_2) = \mathbf{r}_2\,.$$

(Note that in this example u plays the role of x and $\mathbf{r}$ that of $\mathbf{y}$.) The Euler–Lagrange equations for this functional are

$$\frac{\mathrm{d}}{\mathrm{d}u}\left(\frac{\partial}{\partial x_i'}\left(n\sqrt{\mathbf{r}'^2}\right)\right) - \sqrt{\mathbf{r}'^2}\,\frac{\partial n}{\partial x_i} = \frac{\mathrm{d}}{\mathrm{d}u}\left(\frac{n\,x_i'}{\sqrt{\mathbf{r}'^2}}\right) - \sqrt{\mathbf{r}'^2}\,\frac{\partial n}{\partial x_i} = 0\,, \qquad i = 1,2,3\,,$$

where $\mathbf{r} = (x_1, x_2, x_3)$. Taking into account that

$$\frac{1}{\sqrt{\mathbf{r}'^2}}\,\frac{\mathrm{d}}{\mathrm{d}u} = \frac{\mathrm{d}}{\mathrm{d}s}\,,$$

the previous equations can be written in the following more geometric fashion:

$$\boxed{\frac{\mathrm{d}}{\mathrm{d}s}\left(n\,\frac{\mathrm{d}\mathbf{r}}{\mathrm{d}s}\right) = \frac{\partial n}{\partial \mathbf{r}}\,.} \tag{4.19}$$

This is the fundamental equation of geometric optics.

For example, if the index of refraction depends only on r (that is, if the optical medium is *spherically symmetric*), then

$$\frac{\mathrm{d}}{\mathrm{d}s}\left(\mathbf{r}\times n\frac{\mathrm{d}\mathbf{r}}{\mathrm{d}s}\right)=\mathbf{r}\times\frac{\partial n}{\partial\mathbf{r}}=n'(r)\,\mathbf{r}\times\frac{\mathbf{r}}{r}=0\,.$$

Hence in this case *the path of the light ray is contained in a plane passing through the origin of coordinates* perpendicular to the constant vector $n\mathbf{r}\times\frac{\mathrm{d}\mathbf{r}}{\mathrm{d}s}$.

Exercise 4.1. The *eikonal equation* in geometric optics is the partial differential equation

$$\left(\nabla S(\mathbf{r})\right)^2=n^2(\mathbf{r}),$$

where $n(\mathbf{r})$ is the refraction index of the optical medium. Show that if $S(\mathbf{r})$ is a solution of the eikonal equation the orthogonal trajectories to the surfaces of constant S are possible trajectories for light rays in the medium.

Solution. Since $\left|\frac{\mathrm{d}\mathbf{r}}{\mathrm{d}s}\right|=1$, the differential equation of the orthogonal trajectories to the surfaces of constant S is

$$\frac{\mathrm{d}\mathbf{r}}{\mathrm{d}s}=\frac{\nabla S}{|\nabla S|}=\frac{\nabla S}{n}\quad\Longleftrightarrow\quad n\frac{\mathrm{d}\mathbf{r}}{\mathrm{d}s}=\nabla S. \tag{4.20}$$

We must show that these trajectories satisfy the differential equation (4.19) of the light rays deduced in the previous example, i.e.,

$$\frac{\mathrm{d}}{\mathrm{d}s}\left(n\frac{\mathrm{d}\mathbf{r}}{\mathrm{d}s}\right)=\nabla n.$$

To this end, note that Eq. (4.20) implies that

$$\frac{\mathrm{d}}{\mathrm{d}s}\left(n\frac{\mathrm{d}\mathbf{r}}{\mathrm{d}s}\right)=\frac{\mathrm{d}}{\mathrm{d}s}\nabla S=\left(\frac{\mathrm{d}\mathbf{r}}{\mathrm{d}s}\cdot\nabla\right)\nabla S=\frac{1}{n}\left(\nabla S\cdot\nabla\right)\nabla S. \tag{4.21}$$

On the other hand, from the equality

$$(\nabla S\cdot\nabla)\frac{\partial S}{\partial x_i}=\left(\sum_{j=1}^{3}\frac{\partial S}{\partial x_j}\frac{\partial}{\partial x_j}\right)\frac{\partial S}{\partial x_i}=\sum_{j=1}^{3}\frac{\partial S}{\partial x_j}\frac{\partial^2 S}{\partial x_i\partial x_j}=\frac{1}{2}\frac{\partial}{\partial x_i}\sum_{j=1}^{3}\left(\frac{\partial S}{\partial x_j}\right)^2$$

we deduce that

$$(\nabla S\cdot\nabla)\,\nabla S=\frac{1}{2}\nabla\left((\nabla S)^2\right)=\frac{1}{2}\nabla\left(n^2\right)=n\nabla n.$$

Hence equation (4.21) simplifies to

$$\frac{\mathrm{d}}{\mathrm{d}s}\left(n\frac{\mathrm{d}\mathbf{r}}{\mathrm{d}s}\right) = \frac{1}{n}\cdot n\nabla n = \nabla n,$$

as claimed.

4.1.3 Variation and variational derivative

Let uus denote by

$$\boxed{\delta\mathbf{y}(x) := \mathbf{y}_\varepsilon(x) - \mathbf{y}(x) = \varepsilon\boldsymbol{\eta}(x)}$$

the **variation** of the function $\mathbf{y}(x)$, where the vector-valued function $\boldsymbol{\eta}$ satisfies the boundary conditions (4.15). The change in the functional F when its argument $\mathbf{y}$ is incremented by $\delta\mathbf{y}$ is then (using the notation of Eq. (4.16))

$$F[\mathbf{y} + \delta\mathbf{y}] - F[\mathbf{y}] = F[\mathbf{y}_\varepsilon] - F[\mathbf{y}] = g(\varepsilon) - g(0)\,.$$

To first order in the small parameter ε, this change is given by

$$\varepsilon g'(0) = \varepsilon \left.\frac{\mathrm{d}}{\mathrm{d}\varepsilon}\right|_{\varepsilon=0} F[\mathbf{y}_\varepsilon] =: \delta F[\mathbf{y}]\,, \tag{4.22}$$

so that (by definition of derivative) we have

$$F[\mathbf{y} + \delta\mathbf{y}] - F[\mathbf{y}] = \varepsilon\left(\left.\frac{\mathrm{d}}{\mathrm{d}\varepsilon}\right|_{\varepsilon=0} F[\mathbf{y}_\varepsilon]\right) + o(\varepsilon) = \delta F(\mathbf{y}) + o(\varepsilon)\,.$$

The functional $\delta F[\mathbf{y}]$ is called the **variation** of F at $\mathbf{y}$. From Eqs. (4.17) and (4.22) it follows that we can write this variation as

$$\boxed{\delta F[\mathbf{y}] = \int_{x_1}^{x_2} \frac{\delta f}{\delta\mathbf{y}}\big(x, \mathbf{y}(x), \mathbf{y}'(x)\big)\cdot\delta\mathbf{y}(x)\,\mathrm{d}x\,,} \tag{4.23}$$

where

$$\boxed{\frac{\delta f}{\delta\mathbf{y}} := \frac{\partial f}{\partial\mathbf{y}} - \frac{\mathrm{d}}{\mathrm{d}x}\frac{\partial f}{\partial\mathbf{y}'}\,.} \tag{4.24}$$

The (n-component) vector-valued function $\frac{\delta f}{\delta\mathbf{y}}(x, \mathbf{y}, \mathbf{y}')$ is called the **variational derivative** of the density f with respect to the function $\mathbf{y}(x)$. In particular, with this notation the Euler–Lagrange equations (4.18) of the functional (4.13) simply express the *vanishing of the variational derivative of its density* f:

$$\boxed{\frac{\delta f}{\delta\mathbf{y}} = 0\,.}$$

By Eq. (4.23), this is equivalent to the *vanishing of the variation of the functional* F:

$$\boxed{\frac{\delta f}{\delta \mathbf{y}} = 0 \iff \delta F[\mathbf{y}] = 0\,.} \tag{4.25}$$

In other words, $\mathbf{y}(x)$ satisfies the Euler–Lagrange equations for a density $f(x, \mathbf{y}, \mathbf{y}')$ if the increment $F(\mathbf{y} + \delta\mathbf{y}) - F(\mathbf{y})$ of the corresponding functional F is $o(\varepsilon)$, so that F is "stationary" at $\mathbf{y}(x)$ (i.e., does not vary appreciably near $\mathbf{y}(x)$).

Consider two functionals of the form (4.13)–(4.14) with densities f_1 and f_2 differing by the **total derivative** with respect to x of a function $g(x, \mathbf{y})$:

$$f_2(x, \mathbf{y}, \mathbf{y}') = f_1(x, \mathbf{y}, \mathbf{y}') + \frac{\mathrm{d}}{\mathrm{d}x}\, g(x, \mathbf{y}), \qquad \frac{\mathrm{d}}{\mathrm{d}x}\, g(x, \mathbf{y}) := \frac{\partial g(x, \mathbf{y})}{\partial x} + \frac{\partial g(x, \mathbf{y})}{\partial \mathbf{y}}\, \mathbf{y}'\,.$$

We then have

$$\begin{aligned} F_2[\mathbf{y}] - F_1[\mathbf{y}] &= \int_{x_1}^{x_2} \frac{\mathrm{d}}{\mathrm{d}x} g(x, \mathbf{y}(x))\, \mathrm{d}x = g(x_1, \mathbf{y}(x_1)) - g(x_2, \mathbf{y}(x_2)) \\ &= g(x_1, \mathbf{y}_1) - g(x_2, \mathbf{y}_2)\,, \end{aligned}$$

on account of the boundary conditions (4.14) satisfied by the functions $\mathbf{y}(x)$ in the domain of the functionals F_1 and F_2. Hence these functionals *differ by a constant*, and therefore *they have the same variational derivative* (as they have the same variation). It follows that the Euler–Lagrange equations of the densities f_1 and f_2 must be exactly the *same*, as can also be checked by direct differentiation (exercise). In other words:

> Two densities differing by a total derivative give rise to the *same* Euler–Lagrange equations.

Conversely, *if two densities f_1 and f_2 give rise to the* same *Euler–Lagrange equations, they must necessarily differ by a total derivative.* Indeed, by hypothesis

$$\frac{\delta f_1}{\delta \mathbf{y}} = \frac{\delta f_2}{\delta \mathbf{y}}\,,$$

and thus the function $f = f_1 - f_2$ has zero variational derivative. It follows that f is a total derivative (see the following exercise for a proof), as claimed.

Exercise 4.2. Show that if the variational derivative of a function $f(x, \mathbf{y}, \mathbf{y}')$ vanishes identically then f is the total derivative of a function $g(x, \mathbf{y})$.

Solution. Indeed, if the variational derivative of $f(x, \mathbf{y}, \mathbf{y}')$ vanishes identically we have

$$\frac{\delta f}{\delta y_i} := \frac{\partial f}{\partial y_i} - \frac{\mathrm{d}}{\mathrm{d}x}\frac{\partial f}{\partial y_i'} = -\sum_{j=1}^{n} \frac{\partial^2 f}{\partial y_i' \partial y_j'}\, y_j'' - \sum_{j=1}^{n} \frac{\partial^2 f}{\partial y_i' \partial y_j}\, y_j' - \frac{\partial^2 f}{\partial x \partial y_i'} + \frac{\partial f}{\partial y_i} = 0 \tag{4.26}$$

for $i = 1, \ldots, n$ and *all* $(x, \mathbf{y}, \mathbf{y}', \mathbf{y}'')$. Since none of the partial derivatives appearing in the previous identity depend on $\mathbf{y}''$, the coefficient of y_j'' must vanish identically. We thus have

$$\frac{\partial^2 f}{\partial y_i' \partial y_j'} = 0, \qquad i, j = 1, \ldots, n,$$

i.e., $\dfrac{\partial f}{\partial y_i'}$ is independent of $\mathbf{y}'$ for all i. Hence

$$\frac{\partial f}{\partial y_i'} = g_i(x, \mathbf{y}) \quad (i = 1, \ldots, n) \quad \Longrightarrow \quad f = \sum_{i=1}^{n} g_i(x, \mathbf{y}) y_i' + g_0(x, \mathbf{y})$$

for certain functions $g_i(x, \mathbf{y})$, $g_0(x, \mathbf{y})$. Substituting into Eq. (4.26) we then obtain

$$-\sum_{j=1}^{n} \frac{\partial g_i}{\partial y_j} y_j' - \frac{\partial g_i}{\partial x} + \sum_{j=1}^{n} \frac{\partial g_j}{\partial y_i} y_j' + \frac{\partial g_0}{\partial y_i} = 0 . \tag{4.27}$$

Since none of the partial derivatives in Eq. (4.27) depend on $\mathbf{y}'$, equating to zero the coefficient of y_j' in the previous identity we deduce that

$$\frac{\partial g_i}{\partial y_j} = \frac{\partial g_j}{\partial y_i}, \qquad i, j = 1, \ldots, n .$$

It can be shown that these equations imply that there is a function $k(x, \mathbf{y})$ such that

$$g_i = \frac{\partial k}{\partial y_i}, \qquad i = 1, \ldots, n .$$

Equation (4.27) then reduces to

$$\frac{\partial}{\partial y_i}\left(g_0 - \frac{\partial k}{\partial x}\right) = 0 \quad (i = 1, \ldots, n) \quad \Longrightarrow \quad g_0 - \frac{\partial k}{\partial x} = l(x)$$

for some function $l(x)$. We then have

$$f = \sum_{i=1}^{n} \frac{\partial k(x, \mathbf{y})}{\partial y_i} y_i' + \frac{\partial k(x, \mathbf{y})}{\partial x} + l(x) = \frac{\mathrm{d}}{\mathrm{d}x}\left(k(x, \mathbf{y}) + \int l(x)\, \mathrm{d}x\right).$$

4.2 HAMILTON'S PRINCIPLE FOR UNCONSTRAINED SYSTEMS

4.2.1 Hamilton's principle for a single particle

Consider, first, the motion of a particle of mass m subject to an *irrotational* force

$$\mathbf{F}(t, \mathbf{r}) = -\frac{\partial V(t, \mathbf{r})}{\partial \mathbf{r}} . \tag{4.28}$$

Newton's equations of motion are in this case

$$m\ddot{x}_i = -\frac{\partial V}{\partial x_i}\,, \qquad i = 1, 2, 3\,, \tag{4.29}$$

where we have again denoted by x_i the i-th component of the particle's position vector $\mathbf{r}$. We ask ourselves if Eqs. (4.29) are the Euler–Lagrange equations of some functional

$$\int_{t_1}^{t_2} L\big(t, \mathbf{r}(t), \dot{\mathbf{r}}(t)\big)\,\mathrm{d}t\,.$$

(Note, again, that in this case t, $\mathbf{r}$, and L respectively play the roles of x, $\mathbf{y}$, and f in the previous sections.) Although it is not difficult to answer this question in the affirmative simply by inspection, we can proceed more systematically as follows. Writing Eqs. (4.29) in the form

$$\frac{\mathrm{d}}{\mathrm{d}t}(m\dot{x}_i) + \frac{\partial V}{\partial x_i} = 0\,, \qquad i = 1, 2, 3\,,$$

we see that it suffices to find a function $L(t, \mathbf{r}, \dot{\mathbf{r}})$ verifying the equations

$$\frac{\partial L}{\partial \dot{x}_i} = m\dot{x}_i\,, \qquad \frac{\partial L}{\partial x_i} = -\frac{\partial V}{\partial x_i}\,; \qquad i = 1, 2, 3\,.$$

Integrating first the three equations for $\dfrac{\partial L}{\partial x_i}$ we obtain

$$L = -V + g(t, \dot{\mathbf{r}})\,,$$

and substituting back into the remaining equations we have

$$\frac{\partial g}{\partial \dot{x}_i} = m\dot{x}_i\,, \qquad i = 1, 2, 3\,,$$

which determines the function g:

$$g = \frac{1}{2}m\dot{\mathbf{r}}^2 + h(t)\,.$$

Thus the simplest function with the desired property is[2]

$$\boxed{L = \frac{1}{2}m\dot{\mathbf{r}}^2 - V(t, \mathbf{r}) = T - V(t, \mathbf{r})\,.} \tag{4.30}$$

The function L is called the system's **Lagrangian**. We have therefore proved **Hamilton's principle** in its simplest form:

---

[2]Note that

$$h(t) = \frac{\mathrm{d}}{\mathrm{d}t}\int h(s)\,\mathrm{d}s$$

is obviously a total derivative. Therefore adding it to the Lagrangian (4.30) does not change its Euler–Lagrange equations (4.29), as we saw in Section (4.1.3).

The trajectory followed by a particle of mass m subject to the irrotational force (4.28) as it moves from a point $\mathbf{r}_1$ at time t_1 to another point $\mathbf{r}_2$ at time t_2 is a *stationary point* of the functional

$$S[\mathbf{r}] = \int_{t_1}^{t_2} L\big(t, \mathbf{r}(t), \dot{\mathbf{r}}(t)\big)\,\mathrm{d}t \qquad (\text{with} \quad \mathbf{r}(t_1) = \mathbf{r}_1, \ \ \mathbf{r}(t_2) = \mathbf{r}_2), \tag{4.31}$$

where $L = T - V(t, \mathbf{r})$. In other words, Newton's equations of motion are equivalent to the *Euler–Lagrange equations*

$$\frac{\delta L}{\delta \mathbf{r}} = 0,$$

i.e., to the *vanishing of the variation of the functional S:*

$$\delta S[\mathbf{r}] = 0.$$

The functional (4.31) is called the **action**[3]. Note that the action has dimensions of energy × time or length × momentum, since L (like T or V) has dimensions of energy.

4.2.2 Hamilton's principle for a system of particles

Hamilton's principle is extended without difficulty to a system of N particles, provided that the total forces $\mathbf{F}_i$ acting on each particle are *irrotational*, i.e.,

$$\mathbf{F}_i(t, \mathbf{r}_1, \ldots, \mathbf{r}_N) = -\frac{\partial V(t, \mathbf{r}_1, \ldots, \mathbf{r}_N)}{\partial \mathbf{r}_i}, \qquad i = 1, \ldots, N. \tag{4.32}$$

Indeed, it is easy to check that Newton's equations of motion for the system,

$$m_i \ddot{\mathbf{r}}_i = -\frac{\partial V(t, \mathbf{r}_1, \ldots, \mathbf{r}_N)}{\partial \mathbf{r}_i}, \qquad i = 1, \ldots, N,$$

are the Euler–Lagrange equations of the action functional

$$\boxed{S[\mathbf{r}_1, \ldots, \mathbf{r}_N] = \int_{t_1}^{t_2} L\big(t, \mathbf{r}_1(t), \ldots, \mathbf{r}_N(t), \dot{\mathbf{r}}_1(t), \ldots, \dot{\mathbf{r}}_N(t)\big)\,\mathrm{d}t} \tag{4.33}$$

with boundary conditions

$$\mathbf{r}_i(t_1) = \mathbf{r}_{i1}, \qquad \mathbf{r}_i(t_2) = \mathbf{r}_{i2}; \qquad i = 1, \ldots, N, \tag{4.34}$$

where in this case the Lagrangian L is given by

$$\begin{aligned} L(t, \mathbf{r}_1, \ldots, \mathbf{r}_N, \dot{\mathbf{r}}_1, \ldots, \dot{\mathbf{r}}_N) &= T - V(t, \mathbf{r}_1, \ldots, \mathbf{r}_N) \\ &= \frac{1}{2}\sum_{i=1}^{N} m_i \dot{\mathbf{r}}_i^2 - V(t, \mathbf{r}_1, \ldots, \mathbf{r}_N). \end{aligned} \tag{4.35}$$

[3]The name "action" is due to Feynman. In the classical literature, the action computed along a trajectory (i.e., a solution of the Euler–Lagrange equations) is called *Hamilton's principal function.*

To check this statement, we write the three Euler–Lagrange equations for the i-th particle (i.e., one for each of the components of its position vector $\mathbf{r}_i$) in vector form :

$$0 = \frac{\mathrm{d}}{\mathrm{d}t}\frac{\partial L}{\partial \dot{\mathbf{r}}_i} - \frac{\partial L}{\partial \mathbf{r}_i} = \frac{\mathrm{d}}{\mathrm{d}t}\frac{\partial T}{\partial \dot{\mathbf{r}}_i} + \frac{\partial V}{\partial \mathbf{r}_i} = \frac{\mathrm{d}}{\mathrm{d}t}(m_i\dot{\mathbf{r}}_i) + \frac{\partial V}{\partial \mathbf{r}_i} = m_i\ddot{\mathbf{r}}_i - \mathbf{F}_i\,,$$

and observe that this is precisely the equation of motion of $\mathbf{r}_i$. In other words, the following more general version of Hamilton's principle holds:

The trajectory followed by a system of N particles subject to the irrotational forces (4.32) is a *stationary point* of the action functional (4.33)–(4.34) with Lagrangian $L = T - V(t, \mathbf{r}_1, \ldots, \mathbf{r}_N)$. In other words, the system's equations of motion are again the *Euler–Lagrange equations*

$$\frac{\delta L}{\delta \mathbf{r}_i} = 0\,, \qquad i = 1, \ldots, N\,,$$

which express the *vanishing of the variation of the action*:

$$\delta S[\mathbf{r}_1, \ldots, \mathbf{r}_N] = 0\,.$$

• Hamilton's variational principle is sometimes called **principle of least action**, since in many cases the trajectories of a mechanical system turn out to be *minima* of the action (at least *locally*). More properly, this principle should be called **principle of stationary action**, since as we know the Euler–Lagrange equations only guarantee the stationary character of the action.

• From Hamilton's principle and the conservation laws of the Euler–Lagrange equations derived in the previous subsection, it follows that if the Lagrangian (4.35) does not depend explicitly on time the energy integral

$$h = \sum_{i=1}^{N} \dot{\mathbf{r}}_i \frac{\partial L}{\partial \dot{\mathbf{r}}_i} - L = \sum_{i=1}^{N} m_i \dot{\mathbf{r}}_i^2 - L = 2T - (T - V) = T + V,$$

which in this case coincides with the system's *total energy*, is conserved. This result is consistent with the one obtained in Section 2.6.2, since L is independent of time if and only if the potential V is time-independent, in which case the forces (4.32) acting on the system are not only irrotational but *conservative.*

• Likewise, if the Lagrangian L does not depend (for instance) on the x coordinate of the i-th particle, i.e., if $\frac{\partial L}{\partial x_i} = 0$, then $\frac{\partial L}{\partial \dot{x}_i}$ is conserved:

$$\frac{\partial L}{\partial x_i} = 0 \quad \Longrightarrow \quad \frac{\partial L}{\partial \dot{x}_i} = \text{const.}$$

Note that this result is nothing more than the conservation law for the x component of the i-th particle's momentum, since

$$\frac{\partial L}{\partial \dot{x}_i} = \frac{\partial T}{\partial \dot{x}_i} = m_i \dot{x}_i = (\mathbf{p}_i)_x .$$

This is in agreement with the discussion in Section 2.6.2, since

$$\frac{\partial L}{\partial x_i} = -\frac{\partial V}{\partial x_i} = \left(\mathbf{F}_i\right)_x .$$

Obviously, the same result holds for the coordinates y_i or z_i.

4.2.3 Covariance of the Lagrangian formulation

One of the great advantages of the Lagrangian formulation of mechanics is its ***covariance*** *under coordinate changes*, i.e., that it treats all systems of curvilinear coordinates on the same footing.

More specifically, consider a particle of mass m subject to an irrotational force, whose trajectories are the critical points of the action (4.31). Let $(q_1, q_2, q_3) =: \mathbf{q}$ be a system of *curvilinear coordinates*, and denote by $\mathbf{r}(\mathbf{q})$ the function expressing the Cartesian coordinates $\mathbf{r}$ in terms of the curvilinear ones $\mathbf{q}$. Suppose that the particle's trajectory in the coordinates q_i is given by a certain function $\mathbf{q}(t) = (q_1(t), q_2(t), q_3(t))$. In Cartesian coordinates the trajectory is then $\mathbf{r} = \mathbf{r}(\mathbf{q}(t))$, which (with a slight abuse of notation) we shall denote by $\mathbf{r}(t)$. We can then express the Lagrangian $L(t, \mathbf{r}(t), \dot{\mathbf{r}}(t))$ in terms of $\mathbf{q}(t)$ and its time derivatives $\dot{\mathbf{q}}(t)$ using the change of coordinates formula $\mathbf{r} = \mathbf{r}(\mathbf{q})$ and its time derivative

$$\dot{\mathbf{r}} = \sum_{i=1}^{3} \frac{\partial \mathbf{r}(\mathbf{q})}{\partial q_i} \dot{q}_i =: \frac{\partial \mathbf{r}(\mathbf{q})}{\partial \mathbf{q}} \dot{\mathbf{q}} ,$$

so that

$$L\big(t, \mathbf{r}, \dot{\mathbf{r}}\big) = L\left(t, \mathbf{r}(\mathbf{q}), \frac{\partial \mathbf{r}(\mathbf{q})}{\partial \mathbf{q}} \dot{\mathbf{q}}\right) =: \widetilde{L}\big(t, \mathbf{q}, \dot{\mathbf{q}}\big) . \tag{4.36}$$

The *action* of the trajectory $\mathbf{r}(t)$ is then given by

$$S[\mathbf{r}] = \int_{t_1}^{t_2} L\big(t, \mathbf{r}(t), \dot{\mathbf{r}}(t)\big)\, \mathrm{d}t = \int_{t_1}^{t_2} \widetilde{L}\big(t, \mathbf{q}(t), \dot{\mathbf{q}}(t)\big)\, \mathrm{d}t =: \widetilde{S}[\mathbf{q}] .$$

By Hamilton's principle, the equations of motion are obtained from the condition $\delta S[\mathbf{r}] = 0$, i.e., $\delta \widetilde{S}[\mathbf{q}] = 0$, which is in turn equivalent to the Euler–Lagrange equations $\dfrac{\delta \widetilde{L}}{\delta \mathbf{q}} = 0$. Hence:

The equations of motion of the particle in curvilinear coordinates (q_1, q_2, q_3) are the Euler–Lagrange equations of the Lagrangian $\widetilde{L}$ in Eq. (4.36), that is

$$\frac{\mathrm{d}}{\mathrm{d}t}\frac{\partial \widetilde{L}}{\partial \dot{q}_i} - \frac{\partial \widetilde{L}}{\partial q_i} = 0\,, \qquad i = 1, 2, 3\,. \tag{4.37}$$

Note that $\widetilde{L}$ is nothing but the expression of the Lagrangian L in terms of the curvilinear coordinates $\mathbf{q}$ and their time derivatives. With this understanding the tilde can be dropped, and Eqs. (4.37) can be simply written as

$$\boxed{\frac{\mathrm{d}}{\mathrm{d}t}\frac{\partial L}{\partial \dot{q}_i} - \frac{\partial L}{\partial q_i} = 0\,, \qquad i = 1, 2, 3\,.} \tag{4.38}$$

Example 4.7. *Equations of motion in spherical coordinates.*

The kinetic energy of a particle of mass m in spherical coordinates (r, θ, φ) is given by

$$\frac{1}{2}m\dot{\mathbf{r}}^2 = \frac{m}{2}\,(\dot{r}^2 + r^2\dot{\theta}^2 + r^2\sin^2\theta\,\dot{\varphi}^2)$$

(cf. Eq. (2.11)). Thus the Lagrangian in these coordinates is given by

$$\boxed{L = T - V(t, r, \theta, \varphi) = \frac{m}{2}\,(\dot{r}^2 + r^2\dot{\theta}^2 + r^2\sin^2\theta\,\dot{\varphi}^2) - V(t, r, \theta, \varphi)\,.} \tag{4.39}$$

The particle's *equations of motion in spherical coordinates* are thus

$$\boxed{\begin{aligned}
&\frac{\mathrm{d}}{\mathrm{d}t}\frac{\partial L}{\partial \dot{r}} - \frac{\partial L}{\partial r} = m\ddot{r} - mr(\dot{\theta}^2 + \sin^2\theta\,\dot{\varphi}^2) + \frac{\partial V}{\partial r} = 0\,,\\
&\frac{\mathrm{d}}{\mathrm{d}t}\frac{\partial L}{\partial \dot{\theta}} - \frac{\partial L}{\partial \theta} = m\frac{\mathrm{d}}{\mathrm{d}t}(r^2\dot{\theta}) - mr^2\sin\theta\cos\theta\,\dot{\varphi}^2 + \frac{\partial V}{\partial \theta} = 0\,,\\
&\frac{\mathrm{d}}{\mathrm{d}t}\frac{\partial L}{\partial \dot{\varphi}} - \frac{\partial L}{\partial \varphi} = m\frac{\mathrm{d}}{\mathrm{d}t}(r^2\sin^2\theta\,\dot{\varphi}) + \frac{\partial V}{\partial \varphi} = 0\,.
\end{aligned}} \tag{4.40}$$

If the potential V does not depend on the azimuthal angle φ, then the quantity

$$\boxed{\frac{\partial L}{\partial \dot{\varphi}} = mr^2\sin^2\theta\,\dot{\varphi}}$$

is conserved. This conserved quantity is nothing but the z component of the angular momentum $\mathbf{J}$, since[a],

$$\mathbf{J} = r\mathbf{e}_r \times m(\dot{r}\mathbf{e}_r + r\dot{\theta}\,\mathbf{e}_\theta + r\sin\theta\dot{\varphi}\,\mathbf{e}_\varphi) = mr^2\dot{\theta}\mathbf{e}_\varphi - mr^2\sin\theta\dot{\varphi}\,\mathbf{e}_\theta\,,$$
$$\mathbf{e}_z = \cos\theta\,\mathbf{e}_r - \sin\theta\,\mathbf{e}_\theta \quad\Longrightarrow\quad J_z = \mathbf{J}\cdot\mathbf{e}_z = mr^2\sin^2\theta\,\dot{\varphi}\,.$$

Likewise, if the potential V is independent of t, i.e., if V is a function of (r,θ,φ) only, then $\dfrac{\partial L}{\partial t} = 0$, and hence the quantity

$$h = \dot{r}\frac{\partial L}{\partial \dot{r}} + \dot{\theta}\frac{\partial L}{\partial \dot{\theta}} + \dot{\varphi}\frac{\partial L}{\partial \dot{\varphi}} - L = m\,(\dot{r}^2 + r^2\dot{\theta}^2 + r^2\sin^2\theta\,\dot{\varphi}^2) - L = 2T - (T-V) = T+V$$

is conserved. This is, of course, the law of *conservation of energy* discussed in the previous chapters. Note, finally, that even if the potential is independent of θ the function $\dfrac{\partial L}{\partial \dot{\theta}}$ is *not* conserved, since the angle θ appears explicitly in the kinetic energy and as a consequence $\dfrac{\partial L}{\partial \theta}$ never vanishes.

[a]To avoid confusion with the Lagrangian L, throughout this chapter we shall denote by $\mathbf{J}$ the angular momentum.

Example 4.8. *Equations of motion in polar coordinates.*

If a particle moves on a plane subject to an irrotational force with potential V, its Lagrangian in polar coordinates (r,φ) is given by

$$\boxed{L = T - V(r,\varphi) = \frac{m}{2}\,(\dot{r}^2 + r^2\dot{\varphi}^2) - V(t,r,\varphi).}$$

Its corresponding Euler–Lagrange equations are

$$\boxed{\begin{aligned} \frac{\mathrm{d}}{\mathrm{d}t}\frac{\partial L}{\partial \dot{r}} - \frac{\partial L}{\partial r} &= m\ddot{r} - mr\dot{\varphi}^2 + \frac{\partial V}{\partial r} = 0\,,\\ \frac{\mathrm{d}}{\mathrm{d}t}\frac{\partial L}{\partial \dot{\varphi}} - \frac{\partial L}{\partial \varphi} &= m\frac{\mathrm{d}}{\mathrm{d}t}(r^2\dot{\varphi}) + \frac{\partial V}{\partial \varphi} = 0\,. \end{aligned}} \tag{4.41}$$

Taking into account that

$$\nabla V = \frac{\partial V}{\partial r}\,\mathbf{e}_r + \frac{1}{r}\frac{\partial V}{\partial \varphi}\,\mathbf{e}_\varphi = -\mathbf{F} = -F_r\mathbf{e}_r - F_\varphi\mathbf{e}_\varphi\,,$$

we can write the previous equations as

$$\ddot{r} - r\dot{\varphi}^2 = \frac{F_r}{m}\,, \qquad \frac{1}{r}\frac{\mathrm{d}}{\mathrm{d}t}(r^2\dot{\varphi}) = \frac{F_\varphi}{m}\,.$$

Thus the left-hand sides of the previous equations are nothing but the radial and angular components of the acceleration, a_r and a_φ. When the potential V is independent of φ (in which case the force is *central*), the second equation of motion yields the law of conservation of angular momentum

$$mr^2\dot{\varphi} = \text{const.} = J\,,$$

while the equation of motion for the radial coordinate can be written as

$$m\ddot{r} - \frac{J^2}{mr^3} + \frac{\partial V(t,r)}{\partial r} = 0\,.$$

Finally, when V does not depend explicitly on time (i.e., when the force is not just irrotational but conservative) we have $\dfrac{\partial L}{\partial t} = 0$, which implies that the energy integral

$$h = \dot{r}\frac{\partial L}{\partial \dot{r}} + \dot{\varphi}\frac{\partial L}{\partial \dot{\varphi}} - L = m\,(\dot{r}^2 + r^2\dot{\varphi}^2) - L = 2T - (T - V) = T + V$$

is conserved. This is nothing but the law of conservation of energy discussed in the previous two chapters.

Exercise 4.3. Compare Eqs. (4.40) with Newton's second law written in spherical coordinates.

Solution. From the expression (2.18) of the gradient in spherical coordinates

$$\frac{\partial V}{\partial \mathbf{r}} = \frac{\partial V}{\partial r}\mathbf{e}_r + \frac{1}{r}\frac{\partial V}{\partial \theta}\mathbf{e}_\theta + \frac{1}{r\sin\theta}\frac{\partial V}{\partial \varphi}\mathbf{e}_\varphi\,,$$

it follows that Newton's equations of motion $m\mathbf{a} = -\dfrac{\partial V}{\partial \mathbf{r}}$ can be written as

$$ma_r = -\frac{\partial V}{\partial r}, \qquad ma_\theta = -\frac{1}{r}\frac{\partial V}{\partial \theta}, \qquad ma_\varphi = -\frac{1}{r\sin\theta}\frac{\partial V}{\partial \varphi}.$$

Using Eqs. (2.12) for the acceleration in spherical coordinates we immediately obtain the Euler–Lagrange equations (4.40).

Exercise 4.4. i) Write down the equations of motion of a particle subject to an irrotational force in an arbitrary system of orthogonal curvilinear coordinates $\mathbf{q} = (q_1, q_2, q_3)$. ii) Use these equations to find an expression for the components of the acceleration in the curvilinear coordinates $\mathbf{q}$.

Solution. i) In any orthogonal curvilinear coordinate system $\mathbf{q}$ we have

$$\dot{\mathbf{r}}^2 = \sum_{i=1}^{3} h_i^2(\mathbf{q})\dot{q}_i^2, \qquad \frac{\partial V}{\partial \mathbf{r}} = \sum_{i=1}^{3} \frac{1}{h_i(\mathbf{q})} \frac{\partial V}{\partial q_i} \mathbf{e}_{q_i},$$

where the metric coefficients $h_i(\mathbf{q})$ are defined in Eq. (2.3). From the expression for $\dot{\mathbf{r}}^2$ it is straightforward to compute the Lagrangian in the curvilinear coordinates $\mathbf{q}$:

$$L = T - V(\mathbf{q}) = \frac{m}{2}\dot{\mathbf{r}}^2 - V(\mathbf{q}) = \frac{m}{2}\sum_{i=1}^{3} h_i^2(\mathbf{q})\dot{q}_i^2 - V(\mathbf{q}).$$

The canonical momenta are then

$$p_i = \frac{\partial L}{\partial \dot{\mathbf{q}}_i} = m h_i^2(\mathbf{q})\dot{q}_i, \qquad i = 1, 2, 3,$$

from which we easily obtain the Euler–Lagrange equations of motion

$$\dot{p}_i = m h_i^2 \ddot{q}_i + 2m h_i \dot{q}_i \sum_{j=1}^{3} \frac{\partial h_i}{\partial q_j}\dot{q}_j = \frac{\partial L}{\partial q_i} = m \sum_{j=1}^{3} h_j \frac{\partial h_j}{\partial q_i}\dot{q}_j^2 - \frac{\partial V}{\partial q_i}, \qquad i = 1, 2, 3,$$

or equivalently

$$\boxed{h_i \ddot{q}_i + 2\dot{q}_i \sum_{j=1}^{3} \frac{\partial h_i}{\partial q_j}\dot{q}_j - \sum_{j=1}^{3} \frac{h_j}{h_i}\frac{\partial h_j}{\partial q_i}\dot{q}_j^2 = -\frac{1}{m h_i}\frac{\partial V}{\partial q_i}, \qquad i = 1, 2, 3.}$$

ii) The components of the acceleration in the $\mathbf{q}$ orthogonal coordinate system can be found comparing the previous equations with Newton's equations of motion

$$a_{q_i} = \frac{F_{q_i}}{m} = -\frac{1}{m h_i}\frac{\partial V}{\partial q_i}, \qquad i = 1, 2, 3.$$

We thus obtain

$$\boxed{a_{q_i} = h_i \ddot{q}_i + 2\dot{q}_i \sum_{j=1}^{3} \frac{\partial h_i}{\partial q_j}\dot{q}_j - \sum_{j=1}^{3} \frac{h_j}{h_i}\frac{\partial h_j}{\partial q_i}\dot{q}_j^2, \qquad i = 1, 2, 3.} \tag{4.42}$$

The previous expression for the components of the acceleration in the (orthogonal) curvilinear coordinates $\mathbf{q}$ coincides with the one obtained in Exercise 2.3.

Indeed, by Eq. (2.16) in the previous exercise we have

$$
\begin{aligned}
h_k \sum_{i,j=1}^{3} \Gamma_k^{ij} \dot{q}_i \dot{q}_j &= \frac{1}{2h_k} \sum_{i,j=1}^{3} \left[\frac{\partial}{\partial q_i}(h_j^2)\delta_{jk} + \frac{\partial}{\partial q_j}(h_i^2)\delta_{ik} - \frac{\partial}{\partial q_k}(h_j^2)\delta_{ij} \right] \dot{q}_i \dot{q}_j \\
&= \frac{1}{2h_k} \left(2 \sum_{j=1}^{3} \frac{\partial}{\partial q_j}(h_k^2) \dot{q}_k \dot{q}_j - \sum_{j=1}^{3} \frac{\partial}{\partial q_k}(h_j^2) \dot{q}_j^2 \right) \\
&= 2\dot{q}_k \sum_{j=1}^{3} \frac{\partial h_k}{\partial q_j} \dot{q}_j - \sum_{j=1}^{3} \frac{h_j}{h_k} \frac{\partial h_j}{\partial q_k} \dot{q}_j^2 .
\end{aligned}
$$

From Eq. (2.15) it then follows that

$$
a_{q_k} = h_k \ddot{q}_k + 2\dot{q}_k \sum_{j=1}^{3} \frac{\partial h_k}{\partial q_j} \dot{q}_j - \sum_{j=1}^{3} \frac{h_j}{h_k} \frac{\partial h_j}{\partial q_j} \dot{q}_j^2 ,
$$

which is Eq. (4.42) with i replaced by k.

4.2.4 Lagrangian of a charged particle in an electromagnetic field

We shall next derive the Euler–Lagrange equations of motion of a particle of mass m and charge e in the electromagnetic field generated by the potentials $\Phi(t, \mathbf{r})$ and $\mathbf{A}(t, \mathbf{r})$. As we saw in Chapter 2, the Newtonian equations of motion are

$$
m\ddot{\mathbf{r}} = -e\left(\frac{\partial \Phi}{\partial \mathbf{r}} + \frac{\partial \mathbf{A}}{\partial t} \right) + e\dot{\mathbf{r}} \times (\nabla \times \mathbf{A}) .
$$

To simplify these equations, we recall the identity

$$
\mathbf{a} \times (\nabla \times \mathbf{b}) = \nabla(\mathbf{a} \cdot \mathbf{b}) - (\mathbf{a} \cdot \nabla)\mathbf{b}, \tag{4.43}
$$

where $\mathbf{a}(\mathbf{r})$, $\mathbf{b}(\mathbf{r})$ are vector fields and $\mathbf{a} \cdot \nabla$ is the differential operator defined by

$$
\mathbf{a} \cdot \nabla = \sum_{i=1}^{3} a_i \frac{\partial}{\partial x_i} .
$$

We shall also use in what follows the alternative notation

$$
(\mathbf{a} \cdot \nabla)\mathbf{b} \equiv \sum_{i=1}^{3} a_i \frac{\partial \mathbf{b}}{\partial x_i} =: \frac{\partial \mathbf{b}}{\partial \mathbf{r}} \mathbf{a} ,
$$

where $\frac{\partial \mathbf{b}}{\partial \mathbf{r}}$ is the matrix with elements $\frac{\partial b_i}{\partial x_j}$. From Eq. (4.43) it follows that

$$
-\frac{\partial \mathbf{A}}{\partial t} + \dot{\mathbf{r}} \times (\nabla \times \mathbf{A}) = -\frac{\partial \mathbf{A}}{\partial t} + \frac{\partial}{\partial \mathbf{r}}(\dot{\mathbf{r}} \cdot \mathbf{A}) - \frac{\partial \mathbf{A}}{\partial \mathbf{r}} \cdot \dot{\mathbf{r}} = \frac{\partial}{\partial \mathbf{r}}(\dot{\mathbf{r}} \cdot \mathbf{A}) - \frac{\mathrm{d}\mathbf{A}}{\mathrm{d}t} , \tag{4.44}
$$

and we can therefore rewrite the equation of motion as

$$\frac{\mathrm{d}}{\mathrm{d}t}(m\dot{\mathbf{r}} + e\mathbf{A}) + e\frac{\partial}{\partial \mathbf{r}}(\Phi - \dot{\mathbf{r}} \cdot \mathbf{A}) = 0 .$$

These are the Euler–Lagrange equations of a Lagrangian $L(t, \mathbf{r}, \dot{\mathbf{r}})$ provided that

$$\frac{\partial L}{\partial \dot{\mathbf{r}}} = m\dot{\mathbf{r}} + e\mathbf{A} , \qquad \frac{\partial L}{\partial \mathbf{r}} = -e\frac{\partial}{\partial \mathbf{r}}(\Phi - \dot{\mathbf{r}} \cdot \mathbf{A}) .$$

Integrating the second equation we obtain

$$L = e(\dot{\mathbf{r}} \cdot \mathbf{A} - \Phi) + g(t, \dot{\mathbf{r}}) ,$$

and substituting back into the first one we have

$$\frac{\partial g}{\partial \dot{\mathbf{r}}} = m\dot{\mathbf{r}} \quad \Longrightarrow \quad g = \frac{1}{2}m\dot{\mathbf{r}}^2 ,$$

up to an arbitrary (and inessential) function of t. We have thus obtained the following result:

The equations of motion of a particle of mass m and charge e in the electromagnetic field generated by the potentials $\Phi(t, \mathbf{r})$ and $\mathbf{A}(t, \mathbf{r})$ are the Euler–Lagrange equations of the Lagrangian

$$L(t, \mathbf{r}, \dot{\mathbf{r}}) = \frac{1}{2}m\dot{\mathbf{r}}^2 - e\Phi(t, \mathbf{r}) + e\dot{\mathbf{r}} \cdot \mathbf{A}(t, \mathbf{r}) . \tag{4.45}$$

Note that we can express the previous Lagrangian as

$$L = T - U ,$$

where the potential U is given by

$$U(t, \mathbf{r}, \dot{\mathbf{r}}) = e\left[\Phi(t, \mathbf{r}) - \dot{\mathbf{r}} \cdot \mathbf{A}(t, \mathbf{r})\right] \tag{4.46}$$

and is thus *velocity dependent.* If the fields are *static*, i.e., if

$$\frac{\partial \Phi}{\partial t} = 0 , \qquad \frac{\partial \mathbf{A}}{\partial t} = 0 ,$$

then L is independent of t and therefore

$$h = \dot{\mathbf{r}} \cdot \frac{\partial L}{\partial \dot{\mathbf{r}}} - L = m\dot{\mathbf{r}}^2 + e\dot{\mathbf{r}} \cdot \mathbf{A} - L = \frac{1}{2}m\dot{\mathbf{r}}^2 + e\Phi(\mathbf{r})$$

is conserved. This is the conservation law of the particle's electromechanical energy discussed in Section (2.4.4).

4.3 SYSTEMS WITH CONSTRAINTS

4.3.1 Motion of a particle on a smooth surface

The simplest case of a mechanical system with constraints is that of a particle of mass m subject to an external irrotational force with potential $V(t, \mathbf{r})$, whose coordinates $\mathbf{r}$ satisfy at every instant t the **constraint** (i.e., restriction)

$$\boxed{\phi(t, \mathbf{r}) = 0\,.} \tag{4.47}$$

In general, (4.47) is the equation of a moving *surface*. In particular, if ϕ does not depend[4] on t the particle is forced to move on the fixed surface of equation $\phi(\mathbf{r}) = 0$. Although the external force is irrotational, it is essential to take also into account the **reaction** (or **constraint**) **force** $\mathbf{F}^{(c)}(t, \mathbf{r}, \dot{\mathbf{r}})$ exerted by the constraint surface (4.47) on the particle at each instant t, so that Newton's equations of motion are in this case

$$\boxed{m\ddot{\mathbf{r}} + \frac{\partial V(t, \mathbf{r})}{\partial \mathbf{r}} = \mathbf{F}^{(c)}(t, \mathbf{r}, \dot{\mathbf{r}})\,.} \tag{4.48}$$

We ask ourselves whether Eqs. (4.48) are the Euler–Lagrange equations of some action functional. In order to answer this question, let us introduce two *independent coordinates* $(q_1, q_2) = \mathbf{q}$ parametrizing the surface (4.47). For instance, if

$$\phi(t, \mathbf{r}) = \mathbf{r}^2 - a(t)^2\,, \tag{4.49}$$

which is the equation of a sphere centered at the origin with variable radius $a(t) \geqslant 0$, we can use spherical coordinates $q_1 = \theta$, $q_2 = \varphi$. We shall express (with a slight abuse of notation) the relation between the generalized coordinates $\mathbf{q}$ and the Cartesian ones $\mathbf{r}$ in the general form

$$\mathbf{r} = \mathbf{r}(t, \mathbf{q})\,, \tag{4.50}$$

where for each fixed t the mapping $\mathbf{q} \mapsto \mathbf{r}$ must be *bijective* (from an open subset of $\mathbb{R}^2$ to an open subset of the constraint surface at time t). For instance, for the constraint (4.49) the function $\mathbf{r}(t, \mathbf{q}) = \mathbf{r}(t, \theta, \varphi)$ is given by

$$\mathbf{r}(t, \theta, \varphi) = a(t)(\sin\theta\cos\varphi, \sin\theta\sin\varphi, \cos\theta)\,.$$

In general, we can specify the position of the particle at each instant t using the value $\mathbf{q}(t)$ taken by its **generalized coordinates** q_i at that time: indeed, $\mathbf{r} = \mathbf{r}(t, \mathbf{q}(t))$. It is important to note that, while the three Cartesian coordinates x_i are *not* independent, since they are connected by the relation (4.47), the two generalized coordinates q_i are by construction independent variables (i.e., can take *arbitrary* values in some open subset of $\mathbb{R}^2$). For this reason, it is easy to convince oneself that only two of the three (scalar) equations of motion (4.48) can actually be independent.

[4]Time-independent constraints are called *scleronomic* in the specialized literature, while time-dependent constraints are termed *rheonomic*. This terminology originates from the Greek words σκλερός (skleros), meaning "hard, rigid," ῥέο (rheo), meaning "to flow," and νόμος (nomos), meaning "law or rule."

We shall assume that the constraint surface (4.47) is *smooth*, i.e., that there is no *friction*. If this is the case, *the constraint force at each instant t is* perpendicular *to the corresponding* instantaneous *constraint surface* $\phi(t, \mathbf{r}) = 0$. When this happens we shall say that the constraint (4.47) is **ideal**. To formulate analytically the condition of ideal constraint, note that for each t the two vectors

$$\frac{\partial \mathbf{r}(t, \mathbf{q})}{\partial q_i}, \qquad i = 1, 2, \tag{4.51}$$

are *tangent to the constraint surface* at the point with generalized coordinates $\mathbf{q}$, and in fact are a *basis* of the tangent plane to the instantaneous surface $\phi(t, \mathbf{r}) = 0$ at this point. Hence *the constraint is ideal if the constraint force verifies the condition*

$$\boxed{\mathbf{F}^{(c)} \cdot \frac{\partial \mathbf{r}}{\partial q_i} = 0, \qquad i = 1, 2,} \tag{4.52}$$

at each point on the constraint surface. Projecting the equation of motion (4.48) onto the tangent plane to the constraint surface (i.e., multiplying scalarly by each of the two vectors (4.51)) we obtain the two *independent* equations

$$m\ddot{\mathbf{r}} \frac{\partial \mathbf{r}}{\partial q_i} + \frac{\partial V}{\partial \mathbf{r}} \frac{\partial \mathbf{r}}{\partial q_i} = 0, \qquad i = 1, 2,$$

or equivalently

$$\boxed{m\ddot{\mathbf{r}} \frac{\partial \mathbf{r}}{\partial q_i} + \frac{\partial V}{\partial q_i} = 0, \qquad i = 1, 2.} \tag{4.53}$$

Differentiating the relation (4.50) with respect to t, q_i, and $\dot{q}_i$ we obtain the identities[5]

$$\boxed{\begin{aligned} &\dot{\mathbf{r}} = \frac{\partial \mathbf{r}}{\partial t} + \frac{\partial \mathbf{r}}{\partial \mathbf{q}} \dot{\mathbf{q}} = \dot{\mathbf{r}}(t, \mathbf{q}, \dot{\mathbf{q}}) \quad \Longrightarrow \quad \frac{\partial \dot{\mathbf{r}}}{\partial \dot{q}_i} = \frac{\partial \mathbf{r}}{\partial q_i}, \\ &\frac{\partial \dot{\mathbf{r}}}{\partial q_i} = \frac{\partial^2 \mathbf{r}}{\partial t \partial q_i} + \frac{\partial^2 \mathbf{r}}{\partial q_i \partial \mathbf{q}} \dot{\mathbf{q}} = \frac{\partial}{\partial t}\left(\frac{\partial \mathbf{r}}{\partial q_i}\right) + \frac{\partial}{\partial \mathbf{q}}\left(\frac{\partial \mathbf{r}}{\partial q_i}\right) \dot{\mathbf{q}} = \frac{\mathrm{d}}{\mathrm{d}t} \frac{\partial \mathbf{r}}{\partial q_i}, \end{aligned}} \tag{4.54}$$

and hence

$$\ddot{\mathbf{r}} \frac{\partial \mathbf{r}}{\partial q_i} = \frac{\mathrm{d}}{\mathrm{d}t}\left(\dot{\mathbf{r}} \frac{\partial \mathbf{r}}{\partial q_i}\right) - \dot{\mathbf{r}} \frac{\mathrm{d}}{\mathrm{d}t}\left(\frac{\partial \mathbf{r}}{\partial q_i}\right) = \frac{\mathrm{d}}{\mathrm{d}t}\left(\dot{\mathbf{r}} \frac{\partial \dot{\mathbf{r}}}{\partial \dot{q}_i}\right) - \dot{\mathbf{r}} \frac{\partial \dot{\mathbf{r}}}{\partial q_i} = \frac{\mathrm{d}}{\mathrm{d}t} \frac{\partial}{\partial \dot{q}_i}\left(\frac{1}{2} \dot{\mathbf{r}}^2\right) - \frac{\partial}{\partial q_i}\left(\frac{1}{2} \dot{\mathbf{r}}^2\right). \tag{4.55}$$

Thus Eqs. (4.53) can be written as

$$\frac{\mathrm{d}}{\mathrm{d}t} \frac{\partial T}{\partial \dot{q}_i} - \frac{\partial}{\partial q_i}(T - V) = 0, \qquad i = 1, 2,$$

---

[5]We are again using the notation

$$\frac{\partial \mathbf{r}}{\partial \mathbf{q}} \dot{\mathbf{q}} := \sum_{j=1,2} \frac{\partial \mathbf{r}}{\partial q_j} \dot{q}_j, \qquad \frac{\partial^2 \mathbf{r}}{\partial q_i \partial \mathbf{q}} \dot{\mathbf{q}} := \sum_{j=1,2} \frac{\partial^2 \mathbf{r}}{\partial q_i \partial q_j} \dot{q}_j.$$

or, taking into account that V does not depend on $\dot{\mathbf{q}}$,

$$\frac{\mathrm{d}}{\mathrm{d}t}\frac{\partial}{\partial \dot{q}_i}(T-V)-\frac{\partial}{\partial q_i}(T-V)=0\,, \qquad i=1,2\,. \tag{4.56}$$

These are the Euler–Lagrange equations of the Lagrangian $L = T - V$, where *it is understood that the kinetic energy T and the potential V must be expressed in terms of the independent variables* $(t, \mathbf{q}, \dot{\mathbf{q}})$ *using Eq.* (4.50) *and its derivative with respect to t* (i.e., the first Eq. (4.54)). We have thus proved the following fundamental result:

The trajectory $\mathbf{q}(t)$ followed by a particle as it moves from a point with generalized coordinates $\mathbf{q}_1$ (at $t = t_1$) to a second point with generalized coordinates $\mathbf{q}_2$ (at $t = t_2$) obeying an ideal constraint (4.47) at all times is a *stationary point* of the *action*

$$S[\mathbf{q}] = \int_{t_1}^{t_2} L\big(t, \mathbf{q}(t), \dot{\mathbf{q}}(t)\big)\,\mathrm{d}t \qquad (\text{with} \quad \mathbf{q}(t_1)=\mathbf{q}_1, \ \ \mathbf{q}(t_2)=\mathbf{q}_2)\,,$$

where the Lagrangian L equals $T - V$ expressed in terms of the independent variables $(t, \mathbf{q}, \dot{\mathbf{q}})$. The equations of motion are therefore the Euler–Lagrange equations of L,

$$\frac{\delta L}{\delta \mathbf{q}} = 0\,,$$

expressing the vanishing of the variation of the action functional:

$$\delta S[\mathbf{q}] = 0\,.$$

In other words:

Hamilton's principle remains valid in this case if the constraint is *ideal*, i.e., if the constraint force $\mathbf{F}^{(c)}$ satisfies condition (4.52). Moreover, when this is the case the Lagrangian L is equal to $T - V$ expressed in terms of the generalized coordinates q_i and their time derivatives $\dot{q}_i$.

• A **virtual displacement** is a curve $\mathbf{r} = \mathbf{r}(u)$ (where $u \in [u_1, u_2]$ is an arbitrary parameter) entirely contained in the *instantaneous* constraint surface $\phi(t, \mathbf{r}) = 0$ at a certain *fixed* time t, i.e., such that

$$\phi(t, \mathbf{r}(u)) = 0\,, \qquad \forall u \in [u_1, u_2]\,.$$

In other words, a virtual displacement is a succession of *possible* particle positions at the *same* instant t. Note that if the constraint is time-dependent the particle's trajectories are *not* virtual displacements, since for $t' \neq t$ the vector $\mathbf{r}(t')$ belongs to the surface $\phi(t', \mathbf{r}) = 0$, in general different from $\phi(t, \mathbf{r}) = 0$. On the other hand, since a virtual displacement $\mathbf{r}(u)$ is contained in the instantaneous constraint surface $\phi(t, \mathbf{r}) = 0$ for all $u \in [u_1, u_2]$, its tangent vector $\mathbf{r}'(u)$ is *tangent* to this surface at the point $\mathbf{r}(u)$. Hence the ideal constraint condition implies in this case that

$$\mathbf{F}^{(c)}\big(t, \mathbf{r}(u), \mathbf{v}(u)\big) \cdot \mathbf{r}'(u) = 0\,, \qquad \forall u \in [u_1, u_2]\,, \tag{4.57}$$

where the prime denotes derivative with respect to u and $\mathbf{v}(u)$ is a possible velocity[6] for the particle at the point $\mathbf{r}(u)$. Thus *the work W_{12} done by the constraint force along the virtual displacement $\mathbf{r}(u)$ vanishes:*

$$W_{12} = \int_{u_1}^{u_2} \mathbf{F}^{(c)}\big(t, \mathbf{r}(u), \mathbf{v}(u)\big) \cdot \mathbf{r}'(u)\, du = 0\,, \tag{4.58}$$

for *arbitrary* $\mathbf{v}(u)$. Conversely, if the constraint force satisfies (4.58) for *any* virtual displacement $\mathbf{r}(u)$ and allowed velocity $\mathbf{v}(u)$, then Eq. (4.57) holds. This implies that the constraint force is perpendicular to the surface $\phi(t, \mathbf{r}) = 0$ at each point, since any vector tangent to this surface can be obtained as the tangent vector $\mathbf{r}'(u)$ to a curve $\mathbf{r}(u)$ contained in it, i.e., to a virtual displacement. We have thus proved the following result, known as the **principle of virtual work**:

The constraint is ideal—and, thus, *Hamilton's principle* holds—if and only if the constraint force does no work along any *virtual* displacement of the particle.

If the constraint equation (4.47) is *independent of t* (which is the most common case in practice), then the particle's trajectories are virtual displacements, and the principle of virtual work simply states that *the constraint is ideal if and only if the constraint force does no work along* any *trajectory.*

[6]Differentiating the constraint equation $\phi(t, \mathbf{r}) = 0$ with respect to time we obtain

$$\frac{\partial \phi}{\partial t}(t, \mathbf{r}) + \frac{\partial \phi}{\partial \mathbf{r}}(t, \mathbf{r})\dot{\mathbf{r}} = 0\,,$$

which is the constraint satisfied by the particle's velocity $\dot{\mathbf{r}}$ if the particle is at the point $\mathbf{r}$ at time t. Thus the vector field $\mathbf{v}(u)$ must satisfy the condition

$$\frac{\partial \phi}{\partial t}(t, \mathbf{r}(u)) + \frac{\partial \phi}{\partial \mathbf{r}}(t, \mathbf{r}(u))\mathbf{v}(u) = 0\,, \qquad \forall u \in [u_1, u_2].$$

In particular, if the constraint is time-independent then $\mathbf{v}(u)$ must simply be orthogonal to the gradient $\dfrac{\partial \phi}{\partial \mathbf{r}}(\mathbf{r}(u))$, and thus tangent to the constraint surface at each point $\mathbf{r}(u)$.

4.3.2 System of N particles with constraints

Consider next the more general case of a system of N particles subject to the irrotational forces (4.32) and to the $l < 3N$ *independent* constraints[7]

$$\boxed{\phi_i(t, \mathbf{r}_1, \ldots, \mathbf{r}_N) = 0\,, \qquad i = 1, \ldots, l\,.} \tag{4.59}$$

Constraints of this type, which are independent of the particles' *velocities*, are called *holonomic*[8]. The vector

$$\mathbf{x} := (\mathbf{r}_1, \ldots, \mathbf{r}_N) \in \mathbb{R}^{3N}$$

representing the state of the system must belong at each instant t to the surface in $\mathbb{R}^{3N}$—or *manifold*, in a more mathematical language—specified by Eqs. (4.59). Since this manifold has dimension $3N - l = n$, in general it can be parametrized by n independent coordinates $(q_1, \ldots, q_n) =: \mathbf{q}$, in terms of which the vector $\mathbf{x}$ will be expressed by a certain function $\mathbf{x}(t, \mathbf{q})$:

$$\mathbf{x} = \mathbf{x}(t, \mathbf{q})\,. \tag{4.60}$$

In other words, *the state of the system at each instant is uniquely determined by the value of the n* ***generalized coordinates*** *q_i at that instant.* We shall accordingly say that the system possesses n **degrees of freedom**. In particular, the system's trajectory in the space $\mathbb{R}^{3N}$ can be specified by a curve $\mathbf{q}(t)$ in the open subset of $\mathbb{R}^n$ on which the generalized coordinates $\mathbf{q}$ vary, called **configuration space**, through the equation

$$\mathbf{x} = \mathbf{x}(t, \mathbf{q}(t)).$$

It is important to note that, while the *Cartesian* coordinates $\mathbf{x}$ are *not* independent (since they are related by the constraint equations (4.59)), the *generalized* coordinates $\mathbf{q}$ are by construction independent variables.

Again, *we shall suppose that the constraints are* ***ideal,*** in the sense that the constraint force acting on the point $\mathbf{x}$ representing the state of the system, i.e., the vector

$$\mathbf{F}^{(c)}(t, \mathbf{x}, \dot{\mathbf{x}}) := \left(\mathbf{F}_1^{(c)}(t, \mathbf{x}, \dot{\mathbf{x}}), \ldots, \mathbf{F}_N^{(c)}(t, \mathbf{x}, \dot{\mathbf{x}})\right) \in \mathbb{R}^{3N},$$

is *orthogonal to the constraint manifold* defined by Eqs. (4.59) at all times. Since the n vectors

$$\frac{\partial \mathbf{x}(t, \mathbf{q})}{\partial q_i}\,, \qquad i = 1, \ldots, n\,,$$

[7]Mathematically, the independence of the constraints (4.59) is equivalent to the condition that the Jacobian matrix of the vector-valued function $\boldsymbol{\phi} := (\phi_1, \ldots, \phi_l)$ with respect to the $3N$ variables $\mathbf{x} := (\mathbf{r}_1, \ldots, \mathbf{r}_N)$ be of *maximal rank* (equal to l) at all points:

$$\operatorname{rank}\left(\frac{\partial \phi_i}{\partial x_j}\right)_{\substack{1 \leqslant i \leqslant l \\ 1 \leqslant j \leqslant 3N}} = l\,.$$

[8]From the Greek ὅλος (holos), meaning "whole," and νόμος (nomos), meaning "law, rule," probably referring to the fact that holonomic constraints apply to the whole configuration space of the system independently of its velocity.

are a basis of the tangent space to the constraint manifold at each point, the previous condition is equivalent to the relations

$$\mathbf{F}^{(c)} \cdot \frac{\partial \mathbf{x}}{\partial q_i} = \left(\mathbf{F}_1^{(c)}, \dots, \mathbf{F}_N^{(c)}\right) \cdot \left(\frac{\partial \mathbf{r}_1}{\partial q_i}, \dots, \frac{\partial \mathbf{r}_N}{\partial q_i}\right) = \sum_{j=1}^{N} \mathbf{F}_j^{(c)} \cdot \frac{\partial \mathbf{r}_j}{\partial q_i} = 0, \quad i = 1, \dots, n. \tag{4.61}$$

As in the case of a single particle treated above, Eq. (4.61) is equivalent to the *principle of virtual work*, according to which *the constraint forces do no work along any virtual displacement of the system.* By definition, a virtual displacement in this more general context is any curve $\mathbf{x}(u)$ (with $u \in [u_1, u_2]$ an arbitrary parameter) entirely contained in an *instantaneous* constraint surface $\phi_i(t, \mathbf{x}) = 0$ $(i = 1, \dots, l)$ at a *fixed* instant t. Indeed, the work W_{12} done by the constraint forces acting on the system along the virtual displacement $\mathbf{x}(u) = \big(\mathbf{r}_1(u), \dots, \mathbf{r}_N(u)\big)$ is given by

$$W_{12} = \sum_{i=1}^{N} \int_{u_1}^{u_2} F_i^{(c)}\big(t, \mathbf{x}(u), \mathbf{v}(u)\big) \cdot \mathbf{r}_i'(u)\, \mathrm{d}u = \int_{u_1}^{u_2} \mathbf{F}^{(c)}\big(t, \mathbf{x}(u), \mathbf{v}(u)\big) \cdot \mathbf{x}'(u)\, \mathrm{d}u\,,$$

where $\mathbf{v}(u) = (\mathbf{v}_1(u), \dots, \mathbf{v}_N(u)) \in \mathbb{R}^{3N}$ is a possible velocity[9] for the system at the point $\mathbf{x}(u)$. Hence the principle of virtual work, i.e., the requirement that $W_{12} = 0$ for an *arbitrary* virtual displacement $\mathbf{x}(u)$, is equivalent to requiring that

$$\mathbf{F}^{(c)}\big(t, \mathbf{x}(u), \mathbf{v}(u)\big) \cdot \mathbf{x}'(u) = 0$$

for every possible velocity vector $\mathbf{v}(u)$ and every curve $\mathbf{x}(u)$ contained in the instantaneous constraint manifold at time t. This is equivalent to Eq. (4.61), since $\mathbf{x}'(u)$ is an arbitrary tangent vector to the instantaneous constraint manifold at the point $\mathbf{x}(u)$.

Under these conditions—that is, if the constraints are *ideal* and the applied forces acting on the system are *irrotational*—proceeding as in the previous subsection one can prove that *Hamilton's principle is still valid*:

The trajectory $\mathbf{q}(t)$ joining two states $\mathbf{q}_1$ (at $t = t_1$) and $\mathbf{q}_2$ (at $t = t_2$) of a system of particles subject to irrotational forces and *ideal* holonomic constraints is a *stationary point* of the action

$$S[\mathbf{q}] = \int_{t_1}^{t_2} L\big(t, \mathbf{q}(t), \dot{\mathbf{q}}(t)\big)\, \mathrm{d}t\,,$$

[9]Proceeding as in the case of a single particle, it is straightforward to show that the possible velocity vectors $\mathbf{v}(u)$ at a point $\mathbf{x}(u)$ in the instantaneous constraint manifold (4.59) are determined by the linear system

$$\frac{\partial \phi_i}{\partial t}(t, \mathbf{x}(u)) + \frac{\partial \phi_i}{\partial \mathbf{x}}(t, \mathbf{x}(u)) \cdot \mathbf{v}(u) = 0, \qquad i = 1, \dots, l.$$

where the Lagrangian L equals $T-V$ expressed in terms of the independent variables $(t, \mathbf{q}, \dot{\mathbf{q}})$. The equations of motion are thus the Euler–Lagrange equations

$$\frac{\delta L}{\delta \mathbf{q}} = 0\,,$$

expressing the vanishing of the variation of the action functional:

$$\delta S[\mathbf{q}] = 0\,.$$

Exercise 4.5. Give a detailed proof of the previous result.

Solution. The system's equations of motion can be written in vector form as

$$(m_1\ddot{\mathbf{r}}_1, \ldots, m_N\ddot{\mathbf{r}}_N) + \frac{\partial V}{\partial \mathbf{x}} = \mathbf{F}^{(c)}\,.$$

Projecting onto the direction of the vector $\dfrac{\partial \mathbf{x}}{\partial q_i}$ and taking into account Eq. (4.61) we obtain

$$(m_1\ddot{\mathbf{r}}_1, \ldots, m_N\ddot{\mathbf{r}}_N)\cdot\frac{\partial \mathbf{x}}{\partial q_i} + \frac{\partial V}{\partial \mathbf{x}}\frac{\partial \mathbf{x}}{\partial q_i} = \sum_{j=1}^{N} m_j\ddot{\mathbf{r}}_j\frac{\partial \mathbf{r}_j}{\partial q_i} + \frac{\partial V}{\partial q_i} = 0\,, \quad i = 1, \ldots, n\,.$$

From Eqs. (4.55) (with $\mathbf{r}_j$ instead of $\mathbf{r}$) it then follows that

$$\sum_{j=1}^{N} m_j\left[\frac{\mathrm{d}}{\mathrm{d}t}\frac{\partial}{\partial \dot{q}_i}\left(\frac{1}{2}\dot{\mathbf{r}}_j^2\right) - \frac{\partial}{\partial q_i}\left(\frac{1}{2}\dot{\mathbf{r}}_j^2\right)\right] + \frac{\partial V}{\partial q_i} = \frac{\mathrm{d}}{\mathrm{d}t}\frac{\partial T}{\partial \dot{q}_i} - \frac{\partial}{\partial q_i}(T - V) = 0\,,$$

$$i = 1, \ldots, n\,.$$

Since V is independent of $\dot{\mathbf{q}}$, the previous equations can be written in the form

$$\frac{\mathrm{d}}{\mathrm{d}t}\frac{\partial L}{\partial \dot{q}_i} - \frac{\partial L}{\partial q_i} = 0\,, \qquad i = 1, \ldots, n\,,$$

which are the Euler–Lagrange equations of the Lagrangian $L = T - V$.

From what we have just seen, to write down the equations of motion of a mechanical system of N particles subject to l independent ideal holonomic constraints, the remaining (applied) forces being irrotational, we can proceed as follows:

1) Introduce $n = 3N - l$ independent generalized coordinates $(q_1, \dots, q_n) = \mathbf{q}$ parametrizing the constraint manifold (4.59).
2) Express the kinetic energy

$$T = \frac{1}{2} \sum_{i=1}^{N} m_i \dot{\mathbf{r}}_i^2$$

and the potential V of the irrotational forces in terms of $(t, \mathbf{q}, \dot{\mathbf{q}})$, thus obtaining the two functions $T(t, \mathbf{q}, \dot{\mathbf{q}})$ and $V(t, \mathbf{q})$.
3) The system's equations of motion in the generalized coordinates q_i are the Euler–Lagrange equations of the Lagrangian

$$L(t, \mathbf{q}, \dot{\mathbf{q}}) = T(t, \mathbf{q}, \dot{\mathbf{q}}) - V(t, \mathbf{q}),$$

i.e.,

$$\frac{\mathrm{d}}{\mathrm{d}t} \frac{\partial L}{\partial \dot{q}_i} - \frac{\partial L}{\partial q_i} = 0, \qquad i = 1, \dots, n. \tag{4.62}$$

Notation. In classical mechanics textbooks, Eqs. (4.62) are often referred to simply as **Lagrange's equations** for the Lagrangian L.

• One of the advantages of the Lagrangian formulation for systems with constraints is that, as we have just seen, *in order to find the equations of motion it is not necessary to know the constraint forces* (all that is needed is to check that the constraints are *ideal*). In fact, *once these equations have been found* the constraint forces can always be computed using the formula[10]

$$\boxed{\mathbf{F}_i^{(c)} = m_i \ddot{\mathbf{r}}_i + \frac{\partial V}{\partial \mathbf{r}_i}, \qquad 1 \leqslant i \leqslant N,} \tag{4.63}$$

which is nothing but Newton's second law applied to the i-th particle.

Remark 4.1. Hamilton's principle is key to understanding in what sense classical mechanics is the $\hbar \to 0$ limit of quantum mechanics, with the help of Feynman's *path integral* formulation of the latter theory. According to this formulation, the probability $P(t_1, \mathbf{q}_1; t_2, \mathbf{q}_2)$ that a mechanical system with classical Lagrangian $L(t, \mathbf{q}, \dot{\mathbf{q}})$, whose generalized coordinates take the value $\mathbf{q}_1$ at a certain time t_1, is found to have generalized coordinates $\mathbf{q}_2$ at a different time t_2 is given by

$$P(t_1, \mathbf{q}_1; t_2, \mathbf{q}_2) = \left|\Phi(t_1, \mathbf{q}_1; t_2, \mathbf{q}_2)\right|^2,$$

[10]Indeed, $\ddot{\mathbf{r}}_i$ can be computed in terms of $(t, \mathbf{q}, \dot{\mathbf{q}}, \ddot{\mathbf{q}})$ by differentiating $\mathbf{r}_i(t, \mathbf{q})$ twice with respect to time. Once the equations of motion have been found using the Lagrangian formalism, the generalized accelerations $\ddot{\mathbf{q}}$, and hence the accelerations $\ddot{\mathbf{r}}_i$ and the constraint forces $\mathbf{F}_i^{(c)}$, can be expressed in terms of $(t, \mathbf{q}, \dot{\mathbf{q}})$. Note that, in general, the constraint force will depend (usually in a complicated way) on the velocity of the particles.

where the *probability amplitude* $\Phi(t_1, \mathbf{q}_1; t_2, \mathbf{q}_2)$ (in general complex) is given by

$$\Phi(t_1, \mathbf{q}_1; t_2, \mathbf{q}_2) = \text{const.} \sum_{\mathbf{q}} e^{\frac{i}{\hbar} S[\mathbf{q}]} . \tag{4.64}$$

The last "sum"—technically an integral, usually called the *path integral*—is extended to *all* paths $\mathbf{q}(t)$ satisfying the boundary conditions $\mathbf{q}(t_i) = \mathbf{q}_i$, $i = 1, 2$, and

$$S[\mathbf{q}] = \int_{t_1}^{t_2} L\big(t, \mathbf{q}(t), \dot{\mathbf{q}}(t)\big)\, dt$$

is the classical action of the path $\mathbf{q}(t)$. Thus all paths contribute to the probability amplitude $\Phi(t_1, \mathbf{q}_1; t_2, \mathbf{q}_2)$ with the same absolute magnitude, but with different *phases* proportional to their classical action. In the classical limit $\hbar \to 0$ we have $S[\mathbf{q}] \gg \hbar$, and thus the term $e^{iS[\mathbf{q}]/\hbar}$ in the path integral (4.64) is *highly oscillatory* near paths satisfying $\delta S[\mathbf{q}] \neq 0$. As a consequence, the contributions to the sum coming from such paths are vanishingly small as $\hbar \to 0$, since very close to a path $\mathbf{q}$ with $\delta S[\mathbf{q}] \neq 0$ there is a neighboring path whose phase differs by an odd multiple of π from that of $\mathbf{q}$. Thus in the limit $\hbar \to 0$ the overwhelming contribution to the sum (4.64) comes from the path[ll] satisfying $\delta S[\mathbf{q}] = 0$, i.e., from the classical trajectory. In other words, the validity of Hamilton's principle (when $\hbar \to 0$, i.e., in the classical limit) hinges on the fact that in this limit the path with the largest contribution to the probability amplitude $\Phi(t_1, \mathbf{q}_1; t_2, \mathbf{q}_2)$ is the one making the classical action stationary. ■

Example 4.9. *The spherical pendulum.* A spherical pendulum consists of a particle of mass m attached to a rigid rod of length l and negligible mass whose other end is fixed, subject only to Earth's gravitational field $\mathbf{g} = -g\mathbf{e}_z$. In this case there is only one (time-independent) constraint

$$\phi(t, \mathbf{r}) = \mathbf{r}^2 - l^2 = 0\,, \tag{4.65}$$

(if we place the origin at the pendulum's anchor point), and there are therefore $3 - 1 = 2$ degrees of freedom. We shall take as generalized coordinates the polar and azimuthal angles $\theta \in [0, \pi]$, $\varphi \in [0, 2\pi)$ of the spherical coordinate system, in terms of which

$$\mathbf{r}(\theta, \varphi) = l\,\big(\sin\theta\cos\varphi\,\mathbf{e}_1 + \sin\theta\sin\varphi\,\mathbf{e}_2 + \cos\theta\,\mathbf{e}_z\big).$$

(Of course, if the pendulum's pivot is fixed to the ceiling the angle θ must be restricted to the range $[\pi/2, \pi]$.) The constraint force $\mathbf{F}^{(c)}$ (in this case, the rod's reaction) is directed along the rod (toward the origin), and is thus perpendicular to the constraint surface (4.65). Hence the constraint is ideal, and we can apply the

[ll]We are assuming for the sake of simplicity that, as is usually the case, there is a unique classical trajectory satisfying the boundary conditions $\mathbf{q}(t_i) = \mathbf{q}_i$, $i = 1, 2$.

Lagrangian formalism. The potential of the external force $-mg\mathbf{e}_z$ is simply

$$V = mgz = mgl\cos\theta\,,$$

and the kinetic energy is given by

$$T = \frac{1}{2}m\dot{\mathbf{r}}^2 = \frac{1}{2}ml^2(\dot\theta^2 + \sin^2\theta\,\dot\varphi^2).$$

We thus have

$$\boxed{L = ml^2\left[\frac{1}{2}(\dot\theta^2 + \sin^2\theta\,\dot\varphi^2) - k\cos\theta\right], \qquad k := \frac{g}{l},}$$

and Lagrange's equations read

$$\boxed{\ddot\theta = \sin\theta\cos\theta\,\dot\varphi^2 + k\sin\theta\,, \qquad \frac{\mathrm{d}}{\mathrm{d}t}(\sin^2\theta\,\dot\varphi) = 0\,.}$$

From the second Lagrange equation we obtain

$$\sin^2\theta\,\dot\varphi = \frac{J_z}{ml^2} =: c = \text{const.}\,,$$

where $\mathbf{J}$ is the particle's angular momentum. This was to be expected, since

$$\dot{\mathbf{J}} = \mathbf{N} = \mathbf{r}\times(\mathbf{F} + \mathbf{F}^{(c)}) = \mathbf{r}\times\mathbf{F} = -mg\mathbf{r}\times\mathbf{e}_z \quad\Longrightarrow\quad \dot J_z = 0\,.$$

Substituting into the first Lagrange equation we obtain the following second-order differential equation for the angle θ:

$$\boxed{\ddot\theta = c^2\frac{\cos\theta}{\sin^3\theta} + k\sin\theta\,.} \tag{4.66}$$

If $c = 0$ (i.e., $J_z = 0$), the particle moves along a meridian $\varphi = \text{const.}$ (since $\dot\varphi = 0$) and Eq. (4.66) becomes the equation of motion of the simple pendulum

$$\ddot\alpha + k\sin\alpha = 0, \qquad \text{with} \quad \alpha = \pi - \theta.$$

(In fact, $c = 0$ is also possible if either $\theta = 0$ or $\theta = \pi$, but these are just the two equilibria of the motion in a meridian $\varphi = \text{const.}$) Let us see next what happens in the more interesting case $c \neq 0$. Equation (4.66) is then formally the equation of motion of a particle of unit mass moving in the effective one-dimensional potential

$$\boxed{U(\theta) = -\int\left(c^2\frac{\cos\theta}{\sin^3\theta} + k\sin\theta\right)\mathrm{d}\theta = k\cos\theta + \frac{c^2}{2\sin^2\theta}}$$

plotted in Fig. 4.2. The shape of the potential $U(\theta)$ can be determined by taking into account the following facts:

i. $U(\theta)$ diverges as $(\sin\theta)^{-2}$ as $\theta \to 0, \pi$.
ii. The derivative $U'(\theta)$ has the sign of $\theta - \theta_0$, for some $\theta_0 \in (\pi/2, \pi)$.

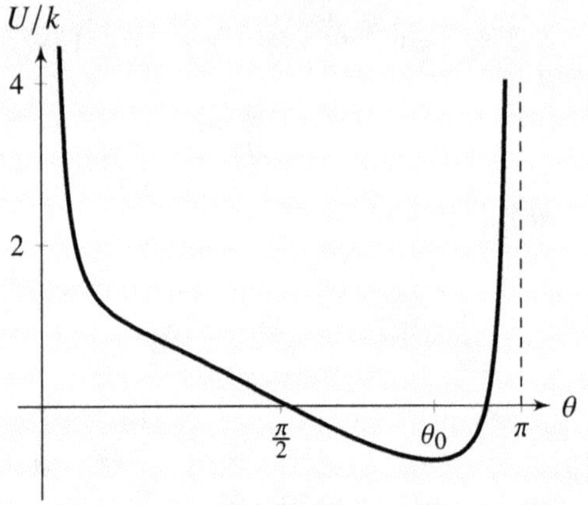

Figure 4.2. Effective potential $U(\theta)$ for $k = 10c^2$.

To prove the last statement, note that

$$U'(\theta) = -(\sin\theta)^{-3}(c^2\cos\theta + k\sin^4\theta)$$

has the sign of $f(\theta) := -(c^2\cos\theta + k\sin^4\theta)$. The function f, and hence U', is clearly negative for $\theta \leqslant \pi/2$. On the other hand,

$$f'(\theta) = c^2\sin\theta - 4k\sin^3\theta\cos\theta$$

is positive over the interval $[\pi/2, \pi)$, so that f is monotonically increasing on $[\pi/2, \pi]$ from $f(\pi/2) = -k < 0$ to $f(\pi) = c^2 > 0$. It follows that there is a unique $\theta_0 \in (\pi/2, \pi)$ such that $f(\theta_0) = 0$, with $f(\theta) < 0$ for $\pi/2 \leqslant \theta < \theta_0$ and $f(\theta) > 0$ for $\theta_0 < \theta \leqslant \pi$. Thus $f(\theta)$, and hence $U'(\theta)$, has the sign of $\theta - \theta_0$, as stated.

Moreover, since L does not depend explicitly on time the energy integral

$$h = \dot\theta\frac{\partial L}{\partial\dot\theta} + \dot\varphi\frac{\partial L}{\partial\dot\varphi} - L = T + V = ml^2\left[\frac{1}{2}\,(\dot\theta^2 + \sin^2\theta\,\dot\varphi^2) + k\cos\theta\right]$$

is conserved. Thus h is the particle's total energy E, which using the conservation of J_z can be expressed as

$$ml^2\left(\frac{1}{2}\dot\theta^2 + U(\theta)\right) = E\,.$$

The motion of the angular coordinate θ is easily determined integrating the previous equation:

$$\boxed{t = \pm \int \frac{\mathrm{d}\theta}{\sqrt{2\left(\frac{E}{ml^2} - U(\theta)\right)}}},$$

while the variation of the azimuthal angle φ then follows from the conservation of J_z:

$$\boxed{\varphi = c \int \frac{\mathrm{d}t}{\sin^2\theta(t)}}.$$

Finally, the equation of the trajectory (θ as a function of φ, or vice versa) is obtained combining the previous equations:

$$\dot{\theta} = \frac{\mathrm{d}\theta}{\mathrm{d}\varphi}\dot{\varphi} = \frac{c}{\sin^2\theta}\frac{\mathrm{d}\theta}{\mathrm{d}\varphi} = \pm\sqrt{2\left(\frac{E}{ml^2} - U(\theta)\right)}$$

$$\Longrightarrow \quad \boxed{\varphi = \pm c \int \frac{\mathrm{d}\theta}{\sin^2\theta\sqrt{2\left(\frac{E}{ml^2} - U(\theta)\right)}}}.$$

From the form of the effective potential $U(\theta)$ it follows that the motion of the coordinate θ is always *periodic*. The period of this motion is given by

$$\tau_\theta = 2\int_{\theta_1}^{\theta_2} \frac{\mathrm{d}\theta}{\sqrt{2\left(\frac{E}{ml^2} - U(\theta)\right)}},$$

where $\theta_1 < \theta_2$ are the two roots of the equation $E/(ml^2) - U(\theta) = 0$ in the interval $(0, \pi)$. The pendulum's motion, however, is *not* periodic in general, since when the coordinate θ returns to its initial value after a period the azimuthal angle φ does not necessarily increase by a multiple of 2π. More precisely, from the equation of the trajectory it follows that in a period of θ the angle φ increases by

$$\Delta\varphi = 2c\int_{\theta_1}^{\theta_2} \frac{\mathrm{d}\theta}{\sin^2\theta\sqrt{2\left(\frac{E}{ml^2} - U(\theta)\right)}}$$

$$= \sqrt{2}\int_{\theta_1}^{\theta_2} \frac{\mathrm{d}\theta}{\sin\theta\sqrt{\frac{E}{mlc^2}\sin^2\theta - \frac{k}{c^2}\sin^2\theta\cos\theta - \frac{1}{2}}}.$$

Hence *the motion is periodic if $\Delta\varphi$ is a rational multiple of 2π.*

Equation (4.66) admits the *constant solution* $\theta = \theta_0$, with $\theta_0 \in (\pi/2, \pi)$ the unique solution of the equation

$$c^2 \cos\theta_0 + k \sin^4\theta_0 = 0,$$

corresponding to a rotation around the z axis with constant angular velocity $\dot{\varphi} = c/\sin^2\theta_0$. The frequency ω of the small oscillations of the angle θ about the solution $\theta = \theta_0$ is given by

$$\begin{aligned}\omega^2 = U''(\theta_0) &= -k\cos\theta_0 + \frac{c^2}{\sin^2\theta_0} + 3c^2\frac{\cos^2\theta_0}{\sin^4\theta_0} \\ &= -k\cos\theta_0 - k\frac{1-\cos^2\theta_0 + 3\cos^2\theta_0}{\cos\theta_0} = k\frac{1+3\cos^2\theta_0}{|\cos\theta_0|},\end{aligned}$$

where we have taken into account that $c^2/\sin^4\theta_0 = -k/\cos\theta_0$.

Finally, the constraint force is easily computed using Eq. (4.63) and noting that in this case $\mathbf{F}^{(c)}$ is directed along $\mathbf{e}_r$ (perpendicular to the constraint surface). Therefore $\mathbf{F}^{(c)} = R\mathbf{e}_r$, with

$$R = m\ddot{\mathbf{r}} \cdot \mathbf{e}_r - m\mathbf{g} \cdot \mathbf{e}_r = ma_r + mg\mathbf{e}_z \cdot \mathbf{e}_r = mg\cos\theta - ml(\dot{\theta}^2 + \sin^2\theta\dot{\varphi}^2),$$

where we have applied Eq. (2.12) with $r = l$ and $\dot{r} = \ddot{r} = 0$. Using the law of conservation of energy we finally obtain

$$\boxed{R = 3mg\cos\theta - \frac{2E}{l}.}$$

Exercise 4.6. Show that $R < 0$ for $\theta \geqslant \pi/2$.

Solution. Indeed, we have

$$\begin{aligned}E = ml^2U(\theta_{1,2}) &= mgl\cos\theta_{1,2} + \frac{ml^2c^2}{2\sin^2\theta_{1,2}} \\ \implies \quad R &= mg\cos\theta + 2mg(\cos\theta - \cos\theta_1) - \frac{mlc^2}{2\sin^2\theta_1},\end{aligned}$$

where $\cos\theta \leqslant \cos\theta_1$ (since $\theta \geqslant \theta_1$) and $\cos\theta \leqslant 0$ for $\theta \in [\pi/2, \pi]$.

4.4 NOETHER'S THEOREM

Consider a mechanical system with Lagrangian $L(t, \mathbf{q}, \dot{\mathbf{q}})$, where $\mathbf{q} = (q_1, \ldots, q_n)$ are the n generalized coordinates. We define the **canonical momentum** associated with the generalized coordinate q_i as the partial derivative of L with respect to the corresponding **generalized velocity** $\dot{q}_i$:

$$p_i := \frac{\partial L}{\partial \dot{q}_i} \, . \tag{4.67}$$

Lagrange's equation of motion for the coordinate q_i is then

$$\dot{p}_i = \frac{\partial L}{\partial q_i} \, . \tag{4.68}$$

We shall say that the coordinate q_i is **cyclic** (or **ignorable**) if L is independent of q_i, i.e.,

$$\frac{\partial L}{\partial q_i} = 0 \, .$$

From Eq. (4.68) we then obtain the following conservation law:

If the coordinate q_i is cyclic, its corresponding canonical momentum p_i is conserved.

Likewise, if L does not explicitly depend on t, we saw in Section 4.1.2 that the energy integral

$$h = \sum_{i=1}^{n} \dot{q}_i \frac{\partial L}{\partial \dot{q}_i} - L = \sum_{i=1}^{n} p_i \dot{q}_i - L \tag{4.69}$$

is conserved. In many mechanical systems the kinetic energy is a *quadratic form* in the generalized velocities, i.e.,

$$T = \frac{1}{2} \sum_{i,j=1}^{n} a_{ij}(t, \mathbf{q}) \, \dot{q}_i \dot{q}_j \, , \qquad \text{with } a_{ij} = a_{ji} \, ,$$

and $L = T - V(t, \mathbf{q})$. A mechanical system of this type is called **natural**. In fact, most of the systems considered so far—with the important exception of the Lagrangian of a charged particle in an electromagnetic field (4.45)—are natural. In a natural mechanical system, the generalized momenta

$$p_i = \frac{\partial T}{\partial \dot{q}_i} = \sum_{j=1}^{n} a_{ij}(t, \mathbf{q}) \, \dot{q}_j \, , \qquad i = 1, \ldots, n \, ,$$

are linear in the generalized velocities $\dot{q}_i$, and the energy integral is simply

$$h = \sum_{i,j=1}^{n} a_{ij}(t, \mathbf{q}) \, \dot{q}_i \dot{q}_j - L = 2T - (T - V) = T + V \, .$$

Hence:

In a natural mechanical system the energy integral is equal to the total energy. In particular, in natural mechanical systems the conservation of h, which will occur if the coefficients a_{ij} and V are both independent of t, is nothing but the law of conservation of energy.

Example 4.10. Consider, first, the Lagrangian of a particle of mass m in Cartesian coordinates $\mathbf{r} = (x_1, x_2, x_3)$, given by

$$\boxed{L = \frac{1}{2}m\dot{\mathbf{r}}^2 - V(t, \mathbf{r})\,.}$$

In this case

$$p_i = \frac{\partial L}{\partial \dot{x}_i} = m\dot{x}_i\,,$$

and thus the canonical momentum corresponding to the coordinate x_i is the i-th component of the linear momentum. Moreover, L is clearly natural and therefore the energy integral h coincides with the particle's energy.

Consider next the Lagrangian of a particle of mass m in spherical coordinates (r, θ, φ):

$$L = \frac{m}{2}\left(\dot{r}^2 + r^2\dot{\theta}^2 + r^2\sin^2\theta\,\dot{\varphi}^2\right) - V(t, r, \theta, \varphi)\,, \tag{4.70}$$

for which

$$p_r = m\dot{r}\,, \qquad p_\theta = mr^2\dot{\theta}\,, \qquad p_\varphi = mr^2\sin^2\theta\,\dot{\varphi}\,. \tag{4.71}$$

In this case the kinetic energy (the first term in the Lagrangian) depends on r and θ, so that p_r and p_θ are *not* conserved even if V is independent of r or θ. On the other hand, if V does not depend on φ then L is independent of this coordinate, and hence p_φ is conserved:

$$\frac{\partial V}{\partial \varphi} = 0 \quad \Longrightarrow \quad p_\varphi = \text{const.}$$

As we know, p_φ is the z component of the particle's angular momentum. Moreover, since the kinetic energy is quadratic in the generalized velocities the Lagrangian is natural, and hence the energy integral coincides with the energy $T + V$, as we saw in Example 4.7. Hence if L does not depend on t—i.e., if the potential V is independent of time—energy is conserved.

Consider, finally, the Lagrangian (4.45) of a particle of mass m and charge e moving in an electromagnetic field with potentials $\Phi(t, \mathbf{r})$ and $\mathbf{A}(t, \mathbf{r})$. The canonical momentum corresponding to the coordinate x_i is now

$$p_i = \frac{\partial L}{\partial \dot{x}_i} = m\dot{x}_i + eA_i(t, \mathbf{r})\,.$$

Thus in this case *the canonical and the linear momenta are in general different.* In particular, if L does not depend on the coordinate x_i, i.e., if

$$\frac{\partial \Phi}{\partial x_i} = 0\,, \qquad \frac{\partial \mathbf{A}}{\partial x_i} = 0\,,$$

p_i is conserved but $m\dot{x}_i$ is not conserved in general. The energy integral is given by

$$h = \sum_{i=1}^{3} p_i \dot{x}_i - L = (m\dot{\mathbf{r}} + e\mathbf{A}) \cdot \dot{\mathbf{r}} - L = \frac{1}{2} m\dot{\mathbf{r}}^2 + e\Phi\,.$$

Therefore in this case h is the sum of the particle's kinetic energy and its potential electric energy. If L does not depend on t, that is if

$$\frac{\partial \Phi}{\partial t} = 0\,, \qquad \frac{\partial \mathbf{A}}{\partial t} = 0\,,$$

then h is conserved. Although the system is *not* natural, we can also interpret h in this case as the total energy. Indeed, if Φ and $\mathbf{A}$ do not depend on t the electric force is *conservative* with potential $e\Phi(\mathbf{r})$, and therefore h is the sum of the kinetic energy and the potential energy of the electric force. This is indeed the total energy of the particle, since the magnetic force does no work as it is always perpendicular to the particle's velocity.

The conservation of the canonical momentum p_i and the energy integral h are clearly a consequence of the *invariance* of the Lagrangian under translations in the coordinate q_i ($q_i \mapsto q_i + \varepsilon$) or the time t ($t \mapsto t + \varepsilon$), respectively, where $\varepsilon \in \mathbb{R}$ is a *continuous parameter*. In fact, one of the fundamental principles of modern physics is that continuous transformations leaving invariant the Lagrangian—or, more generally, the action—give rise to conserved quantities. This is precisely the import of **Noether's theorem**:

Suppose that the *action* of a mechanical system with Lagrangian $L(t, \mathbf{q}, \dot{\mathbf{q}})$ is *invariant* under a *one-parameter family* of invertible transformations

$$\tilde{t} = t + \varepsilon\tau(t, \mathbf{q}) + O(\varepsilon^2)\,, \qquad \tilde{\mathbf{q}} = \mathbf{q} + \varepsilon\eta(t, \mathbf{q}) + O(\varepsilon^2)\,, \tag{4.72}$$

i.e., that

$$\int_{\tilde{t}_1}^{\tilde{t}_2} L\left(\tilde{t}, \tilde{\mathbf{q}}, \frac{d\tilde{\mathbf{q}}}{d\tilde{t}}\right) d\tilde{t} = \int_{t_1}^{t_2} L(t, \mathbf{q}, \dot{\mathbf{q}})\, dt\,, \qquad \forall t_1, t_2\,. \tag{4.73}$$

Then the function

$$I(t, \mathbf{q}, \dot{\mathbf{q}}) := \mathbf{p}\,\eta - h\tau\,,$$

where $\mathbf{p} = \dfrac{\partial L}{\partial \dot{\mathbf{q}}}$ and $h = \mathbf{p}\dot{\mathbf{q}} - L$, is conserved.

Proof. We begin by computing the derivatives of $\frac{\mathrm{d}\widetilde{\mathbf{q}}}{\mathrm{d}\tilde{t}}$ and $\frac{\mathrm{d}\tilde{t}}{\mathrm{d}t}$ with respect to ε at $\varepsilon = 0$, that we shall need in the sequel:

$$\frac{\mathrm{d}\tilde{t}}{\mathrm{d}t} = 1 + \varepsilon\,\dot{\tau} + O(\varepsilon^2) \qquad \Longrightarrow \qquad \left.\frac{\mathrm{d}\tilde{t}}{\mathrm{d}t}\right|_{\varepsilon=0} = 1\,, \quad \left.\frac{\partial}{\partial\varepsilon}\right|_{\varepsilon=0} \frac{\mathrm{d}\tilde{t}}{\mathrm{d}t} = \dot{\tau}\,.$$

$$\begin{aligned}\frac{\mathrm{d}\widetilde{\mathbf{q}}}{\mathrm{d}\tilde{t}} &= \left(\frac{\mathrm{d}\tilde{t}}{\mathrm{d}t}\right)^{-1}\frac{\mathrm{d}\widetilde{\mathbf{q}}}{\mathrm{d}t} = \Big(1 + \varepsilon\,\dot{\tau} + O(\varepsilon^2)\Big)^{-1}\Big(\dot{\mathbf{q}} + \varepsilon\,\dot{\eta} + O(\varepsilon^2)\Big) \\ &= \Big(1 - \varepsilon\,\dot{\tau} + O(\varepsilon^2)\Big)\Big(\dot{\mathbf{q}} + \varepsilon\,\dot{\eta} + O(\varepsilon^2)\Big) = \dot{\mathbf{q}} + \varepsilon(\dot{\eta} - \dot{\mathbf{q}}\dot{\tau}) + O(\varepsilon^2) \\ &\Longrightarrow \quad \left.\frac{\partial}{\partial\varepsilon}\right|_{\varepsilon=0} \frac{\mathrm{d}\widetilde{\mathbf{q}}}{\mathrm{d}\tilde{t}} = \dot{\eta} - \dot{\mathbf{q}}\dot{\tau}\,.\end{aligned}$$

The invariance of the action under the transformation $(t, \mathbf{q}) \mapsto (\tilde{t}, \widetilde{\mathbf{q}})$ can be expressed in the equivalent form[12]

$$L\left(\tilde{t}, \widetilde{\mathbf{q}}, \frac{\mathrm{d}\widetilde{\mathbf{q}}}{\mathrm{d}\tilde{t}}\right)\frac{\mathrm{d}\tilde{t}}{\mathrm{d}t} = L(t, \mathbf{q}, \dot{\mathbf{q}})\,. \tag{4.74}$$

Differentiating (4.74) with respect to ε and setting $\varepsilon = 0$ we then obtain

$$\left.\frac{\mathrm{d}\tilde{t}}{\mathrm{d}t}\right|_{\varepsilon=0} \left.\frac{\partial}{\partial\varepsilon}\right|_{\varepsilon=0} L\left(\tilde{t}, \widetilde{\mathbf{q}}, \frac{\mathrm{d}\widetilde{\mathbf{q}}}{\mathrm{d}\tilde{t}}\right) + L\left.\frac{\partial}{\partial\varepsilon}\right|_{\varepsilon=0} \frac{\mathrm{d}\tilde{t}}{\mathrm{d}t} = \frac{\partial L}{\partial t}\tau + \frac{\partial L}{\partial \mathbf{q}}\eta + \frac{\partial L}{\partial \dot{\mathbf{q}}}(\dot{\eta} - \dot{\mathbf{q}}\dot{\tau}) + L\dot{\tau} = 0\,. \tag{4.75}$$

Using Lagrange's equations we can rewrite the previous equality as follows:

$$0 = \frac{\partial L}{\partial t}\tau + \dot{\mathbf{p}}\,\eta + \mathbf{p}\,\dot{\eta} - \mathbf{p}\,\dot{\mathbf{q}}\,\dot{\tau} + L\dot{\tau} = \frac{\partial L}{\partial t}\tau + \frac{\mathrm{d}}{\mathrm{d}t}(\mathbf{p}\,\eta) - h\dot{\tau} = \frac{\mathrm{d}}{\mathrm{d}t}(\mathbf{p}\,\eta - h\tau) + \left(\dot{h} + \frac{\partial L}{\partial t}\right)\tau\,.$$

It is straightforward to check that the last term vanishes identically on account of Lagrange's equations:

$$\dot{h} = \dot{\mathbf{p}}\,\dot{\mathbf{q}} + \mathbf{p}\,\ddot{\mathbf{q}} - \frac{\partial L}{\partial t} - \frac{\partial L}{\partial \mathbf{q}}\dot{\mathbf{q}} - \mathbf{p}\,\ddot{\mathbf{q}} = \left(\dot{\mathbf{p}} - \frac{\partial L}{\partial \mathbf{q}}\right)\dot{\mathbf{q}} - \frac{\partial L}{\partial t} = -\frac{\partial L}{\partial t}\,. \qquad \blacksquare$$

Generally speaking, a *symmetry* of an object is any transformation leaving the object invariant. The set of all symmetries of an object is a *group* (with composition as group multiplication), since i) the composition of two symmetries is clearly a symmetry, ii) the inverse of a symmetry is also a symmetry (why?), and iii) the identity transformation is obviously a symmetry. Thus the family of transformations (4.72) are a *one-parameter group* of symmetries of the *action*. Families of symmetries of an object depending on one or more continuous parameters—like the transformations (4.72)—are usually called *continuous symmetries*. Thus the import of Noether's theorem is that *every continuous symmetry of the action yields a conservation law.* As mentioned before, this is in fact one of the most fundamental principles in modern physics, which actually holds in much more general settings like classical or quantum field theory.

---

[12]Indeed, integrating (4.74) between t_1 and t_2 we obtain Eq. (4.73). Conversely, Eq. (4.73) implies (4.74), since the times t_1 and t_2 are arbitrary.

Example 4.11. Consider a system of N particles subject only to irrotational forces generated by a potential $V(t, \mathbf{r}_1, \ldots, \mathbf{r}_N)$. We can then take the Cartesian coordinates $\mathbf{q} = (\mathbf{r}_1, \ldots, \mathbf{r}_N)$ as generalized coordinates, and $L = T - V$ as the system's Lagrangian. The kinetic energy

$$T = \frac{1}{2}\sum_{i=1}^{N} m_i \dot{\mathbf{r}}_i^2$$

is then invariant under two types of transformations:

i. *Translations* of the particles' coordinates in the direction of a unit vector $\mathbf{n}$:

$$\tilde{t} = t\,, \qquad \widetilde{\mathbf{r}}_i = \mathbf{r}_i = \mathbf{r}_i + \varepsilon \mathbf{n} \qquad (1 \leqslant i \leqslant N, \quad \varepsilon \in \mathbb{R}); \tag{4.76}$$

indeed, $\dot{\widetilde{\mathbf{r}}}_i = \dot{\mathbf{r}}_i$.

ii. *Rotations* of the particles' coordinates about an axis $\mathbf{n}$:

$$\tilde{t} = t\,, \qquad \widetilde{\mathbf{r}}_i = R(\varepsilon)\mathbf{r}_i \qquad (1 \leqslant i \leqslant N, \quad \varepsilon \in \mathbb{R}); \tag{4.77}$$

indeed, $\dot{\widetilde{\mathbf{r}}}_i = R(\varepsilon)\dot{\mathbf{r}}_i$ and hence $\dot{\widetilde{\mathbf{r}}}_i^2 = \dot{\mathbf{r}}_i^2$.

Obviously, the Lagrangian (and hence the action, since $\tilde{t} = t$) will be invariant under the previous transformations if and only if the potential $V(t, \mathbf{r}_1, \ldots, \mathbf{r}_N)$ is, i.e., provided that

$$V(t, \widetilde{\mathbf{r}}_1, \ldots, \widetilde{\mathbf{r}}_N) = V(t, \mathbf{r}_1, \ldots, \mathbf{r}_N)\,.$$

Suppose, first, that the potential is invariant under the translations (4.76), for which

$$\tau = 0\,, \qquad \eta_i = \mathbf{n} \qquad (1 \leqslant i \leqslant N)\,.$$

The corresponding *conserved quantity* is then

$$I = \sum_{i=1}^{N} \frac{\partial L}{\partial \dot{\mathbf{r}}_i}\eta_i - 0 \cdot h = \sum_{i=1}^{N} m_i \dot{\mathbf{r}}_i \cdot \mathbf{n} = \mathbf{n} \cdot \sum_{i=1}^{N} m_i \dot{\mathbf{r}}_i = \boxed{\mathbf{P} \cdot \mathbf{n}}\,,$$

i.e., the component of the system's *total linear momentum* along the direction of the vector $\mathbf{n}$.

Suppose next that the potential is invariant under the rotations (4.77). What is the conserved quantity associated with this invariance of the action? To answer this question, let us take the z axis in the direction of the vector $\mathbf{n}$, so that

$$R(\varepsilon) = \begin{pmatrix} \cos\varepsilon & -\sin\varepsilon & 0 \\ \sin\varepsilon & \cos\varepsilon & 0 \\ 0 & 0 & 1 \end{pmatrix}.$$

Expanding $R(\varepsilon)$ in powers of ε we obtain

$$\widetilde{\mathbf{r}}_i = R(\varepsilon)\mathbf{r}_i = \mathbf{r}_i + \varepsilon A\mathbf{r}_i + O(\varepsilon^2)\,, \qquad 1 \leqslant i \leqslant N\,, \tag{4.78}$$

with

$$A = R'(0) = \begin{pmatrix} 0 & -1 & 0 \\ 1 & 0 & 0 \\ 0 & 0 & 0 \end{pmatrix}.$$

Since

$$A\mathbf{r}_i = (-y_i, x_i, 0) = \mathbf{e}_3 \times \mathbf{r}_i = \mathbf{n} \times \mathbf{r}_i\,,$$

we can rewrite Eq. (4.78) in vector form as

$$\widetilde{\mathbf{r}}_i = \mathbf{r}_i + \varepsilon\,\mathbf{n} \times \mathbf{r}_i + O(\varepsilon^2)\,, \qquad 1 \leqslant i \leqslant N\,.$$

Hence in this case

$$\tau = 0\,, \qquad \eta_i = \mathbf{n} \times \mathbf{r}_i\,, \qquad 1 \leqslant i \leqslant N\,,$$

and the conserved quantity associated with the invariance of the action under rotations about the **n** axis is therefore

$$I = \sum_{i=1}^{N} \frac{\partial L}{\partial \dot{\mathbf{r}}_i}\eta_i - 0 \cdot h = \sum_{i=1}^{N} m_i \dot{\mathbf{r}}_i \cdot (\mathbf{n} \times \mathbf{r}_i) = \mathbf{n} \cdot \sum_{i=1}^{N} m_i \mathbf{r}_i \times \dot{\mathbf{r}}_i = \boxed{\mathbf{J} \cdot \mathbf{n}\,,}$$

where **J** is the system's total angular momentum.

Exercise 4.7. Determine the conserved quantity $I(t, \mathbf{q}, \dot{\mathbf{q}})$ associated with the invariance of the action under the *space-time dilations*

$$\tilde{t} = \lambda^\alpha t\,, \qquad \widetilde{\mathbf{q}} = \lambda \mathbf{q}, \qquad \forall \lambda > 0, \tag{4.79}$$

where $\alpha \in \mathbb{R}$ is fixed.

Solution. As in the formulation of Noether's theorem the parameter $\varepsilon = 0$ corresponds to the identity transformation, we set (for instance) $\lambda = \mathrm{e}^\varepsilon$ in Eq. (4.79). Expanding to first order in ε we then obtain

$$\tilde{t} = \mathrm{e}^{\alpha\varepsilon} t = t + \varepsilon\alpha t + O(\varepsilon^2)\,, \qquad \widetilde{\mathbf{q}} = \mathrm{e}^\varepsilon \mathbf{q} = \mathbf{q} + \varepsilon \mathbf{q} + O(\varepsilon)^2\,.$$

Hence

$$\tau = \alpha t\,, \qquad \eta = \mathbf{q}\,,$$

and thus the conserved quantity associated with the invariance of the action under the transformations (4.79) is given by

$$I(t, \mathbf{q}, \dot{\mathbf{q}}) = \mathbf{q}\,\frac{\partial L}{\partial \dot{\mathbf{q}}} - \alpha t h\,.$$

Note that the action is invariant under the dilations (4.79) if the Lagrangian L verifies the condition

$$L\left(\tilde{t}, \tilde{\mathbf{q}}, \frac{\mathrm{d}\tilde{\mathbf{q}}}{\mathrm{d}\tilde{t}}\right) \mathrm{d}\tilde{t} = L(\lambda^\alpha t, \lambda \mathbf{q}, \lambda^{1-\alpha}\dot{\mathbf{q}})\lambda^\alpha \,\mathrm{d}t = L(t, \mathbf{q}, \dot{\mathbf{q}})\,\mathrm{d}t\,,$$

i.e., if L transforms under dilations as

$$L(\lambda^\alpha t, \lambda \mathbf{q}, \lambda^{1-\alpha}\dot{\mathbf{q}}) = \lambda^{-\alpha} L(t, \mathbf{q}, \dot{\mathbf{q}})\,.$$

Suppose, for instance, that the system is *natural.* In this case the previous condition becomes

$$\frac{1}{2}\lambda^{2-2\alpha}\sum_{i,j=1}^{n} a_{ij}(\lambda^\alpha t, \lambda\mathbf{q})\dot{q}_i\dot{q}_j - V(\lambda^\alpha t, \lambda\mathbf{q})$$
$$= \frac{1}{2}\lambda^{-\alpha}\sum_{i,j=1}^{n} a_{ij}(t, \mathbf{q})\dot{q}_i\dot{q}_j - \lambda^{-\alpha}V(t, \mathbf{q})\,.$$

Equating the coefficient of $\dot{q}_i\dot{q}_j$ in both sides of this equality we obtain

$$\lambda^{2-2\alpha}a_{ij}(\lambda^\alpha t, \lambda\mathbf{q}) = \lambda^{-\alpha}a_{ij}(t, \mathbf{q}) \iff a_{ij}(\lambda^\alpha t, \lambda\mathbf{q}) = \lambda^{\alpha-2}a_{ij}(t, \mathbf{q})\,,$$

and hence

$$V(\lambda^\alpha t, \lambda\mathbf{q}) = \lambda^{-\alpha}V(t, \mathbf{q})\,.$$

For example, if the matrix a_{ij} is constant then we must have $\alpha = 2$, and therefore

$$V(\lambda^2 t, \lambda\mathbf{q}) = \lambda^{-2}V(t, \mathbf{q})\,.$$

Consider, for instance, the case of a particle of mass m that moves subject to the central potential $V(r) = k/(2r^2)$, with $k \neq 0$. From the previous discussion it easily follows that in this case the action is invariant under the dilations (4.79) with $\alpha = 2$. In this case the energy $T + V = E$ and the function

$$I = m\mathbf{r}\dot{\mathbf{r}} - 2ht = mr\dot{r} - 2Et = \frac{\mathrm{d}}{\mathrm{d}t}\left(\frac{1}{2}mr^2 - Et^2\right) = \text{const.} \tag{4.80}$$

is conserved. Note that the value of the conserved quantity I can be expressed in terms of the initial data $r_0 = r(0)$ and $\dot{r}_0 = \dot{r}(0)$ by evaluating it at $t = 0$:

$$I = mr\dot{r} - 2Et\Big|_{t=0} = mr_0\dot{r}_0 \,.$$

Integrating Eq. (4.80) we can easily determine the motion of the r coordinate:

$$It = mr_0\dot{r}_0 t = \frac{1}{2}mr^2 - Et^2 - \frac{1}{2}mr_0^2 \quad \Longrightarrow \quad \boxed{r = \sqrt{r_0^2 + 2r_0\dot{r}_0 t + \frac{2E}{m}t^2}\,.}$$

The motion of the angular coordinate φ (in the plane of motion) is obtained integrating the law of conservation of angular momentum $mr^2\dot{\varphi} = J$:

$$\boxed{\varphi = \varphi_0 + \frac{J}{m}\int_0^t \frac{\mathrm{d}s}{r^2(s)} = \varphi_0 + \frac{J}{m}\int_0^t \frac{\mathrm{d}s}{r_0^2 + 2r_0\dot{r}_0 s + \frac{2E}{m}s^2}\,, \qquad \varphi_0 := \varphi(0)\,.}$$

If $E = 0$ the integral is elementary:

$$\varphi = \begin{cases} \varphi_0 + \frac{Jt}{mr_0^2}\,, & \dot{r}_0 = 0 \\ \varphi_0 + \frac{J}{2mr_0\dot{r}_0}\log\left(1 + \frac{2\dot{r}_0 t}{r_0}\right), & \dot{r}_0 \neq 0\,. \end{cases}$$

In the more general case $E \neq 0$ the integral can be evaluated in terms of hyperbolic, rational or trigonometric functions depending on whether the discriminant of the polynomial in the denominator

$$4r_0^2\left(\dot{r}_0^2 - 2\frac{E}{m}\right) = -\frac{4}{m^2}\left(J^2 + k\,m\right)$$

is respectively positive, zero or negative (exercise).

4.5 INTRODUCTION TO HAMILTONIAN MECHANICS

4.5.1 Hamilton's canonical equations

Lagrange's equations of motion of a mechanical system:

$$\frac{\mathrm{d}}{\mathrm{d}t}\frac{\partial L}{\partial \dot{\mathbf{q}}} - \frac{\partial L}{\partial \mathbf{q}} = 0\,, \tag{4.81}$$

although more versatile than Newton's, suffer from two main drawbacks. First of all, Eqs. (4.81) *are not in normal form*, i.e., the second derivatives $\ddot{q}_i$ are not expressed in terms of $(t, \mathbf{q}, \dot{\mathbf{q}})$. Secondly, they are *second-order* equations, so that the graphs of two solutions $\mathbf{q}_1(t)$ and $\mathbf{q}_2(t)$—i.e., two system trajectories—can intersect in the extended configuration space $\mathbb{R} \times \mathbb{R}^n$ of the variables $(t, \mathbf{q})$ without violating the existence and uniqueness theorem for systems of ordinary differential equations. Both problems can be solved if we are able to express Eqs. (4.81) as a *normal* system of *first-order* differential equations. Since Lagrange's equations are first-order in the canonical momenta p_i, the most natural way to achieve this is to use as dependent variables $\mathbf{q} = (q_1, \ldots, q_n)$ and $\mathbf{p} := (p_1, \ldots, p_n)$, in terms of which Eqs. (4.81) can be rewritten as

$$\frac{\mathrm{d}\mathbf{q}}{\mathrm{d}t} = \dot{\mathbf{q}}\,, \qquad \frac{\mathrm{d}\mathbf{p}}{\mathrm{d}t} = \frac{\partial L}{\partial \mathbf{q}}(t, \mathbf{q}, \dot{\mathbf{q}})\,. \tag{4.82}$$

The problem is that $\dot{\mathbf{q}}$, which appears in the RHS of these equations, *must be expressed as a function of* $(t, \mathbf{q}, \mathbf{p})$ *using the relation*

$$\mathbf{p} = \frac{\partial L}{\partial \dot{\mathbf{q}}}(t, \mathbf{q}, \dot{\mathbf{q}})\,. \tag{4.83}$$

By the inverse function theorem, for this to be possible (at least locally) we must have

$$\det\left(\frac{\partial p_i}{\partial \dot{q}_j}\right)_{1\leqslant i,j\leqslant n} = \det\left(\frac{\partial^2 L}{\partial \dot{q}_i \partial \dot{q}_j}\right)_{1\leqslant i,j\leqslant n} \neq 0\,. \tag{4.84}$$

For instance, it can be shown that this condition automatically holds in a natural mechanical system. Indeed, in such a system we have

$$\frac{\partial^2 L}{\partial \dot{q}_i \partial \dot{q}_j} = a_{ij}(t, \mathbf{q})$$

(cf. Section 4.4). Since $T > 0$ for $\dot{\mathbf{q}} \neq 0$, the matrix $\big(a_{ij}(t, \mathbf{q})\big)_{1\leqslant i,j\leqslant n}$ is positive definite, and therefore invertible.

As we have just remarked, (4.82) should be more precisely written as

$$\frac{\mathrm{d}\mathbf{q}}{\mathrm{d}t} = \dot{\mathbf{q}}(t, \mathbf{q}, \mathbf{p})\,, \qquad \frac{\mathrm{d}\mathbf{p}}{\mathrm{d}t} = \frac{\partial L}{\partial \mathbf{q}}\big(t, \mathbf{q}, \dot{\mathbf{q}}(t, \mathbf{q}, \mathbf{p})\big)\,,$$

where $\dot{\mathbf{q}}(t, \mathbf{q}, \mathbf{p})$ is the (vector-valued) function obtained by solving Eq. (4.83) for $\dot{\mathbf{q}}$ in terms of $(t, \mathbf{q}, \mathbf{p})$. In order to recast the previous system in a more symmetric form, it is essential to study how the Lagrangian L depends on the variables $(t, \mathbf{q}, \dot{\mathbf{q}})$. The differential of L, considered as a function of these variables, is given by

$$\mathrm{d}L = \frac{\partial L}{\partial t}\,\mathrm{d}t + \frac{\partial L}{\partial \mathbf{q}}\,\mathrm{d}\mathbf{q} + \frac{\partial L}{\partial \dot{\mathbf{q}}}\,\mathrm{d}\dot{\mathbf{q}} = \frac{\partial L}{\partial t}\,\mathrm{d}t + \frac{\partial L}{\partial \mathbf{q}}\,\mathrm{d}\mathbf{q} + \mathbf{p}\,\mathrm{d}\dot{\mathbf{q}}\,. \tag{4.85}$$

Taking into account that

$$\mathbf{p}\,\mathrm{d}\dot{\mathbf{q}} = \mathrm{d}(\mathbf{p}\,\dot{\mathbf{q}}) - \dot{\mathbf{q}}\,\mathrm{d}\mathbf{p}\,,$$

from Eq. (4.85) we obtain

$$\mathrm{d}(\mathbf{p}\,\dot{\mathbf{q}} - L) = \mathrm{d}h = -\frac{\partial L}{\partial t}\,\mathrm{d}t - \frac{\partial L}{\partial \mathbf{q}}\,\mathrm{d}\mathbf{q} + \dot{\mathbf{q}}\,\mathrm{d}\mathbf{p}\,. \tag{4.86}$$

If in the previous formula we consider $\dot{\mathbf{q}}$ as a function of the variables $(t, \mathbf{q}, \mathbf{p})$ the energy integral h becomes a function

$$H(t, \mathbf{q}, \mathbf{p}) := h(t, \mathbf{q}, \dot{\mathbf{q}}(t, \mathbf{q}, \mathbf{p})) \tag{4.87}$$

of these variables called the system's **Hamiltonian**. Since $dH = dh$ is given by the RHS of Eq. (4.86), the partial derivatives of $H(t, \mathbf{q}, \mathbf{p})$ with respect to the independent variables $(t, \mathbf{q}, \mathbf{p})$ are simply the coefficients of dt, $d\mathbf{q}$, and $d\mathbf{p}$ in the previous equation, i.e.,

$$\frac{\partial H}{\partial t} = -\frac{\partial L}{\partial t}, \qquad \frac{\partial H}{\partial \mathbf{q}} = -\frac{\partial L}{\partial \mathbf{q}}, \qquad \frac{\partial H}{\partial \mathbf{p}} = \dot{\mathbf{q}}. \tag{4.88}$$

It is understood that in the RHS of these equations $\dot{\mathbf{q}}$ *must be expressed in terms of* $(t, \mathbf{q}, \mathbf{p})$ *inverting* Eq. (4.83). From Eqs. (4.88) it then follows that Lagrange's equations of motion (4.82) are equivalent to the following system of first-order ordinary differential equations in the *independent variables* $(\mathbf{q}, \mathbf{p})$:

$$\frac{d\mathbf{q}}{dt} = \frac{\partial H}{\partial \mathbf{p}}(t, \mathbf{q}, \mathbf{p}), \qquad \frac{d\mathbf{p}}{dt} = -\frac{\partial H}{\partial \mathbf{q}}(t, \mathbf{q}, \mathbf{p}). \tag{4.89}$$

Equations (4.89) are known as **Hamilton's canonical equations**.

Remark 4.2. In mathematics, the passage from the generalized coordinates and velocities $(\mathbf{q}, \dot{\mathbf{q}})$ to the canonical variables $(\mathbf{q}, \mathbf{p})$, where $\dot{\mathbf{q}}$ and $\mathbf{p}$ are related through (4.83), is called a **Legendre transformation**. This type of transformation is widely used, among other areas of physics, in thermodynamics. ■

- In order to write Hamilton's canonical equations of a mechanical system with Lagrangian $L(t, \mathbf{q}, \dot{\mathbf{q}})$ we can proceed as follows:

1) Find the canonical momenta

$$p_i = \frac{\partial L}{\partial \dot{q}_i}(t, \mathbf{q}, \dot{\mathbf{q}}), \qquad i = 1, \ldots, n.$$

2) Use the above equations to solve for the generalized velocities $\dot{q}_i$ in terms of the canonical momenta p_j:

$$\dot{q}_i = \dot{q}_i(t, \mathbf{q}, \mathbf{p}), \qquad i = 1, \ldots, n. \tag{4.90}$$

3) Compute the system's Hamiltonian

$$H(t, \mathbf{q}, \mathbf{p}) = \mathbf{p} \cdot \dot{\mathbf{q}} - L$$

using Eqs. (4.90) to express $\dot{\mathbf{q}}$ as a function of the variables $(t, \mathbf{q}, \mathbf{p})$.

4) Hamilton's canonical equations (4.89) can then be written down by computing the partial derivatives of H with respect to the canonical variables $\mathbf{q}$ and $\mathbf{p}$. In fact, the first n equations

$$\dot{q}_i = \frac{\partial H}{\partial p_i}, \qquad i = 1, \ldots, n,$$

are actually equations (4.90), so that in practice it is only necessary to find the n remaining equations

$$\dot{p}_i = -\frac{\partial H}{\partial q_i}, \qquad i = 1, \ldots, n.$$

- Recall that in a *natural* mechanical system

$$\boxed{h = \dot{\mathbf{q}}\,\frac{\partial L}{\partial \dot{\mathbf{q}}} - L = T + V,}$$

and therefore:

The Hamiltonian of a *natural* mechanical system is the energy $T + V$ *expressed in terms of the variables* $(t, \mathbf{q}, \mathbf{p})$.

4.5.2 Basic conservation laws

First of all, from Hamilton's equations it follows that if the Hamiltonian H is independent of a coordinate q_i the corresponding momentum p_i is conserved:

$$\boxed{\frac{\partial H}{\partial q_i} = 0 \quad \Longrightarrow \quad p_i = \text{const.}}$$

Likewise, if H is independent of the momentum p_i its corresponding coordinate q_i is conserved:

$$\boxed{\frac{\partial H}{\partial p_i} = 0 \quad \Longrightarrow \quad q_i = \text{const.}}$$

This example illustrates the great *symmetry* between the generalized coordinates q_i and their associated momenta p_i, which is in fact one of the distinctive advantages of the Hamiltonian formulation of mechanics.

From Hamilton's equations we also deduce that

$$\frac{\mathrm{d}H}{\mathrm{d}t} = \frac{\partial H}{\partial t} + \frac{\partial H}{\partial \mathbf{q}}\dot{\mathbf{q}} + \frac{\partial H}{\partial \mathbf{p}}\dot{\mathbf{p}} = \frac{\partial H}{\partial t} + \frac{\partial H}{\partial \mathbf{q}}\frac{\partial H}{\partial \mathbf{p}} - \frac{\partial H}{\partial \mathbf{p}}\frac{\partial H}{\partial \mathbf{q}} = \frac{\partial H}{\partial t}.$$

Hence the Hamiltonian is conserved if it does not depend explicitly on t:

$$\boxed{\frac{\partial H}{\partial t} = 0 \quad \Longrightarrow \quad H = \text{const.}}$$

Note that from the first Eq. (4.88), namely

$$\boxed{\frac{\partial H}{\partial t} = -\frac{\partial L}{\partial t}\,,}$$

it follows that H is conserved if and only if L is independent of t. Since $H(t, q, \mathbf{p}) = h(t, \mathbf{q}, \dot{\mathbf{q}})$, this is the conservation of the energy integral deduced in the Lagrangian formalism (cf. Section 4.4).

• Another advantage of the Hamiltonian formulation of mechanics over the Lagrangian one consists of the following fact: *if the coordinate q_i is* cyclic, *it is possible to* eliminate *from Hamilton's equations the degree of freedom corresponding to this coordinate and its associated momentum p_i, reducing these equations to a system of $2(n-1)$ canonical equations.*

Indeed, suppose that

$$\frac{\partial H}{\partial q_i} = 0\,,$$

so that $p_i(t) = c$ for all t. It is then immediate to check that *the equations of motion of the remaining coordinates and momenta are Hamilton's canonical equations for the Hamiltonian*

$$H\big|_{p_i=c} = H(t, q_1, \ldots, q_{i-1}, q_{i+1}, \ldots, q_n, p_1, \ldots, p_{i-1}, c, p_{i+1}, \ldots, p_n)\,,$$

which depends only on the $2(n-1)$ canonical variables (q_j, p_j) with $j \neq i$. Indeed, if $j \neq i$ we have

$$\dot{q}_j = \frac{\partial H}{\partial p_j}\bigg|_{p_i=c} = \frac{\partial}{\partial p_j}\left(H\big|_{p_i=c}\right), \qquad \dot{p}_j = -\frac{\partial H}{\partial q_j}\bigg|_{p_i=c} = -\frac{\partial}{\partial q_j}\left(H\big|_{p_i=c}\right).$$

Once these equations are solved, the motion of the cyclic coordinate q_i is determined simply by integrating its corresponding canonical equation

$$\dot{q}_i = \frac{\partial H}{\partial p_i}\bigg|_{p_i=c} = \frac{\partial H\big|_{p_i=c}}{\partial c}\,,$$

i.e.,

$$q_i(t) = \int \frac{\partial H\big|_{p_i=c}}{\partial c}\big(t, q_1(t), \ldots, q_{i-1}(t), q_{i+1}(t), \ldots, q_n(t),$$
$$p_1(t), \ldots, p_{i-1}(t), c, p_{i+1}(t), \ldots, p_n(t)\big)\, \mathrm{d}t\,.$$

Example 4.12. *Hamiltonian of a particle in Cartesian coordinates.*

As we saw in Section 4.2.1, in this case the Lagrangian is given by

$$L = \frac{1}{2}m\dot{\mathbf{r}}^2 - V(t, \mathbf{r})\,,$$

and the canonical momentum coincides with the linear one:

$$\mathbf{p} = \frac{\partial L}{\partial \dot{\mathbf{r}}} = m\dot{\mathbf{r}} \iff \dot{\mathbf{r}} = \frac{\mathbf{p}}{m}.$$

Since the Lagrangian is natural, the Hamiltonian is the total energy $T + V$ expressed in terms of $(t, \mathbf{r}, \mathbf{p})$:

$$H(t, \mathbf{r}, \mathbf{p}) = \frac{1}{2} m\dot{\mathbf{r}}^2 + V(t, \mathbf{r}) = \boxed{\frac{\mathbf{p}^2}{2m} + V(t, \mathbf{r})}.$$

If H does not depend on the coordinate x_i (i.e., if V is independent of x_i), the corresponding momentum $p_i = m\dot{x}_i$ is conserved, whereas if H is time-independent (equivalently, if V does not depend on t) H itself is conserved. These are nothing but the laws of conservation of the i-th component of the linear momentum and the total energy that we already knew.

Example 4.13. *Hamiltonian of a particle in spherical coordinates.*

As we saw in Example 4.7, the Lagrangian of a particle of mass m in spherical coordinates is given by Eq. (4.70), namely

$$L = \frac{m}{2} (\dot{r}^2 + r^2\dot{\theta}^2 + r^2 \sin^2\theta\, \dot{\varphi}^2) - V(t, r, \theta, \varphi).$$

This Lagrangian is clearly *natural*, so that its corresponding Hamiltonian is simply the total energy

$$T + V = \frac{m}{2} (\dot{r}^2 + r^2\dot{\theta}^2 + r^2 \sin^2\theta\, \dot{\varphi}^2) + V(t, r, \theta, \varphi),$$

expressed in terms of the canonical momenta (4.71):

$$p_r = m\dot{r}, \qquad p_\theta = mr^2\dot{\theta}, \qquad p_\varphi = mr^2 \sin^2\theta\, \dot{\varphi}.$$

From these equations we obtain

$$\boxed{\dot{r} = \frac{p_r}{m}, \qquad \dot{\theta} = \frac{p_\theta}{mr^2}, \qquad \dot{\varphi} = \frac{p_\varphi}{mr^2 \sin^2\theta},} \tag{4.91}$$

which yields the following expression for the Hamiltonian:

$$\boxed{H(t, r, \theta, \varphi, p_r, p_\theta, p_\varphi) = \frac{1}{2m}\left(p_r^2 + \frac{p_\theta^2}{r^2} + \frac{p_\varphi^2}{r^2 \sin^2\theta}\right) + V(t, r, \theta, \varphi).} \tag{4.92}$$

Hamilton's canonical equations are in this case the three equations (4.91), along with the three equations for the time derivatives of the momenta:

$$\boxed{\begin{aligned}\dot{p}_r &= -\frac{\partial H}{\partial r} = -\frac{\partial V}{\partial r} + \frac{1}{mr^3}\left(p_\theta^2 + \frac{p_\varphi^2}{\sin^2\theta}\right),\\ \dot{p}_\theta &= -\frac{\partial H}{\partial \theta} = -\frac{\partial V}{\partial \theta} + \frac{p_\varphi^2}{mr^2}\frac{\cos\theta}{\sin^3\theta},\\ \dot{p}_\varphi &= -\frac{\partial H}{\partial \varphi} = -\frac{\partial V}{\partial \varphi}.\end{aligned}}$$

As we already knew, from the last of these equations it follows that p_φ (which is equal to the z component of the angular momentum) is conserved if the potential does not depend on φ. Similarly, since

$$\frac{\partial H}{\partial t} = \frac{\partial V}{\partial t},$$

if V is independent of t the Hamiltonian H, which coincides with the system's total energy, is conserved.

Example 4.14. *Hamiltonian of a charged particle in an electromagnetic field.*
As we saw in Example 4.10, using Cartesian coordinates $\mathbf{r} = (x_1, x_2, x_3)$ the Lagrangian can be taken as

$$L = \frac{1}{2} m\dot{\mathbf{r}}^2 - e\Phi(t, \mathbf{r}) + e\dot{\mathbf{r}} \cdot \mathbf{A}(t, \mathbf{r}). \tag{4.93}$$

Hence the canonical momenta are given by

$$p_i = m\dot{x}_i + eA_i(t, \mathbf{r}), \qquad i = 1, 2, 3, \tag{4.94}$$

and the Hamiltonian reads

$$H = \frac{1}{2} m\dot{\mathbf{r}}^2 + e\Phi(t, \mathbf{r}),$$

where it is understood that the velocities must be expressed in terms of the canonical momenta. Since

$$\boxed{\dot{x}_i = \frac{1}{m}\big(p_i - eA_i(t, \mathbf{r})\big), \qquad i = 1, 2, 3,} \tag{4.95}$$

substituting into the formula for H we obtain the expression

$$\boxed{H(t, \mathbf{r}, \mathbf{p}) = \frac{1}{2m}\big(\mathbf{p} - e\mathbf{A}(t, \mathbf{r})\big)^2 + e\Phi(t, \mathbf{r}).}$$

Note that in this formula $\mathbf{p}$ does *not* denote the *linear* momentum of the particle, but rather the vector whose three components are the *canonical* momenta p_i given by Eq. (4.94). Hamilton's equations are the three equations (4.95), along with

$$\boxed{\dot{p}_i = -\frac{\partial H}{\partial x_i} = -e\,\frac{\partial \Phi}{\partial x_i}(t,\mathbf{r}) + \frac{e}{m}\,\big(\mathbf{p} - e\mathbf{A}(t,\mathbf{r})\big)\cdot\frac{\partial \mathbf{A}}{\partial x_i}(t,\mathbf{r})\,, \qquad i = 1,2,3\,.}$$

The previous Hamiltonian can also be easily calculated in *spherical coordinates*. In fact, we know that the Lagrangian is *covariant* under coordinate changes, so that in order to obtain the Lagrangian of a charged particle in spherical coordinates it suffices to express the Lagrangian (4.93) in these coordinates. Since

$$\dot{\mathbf{r}}\cdot\mathbf{A} = (\dot{r}\mathbf{e}_r + r\dot{\theta}\mathbf{e}_\theta + r\sin\theta\,\dot{\varphi}\mathbf{e}_\varphi)\cdot(A_r\mathbf{e}_r + A_\theta\mathbf{e}_\theta + A_\varphi\mathbf{e}_\varphi) = \dot{r}A_r + r\dot{\theta}A_\theta + r\sin\theta\,\dot{\varphi}A_\varphi\,,$$

substituting into Eq. (4.45) we obtain

$$\boxed{L = \frac{m}{2}(\dot{r}^2 + r^2\dot{\theta}^2 + r^2\sin^2\theta\,\dot{\varphi}^2) - e\Phi + e\big(\dot{r}A_r + r\dot{\theta}A_\theta + r\sin\theta\,\dot{\varphi}A_\varphi\big)\,.}$$

The canonical momenta are now

$$p_r = m\dot{r} + eA_r\,, \qquad p_\theta = mr^2\dot{\theta} + erA_\theta\,, \qquad p_\varphi = mr^2\sin^2\theta\,\dot{\varphi} + er\sin\theta\,A_\varphi\,,$$

so that

$$\begin{aligned}
\dot{r} &= \frac{1}{m}\,(p_r - eA_r)\,,\\
r\dot{\theta} &= \frac{1}{mr}\,(p_\theta - erA_\theta)\,,\\
r\sin\theta\,\dot{\varphi} &= \frac{1}{mr\sin\theta}\,(p_\varphi - er\sin\theta\,A_\varphi)\,.
\end{aligned}$$

Substituting into the definition of H we finally obtain

$$\begin{aligned}
H &= \dot{r}p_r + \dot{\theta}p_\theta + \dot{\varphi}p_\varphi - \frac{m}{2}(\dot{r}^2 + r^2\dot{\theta}^2 + r^2\sin^2\theta\,\dot{\varphi}^2)\\
&\quad + e\Phi - e\big(\dot{r}A_r + r\dot{\theta}A_\theta + r\sin\theta\,\dot{\varphi}A_\varphi\big) = \frac{m}{2}\,\big(\dot{r}^2 + r^2\dot{\theta}^2 + r^2\sin^2\theta\,\dot{\varphi}^2\big) + e\Phi\\
&\boxed{= \frac{1}{2m}\left[(p_r - eA_r)^2 + \frac{(p_\theta - erA_\theta)^2}{r^2} + \frac{(p_\varphi - er\sin\theta\,A_\varphi)^2}{r^2\sin^2\theta}\right] + e\Phi\,,}
\end{aligned}$$

which can also be expressed as

$$\boxed{H = \frac{1}{2m}\left[(p_r - eA_r)^2 + \left(\frac{p_\theta}{r} - eA_\theta\right)^2 + \left(\frac{p_\varphi}{r\sin\theta} - eA_\varphi\right)^2\right] + e\Phi\,.} \tag{4.96}$$

Exercise 4.8. Write down Hamilton's canonical equations for the Hamiltonian (4.96).

Solution. To begin with, Hamilton's equations for the coordinates r, θ, and φ are simply the expressions for $\dot{r}$, $\dot{\theta}$, and $\dot{\varphi}$ in terms of the canonical momenta p_r, p_θ, and p_φ found above, namely

$$\dot{r} = \frac{1}{m}\,(p_r - eA_r)\,, \quad \dot{\theta} = \frac{1}{mr^2}\,(p_\theta - erA_\theta)\,, \quad \dot{\varphi} = \frac{1}{mr^2\sin^2\theta}\,(p_\varphi - er\sin\theta\, A_\varphi)\,.$$

The remaining equations are easily found differentiating the Hamiltonian H with respect to the coordinates, namely

$$\begin{aligned}
\dot{p}_r &= -\frac{\partial H}{\partial r} = -e\frac{\partial \Phi}{\partial r} + \frac{e}{m}(p_r - eA_r)\frac{\partial A_r}{\partial r} + \frac{1}{m}\left(\frac{p_\theta}{r} - eA_\theta\right)\left(\frac{p_\theta}{r^2} + e\frac{\partial A_\theta}{\partial r}\right) \\
&\quad + \frac{1}{m}\left(\frac{p_\varphi}{r\sin\theta} - eA_\varphi\right)\left(\frac{p_\varphi}{r^2\sin\theta} + e\frac{\partial A_\varphi}{\partial r}\right), \\
\dot{p}_\theta &= -\frac{\partial H}{\partial \theta} = -e\frac{\partial \Phi}{\partial \theta} + \frac{e}{m}(p_r - eA_r)\frac{\partial A_r}{\partial \theta} + \frac{e}{m}\left(\frac{p_\theta}{r} - eA_\theta\right)\frac{\partial A_\theta}{\partial \theta} \\
&\quad + \frac{1}{m}\left(\frac{p_\varphi}{r\sin\theta} - eA_\varphi\right)\left(\frac{p_\varphi\cos\theta}{r\sin^2\theta} + e\frac{\partial A_\varphi}{\partial \theta}\right), \\
\dot{p}_\varphi &= -\frac{\partial H}{\partial \varphi} = -e\frac{\partial \Phi}{\partial \varphi} + \frac{e}{m}(p_r - eA_r)\frac{\partial A_r}{\partial \varphi} + \frac{e}{m}\left(\frac{p_\theta}{r} - eA_\theta\right)\frac{\partial A_\theta}{\partial \varphi} \\
&\quad + \frac{e}{m}\left(\frac{p_\varphi}{r\sin\theta} - eA_\varphi\right)\frac{\partial A_\varphi}{\partial \varphi}.
\end{aligned}$$

4.5.3 Poisson brackets

As we have just seen, in the Hamiltonian formalism the equations of motion are *first-order* in the variables $\mathbf{q}$ (the generalized coordinates of the Lagrangian formalism) and $\mathbf{p}$ (their associated canonical momenta). Thus the motion of the system can be represented by the trajectory of a *single point* in the space $\mathbb{R}^{2n}$ where the canonical variables $(\mathbf{q}, \mathbf{p})$ take values, usually referred to as the system's **phase space**. Note that, by the existence and uniqueness theorem for systems of first-order ordinary differential equations, there is a *unique* trajectory $(\mathbf{q}(t), \mathbf{p}(t))$ passing through any point $(\mathbf{q}_0, \mathbf{p}_0)$ in phase space at a certain initial time t_0, i.e., verifying the initial conditions $\mathbf{q}(t_0) = \mathbf{q}_0$, $\mathbf{p}(t_0) = \mathbf{p}_0$. (We are assuming, as we shall implicitly do in what follows, that the Hamiltonian $H(t, \mathbf{q}, \mathbf{p})$ is of class C^2 in the variables $(t, \mathbf{q}, \mathbf{p})$ for all t.) It can be shown that if H is time-independent this implies that *the system's trajectories in phase space do not intersect.*

The rate of change of a smooth function $f(t, \mathbf{q}, \mathbf{p})$ (usually called a **dynamical variable**) as the canonical variables $\mathbf{q}$ and $\mathbf{p}$ evolve with time through Hamilton's canonical equations (4.89) for a given Hamiltonian $H(t, \mathbf{q}, \mathbf{p})$ is given by

$$\dot{f} = \frac{\partial f}{\partial t} + \frac{\partial f}{\partial \mathbf{q}}\dot{\mathbf{q}} + \frac{\partial f}{\partial \mathbf{p}}\dot{\mathbf{p}} = \frac{\partial f}{\partial t} + \frac{\partial f}{\partial \mathbf{q}}\frac{\partial H}{\partial \mathbf{p}} - \frac{\partial f}{\partial \mathbf{p}}\frac{\partial H}{\partial \mathbf{q}}\,.$$

This suggests defining the **Poisson bracket** of two dynamical variables $f(t, \mathbf{q}, \mathbf{p})$ and $g(t, \mathbf{q}, \mathbf{p})$ as the expression

$$\{f, g\} := \frac{\partial f}{\partial \mathbf{q}} \frac{\partial g}{\partial \mathbf{p}} - \frac{\partial f}{\partial \mathbf{p}} \frac{\partial g}{\partial \mathbf{q}} \equiv \sum_{i=1}^{n} \left(\frac{\partial f}{\partial q_i} \frac{\partial g}{\partial p_i} - \frac{\partial f}{\partial p_i} \frac{\partial g}{\partial q_i} \right). \tag{4.97}$$

Using this definition, the previous formula for $\dot{f}$ can be concisely written as

$$\dot{f} = \frac{\partial f}{\partial t} + \{f, H\}; \tag{4.98}$$

in particular, if f does not depend explicitly on time t we obtain the simpler expression

$$\dot{f} = \{f, H\}.$$

Applying the previous formula to the coordinates $(\mathbf{q}, \mathbf{p})$ in phase space we obtain the following formulation of Hamilton's canonical equations in terms of the Poisson bracket:

$$\dot{q}_i = \{q_i, H\}, \qquad \dot{p}_i = \{p_i, H\}, \qquad i = 1, \dots, n.$$

The Poisson brackets of the canonical coordinates and momenta among themselves are particularly simple:

$$\{q_i, q_j\} = \{p_i, p_j\} = 0, \qquad \{q_i, p_j\} = \delta_{ij}, \qquad i, j = 1, \dots, n, \tag{4.99}$$

where δ_{ij} is Kronecker's delta.

The following properties of the Poisson bracket follow immediately from its definition:

1) *Antisymmetry*: $\{f, g\} = -\{g, f\}$. In particular, $\{f, f\} = 0$.
2) *Bilinearity*: $\{\lambda f + \mu g, h\} = \lambda\{f, h\} + \mu\{g, h\}$, where λ, μ are constant (or, more generally, functions only of t). (By antisymmetry, the analogous property holds for the Poisson bracket $\{f, \lambda g + \mu h\}$.)
3) *Leibniz's rule*: $\{fg, h\} = f\{g, h\} + \{f, h\}g$ (and similarly for $\{f, gh\}$).

On the other hand, a long but straightforward calculation shows that the Poisson bracket satisfies the so-called **Jacobi identity**

$$\{\{f, g\}, h\} + \{\{g, h\}, f\} + \{\{h, f\}, g\} = 0. \tag{4.100}$$

From Leibniz's rule for partial derivatives it easily follows that

$$\frac{\partial}{\partial t} \{f, g\} = \left\{ \frac{\partial f}{\partial t}, g \right\} + \left\{ f, \frac{\partial g}{\partial t} \right\}.$$

Using this relation and the Jacobi identity we can derive an important generalization of the previous result, known as the *Jacobi–Poisson identity*:

$$\boxed{\frac{\mathrm{d}}{\mathrm{d}t}\{f,g\} = \{\dot f,g\} + \{f,\dot g\}\,.} \tag{4.101}$$

Indeed,

$$\begin{aligned}
\frac{\mathrm{d}}{\mathrm{d}t}\{f,g\} &= \frac{\partial}{\partial t}\{f,g\} + \{\{f,g\},H\} \\
&= \left\{\frac{\partial f}{\partial t},g\right\} + \left\{f,\frac{\partial g}{\partial t}\right\} - \{\{g,H\},f\} - \{\{H,f\},g\} \\
&= \left\{\frac{\partial f}{\partial t},g\right\} + \{\{f,H\},g\} + \left\{f,\frac{\partial g}{\partial t}\right\} + \{f,\{g,H\}\} \\
&= \left\{\frac{\partial f}{\partial t} + \{f,H\},g\right\} + \left\{f,\frac{\partial g}{\partial t} + \{g,H\}\right\} \\
&= \{\dot f,g\} + \{f,\dot g\}\,,
\end{aligned}$$

where we have used the Jacobi identity in the second equality. An important corollary of the Jacobi–Poisson identity is the so-called **Jacobi–Poisson theorem**, of fundamental importance for obtaining first integrals of Hamiltonian systems:

> If $f(t,\mathbf{q},\mathbf{p})$ and $g(t,\mathbf{q},\mathbf{p})$ are two first integrals of Hamilton's canonical equations (4.89), so is their Poisson bracket $\{f,g\}$.

Example 4.15. As we saw in Example 4.12, the Hamiltonian of a particle of mass m in Cartesian coordinates is given by

$$H = \frac{\mathbf{p}^2}{2m} + V(t,\mathbf{r})\,,$$

where $\mathbf{r} = (x_1,x_2,x_3)$ plays the role of $\mathbf{q}$ and $\mathbf{p} = m\dot{\mathbf{r}}$ is the linear momentum. We can compute the Poisson bracket of any two components of the angular momentum

$$\mathbf{J} = \mathbf{r}\times\mathbf{p} = \left(x_2p_3 - x_3p_2, x_3p_1 - x_1p_3, x_1p_2 - x_2p_1\right)$$

by applying the properties of the Poisson bracket reviewed above and the fundamental brackets (4.99). For instance,

$$\begin{aligned}
\{J_1,J_2\} &= \{x_2p_3 - x_3p_2, x_3p_1 - x_1p_3\} \\
&= \{x_2p_3,x_3p_1\} - \{x_2p_3,x_1p_3\} - \{x_3p_2,x_3p_1\} + \{x_3p_2,x_1p_3\} \\
&= -x_2p_1 + p_2x_1 = J_3\,.
\end{aligned}$$

Proceeding in this way we obtain the important relations

$$\boxed{\{J_i, J_j\} = J_k\,, \qquad (i, j, k) = \textit{cyclic} \text{ permutation of } (1, 2, 3)\,.}$$

Suppose now that any two components of the angular momentum are conserved, for instance J_1 and J_2. By the Jacobi–Poisson theorem, the remaining component $J_3 = \{J_1, J_2\}$ will also be conserved. In other words, *if any two components of the angular moment* **J** *are conserved then* **J** *is conserved.* Likewise, suppose that the projection of the linear momentum **p** (which in this case coincides with the canonical one) along a certain direction **n** and the angular momentum are conserved. Choosing the coordinates appropriately, we can assume that p_1 and **J** are conserved. The relation

$$\{p_1, J_2\} = \{p_1, x_3 p_1 - x_1 p_3\} = -\{p_1, x_1 p_3\} = p_3$$

then implies, by the Jacobi–Poisson theorem, that p_3 is also conserved. Similarly, from the Poisson bracket

$$\{p_1, J_3\} = \{p_1, x_1 p_2 - x_2 p_1\} = \{p_1, x_1 p_2\} = -p_2$$

we deduce that p_2 is conserved. Hence in this case the linear momentum **p** is conserved.

Remark 4.3. The fundamental Poisson brackets (4.99) make it possible to establish a formal analogy between classical and quantum mechanics. Consider, for the sake of simplicity, the case of a single particle in *Cartesian* coordinates, for which $q_j = x_j$ and $p_j = m\dot{x}_j$. In quantum mechanics the dynamical variables (q_j, p_j) are replaced (in the so-called Schrödinger picture) by the *self-adjoint operators*

$$\boxed{Q_j = q_j\,, \qquad P_j = -\mathrm{i}\hbar\frac{\partial}{\partial q_j}}\,,$$

whose action on a complex valued *wave function* (probability amplitude) $\psi(\mathbf{q})$ is given by

$$(Q_j\psi)(\mathbf{q}) = q_j\psi(\mathbf{q})\,, \qquad (P_j\psi)(\mathbf{q}) = -\mathrm{i}\hbar\,\frac{\partial\psi}{\partial q_j}(\mathbf{q})\,.$$

The operators (Q_i, P_j) satisfy *commutation relations* totally analogous to Eq. (4.99):

$$\boxed{[Q_i, Q_j] = [P_i, P_j] = 0\,, \qquad [Q_i, P_j] = \mathrm{i}\hbar\left[\frac{\partial}{\partial q_j}, q_i\right] = \mathrm{i}\hbar\,\delta_{ij}\,,} \tag{4.102}$$

where the *commutator* of two operators A, B is defined by

$$[A, B] = AB - BA\,.$$

Any other function $f(\mathbf{q}, \mathbf{p})$ is represented in quantum mechanics by a self-adjoint operator $F(\mathbf{Q}, \mathbf{P})$ such that

$$F(\mathbf{q}, \mathbf{p}) = f(\mathbf{q}, \mathbf{p}) .$$

This fact is known as *Bohr's correspondence principle.* It is important to realize in this respect that, since the product of operators is not commutative in general, the operator F determines the classical function f but not vice versa. For instance,

$$F_1(Q, P) = PQ^2P \neq F_2(Q, P) = \frac{1}{2}\,(Q^2P^2 + P^2Q^2)$$

(in fact, $F_1 - F_2 = \hbar^2$), even though $f_1(q, p) = f_2(q, p) = q^2p^2$. This fact is not surprising, since classical mechanics is the limit as $\hbar \to 0$ of quantum mechanics, so the former theory must be determined by the latter. The converse, however, is not necessarily true, as there may exist different theories with the same limit as $\hbar \to 0$.

The commutator $[A, B]$ of two operators A and B has algebraic properties formally analogous to those of the Poisson bracket. Indeed, it is obviously antisymmetric and linear in each of its arguments. In addition, if A, B, and C are three operators then it is immediate to show that

$$[AB, C] = A\,[B, C] + [A, C]\,B . \tag{4.103}$$

This identity is similar to Leibniz's rule satisfied by the Poisson bracket, with the only difference that the *order* in which operators appear in Eq. (4.103) is essential for its validity. Finally, it is straightforward to show that the commutator also verifies the *Jacobi identity*

$$[[A, B], C] + [[B, C], A] + [[C, A], B] = 0 ,$$

where the order is again essential.

If $F(\mathbf{Q}, \mathbf{P})$ and $G(\mathbf{Q}, \mathbf{P})$ are two self-adjoint operators depending *polynomially* on $(\mathbf{Q}, \mathbf{P})$ (and not explicitly dependent on $\hbar$), by repeatedly applying Eq. (4.103) we can always express the commutator $[F, G]$ in terms of the canonical commutators in Eq. (4.102). For instance,

$$[Q, P^2] = [Q, P]P + P[Q, P] = 2i\hbar P. \tag{4.104}$$

It then follows from Leibniz's rule that the Poisson bracket $\{f, g\}$ of the corresponding classical functions $f(\mathbf{q}, \mathbf{p}) = F(\mathbf{q}, \mathbf{p})$, $g(\mathbf{q}, \mathbf{p}) = G(\mathbf{q}, \mathbf{p})$ will satisfy the *same* expression replacing Q_i by q_i, P_i by p_i and the canonical commutators (4.102) by the canonical Poisson brackets (4.99). For instance, the classical analogue of Eq. (4.104) is

$$\{q, p^2\} = \{q, p\}p + p\{q, p\} = 2p.$$

In general, if

$$[F, G] = i\hbar K , \tag{4.105}$$

where $K(\mathbf{Q}, \mathbf{P})$ is a polynomial independent of $\hbar$, the classical Poisson bracket $\{f, g\}$ will be given by

$$\{f, g\} = k \tag{4.106}$$

with $k(\mathbf{q},\mathbf{p}) = K(\mathbf{q},\mathbf{p})$ (see the next exercise for an example). Therefore *the commutator in quantum mechanics determines the Poisson bracket in classical mechanics through the relation*

$$\boxed{\frac{1}{i\hbar}\left[F,G\right] \to \{f,g\}\,.}$$

Note, finally, that the opposite route (from the Poisson bracket in classical mechanics to the commutator in quantum mechanics) *is not well defined in general*, since as we have remarked different self-adjoint operators $F(\mathbf{Q},\mathbf{P})$ can yield the same function $f(\mathbf{q},\mathbf{p})$. ■

Example 4.16. Consider, for example, the commutator $[Q^2, P^2]$. Using repeatedly Eq. (4.103) we obtain

$$\begin{aligned}[Q^2,P^2] &= [Q\cdot Q,P^2] = Q[Q,P^2] + [Q,P^2]Q \\ &= QP[Q,P] + Q[Q,P]P + P[Q,P]Q + [Q,P]PQ = 2i\hbar(QP+PQ)\,.\end{aligned}$$

At the classical level, repeated application of Leibniz's rule to the Poisson bracket $\{q^2,p^2\}$ yields

$$\begin{aligned}\{q^2,p^2\} &= q\{q,p^2\} + \{q,p^2\}q = qp\{q,p\} + q\{q,p\}p + p\{q,p\}q + \{q,p\}pq \\ &= 2(qp+pq) = 4qp\,,\end{aligned}$$

which is indeed obtained from $[Q^2,P^2]/(i\hbar)$ replacing Q by q and P by p.

4.5.4 Canonical transformations

We have already remarked in the previous subsection that in the Hamiltonian formulation the canonical variables $\mathbf{q}$ and $\mathbf{p}$ have identical status. It is therefore reasonable to try to simplify Hamilton's equations (4.89) using general changes of variables of the form

$$\widetilde{\mathbf{q}} = \widetilde{\mathbf{q}}(t,\mathbf{q},\mathbf{p})\,,\qquad \widetilde{\mathbf{p}} = \widetilde{\mathbf{p}}(t,\mathbf{q},\mathbf{p}) \tag{4.107}$$

involving both coordinates and momenta. The problem is that, in general, such a transformation maps the system (4.89) into a first-order system that need *not* be in general of Hamiltonian type, i.e., of the form

$$\dot{\widetilde{\mathbf{q}}} = \frac{\partial \widetilde{H}}{\partial \widetilde{\mathbf{p}}}\,,\qquad \dot{\widetilde{\mathbf{p}}} = -\frac{\partial \widetilde{H}}{\partial \widetilde{\mathbf{q}}} \tag{4.108}$$

for a certain function $\widetilde{H}(t,\widetilde{\mathbf{q}},\widetilde{\mathbf{p}})$. The transformation (4.107) is said to be **canonical** provided that it maps Hamilton's equations of *any* Hamiltonian $H(t,\mathbf{q},\mathbf{p})$ into the canonical equations of another Hamiltonian $\widetilde{H}(t,\widetilde{\mathbf{q}},\widetilde{\mathbf{p}})$.

Example 4.17. The transformation

$$\tilde{\mathbf{q}} = \mathbf{p}, \qquad \tilde{\mathbf{p}} = \mathbf{q}$$

is canonical, since it transforms Hamilton's equations of any Hamiltonian $H(t, \mathbf{q}, \mathbf{p})$ into those of the Hamiltonian $\tilde{H}(t, \tilde{\mathbf{q}}, \tilde{\mathbf{p}}) = -H(t, \mathbf{q}, \mathbf{p})$. Indeed,

$$\dot{\tilde{\mathbf{q}}} = \dot{\mathbf{p}} = -\frac{\partial H}{\partial \mathbf{q}} = \frac{\partial \tilde{H}}{\partial \tilde{\mathbf{p}}}, \qquad \dot{\tilde{\mathbf{p}}} = \dot{\mathbf{q}} = \frac{\partial H}{\partial \mathbf{p}} = -\frac{\partial \tilde{H}}{\partial \tilde{\mathbf{q}}}.$$

The transformation

$$\tilde{\mathbf{q}} = \mathbf{p}, \qquad \tilde{\mathbf{p}} = -\mathbf{q},$$

is also canonical, with $\tilde{H}(t, \tilde{\mathbf{q}}, \tilde{\mathbf{p}}) = H(t, \mathbf{q}, \mathbf{p})$.

Exercise 4.9. Show that the transformation

$$\tilde{q} = p^2, \qquad \tilde{p} = q \tag{4.109}$$

is *not* canonical.

Solution. The equations of motion for the new canonical variables $(\tilde{q}, \tilde{p})$ are

$$\dot{\tilde{q}} = 2p\dot{p} = -2p\frac{\partial H}{\partial q}, \qquad \dot{\tilde{p}} = \dot{q} = \frac{\partial H}{\partial p}.$$

For the change of variables (4.109) to be a canonical transformation, for *any* given $H(t, q, p)$ there must exist a function $\tilde{H}(t, \tilde{q}, \tilde{p})$ such that

$$\frac{\partial \tilde{H}}{\partial \tilde{p}} = -2p\frac{\partial H}{\partial q}, \qquad \frac{\partial \tilde{H}}{\partial \tilde{q}} = -\frac{\partial H}{\partial p}. \tag{4.110}$$

The differential of $\tilde{H}$ is therefore

$$\mathrm{d}\tilde{H} = \frac{\partial \tilde{H}}{\partial t}\,\mathrm{d}t + \frac{\partial \tilde{H}}{\partial \tilde{q}}\,\mathrm{d}\tilde{q} + \frac{\partial \tilde{H}}{\partial \tilde{p}}\,\mathrm{d}\tilde{p} = \frac{\partial \tilde{H}}{\partial t}\,\mathrm{d}t - 2p\frac{\partial H}{\partial p}\,\mathrm{d}p - 2p\frac{\partial H}{\partial q}\,\mathrm{d}q.$$

The partial derivatives of $\tilde{H}$ (considered as a function of the variable (t, q, p)) with respect to the canonical variables (q, p) are thus

$$\frac{\partial \tilde{H}}{\partial q} = -2p\frac{\partial H}{\partial q}, \qquad \frac{\partial \tilde{H}}{\partial p} = -2p\frac{\partial H}{\partial p}.$$

This implies that

$$\frac{\partial^2 \widetilde{H}}{\partial p \partial q} = -2\frac{\partial}{\partial p}\left(p\frac{\partial H}{\partial q}\right) = -2\frac{\partial H}{\partial q} - 2p\frac{\partial^2 H}{\partial p \partial q} = \frac{\partial \widetilde{H}}{\partial q \partial p} = -2p\frac{\partial^2 H}{\partial p \partial q}$$
$$\implies \quad \frac{\partial H}{\partial q} = 0,$$

which shows that if $\frac{\partial H}{\partial q} \neq 0$ there is no Hamiltonian $\widetilde{H}(t, \widetilde{q}, \widetilde{p})$ satisfying Eqs. (4.110). Hence the transformation (4.109) is *not* canonical, as claimed.

An important result in Hamiltonian mechanics states that *the transformation* (4.107) *is canonical if and only if the Poisson brackets of the transformed canonical variables* $\widetilde{\mathbf{q}}$ *and* $\widetilde{\mathbf{p}}$ *verify*

$$\{\widetilde{q}_i, \widetilde{q}_j\} = \{\widetilde{p}_i, \widetilde{p}_j\} = 0\,, \qquad \{\widetilde{q}_i, \widetilde{p}_j\} = \lambda\delta_{ij}\,, \qquad i, j = 1, \ldots, n\,,$$

with $\lambda \neq 0$ *constant*[13]. In particular, if $\lambda = 1$ the variables $\widetilde{\mathbf{q}}$ and $\widetilde{\mathbf{p}}$ are said to be **canonically conjugate**.

Exercise 4.10. Applying the previous result, show that the transformation (4.109) is not canonical.

Solution. Indeed,

$$\{\widetilde{q}, \widetilde{p}\} = \{p^2, q\} = -2p$$

is not constant.

In fact, it can be shown that it is always possible to find a canonical transformation (4.107) mapping Hamilton's equations (4.89) of *any* Hamiltonian $H(t, \mathbf{q}, \mathbf{p})$ into the canonical equations of the Hamiltonian $\widetilde{H} = 0$, that is to say, into the trivial system

$$\dot{\widetilde{\mathbf{q}}} = 0\,, \qquad \dot{\widetilde{\mathbf{p}}} = 0\,.$$

The general solution of this system is obviously

$$\widetilde{\mathbf{q}} = \widetilde{\mathbf{q}}_0\,, \qquad \widetilde{\mathbf{p}} = \widetilde{\mathbf{p}}_0\,,$$

with $\widetilde{\mathbf{q}}_0, \widetilde{\mathbf{p}}_0$ arbitrary constant vectors. The general solution of the canonical equations of the original Hamiltonian $H(t, \mathbf{q}, \mathbf{p})$ is then obtained inverting the relations

$$\widetilde{\mathbf{q}}(t, \mathbf{q}, \mathbf{p}) = \widetilde{\mathbf{q}}_0\,, \qquad \widetilde{\mathbf{p}}(t, \mathbf{q}, \mathbf{p}) = \widetilde{\mathbf{p}}_0$$

to express $\mathbf{q}$ and $\mathbf{p}$ in terms of t and the $2n$ constants $(\widetilde{\mathbf{q}}_0, \widetilde{\mathbf{p}}_0)$. In fact, this is one of the most effective methods for solving Hamilton's equations, using the so-called *Hamilton–Jacobi equation* for finding a canonical transformation mapping the original Hamiltonian H into $\widetilde{H} = 0$.

[13]Usually only canonical transformations with $\lambda = 1$ (called *proper*) are considered. This does not entail any real restriction, since if (4.107) is a canonical transformation with $\lambda \neq 1$ the transformation $(\mathbf{q}, \mathbf{p}) \mapsto (\widetilde{\mathbf{q}}, \widetilde{\mathbf{p}}/\lambda)$ is another canonical transformation with $\lambda = 1$

4.5.5 The Hamilton–Jacobi equation

The **Hamilton–Jacobi equation** is the partial differential equation

$$\boxed{\frac{\partial S}{\partial t} + H\left(t, \mathbf{q}, \frac{\partial S}{\partial \mathbf{q}}\right) = 0} \tag{4.111}$$

in the unknown function $S(t, \mathbf{q})$. A *complete solution* of the Hamilton–Jacobi equation is a solution $S(t, \mathbf{q}, \boldsymbol{\alpha})$ depending on n parameters $(\alpha_1, \dots, \alpha_n) = \boldsymbol{\alpha}$ (where n is the number of degrees of freedom) that satisfies the condition

$$\det\left(\frac{\partial^2 S}{\partial q_i \partial \alpha_j}\right)_{1 \leqslant i,j \leqslant n} \neq 0. \tag{4.112}$$

If $S(t, \mathbf{q}, \boldsymbol{\alpha})$ is a complete solution of the Hamilton–Jacobi equation, it is shown in textbooks on theoretical mechanics that the mapping $(\mathbf{q}, \mathbf{p}) \to (\boldsymbol{\alpha}, \boldsymbol{\beta})$ implicitly defined by[14]

$$\boxed{\mathbf{p} = \frac{\partial S(t, \mathbf{q}, \boldsymbol{\alpha})}{\partial \mathbf{q}}, \qquad \boldsymbol{\beta} = -\frac{\partial S(t, \mathbf{q}, \boldsymbol{\alpha})}{\partial \boldsymbol{\alpha}}} \tag{4.113}$$

is a *canonical transformation* sending Hamilton's equations for H to the corresponding equations for the trivial Hamiltonian $\widetilde{H}(t, \boldsymbol{\alpha}, \boldsymbol{\beta}) = 0$. Hence the *general solution* of the equations of motion in the original coordinates $(\mathbf{q}, \mathbf{p})$ is implicitly given by Eq. (4.113), with $\boldsymbol{\alpha}, \boldsymbol{\beta} \in \mathbb{R}^n$ arbitrary constants.

The problem of finding the general solution of the equations of motion of a Hamiltonian system is thus reduced to finding a complete solution of the Hamilton–Jacobi equation. Although solving a *partial* differential equation is in general more difficult than solving a system of *ordinary differential equations,* in this case it is sufficient to find a *complete* solution rather than the *general* solution of the Hamilton–Jacobi equation, which typically depends on arbitrary functions. In fact, in many cases it is possible to find a complete solution of the Hamilton–Jacobi equation by several standard techniques, such as separation of variables in suitable coordinates.

The Hamilton–Jacobi equation was the starting point used by Schrödinger in his 1926 paper to deduce the equation that now bears his name. Indeed, let us take for simplicity the canonical coordinates $\mathbf{q}$ as the position vector $\mathbf{r}$ of a particle, so that

$$H = \frac{\mathbf{p}^2}{2m} + V(\mathbf{r}).$$

Let us define a function $\psi(t, \mathbf{r})$ through the equation

$$\psi(t, \mathbf{r}) = e^{\frac{i}{\hbar} S(t, \mathbf{r})},$$

[14]Note that, by the inverse function theorem, Eq. (4.112) guarantees that the second equation (4.113) can be locally inverted to express $\mathbf{q}$ as a function of the parameters $\boldsymbol{\alpha}$ and $\boldsymbol{\beta}$. Substituting this expression into the first equation (4.113) we then obtain an expression for $\mathbf{p}$ in terms of $\boldsymbol{\alpha}$ and $\boldsymbol{\beta}$.

where $\hbar$ is a constant[15] with dimensions of *action*. (Note that the introduction of a constant with action dimensions in the previous formula is necessary, since the Hamilton–Jacobi equation implies that S has dimensions of energy times time.) The partial derivatives of ψ with respect to time and position are given by

$$\frac{\partial \psi}{\partial t} = \frac{\mathrm{i}}{\hbar}\frac{\partial S}{\partial t}\psi, \qquad \nabla\psi = \frac{\mathrm{i}}{\hbar}(\nabla S)\psi,$$
$$\nabla^2\psi = \frac{\mathrm{i}}{\hbar}(\nabla^2 S)\psi + \frac{\mathrm{i}}{\hbar}\nabla S\cdot\nabla\psi = \frac{\mathrm{i}}{\hbar}\hbar(\nabla^2 S)\psi - \frac{1}{\hbar^2}\left(\nabla S\right)^2\psi .$$

The first term in the RHS of the last formula can be dropped provided that

$$\frac{\left|\nabla^2 S\right|}{(\nabla S)^2} \ll \frac{1}{\hbar}, \tag{4.114}$$

in which case we have the approximate equality

$$\left(\nabla S\right)^2\psi = -\hbar^2\nabla^2\psi.$$

From the Hamilton–Jacobi equation, which in this case reduces to

$$\frac{\partial S}{\partial t} + \frac{1}{2m}\left(\nabla S\right)^2 + V(\mathbf{r}) = 0,$$

and the previous approximate formula for $\left(\nabla S\right)^2\psi$ we obtain the following differential equation for the *wave function* $\psi(\mathbf{r},t)$:

$$\mathrm{i}\hbar\frac{\partial\psi}{\partial t} = -\frac{\partial S}{\partial t}\psi = \frac{1}{2m}\left(\nabla S\right)^2\psi + V(\mathbf{r})\psi = -\frac{\hbar^2}{2m}\nabla^2\psi + V(\mathbf{r})\psi.$$

This is the celebrated *time-dependent Schrödinger equation*, which governs the time evolution of the wave function (in the absence of measurements) in quantum mechanics.

The physical significance of condition (4.114) can be understood using the first equation (4.113), which in this case reduces to

$$\mathbf{p} = \nabla S,$$

and the formula for the particle's *de Broglie wavelength*

$$\lambda = \frac{2\pi\hbar}{p}.$$

For example, in one dimension Eq. (4.114) reads

$$\frac{p'(x)}{p(x)^2} \ll \frac{1}{\hbar}, \tag{4.115}$$

where

$$p(x) = \sqrt{2m(E - V(x))}.$$

Equation (4.115) can be rewritten as

$$\lambda\,\frac{p'(x)}{p(x)} \ll 2\pi;$$

in other words, the relative variation of the particle's momentum over a de Broglie wavelength must be very small.

---

[15] $\hbar$ is the *reduced Planck constant*, whose value in the SI system is $1.054571817\cdot 10^{-34}$ J s

Exercise 4.11. Using the Hamilton–Jacobi equation, solve the equations of motion for a particle of mass m moving in a plane under the action of a central potential $V(r)$.

Solution. The Hamiltonian

$$H = \frac{1}{2m}\left(p_r^2 + \frac{p_\varphi^2}{r^2}\right) + V(r)$$

does not depend on the time t and the polar angle φ. For this reason, we try a simple solution of the Hamilton–Jacobi equation of the form

$$S(r, \varphi, \alpha_1, \alpha_2) = -\alpha_1 t + W(r, \alpha_1, \alpha_2) + \alpha_2 \varphi.$$

Since

$$\frac{\partial S}{\partial t} = -\alpha_1, \qquad \frac{\partial S}{\partial r} = \frac{\partial W}{\partial r} \equiv W', \qquad \frac{\partial S}{\partial \varphi} = \alpha_2,$$

substituting into the Hamilton–Jacobi equation we obtain the following first-order ordinary differential equation for W considered as a function of r:

$$\frac{W'^2}{2m} + \frac{\alpha_2^2}{2mr^2} + V(r) = \alpha_1,$$

whose general solution is

$$W = \pm\sqrt{2m}\int dr\sqrt{\alpha_1 - V(r) - \frac{\alpha_2^2}{2mr^2}}\,.$$

The general solution of Hamilton's equations of motion is given by Eqs. (4.113), namely

$$p_r = \frac{\partial S}{\partial r} = W'(r) = \pm\sqrt{2m}\sqrt{\alpha_1 - V(r) - \frac{\alpha_2^2}{2mr^2}}, \qquad p_\varphi = \frac{\partial S}{\partial \varphi} = \alpha_2, \tag{4.116}$$

$$\beta_1 = -\frac{\partial S}{\partial \alpha_1} = t - \frac{\partial W}{\partial \alpha_1} = t \mp \sqrt{\frac{m}{2}}\int dr\left(\alpha_1 - V(r) - \frac{\alpha_2^2}{2mr^2}\right)^{-1/2}, \tag{4.117}$$

$$\begin{aligned}\beta_2 &= -\frac{\partial S}{\partial \alpha_2} = -\varphi - \frac{\partial W}{\partial \alpha_2} \\ &= -\varphi \pm \frac{\alpha_2}{\sqrt{2m}}\int \frac{dr}{r^2}\left(\alpha_1 - V(r) - \frac{\alpha_2^2}{2mr^2}\right)^{-1/2},\end{aligned} \tag{4.118}$$

with $\alpha_{1,2}$ and $\beta_{1,2}$ constant. As expected, p_φ is conserved, since the polar angle φ is cyclic. It is also clear that the constant α_1 is the particle's conserved energy E, since from Eq. (4.116) we obtain

$$\alpha_1 = \frac{1}{2m}\left(p_r^2 + \frac{\alpha_2^2}{r^2}\right) + V(r) = \frac{1}{2m}\left(p_r^2 + \frac{p_\varphi^2}{r^2}\right) + V(r).$$

Setting $\beta_1 = t_0$ and $\beta_2 = -\varphi_0$ we can rewrite Eqs. (4.117)–(4.118) as

$$\boxed{\begin{aligned}\varphi &= \varphi_0 \pm \frac{p_\varphi}{\sqrt{2m}} \int \frac{\mathrm{d}r}{r^2}\left(E - V(r) - \frac{p_\varphi^2}{2mr^2}\right)^{-1/2}, \\ t &= t_0 \pm \sqrt{\frac{m}{2}} \int \mathrm{d}r \left(E - V(r) - \frac{p_\varphi^2}{2mr^2}\right)^{-1/2}.\end{aligned}}$$

The first equation is the equation of the orbit (r as a function of φ), while the second one determines the time variation of r. These equations respectively coincide with Eqs. (3.19) and (3.21) deduced in Section 3.2.2 using the Newtonian formalism (as $L = p_\varphi$). Note, finally, that the completeness condition (4.112) is satisfied. Indeed, since

$$\frac{\partial S}{\partial \alpha_1} = -t, \qquad \frac{\partial S}{\partial \alpha_2} = \varphi + \frac{\partial W}{\partial \alpha_2} \quad \Longrightarrow \quad \frac{\partial^2 S}{\partial \varphi \partial \alpha_1} = 0, \qquad \frac{\partial^2 S}{\partial \varphi \partial \alpha_2} = 1,$$

this condition reads

$$\begin{vmatrix} \dfrac{\partial W'}{\partial \alpha_1} & \dfrac{\partial W'}{\partial \alpha_2} \\ 0 & 1 \end{vmatrix} = \frac{\partial W'}{\partial \alpha_1} \neq 0,$$

which clearly holds in this case as

$$\frac{\partial W'}{\partial \alpha_1} = \pm\sqrt{\frac{m}{2}}\left(E - V(r) - \frac{p_\varphi^2}{2mr^2}\right)^{-1/2} \neq 0.$$

CHAPTER 5

Small oscillations

In this chapter we analyze the motion of a conservative mechanical system near a stable equilibrium. We start by linearizing the Lagrangian equations of motion near the equilibrium, and then solve them in terms of two constant symmetric matrices that characterize the quadratic parts of the kinetic and potential energy of the system. In this way we determine the system's characteristic frequencies and its corresponding normal modes of vibration. As applications of this general method, we examine the small oscillations of a double pendulum and the longitudinal vibrations of the CO_2 molecule.

5.1 LINEARIZED EQUATIONS OF MOTION

We shall assume that the constraints are *holonomic* and *time-independent*, so that the position vector of each particle is a function only of the generalized coordinates $\mathbf{q} = (q_1, \ldots, q_n)$ not explicitly depending on time:

$$\mathbf{r}_k = \mathbf{r}_k(\mathbf{q}), \qquad k = 1, \ldots, N.$$

Since

$$\dot{\mathbf{r}}_k = \sum_{j=1}^{n} \frac{\partial \mathbf{r}_k(\mathbf{q})}{\partial q_j} \dot{q}_j, \tag{5.1}$$

the system's kinetic energy is then of the form

$$T(\mathbf{q}, \dot{\mathbf{q}}) = \frac{1}{2} \sum_{i,j=1}^{n} t_{ij}(\mathbf{q}) \dot{q}_i \dot{q}_j$$

with

$$t_{ij}(\mathbf{q}) = \sum_{k=1}^{N} m_k \frac{\partial \mathbf{r}_k(\mathbf{q})}{\partial q_i} \cdot \frac{\partial \mathbf{r}_k(\mathbf{q})}{\partial q_j} = t_{ji}(\mathbf{q}).$$

DOI: 10.1201/9781003600633-5

If the system is conservative, with potential energy $V(\mathbf{q})$, its Lagrangian is given by

$$L = \frac{1}{2}\sum_{i,j=1}^{n} t_{ij}(\mathbf{q})\dot{q}_i\dot{q}_j - V(\mathbf{q}),$$

and Lagrange's equations of motion read:

$$\frac{\mathrm{d}}{\mathrm{d}t}\left(\sum_{j=1}^{n} t_{ij}(\mathbf{q})\dot{q}_j\right) + \frac{\partial V}{\partial q_i}(\mathbf{q})$$
$$= \sum_{j=1}^{n} t_{ij}(\mathbf{q})\ddot{q}_j + \sum_{j,k=1}^{n}\frac{\partial t_{ij}}{\partial q_k}(\mathbf{q})\,\dot{q}_j\dot{q}_k + \frac{\partial V}{\partial q_i}(\mathbf{q}) = 0\,, \qquad i = 1,\dots,n. \quad (5.2)$$

Thus an equilibrium solution $\mathbf{q}(t) = \mathbf{q}_0$ exists if and only if

$$\frac{\partial V}{\partial q_i}(\mathbf{q}_0) = 0\,, \qquad i = 1,\dots,n,$$

i.e., if $\mathbf{q}_0$ is a *critical point* of the potential $V(\mathbf{q})$. As in the one-dimensional case (cf. Section 2.5.3), it can be shown that the equilibrium $\mathbf{q}_0$ is *stable* if and only if $\mathbf{q}_0$ is a *local minimum* of V.

We wish to describe the motion of the system near a stable equilibrium $\mathbf{q}_0$. To this end, let us assume without loss of generality (choosing $\mathbf{q} - \mathbf{q}_0$ as new generalized coordinates if necessary) that $\mathbf{q}_0 = 0$, and normalize the potential so that $V(0) = 0$. The energy of the equilibrium solution $\mathbf{q} = 0$ is then $E_0 = 0$. Consider now a motion of the system close to the equilibrium solution $\mathbf{q} = 0$, i.e., with $|\mathbf{q}(0)|$ and $|\dot{\mathbf{q}}(0)|$ small. The energy of such a motion is then close to $E_0 = 0$ and verifies

$$E = T + V(\mathbf{q}) \geqslant V(\mathbf{q}) \geqslant V(0) = 0$$

(as $\mathbf{q} = 0$ is by hypothesis a local minimum of V), i.e., E is positive and small. Since $\mathbf{q} = 0$ is by hypothesis a critical point of V with $V(0) = 0$, we can write its Taylor expansion about 0 as[1]

$$V(\mathbf{q}) = \frac{1}{2}\sum_{i,j=1}^{n} b_{ij}q_iq_j + o(|\mathbf{q}|^2)\,, \qquad \text{with} \quad b_{ij} = \frac{\partial^2 V}{\partial q_i q_j}(0) = b_{ji}\,.$$

Furthermore, since 0 is a local minimum of V the quadratic form $\sum_{i,j=1}^{n} b_{ij}q_iq_j$ is positive semidefinite, i.e., the eigenvalues of the symmetric $n \times n$ matrix

$$B = (b_{ij})_{1\leqslant i,j\leqslant n}$$

[1]In the previous formula we are using the standard notation $o(t)$ to denote a function verifying $\lim_{t\to 0}\frac{o(t)}{t} = 0$.

are nonnegative. We shall actually assume that B is *positive definite* (i.e., all its eigenvalues are *strictly positive*), so that

$$\boxed{V(\mathbf{q}) \simeq \frac{1}{2}\sum_{i,j=1}^{n} b_{ij}q_i q_j}$$

near the origin. Similarly, the system's kinetic energy can be expanded near the equilibrium $\mathbf{q} = \dot{\mathbf{q}} = 0$ as

$$T = \frac{1}{2}\sum_{i,j=1}^{n} a_{ij}\dot{q}_i\dot{q}_j + o(|\dot{\mathbf{q}}^2|)\,, \qquad \text{with} \quad a_{ij} = t_{ij}(0) = a_{ji}\,.$$

Note also that the quadratic form

$$\boxed{T_0 := \frac{1}{2}\sum_{i,j=1}^{n} a_{ij}\dot{q}_i\dot{q}_j}$$

is *positive definite.* Indeed, the kinetic energy $T(\mathbf{q}, \dot{\mathbf{q}})$ is nonnegative and vanishes only if $\dot{\mathbf{r}}_k = 0$ for all $k = 1, \ldots, N$. In particular, $T_0 = T(0, \dot{\mathbf{q}})$ is also positive semidefinite, and by Eq. (5.1) it can only vanish when

$$\sum_{j=1}^{n} \frac{\partial \mathbf{r}_k}{\partial q_j}(0)\,\dot{q}_j = 0\,, \qquad k = 1, \ldots, N\,. \tag{5.3}$$

Recall that the vectors

$$\frac{\partial \mathbf{x}}{\partial q_j} := \left(\frac{\partial \mathbf{r}_1}{\partial q_j}, \ldots, \frac{\partial \mathbf{r}_N}{\partial q_j}\right), \qquad j = 1, \ldots, n\,,$$

are linearly independent at each point, since they are a basis of the tangent space to the system's constraint manifold (see Section 4.3.2). Thus Eqs. (5.3), which are equivalent to the single relation

$$\sum_{j=1}^{n} \frac{\partial \mathbf{x}}{\partial q_j}(0)\,\dot{q}_j = 0,$$

imply that $\dot{q}_j = 0$ for all j. This shows that $T_0 = 0$ if and only if $\dot{\mathbf{q}} = 0$, and hence T_0 is positive definite, as claimed. Thus near the equilibrium solution $\mathbf{q} = \dot{\mathbf{q}} = 0$ we also have $T \simeq T_0$. It follows that for small displacements $\mathbf{q}$ and small velocities $\dot{\mathbf{q}}$ the system's Lagrangian can be approximated by the quadratic Lagrangian

$$\boxed{L_0 := \frac{1}{2}\sum_{i,j=1} a_{ij}\dot{q}_i\dot{q}_j - \frac{1}{2}\sum_{i,j=1}^{n} b_{ij}q_i q_j\,.} \tag{5.4}$$

The motion of the system near its stable equilibrium $\mathbf{q} = 0$ is thus approximately governed by the Euler–Lagrange equations of the Lagrangian L_0, namely

$$\sum_{j=1}^{n} \left(a_{ij}\ddot{q}_j + b_{ij}q_j \right) = 0\,, \qquad i = 1, \ldots, n\,,$$

or in matrix form

$$\boxed{A\ddot{\mathbf{q}} + B\mathbf{q} = 0\,,} \tag{5.5}$$

where A is the positive definite $n \times n$ symmetric matrix with matrix elements a_{ij}. Equations (5.5) are a system of n second-order linear homogeneous differential equations with constant coefficients. Note that these equations could have also been derived by *linearizing* the exact equations of motion, i.e., by retaining only the terms linear in $(\mathbf{q}, \dot{\mathbf{q}}, \ddot{\mathbf{q}})$ in Eqs. (5.2).

5.2 NORMAL COORDINATES AND NORMAL MODES

One of the easiest ways of solving the linearized equations (5.5) is by transforming the Lagrangian L_0 into a suitable canonical form. Indeed, since T_0 is positive definite there is a non-singular (in general non-orthogonal) linear change of variables

$$q_i = \sum_{j=1}^{n} M_{ij}\widetilde{q}_j\,, \qquad i = 1, \ldots, n\,,$$

or in matrix form

$$\mathbf{q} = M\widetilde{\mathbf{q}}$$

(and consequently $\dot{\mathbf{q}} = M\dot{\widetilde{\mathbf{q}}}$), transforming the positive definite quadratic form T_0 into $\dot{\widetilde{\mathbf{q}}}^2/2$. In other words, there exists a non-singular matrix M such that

$$M^{\mathsf{T}}AM = \mathbf{1}\,.$$

Since

$$L_0 = T_0 - \frac{1}{2}\mathbf{q}^{\mathsf{T}} \cdot B\mathbf{q}\,,$$

in the new generalized coordinates $\widetilde{\mathbf{q}}$ the Lagrangian L_0 can then be written as

$$L_0 = \frac{1}{2}\dot{\widetilde{\mathbf{q}}}^2 - \frac{1}{2}\sum_{i,j=1}^{n} \widetilde{b}_{ij}\widetilde{q}_i\widetilde{q}_j = \frac{1}{2}\dot{\widetilde{\mathbf{q}}}^2 - \frac{1}{2}\widetilde{\mathbf{q}}^{\mathsf{T}} \cdot \widetilde{B}\widetilde{\mathbf{q}}\,,$$

where $\widetilde{b}_{ij}$ is the (i, j)-th matrix element of the matrix

$$\widetilde{B} = M^{\mathsf{T}}BM\,.$$

Since the matrix $\widetilde{B}$ is still symmetric and positive definite (exercise), it can be diagonalized by an orthogonal transformation. In other words, there is a real *orthogonal* matrix O such that

$$O^{\mathsf{T}}\widetilde{B}O = \begin{pmatrix} \lambda_1 & & & \\ & \lambda_2 & & \\ & & \ddots & \\ & & & \lambda_n \end{pmatrix} =: \Lambda \tag{5.6}$$

is a diagonal matrix. Note that the numbers λ_i are all *positive*, since $\widetilde{B}$ is a positive definite matrix. Defining new generalized coordinates $\mathbf{Q}$ by the linear transformation $\widetilde{\mathbf{q}} = O\mathbf{Q}$, and taking into account that $\dot{\widetilde{\mathbf{q}}}^2 = (O\dot{\mathbf{Q}})^2 = \dot{\mathbf{Q}}^2$ (since O is orthogonal), we easily find

$$\boxed{L_0 = \frac{1}{2}\dot{\mathbf{Q}}^2 - \frac{1}{2}\mathbf{Q}\cdot\Lambda\mathbf{Q} = \frac{1}{2}\sum_{i=1}^{n}(\dot{Q}_i^2 - \lambda_i Q_i^2)\,.}$$

By the *covariance* of Lagrange's equations (cf. Section 4.2.3), the linearized equations of motion in the generalized coordinates $\mathbf{Q}$ are simply the Euler–Lagrange equations of the previous Lagrangian with respect to the variables $(\mathbf{Q}, \dot{\mathbf{Q}})$, namely the *decoupled* system

$$\boxed{\ddot{Q}_i + \lambda_i Q_i = 0\,, \qquad i = 1, \dots, n\,.} \tag{5.7}$$

In other words, in the generalized coordinates $\mathbf{Q}$ the system is equivalent to a collection of n *decoupled harmonic oscillators* with frequencies

$$\omega_i := \sqrt{\lambda_i}\,.$$

The general solution of the system (5.7) is therefore

$$\boxed{Q_i = A_i\cos(\omega_i t + \alpha_i)\,, \qquad i = 1, \dots, n\,,}$$

or in vector form

$$\mathbf{Q} = \sum_{i=1}^{n} A_i\cos(\omega_i t + \alpha_i)\mathbf{e}_i,$$

where $A_i \geqslant 0$, $\alpha_i \in [0, 2\pi)$ are arbitrary constants, and $\mathbf{e}_i = (0, \dots, 1, \dots, 0)$ is the i-th canonical basis vector. In other words, equations (5.7) possess a *fundamental system of solutions* of the form

$$\boxed{\mathbf{Q}^{(i)}(t) = \cos(\omega_i t + \alpha_i)\mathbf{e}_i\,, \qquad i = 1, \dots, n\,.} \tag{5.8}$$

The generalized coordinates $\mathbf{Q}$ and the n fundamental solutions (5.8) are respectively called the system's **normal coordinates** and **normal modes**. Likewise, the n frequencies ω_i ($i = 1, \dots, n$) are called the system's **normal frequencies**. In terms of the original generalized coordinates

$$\boxed{\mathbf{q} = M\widetilde{\mathbf{q}} = MO\mathbf{Q}} \tag{5.9}$$

the normal modes (5.8) become

$$\boxed{\mathbf{q}^{(i)}(t) = \mathbf{c}_i \cos(\omega_i t + \alpha_i), \qquad i = 1, \dots, n,} \tag{5.10}$$

where the n-dimensional vectors $\mathbf{c}_i$ are given by

$$\boxed{\mathbf{c}_i = (MO)\mathbf{e}_i .} \tag{5.11}$$

In other words, the vectors $\mathbf{c}_i$ are the *columns* of the matrix MO satisfying

$$(MO)^{\mathsf{T}} A (MO) = \mathbf{1}, \qquad (MO)^{\mathsf{T}} B (MO) = \Lambda. \tag{5.12}$$

In particular, since both M and O are invertible, the n vectors $\mathbf{c}_i$ are *linearly independent*. The general solution of the system's linearized equations of motion (5.5)—which, by the previous argument, is an approximate solution of its exact equations of motion (5.2) near the stable equilibrium $\mathbf{q} = 0$—is an arbitrary linear combination

$$\boxed{\mathbf{q}(t) = \sum_{i=1}^{n} a_i \mathbf{q}^{(i)}(t),}$$

with $a_i \in \mathbb{R}$ constant, of the n normal modes $\mathbf{q}^{(i)}(t)$.

Remarks 5.1.

- The matrices M and O, and therefore the vectors $\mathbf{c}_i$, are not unique.
- The vectors $\mathbf{c}_i$ defined by Eq. (5.11) are in general not mutually orthogonal nor of unit length. However, since the n vectors $O\mathbf{e}_i$ are orthonormal (being the columns of an orthogonal matrix), the vectors $\mathbf{c}_i$ satisfy the relations

$$\mathbf{c}_i \cdot A\mathbf{c}_j = \delta_{ij}, \qquad i, j = 1, \dots, n .$$

Indeed, taking into account that $M^{\mathsf{T}} A M = \mathbf{1}$ we have

$$\mathbf{c}_i \cdot A\mathbf{c}_j = (MO\mathbf{e}_i) \cdot (AMO\mathbf{e}_j) = (O\mathbf{e}_i) \cdot (M^{\mathsf{T}} A M O\mathbf{e}_j) = (O\mathbf{e}_i) \cdot (O\mathbf{e}_j) = \delta_{ij} . \qquad \blacksquare$$

How does one find in practice the frequencies ω_i and the corresponding vectors $\mathbf{c}_i$ determining the system's normal modes (5.10) in the original coordinates $\mathbf{q}$? To answer this question it suffices to note that, since $\mathbf{q}^{(i)}(t)$ is a solution of the linearized equations (5.5), the vector $\mathbf{c}_i$ must satisfy the linear system

$$\boxed{(B - \omega_i^2 A)\mathbf{c}_i = 0, \qquad i = 1, \dots, n .} \tag{5.13}$$

As $\mathbf{c}_i$ is nonzero, we must therefore have $\det(B - \omega_i^2 A) = 0$. In other words, the normal frequencies $\omega_i = \sqrt{\lambda_i}$ are the square roots of the n solutions λ_i (counting multiplicities) of the *characteristic equation*

$$\boxed{\det(B - \lambda A) = 0\,.} \tag{5.14}$$

The numbers λ_i are called the eigenvalues of the matrix B relative to the positive definite matrix A; in particular, when $A = \mathbf{1}$ the λ_i's are the ordinary eigenvalues of B. Note that, by Eq. (5.6), the λ_i's are the (ordinary) eigenvalues of the matrix $\widetilde{B}$ in the previous discussion. For each such eigenvalue $\lambda_i = \omega_i^2$, the corresponding (eigen)vector $\mathbf{c}_i$ is then found solving the linear system (5.13).

The previous argument guarantees that the columns $\{\mathbf{c}_1, \dots, \mathbf{c}_n\}$ of the matrix MO constructed above are a basis of $\mathbb{R}^n$ whose elements satisfy Eqs. (5.13). In fact, since M^{T} is non-singular, Eq. (5.13) is equivalent to

$$\left(M^T B - \lambda_i M^{\mathsf{T}} A\right) M (M^{-1}\mathbf{c}_i) = (\widetilde{B} - \lambda_i)\left(M^{-1}\mathbf{c}_i\right) = 0.$$

Hence if $\mathbf{c}_i$ satisfies Eq. (5.13) the vector $M^{-1}\mathbf{c}_i$ is an eigenvector of the matrix $\widetilde{B}$ with eigenvalue $\lambda_i = \omega_i^2$, and conversely. Moreover, the characteristic polynomial of $\widetilde{B}$ is proportional to $\det(B - \lambda A)$:

$$\begin{aligned}\det\left(\widetilde{B} - \lambda\right) &= \det\left(M^{\mathsf{T}} B M - \lambda M^{\mathsf{T}} A M\right) = \det\left(M^{\mathsf{T}}(B - \lambda A) M\right)\\ &= (\det M)^2 \det\left(B - \lambda A\right).\end{aligned}$$

Hence if ω_i^2 is a root of the characteristic equation (5.14) with multiplicity $m \geqslant 1$ it is also a root of the characteristic polynomial of the matrix $\widetilde{B}$ with the same multiplicity. Since $\widetilde{B}$ is *symmetric*, and hence diagonalizable, there are exactly m linearly independent eigenvectors $\mathbf{v}_1, \dots, \mathbf{v}_m$ of $\widetilde{B}$ with eigenvalue ω_i^2. Consequently, the vectors $M\mathbf{v}_1, \dots, M\mathbf{v}_m$ are a basis of solutions of Eq. (5.13). This shows that if ω_i^2 is a root of the characteristic equation (5.14) with multiplicity m, then Eq. (5.13) has exactly m linearly independent solutions.

Remark 5.2. Comparison of Eqs. (5.9) and (5.11) shows that, once the vectors $\mathbf{c}_i$ have been obtained, the normal coordinates $\mathbf{Q}$ can be found through the formula

$$\boxed{\mathbf{Q} = C^{-1}\mathbf{q}\,,} \tag{5.15}$$

where C is the matrix whose columns are the components of the vectors $\mathbf{c}_1, \dots, \mathbf{c}_n$—i.e., the change of basis matrix from the canonical basis of $\mathbb{R}^n$ to the basis $\{\mathbf{c}_1, \dots, \mathbf{c}_n\}$. ■

5.3 THE DOUBLE PENDULUM

Consider the double pendulum schematically represented in Fig. 5.1. Calling $\mathbf{r}_\alpha = (x_\alpha, y_\alpha)$ the position vector of the particle $\alpha = 1, 2$, the system's constraints are

$$\mathbf{r}_1^2 - l_1^2 = 0\,, \qquad (\mathbf{r}_2 - \mathbf{r}_1)^2 - l_2^2 = 0\,.$$

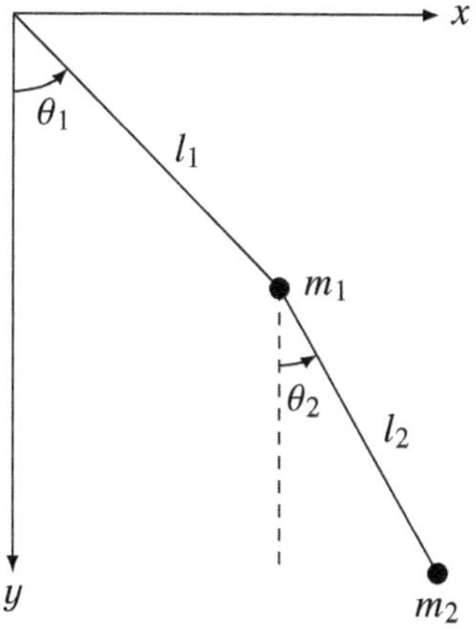

Figure 5.1. Generalized coordinates (θ_1, θ_2) for the double pendulum system.

These constraints are obviously holonomic and time-independent. Moreover, the principle of virtual work clearly holds, since the constraint forces—the tension of the wire or rod connecting the first particle to the anchor point and the second particle to the first one—are respectively parallel to the vectors $\mathbf{r}_1$ and $\mathbf{r} := \mathbf{r}_2 - \mathbf{r}_1$, and thus perpendicular to the particles' infinitesimal displacements (which, as the constraints are time-independent, are also virtual displacements). Hence we can apply the Lagrangian formalism.

We shall use as generalized coordinates the two angles θ_1 and θ_2 between the pendulums' strings and the vertical, and take the y axis *downward* (see Fig. 5.1). We then have

$$\mathbf{r}_1 = l_1(\sin\theta_1, \cos\theta_1), \qquad \mathbf{r} = l_2(\sin\theta_2, \cos\theta_2),$$

and therefore

$$\dot{\mathbf{r}}_1 = l_1\dot{\theta}_1(\cos\theta_1, -\sin\theta_1), \qquad \dot{\mathbf{r}} = l_2\dot{\theta}_2(\cos\theta_2, -\sin\theta_2),$$

so that

$$\begin{aligned} T &= \frac{1}{2}m_1\dot{\mathbf{r}}_1^2 + \frac{1}{2}m_2(\dot{\mathbf{r}}_1 + \dot{\mathbf{r}})^2 = \frac{1}{2}M\dot{\mathbf{r}}_1^2 + \frac{1}{2}m_2\dot{\mathbf{r}}^2 + m_2\dot{\mathbf{r}}_1\cdot\dot{\mathbf{r}} \\ &= \frac{1}{2}Ml_1^2\dot{\theta}_1^2 + \frac{1}{2}m_2l_2^2\dot{\theta}_2^2 + m_2l_1l_2\dot{\theta}_1\dot{\theta}_2(\cos\theta_1\cos\theta_2 + \sin\theta_1\sin\theta_2) \\ &= \frac{1}{2}Ml_1^2\dot{\theta}_1^2 + \frac{1}{2}m_2l_2^2\dot{\theta}_2^2 + m_2l_1l_2\dot{\theta}_1\dot{\theta}_2\cos(\theta_1 - \theta_2), \end{aligned}$$

with $M := m_1 + m_2$. Likewise, if y is the vertical coordinate of the relative position vector $\mathbf{r}$ the potential energy is given by

$$V = -m_1gy_1 - m_2gy_2 = -m_1gy_1 - m_2g(y + y_1) = -Mgl_1\cos\theta_1 - m_2gl_2\cos\theta_2,$$

and therefore the system's Lagrangian can be taken as

$$\boxed{\begin{aligned} L = T - V = \frac{1}{2}\, Ml_1^2\dot\theta_1^2 + \frac{1}{2}\, m_2 l_2^2\dot\theta_2^2 + m_2 l_1 l_2 \dot\theta_1\dot\theta_2 \cos(\theta_1 - \theta_2) \\ + Mgl_1\cos\theta_1 + m_2 g l_2 \cos\theta_2\,. \end{aligned}}$$

The equilibrium positions are determined by the system of equations

$$\frac{\partial V}{\partial \theta_1} = Mgl_1 \sin\theta_1 = 0\,, \qquad \frac{\partial V}{\partial \theta_2} = m_2 g l_2 \sin\theta_2 = 0\,,$$

and are therefore (up to integer multiples of 2π) the four points

$$(0,0)\,, \quad (0,\pi), \quad (\pi,0)\,, \quad (\pi,\pi)\,.$$

It is straightforward to ascertain that the point $(0,0)$ is the unique local (in fact, global) minimum of V, $(\pi,0)$ and $(0,\pi)$ are saddle points, and (π,π) is a global maximum. (In fact, the equilibria $(\pi,0)$ and (π,π) are not possible in practice due to the fact that the first pendulum is anchored to the ceiling, and thus $\theta_1 \in [-\pi/2, \pi/2]$. For the same reason, the equilibrium $(0,\pi)$ is only possible if $l_2 \leqslant l_1$.) The exact equations of motion

$$\begin{aligned} Ml_1^2\ddot\theta_1 + m_2 l_1 l_2 \ddot\theta_2 \cos(\theta_1-\theta_2) + m_2 l_1 l_2 \dot\theta_2^2 \sin(\theta_1-\theta_2) + Mgl_1\sin\theta_1 &= 0\,,\\ m_2 l_2^2\ddot\theta_2 + m_2 l_1 l_2 \ddot\theta_1 \cos(\theta_1-\theta_2) - m_2 l_1 l_2 \dot\theta_1^2 \sin(\theta_1-\theta_2) + m_2 g l_2 \sin\theta_2 &= 0\,, \end{aligned}$$

are a system of nonlinear coupled second-order differential equations that cannot be solved in closed form (i.e., in terms of elementary functions and their primitives). On the other hand, we can easily study the system's motion near the stable equilibrium $\theta_1 = \theta_2 = 0$, i.e., the small oscillations of the two pendulums, through the method explained above.

To begin with, taking into account that

$$\cos\theta = 1 - \frac{\theta^2}{2} + o(\theta^2)$$

we easily obtain

$$\begin{aligned} T_0 = T(0,0,\dot\theta_1,\dot\theta_2) &= \frac{1}{2}\, Ml_1^2\dot\theta_1^2 + \frac{1}{2}\, m_2 l_2^2 \dot\theta_2^2 + m_2 l_1 l_2 \dot\theta_1 \dot\theta_2\,,\\ V = -(Ml_1 + m_2 l_2)g &+ \frac{g}{2}\,(Ml_1\theta_1^2 + m_2 l_2 \theta_2^2) + o(\theta_1^2 + \theta_2^2)\,, \end{aligned}$$

and thus (ignoring the irrelevant constant term in V)

$$\boxed{\begin{aligned} A &= \begin{pmatrix} Ml_1^2 & m_2 l_1 l_2 \\ m_2 l_1 l_2 & m_2 l_2^2 \end{pmatrix} = Ml_1^2 \begin{pmatrix} 1 & \lambda\mu \\ \lambda\mu & \lambda^2\mu \end{pmatrix},\\ B &= g\begin{pmatrix} Ml_1 & 0 \\ 0 & m_2 l_2 \end{pmatrix} = Ml_1^2\omega_0^2 \begin{pmatrix} 1 & 0 \\ 0 & \lambda\mu \end{pmatrix}, \end{aligned}}$$

where

$$\omega_0 := \sqrt{\frac{g}{l_1}}$$

is the natural frequency of the first pendulum and we have set

$$\lambda := \frac{l_2}{l_1}, \qquad \mu := \frac{m_2}{M}.$$

Since B is diagonal, it is convenient to write the characteristic equation (5.14) in the equivalent form

$$\det\left(A - \frac{B}{\omega^2}\right) = M^2 l_1^4 \begin{vmatrix} 1 - \frac{\omega_0^2}{\omega^2} & \lambda\mu \\ \lambda\mu & \lambda\mu\left(\lambda - \frac{\omega_0^2}{\omega^2}\right) \end{vmatrix} = 0$$

$$\Longleftrightarrow \quad \left(1 - \frac{\omega_0^2}{\omega^2}\right)\left(\lambda - \frac{\omega_0^2}{\omega^2}\right) - \lambda\mu = 0,$$

or

$$\left(\frac{\omega_0}{\omega}\right)^4 - (\lambda+1)\left(\frac{\omega_0}{\omega}\right)^2 + \lambda(1-\mu) = 0.$$

The normal frequencies are thus determined by the equation

$$\frac{\omega_0^2}{\omega_\pm^2} = \frac{1}{2}\left(\lambda + 1 \pm \sqrt{(\lambda+1)^2 - 4\lambda(1-\mu)}\right) = \frac{1}{2}\left(\lambda + 1 \pm \sqrt{(\lambda-1)^2 + 4\lambda\mu}\right),$$

whence

$$\boxed{\omega_\pm^2 = \omega_0^2\,\frac{\lambda + 1 \mp \sqrt{(\lambda-1)^2 + 4\lambda\mu}}{2\lambda(1-\mu)}\,.}$$

In particular, when the two pendulums have the same length (i.e., for $\lambda = 1$) we simply have

$$\omega_\pm^2 = \frac{\omega_0^2}{1 \pm \sqrt{\mu}} = \frac{\omega_0^2}{1 \pm \sqrt{\frac{m_2}{m_1+m_2}}}\,.$$

The two normal modes are found by solving the characteristic equation

$$\left(A - \frac{B}{\omega_\pm^2}\right)\mathbf{c}_\pm = 0,$$

i.e.,

$$\left(1 - \frac{\omega_0^2}{\omega_\pm^2}\right)c_{\pm,1} + \lambda\mu c_{\pm,2} = 0.$$

Using the previous formulas for $\omega_\pm^2$ we can rewrite the last equation as

$$\left(1 - \lambda \mp \sqrt{(\lambda-1)^2 + 4\lambda\mu}\right)c_{\pm,1} + 2\lambda\mu c_{\pm,2} = 0,$$

thus obtaining the two (unnormalized) vectors

$$\boxed{\mathbf{c}_\pm = c\left(2\lambda\mu, \lambda - 1 \pm \sqrt{(\lambda-1)^2 + 4\lambda\mu}\right)} \tag{5.16}$$

with arbitrary $c \neq 0$. Hence the two normal mode solutions are

$$(\theta_1^{(\pm)}, \theta_2^{(\pm)}) = \mathbf{c}_\pm \cos(\omega_\pm t + \alpha_\pm),$$

i.e.,

$$\boxed{\begin{aligned} \theta_1^{(\pm)} &= 2c\lambda\mu\cos(\omega_\pm t + \alpha_\pm), \\ \theta_2^{(\pm)} &= c\left(\lambda - 1 \pm \sqrt{(\lambda-1)^2 + 4\lambda\mu}\right)\cos(\omega_\pm t + \alpha_\pm). \end{aligned}}$$

Note that the quotient of the amplitudes of the oscillations of the angles θ_2 and θ_1 in these normal modes, given by

$$\boxed{\frac{\theta_2^{(\pm)}}{\theta_1^{(\pm)}} = \frac{\lambda - 1 \pm \sqrt{(\lambda-1)^2 + 4\lambda\mu}}{2\lambda\mu},}$$

is positive (resp. negative) for the normal mode with the smaller frequency ω_+ (resp. the larger frequency ω_-). Thus in the normal mode with frequency ω_+ the pendulums oscillate *in phase*, whereas in the one with frequency ω_- they oscillate completely *out of phase* (i.e., θ_1 is maximum when θ_2 is minimum, and vice versa). Again, in the particular case in which $l_1 = l_2$ the quotient $\theta_2^{(\pm)}/\theta_1^{(\pm)}$ simplifies to

$$\frac{\theta_2^{(\pm)}}{\theta_1^{(\pm)}} = \pm\frac{1}{\sqrt{\mu}} = \pm\sqrt{1 + \frac{m_1}{m_2}}.$$

Exercise 5.1. Show that for all (positive) values of $\lambda = l_2/l_1$ and $\mu = m_2/M$ we have $\omega_+ < \omega_0 < \omega_-$.

Solution. We need to show that

$$\lambda + 1 - \sqrt{(\lambda-1)^2 + 4\lambda\mu} < 2\lambda(1-\mu) < \lambda + 1 + \sqrt{(\lambda-1)^2 + 4\lambda\mu},$$

or equivalently

$$-\sqrt{(\lambda-1)^2 + 4\lambda\mu} < \lambda - 1 - 2\lambda\mu < \sqrt{(\lambda-1)^2 + 4\lambda\mu}.$$

The previous inequalities are obviously true, since

$$(\lambda - 1 - 2\lambda\mu)^2 = (\lambda-1)^2 + 4\lambda\mu - 4\lambda^2\mu(1-\mu) < (\lambda-1)^2 + 4\lambda\mu$$

(recall that $\mu = m_2/(m_1 + m_2) < 1$).

Exercise 5.2. Find the normal coordinates $\theta_\pm$ for the double pendulum system.

Solution. From Eqs. (5.15)–(5.16) we have

$$\begin{pmatrix}\theta_+\\ \theta_-\end{pmatrix} = \begin{pmatrix} \dfrac{2\lambda\mu}{\lambda-1+\sqrt{(\lambda-1)^2+4\lambda\mu}} & \dfrac{2\lambda\mu}{\lambda-1-\sqrt{(\lambda-1)^2+4\lambda\mu}} \end{pmatrix}^{-1} \begin{pmatrix}\theta_1\\ \theta_2\end{pmatrix}$$

$$= -\frac{1}{4\lambda\mu\sqrt{(\lambda-1)^2+4\lambda\mu}} \begin{pmatrix} \lambda-1-\sqrt{(\lambda-1)^2+4\lambda\mu} & -2\lambda\mu \\ 1-\lambda-\sqrt{(\lambda-1)^2+4\lambda\mu} & 2\lambda\mu \end{pmatrix} \begin{pmatrix}\theta_1\\ \theta_2\end{pmatrix}.$$

In fact, since the normal coordinates Q_i are defined up to multiplication by a constant scalar, we can take as normal coordinates

$$\theta_\pm = \left[1-\lambda \pm \sqrt{(\lambda-1)^2+4\lambda\mu}\right]\theta_1 + 2\lambda\mu\theta_2 .$$

This expression simplifies considerably when the two pendulums have the same length, in which case (dropping the inessential constant factor $\pm 2\sqrt{\mu}$) we obtain

$$\theta_\pm = \theta_1 \pm \sqrt{\mu}\,\theta_2 .$$

5.4 LONGITUDINAL VIBRATIONS OF THE CO_2 MOLECULE

Consider a triatomic molecule made up of two identical atoms of mass m_1 and a single atom of mass m_2. We shall also assume that, as is the case with the CO_2 molecule, the molecule's equilibrium configuration is collinear, with the atom of mass m_2 lying between the other two atoms and separated from each of them by the same distance a. Let us choose the x axis along the line joining the equilibrium positions of the three atoms, and place the origin at the equilibrium position of the atom of mass m_2 (cf. Fig. 5.2). We shall next study the longitudinal vibrations of the molecule, i.e., the motions of its atoms along the line of the molecule at equilibrium (the x axis). Calling x_1 and x_3 the x coordinates of the atoms of mass m_1 (from left to right), and x_2 that of the atom of mass m_2, the system's kinetic and potential energies are given by

$$\boxed{T = \frac{1}{2}\left[m_1(\dot{x}_1^2+\dot{x}_3^2)+m_2\dot{x}_2^2\right], \qquad V = U(x_2-x_1)+U(x_3-x_2),}$$

where U is the interaction potential between the atoms of mass m_2 and each of the atoms of mass m_1. We have assumed that the interaction between the two atoms of mass m_1 is negligible compared to their interaction with the atom of mass m_2, since the strength of atomic interactions usually falls off very quickly with the distance. Although the potential U is not known in detail, we are only interested in small vibrations of the atoms about their equilibrium position $x_1 = -a$, $x_2 = 0$, $x_3 = a$. Imposing that the partial derivatives of V vanish at equilibrium we easily deduce that

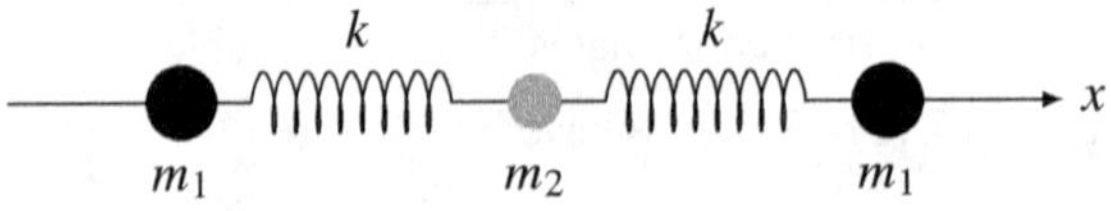

Figure 5.2. Schematic representation of the CO_2 molecule.

$U'(a) = 0$. If we now Taylor expand $U(x)$ about $x = a$ keeping only the lowest order nontrivial term we obtain

$$U(x) \simeq U(a) + \frac{k}{2}(x-a)^2 ,$$

where $k = U''(a) > 0$, and thus (dropping the inessential constant $U(a)$)

$$V \simeq \frac{k}{2}\left[(x_2 - x_1 - a)^2 + (x_3 - x_2 - a)^2\right].$$

Thus in this approximation (i.e., when $|x_2 - x_1|$ and $|x_3 - x_2|$ are both close to a) the molecule behaves as a system of three collinear particles of masses m_1, m_2, and m_1 connected by springs of natural length a and constant k (cf. Fig. 5.2). The system's Lagrangian $L = T - V$ is approximately given by

$$\boxed{L_0 := \frac{1}{2}\left[m_1(\dot{x}_1^2 + \dot{x}_3^2) + m_2\dot{x}_2^2\right] - \frac{k}{2}\left[(x_2 - x_1 - a)^2 + (x_3 - x_2 - a)^2\right].}$$

Since L_0 is invariant under the translation $x_i \mapsto x_i + \varepsilon$ for arbitrary ε, the x component of the linear momentum $P = m_1(\dot{x}_1 + \dot{x}_3) + m_2\dot{x}_2$ is conserved. This suggests, as in the two-body problem, separating the center of mass motion from the particles' relative motion, i.e., to use as generalized coordinates

$$\boxed{X := \frac{1}{M}\left[m_1(x_1 + x_3) + m_2 x_2\right], \qquad q_1 := x_2 - x_1 - a, \qquad q_2 := x_3 - x_2 - a,}$$

where $M = 2m_1 + m_2$ is the molecule's total mass. Inverting these equations we readily find the following expressions for the atoms' physical coordinates in terms of the generalized ones:

$$\boxed{\begin{aligned} &x_1 = X - \frac{m_1 + m_2}{M} q_1 - \frac{m_1}{M} q_2 - a, \qquad x_2 = X + \frac{m_1}{M}(q_1 - q_2), \\ &x_3 = X + \frac{m_1}{M} q_1 + \frac{m_1 + m_2}{M} q_2 + a, \end{aligned}} \tag{5.17}$$

and hence

$$\begin{aligned} \dot{x}_1 &= \dot{X} - \frac{m_1 + m_2}{M}\dot{q}_1 - \frac{m_1}{M}\dot{q}_2, \\ \dot{x}_2 &= \dot{X} + \frac{m_1}{M}(\dot{q}_1 - \dot{q}_2), \\ \dot{x}_3 &= \dot{X} + \frac{m_1}{M}\dot{q}_1 + \frac{m_1 + m_2}{M}\dot{q}_2 . \end{aligned}$$

Substituting these expressions into the Lagrangian L_0 and operating we obtain

$$L_0 = \frac{1}{2}M\dot{X}^2 + L_1(q_1, q_2, \dot{q}_1, \dot{q}_2),$$

where

$$\boxed{L_1 = \frac{m_1(m_1+m_2)}{2M}(\dot{q}_1^2 + \dot{q}_2^2) + \frac{m_1^2}{M}q_1q_2 - \frac{k}{2}(q_1^2 + q_2^2).}$$

Thus in the approximation of small vibrations (i.e., when $|q_1|$ and $|q_2|$ are small) the equation of motion of the center of mass coordinate is $\ddot{X} = 0$, as expected (since there are no external forces), and the motion of the coordinates (q_1, q_2) is governed by the Lagrangian L_1. From the expression of the Lagrangian L_1 we readily obtain the following formulas for the matrices A and B:

$$\boxed{A = \frac{m_1}{M}\begin{pmatrix} m_1+m_2 & m_1 \\ m_1 & m_1+m_2 \end{pmatrix}, \qquad B = k\mathbf{1}.}$$

As in the previous example, we can write the characteristic equation as

$$\det\left(A - \frac{B}{\omega^2}\right) = \frac{m_1^2}{M^2}\begin{vmatrix} m_1+m_2-\frac{kM}{m_1\omega^2} & m_1 \\ m_1 & m_1+m_2-\frac{kM}{m_1\omega^2} \end{vmatrix} = 0$$
$$\iff m_1 + m_2 - \frac{kM}{m_1\omega^2} = \pm m_1,$$

whence we easily obtain

$$\boxed{\omega_\pm = \sqrt{\frac{kM}{m_1(m_1+m_2 \mp m_1)}} = \begin{cases} \sqrt{\frac{kM}{m_1m_2}} \\ \sqrt{\frac{k}{m_1}}. \end{cases}}$$

The normal mode vectors $\mathbf{c}_\pm$ are determined by the eigenvalue equation

$$\left(m_1 + m_2 - \frac{kM}{m_1\omega_\pm^2}\right)c_{\pm,1} + m_1c_{\pm,2} = m_1(\pm c_1 + c_2) = 0,$$

so that

$$\mathbf{c}_\pm = c(1, \mp 1)$$

with $c \neq 0$ constant. Hence the two normal mode solutions are given by

$$\boxed{q_1^{(\pm)} = c\cos(\omega_\pm t + \alpha_\pm), \qquad q_2^{(\pm)} = \mp c\cos(\omega_\pm t + \alpha_\pm),}$$

and the normal mode coordinates are simply

$$\boxed{Q_\pm = q_1 \mp q_2}$$

(why?). The motion of the atoms' physical coordinates x_i in each of these normal modes can be easily obtained from Eqs. (5.17). Note that in the normal mode with the smaller frequency ω_- we have $q_1 = q_2$, or equivalently $x_2 - x_1 = x_3 - x_2$. Hence in this mode the distances between the atom of mass m_2 and each of the atoms of mass m_1 increase or decrease *in step*, oscillating with the same frequency ω_-. Moreover, from Eqs. (5.17) it follows that $x_2 = X$, i.e., the atom of mass m_2 is fixed at the molecule's center of mass (in particular, it is stationary in the CM frame). On the other hand, in the normal mode with the larger frequency ω_+ we have $q_1 = -q_2$, so that $x_2 - x_1 - a$ and $x_3 - x_2 - a$ have opposite signs and oscillate completely *out of phase* with the same frequency ω_+. Thus when the right half of the molecule stretches the left one contracts, and vice versa. Moreover, in this case the x_2 coordinate is given by

$$x_2 = X + \frac{2m_1}{M} q_1 = X + \frac{2cm_1}{M} \cos(\omega_+ t + \alpha_+) ,$$

so that the position of the atom of mass m_2 oscillates with frequency ω_+ in the CM frame.

CHAPTER 6

Motion relative to a non-inertial frame

We examine in this chapter the description of the motion of a dynamical system in a non-inertial (moving) reference frame. To account for the rotation of the axes of the non-inertial frame, we discuss the three-dimensional rotation group and establish the relation between the time derivatives in the fixed (inertial) and moving frames. This allows us to generalize Newton's equations of motion to non-inertial frames, and to explain how several types of non-inertial forces arise. We then apply our results to analyzing the motion of particles near Earth's surface as observed from a terrestrial reference frame, concluding with a discussion of Foucault's pendulum.

6.1 ANGULAR VELOCITY OF A REFERENCE FRAME WITH RESPECT TO ANOTHER

Consider, to begin with, two reference frames S and S' with the same origin, and denote by $\{\mathbf{e}_1, \mathbf{e}_2, \mathbf{e}_3\}$ and $\{\mathbf{e}'_i, \mathbf{e}'_2, \mathbf{e}'_3\}$ the orthonormal positively oriented frames determining the axes of S and S'. We shall always assume in this chapter that the frame S' is *inertial*, and denote by $O(t)$ the linear application relating the vectors $\mathbf{e}'_i$ (**fixed axes**) with the vectors $\mathbf{e}_i$ (**moving axes**):

$$\boxed{\mathbf{e}_i(t) = O(t)\,\mathbf{e}'_i\,, \qquad i = 1, 2, 3\,.} \tag{6.1}$$

We shall often identify in what follows the operator $O(t)$ with its *matrix* in the basis $\{\mathbf{e}'_i\}_{i=1}^{3}$, whose *columns* are the coordinates of the vectors $\mathbf{e}_i(t)$ with respect to the latter basis. Since $O(t)$ transforms a positively oriented orthonormal frame into another such frame, this operator is an element of the **special orthogonal group** $\mathrm{SO}(3)$ of all linear operators $M : \mathbb{R}^3 \to \mathbb{R}^3$ (or, equivalently, 3×3 real matrices M) satisfying the conditions

$$\boxed{M^{\mathsf{T}} M = M M^{\mathsf{T}} = \mathbf{1}\,, \qquad \det M = 1\,.}$$

DOI: 10.1201/9781003600633-6

A theorem first proved by Euler states that *every element M of* SO(3) *is a rotation about a certain axis* $\mathbf{n}$. The proof of this theorem is as follows. First of all, taking the determinant of both members of the equality

$$M^{\mathsf{T}}(M-\mathbf{1}) = \mathbf{1} - M^{\mathsf{T}}$$

and using the elementary identities

$$\det M = \det M^{\mathsf{T}} = 1\,, \qquad \det(\mathbf{1}-M^{\mathsf{T}}) = \det\big((\mathbf{1}-M)^{\mathsf{T}}\big) = \det(\mathbf{1}-M)$$

we obtain

$$\det(M-\mathbf{1}) = \det(\mathbf{1}-M) = -\det(M-\mathbf{1}) \quad\Longrightarrow\quad \det(M-\mathbf{1}) = 0\,.$$

Hence $\lambda = 1$ is an *eigenvalue* of M. In other words, there exists a nonzero vector $\mathbf{n} \in \mathbb{R}^3$ (which we can take w.l.o.g. of unit length) such that $M\mathbf{n} = \mathbf{n}$. Let us next show that M is a rotation about the axis $\mathbf{n}$. Indeed, taking $\mathbf{e}_3' = \mathbf{n}$ the matrix M is of the form

$$M = \begin{pmatrix} a_{11} & a_{12} & 0 \\ a_{21} & a_{22} & 0 \\ 0 & 0 & 1 \end{pmatrix},$$

where $A \equiv (a_{ij})_{1\leqslant i,j\leqslant 2}$ is an orthogonal 2×2 matrix with unit determinant (recall that the columns of an orthogonal matrix are unit vectors that are mutually perpendicular). Since $a_{11}^2 + a_{21}^2 = 1$, we can take

$$a_{11} = \cos\theta, \quad a_{21} = \sin\theta\,, \qquad \text{with } \theta \in [0, 2\pi)\,.$$

Similarly,

$$a_{12} = \cos\psi, \quad a_{22} = \sin\psi\,, \qquad \text{with } \psi \in [0, 2\pi)\,.$$

Imposing the orthogonality of the columns of A we obtain

$$\cos\theta\cos\psi + \sin\theta\sin\psi = \cos(\theta-\psi) = 0\,.$$

Hence $\psi = \theta \pm \frac{\pi}{2}$ (up to an integer multiple of 2π), and

$$M = \begin{pmatrix} \cos\theta & \mp\sin\theta & 0 \\ \sin\theta & \pm\cos\theta & 0 \\ 0 & 0 & 1 \end{pmatrix}.$$

Actually, the solution $\psi = \theta - \frac{\pi}{2}$ is unacceptable, since it implies that $\det M = -1$. Thus $\psi = \theta + \pi/2$, and

$$M = \begin{pmatrix} \cos\theta & -\sin\theta & 0 \\ \sin\theta & \cos\theta & 0 \\ 0 & 0 & 1 \end{pmatrix} =: R_3(\theta) \tag{6.2}$$

is indeed a (counterclockwise) rotation of angle θ about the axis $\mathbf{e}_3' = \mathbf{n}$.

Exercise 6.1. Show that the rotation angle θ of a matrix $M \in \mathrm{SO}(3)$ is determined by the equation $1 + 2\cos\theta = \operatorname{tr} M$, where $\operatorname{tr} M := \sum_{i=1}^{3} M_{ii}$ denotes the **trace** of the matrix M.

Solution. We have just seen that if $M \in \mathrm{SO}(3)$ and the unit vector $\mathbf{n}$ is an eigenvector of M of eigenvalue 1, then M is a rotation about the axis $\mathbf{n}$. To determine the angle of rotation θ, note that $M = UR_3(\theta)U^{-1}$, where U is the change of basis matrix from the original basis $\{\mathbf{e}'_i\}_{i=1}^{3}$ to the basis with $\mathbf{n} = \mathbf{e}'_3$. Taking the trace of this equality and remembering that $\operatorname{tr}(AB) = \operatorname{tr}(BA)$ we obtain

$$\operatorname{tr} M = \operatorname{tr}\left(UR_3(\theta)U^{-1}\right) = \operatorname{tr}\left(U^{-1}UR_3(\theta)\right) = \operatorname{tr} R_3(\theta) = 1 + 2\cos\theta\,.$$

Consider again the rotation matrix $R_3(\theta)$ about the axis $\mathbf{e}'_3$. A direct calculation shows that

$$\frac{\mathrm{d}R_3}{\mathrm{d}\theta}(0) = \begin{pmatrix} 0 & -1 & 0 \\ 1 & 0 & 0 \\ 0 & 0 & 0 \end{pmatrix};$$

hence, if $\mathbf{c} = \sum_{i=1}^{3} c'_i \mathbf{e}'_i \in \mathbb{R}^3$ is an arbitrary vector we have

$$\left.\frac{\mathrm{d}}{\mathrm{d}\theta}\right|_{\theta=0} R_3(\theta)\mathbf{c} = \frac{\mathrm{d}R_3}{\mathrm{d}\theta}(0)\,\mathbf{c} = -c_2\mathbf{e}'_1 + c_1\mathbf{e}'_2 = \mathbf{e}'_3 \times \mathbf{c}\,.$$

In general, if $R_{\mathbf{n}}(\theta)$ denotes the matrix implementing a rotation about the axis $\mathbf{n}$ by an angle θ, we must accordingly have

$$\left.\frac{\mathrm{d}}{\mathrm{d}\theta}\right|_{\theta=0} R_{\mathbf{n}}(\theta)\mathbf{c} = \frac{\mathrm{d}R_{\mathbf{n}}}{\mathrm{d}\theta}(0)\,\mathbf{c} = \mathbf{n} \times \mathbf{c}\,.$$

We can symbolically write

$$\boxed{\frac{\mathrm{d}R_{\mathbf{n}}}{\mathrm{d}\theta}(0) = \mathbf{n}\times\,,} \tag{6.3}$$

with the understanding that both sides are equal when applied to an arbitrary vector $\mathbf{c} \in \mathbb{R}^3$. Another consequence of the previous result is that, since $R_{\mathbf{n}}(0) = \mathbf{1}$, for θ small we have

$$\boxed{R_{\mathbf{n}}(\theta)\mathbf{c} = \mathbf{c} + \theta\,\mathbf{n} \times \mathbf{c} + O(\theta^2)\,.} \tag{6.4}$$

For this reason the transformation

$$\mathbf{c} \mapsto \mathbf{c} + \theta\,\mathbf{n} \times \mathbf{c}$$

is called an *infinitesimal rotation* of angle θ about $\mathbf{n}$.

Exercise 6.2. Show that $R_{\mathbf{n}}(\theta)\mathbf{r} = \cos\theta\,\mathbf{r} + (1-\cos\theta)(\mathbf{n}\cdot\mathbf{r})\mathbf{n} + \sin\theta\,\mathbf{n}\times\mathbf{r}$.

Solution. If $\mathbf{r}$ is parallel to $\mathbf{n}$ the formula is clearly true. On the other hand, if $\mathbf{r}$ is not parallel to $\mathbf{n}$ the vectors $\mathbf{n}$, $\mathbf{n}\times\mathbf{r}$, and

$$(\mathbf{n}\times\mathbf{r})\times\mathbf{n} = \mathbf{r} - (\mathbf{n}\cdot\mathbf{r})\mathbf{n}$$

are mutually orthogonal and nonzero. Moreover, the vectors $\mathbf{n}\times\mathbf{r}$ and $(\mathbf{n}\times\mathbf{r})\times\mathbf{n}$ have the same length $l > 0$, since $(\mathbf{n}\times\mathbf{r})$ is orthogonal to $\mathbf{n}$ and $\mathbf{n}$ is a unit vector. It follows that the vectors

$$\mathbf{e}_1 = \frac{1}{l}\big(\mathbf{r} - (\mathbf{n}\cdot\mathbf{r})\mathbf{n}\big), \quad \mathbf{e}_2 = \frac{1}{l}\,\mathbf{n}\times\mathbf{r}, \quad \mathbf{e}_3 = \mathbf{n}$$

make up a positively oriented orthonormal basis. By construction, $R_{\mathbf{n}}(\theta) = R_3(\theta)$ in this basis. Using Eq. (6.2) for the rotation matrix $R_3(\theta)$ we then obtain

$$\begin{aligned} R_{\mathbf{n}}(\theta)\mathbf{r} &= R_{\mathbf{n}}(\theta)\big(l\mathbf{e}_1 + (\mathbf{n}\cdot\mathbf{r})\mathbf{n}\big) = lR_{\mathbf{n}}(\theta)\mathbf{e}_1 + (\mathbf{n}\cdot\mathbf{r})\mathbf{n} \\ &= l\cos\theta\,\mathbf{e}_1 + l\sin\theta\,\mathbf{e}_2 + (\mathbf{n}\cdot\mathbf{r})\mathbf{n} \\ &= \cos\theta\big(\mathbf{r} - (\mathbf{n}\cdot\mathbf{r})\mathbf{n}\big) + \sin\theta\,\mathbf{n}\times\mathbf{r} + (\mathbf{n}\cdot\mathbf{r})\mathbf{n}, \end{aligned}$$

which is equivalent to the proposed formula.

Let now $O(t) \in \mathrm{SO}(3)$ for all t, and suppose that O is of class C^1 (i.e., that the matrix elements of O are continuously differentiable functions of t). We shall next compute the derivative $\dot{O}(t)$ at an arbitrary time t. To this end, we differentiate with respect to t the identity

$$O(t)O(t)^{\mathsf{T}} = \mathbf{1},$$

obtaining

$$0 = \dot{O}(t)O(t)^{\mathsf{T}} + O(t)\dot{O}(t)^{\mathsf{T}} = \dot{O}(t)O(t)^{\mathsf{T}} + \big[\dot{O}(t)O(t)^{\mathsf{T}}\big]^{\mathsf{T}}.$$

Thus

$$\Omega(t) := \dot{O}(t)O(t)^{\mathsf{T}}$$

is an *antisymmetric* 3×3 matrix. Since $O(t)^{\mathsf{T}} = O(t)^{-1}$, from the previous relation we obtain

$$\boxed{\dot{O}(t) = \Omega(t)O(t)\,.} \tag{6.5}$$

The antisymmetric matrix $\Omega(t)$ can be written as

$$\Omega(t) = \begin{pmatrix} 0 & -\omega_3(t) & \omega_2(t) \\ \omega_3(t) & 0 & -\omega_1(t) \\ -\omega_2(t) & \omega_1(t) & 0 \end{pmatrix} \tag{6.6}$$

for appropriate real numbers $\omega_i(t)$. It is then straightforward to check that if $\mathbf{c} \in \mathbb{R}^3$ is any vector then

$$\Omega(t)\mathbf{c} = \boldsymbol{\omega}(t)\times\mathbf{c},$$

where $\omega(t) \in \mathbb{R}^3$ is the vector with components $\omega_i(t)$. From Eq. (6.5) and the previous identity (with $O(t)\mathbf{c}$ instead of $\mathbf{c}$) we finally deduce that

$$\boxed{\dot{O}(t)\mathbf{c} = \omega(t) \times O(t)\mathbf{c}, \qquad \forall t \in \mathbb{R}.} \tag{6.7}$$

The vector $\omega(t) \in \mathbb{R}^3$, which is in general time-dependent, is determined by the relation

$$\omega(t)\times = \Omega(t) = \dot{O}(t)O(t)^T = \dot{O}(t)O(t)^{-1},$$

which can also be written as

$$\omega(t)\times = \left.\frac{\mathrm{d}}{\mathrm{d}s}\right|_{s=t} O(s)O(t)^{-1}. \tag{6.8}$$

Exercise 6.3. Show that the matrix elements Ω_{ij} of the antisymmetric matrix Ω and the components ω_k of the vector ω are related by

$$\Omega_{ij} = -\sum_{k=1}^{3} \varepsilon_{ijk}\omega_k, \qquad \omega_k = -\frac{1}{2}\sum_{i,j=1}^{3} \varepsilon_{ijk}\Omega_{ij}, \tag{6.9}$$

where ε_{ijk} is Levi-Civita's completely antisymmetric tensor defined in Eq. (1.6).

Solution. Due to the antisymmetry of the Levi-Civita tensor in its first two indices (i.e., the identity $\varepsilon_{ijk} = -\varepsilon_{jik}$), to check the first relation (6.9) it suffices to prove the equalities

$$\Omega_{12} = -\omega_3, \quad \Omega_{13} = \omega_2, \quad \Omega_{23} = -\omega_1,$$

which clearly hold on account of Eq. (6.6). The second relation follows from the first, as it is equivalent to the equalities

$$\omega_1 = -\frac{1}{2}(\varepsilon_{123}\Omega_{23} + \varepsilon_{132}\Omega_{32}) = -\frac{1}{2}(\Omega_{23} - \Omega_{32}) = -\Omega_{23},$$

and similarly

$$\omega_2 = -\Omega_{31} = \Omega_{13}, \qquad \omega_3 = -\Omega_{12},$$

where we have used the antisymmetric character of the matrix Ω.

Exercise 6.4. If $O(t)$ is a rotation about a fixed axis $\mathbf{n}$ by a time-dependent angle $\alpha(t)$, show that

$$\omega(t) = \dot{\alpha}(t)\mathbf{n}.$$

Solution. Indeed, from Eq. (6.8) we obtain

$$\begin{aligned} &O(t) = R_{\mathbf{n}}(\alpha(t)) \\ &\qquad \Longrightarrow \quad O(s)O(t)^{-1} = R_{\mathbf{n}}(\alpha(s))R_{\mathbf{n}}(\alpha(t))^{-1} = R_{\mathbf{n}}(\alpha(s) - \alpha(t)), \end{aligned}$$

and thus, by Eq. (6.3),

$$\omega(t)\times = \frac{\mathrm{d}}{\mathrm{d}s}\bigg|_{s=t} R_{\mathbf{n}}\big(\alpha(s)-\alpha(t)\big) = \dot{\alpha}(t)\frac{\mathrm{d}}{\mathrm{d}\theta}\bigg|_{\theta=0} R_{\mathbf{n}}(\theta) = \dot{\alpha}(t)\mathbf{n}\times$$
$$\Longrightarrow \quad \omega(t) = \dot{\alpha}(t)\mathbf{n}\,,$$

as was to be shown.

Applying Eq. (6.7) to Eq. (6.1), which relates the moving axes unit vectors $\mathbf{e}_i(t) = O(t)\mathbf{e}'_i$ with the fixed ones $\mathbf{e}'_i$, we obtain the important formula

$$\boxed{\dot{\mathbf{e}}_i(t) = \omega(t)\times O(t)\mathbf{e}'_i = \omega(t)\times \mathbf{e}_i(t)\,,} \tag{6.10}$$

where $\dot{\mathbf{e}}_i(t)$ denotes the time derivative of the vector $\mathbf{e}_i(t)$ with respect to the inertial (fixed) frame S'. We thus have

$$\mathbf{e}_i(t+\Delta t) = \mathbf{e}_i(t) + \omega(t)\Delta t\times \mathbf{e}_i(t) + O(\Delta t^2)\,.$$

Comparing with Eq. (6.4), we deduce that to first order in Δt each vector $\mathbf{e}_i(t+\Delta t)$ is obtained from $\mathbf{e}_i(t)$ by applying an *infinitesimal rotation of angle* $\Delta\theta$ *and axis* $\mathbf{n}(t)$ such that

$$\Delta\theta\,\mathbf{n}(t) = \Delta t\,\omega(t)\,.$$

Letting Δt tend to zero in the previous equation we obtain the relation

$$\boxed{\omega(t) = \dot{\theta}(t)\,\mathbf{n}(t)\,.}$$

In other words:

> The direction and the magnitude of the vector $\omega(t)$ are respectively equal to the *instantaneous axis of rotation* and the magnitude of the *instantaneous angular velocity* of the moving axes $\{\mathbf{e}_i\}_{i=1}^3$ with respect to the fixed ones $\{\mathbf{e}'_i\}_{i=1}^3$.

For this reason, the vector $\omega(t)$ is called the **instantaneous angular velocity vector** (at time t) of the moving axes $\{\mathbf{e}_i\}_{i=1}^3$ with respect to the fixed ones $\{\mathbf{e}'_i\}_{i=1}^3$.

6.2 TIME DERIVATIVE IN THE FIXED AND MOVING FRAMES

To study the relation between the description of dynamics in the inertial frame S' (**fixed frame**) and its non-inertial counterpart S (**moving frame**), we shall analyze in this section how the time derivative of a vector $\mathbf{A}(t)$ is expressed in each of these reference frames. To this end, let us start by expanding $\mathbf{A}(t)$ in the *moving* frame $\{\mathbf{e}_i\}_{i=1}^3$:

$$\mathbf{A}(t) = \sum_{i=1}^{3} A_i(t)\,\mathbf{e}_i\,.$$

From now on, to avoid confusion we shall respectively denote by $\left(\frac{\mathrm{d}}{\mathrm{d}t}\right)_{\mathrm{f}}$ and $\left(\frac{\mathrm{d}}{\mathrm{d}t}\right)_{\mathrm{m}}$ the time derivatives with respect to the fixed and moving reference frames. Differentiating the previous equation *in the fixed frame* and using this notation we obtain the identity

$$\left(\frac{\mathrm{d}\mathbf{A}(t)}{\mathrm{d}t}\right)_{\mathrm{f}} = \sum_{i=1}^{3} \dot{A}_i(t)\,\mathbf{e}_i + \sum_{i=1}^{3} A_i(t)\left(\frac{\mathrm{d}\mathbf{e}_i}{\mathrm{d}t}\right)_{\mathrm{f}}\,, \tag{6.11}$$

where we have taken into account that the functions $A_i(t)$ are *scalars*, and therefore their time derivative is the same in any frame. Using Eq. (6.10), which in the notation just introduced reads

$$\boxed{\left(\frac{\mathrm{d}\mathbf{e}_i}{\mathrm{d}t}\right)_{\mathrm{f}} = \boldsymbol{\omega}(t) \times \mathbf{e}_i\,,} \tag{6.12}$$

Eq. (6.11) becomes

$$\left(\frac{\mathrm{d}\mathbf{A}(t)}{\mathrm{d}t}\right)_{\mathrm{f}} = \sum_{i=1}^{3} \dot{A}_i(t)\,\mathbf{e}_i + \boldsymbol{\omega}(t) \times \mathbf{A}(t)\,. \tag{6.13}$$

On the other hand, in the moving frame the vectors $\mathbf{e}_i$ are *constant*, so that $\left(\frac{\mathrm{d}\mathbf{A}}{\mathrm{d}t}\right)_{\mathrm{m}}$ is simply given by

$$\left(\frac{\mathrm{d}\mathbf{A}(t)}{\mathrm{d}t}\right)_{\mathrm{m}} = \sum_{i=1}^{3} \dot{A}_i(t)\,\mathbf{e}_i\,.$$

Comparing the last two equations we obtain the important relation

$$\boxed{\left(\frac{\mathrm{d}\mathbf{A}(t)}{\mathrm{d}t}\right)_{\mathrm{f}} = \left(\frac{\mathrm{d}\mathbf{A}(t)}{\mathrm{d}t}\right)_{\mathrm{m}} + \boldsymbol{\omega}(t) \times \mathbf{A}(t)\,.} \tag{6.14}$$

The above expression is valid for *any* time-dependent vector $\mathbf{A}(t)$; in particular, if we apply it to the instantaneous angular velocity $\boldsymbol{\omega}(t)$ we obtain

$$\boxed{\left(\frac{\mathrm{d}\boldsymbol{\omega}(t)}{\mathrm{d}t}\right)_{\mathrm{f}} = \left(\frac{\mathrm{d}\boldsymbol{\omega}(t)}{\mathrm{d}t}\right)_{\mathrm{m}} =: \dot{\boldsymbol{\omega}}(t)\,.} \tag{6.15}$$

6.3 DYNAMICS IN A NON-INERTIAL REFERENCE FRAME

Consider next the most general situation in which the origin of S is displaced from that of S' by a time-dependent vector $\mathbf{R}(t)$. If $\mathbf{r}$ is the position vector of a particle with respect to the non-inertial frame S, its position vector in the inertial frame S' will be given by

$$\mathbf{r}' = \mathbf{r} + \mathbf{R}.$$

Differentiating this equality with respect to the fixed (inertial) reference frame we obtain

$$\left(\frac{\mathrm{d}\mathbf{r}'}{\mathrm{d}t}\right)_{\mathrm{f}} = \left(\frac{\mathrm{d}\mathbf{r}}{\mathrm{d}t}\right)_{\mathrm{m}} + \omega \times \mathbf{r} + \mathbf{V}, \tag{6.16}$$

where

$$\boxed{\mathbf{V} := \left(\frac{\mathrm{d}\mathbf{R}}{\mathrm{d}t}\right)_{\mathrm{f}}} \tag{6.17}$$

is the velocity of the origin of S measured in the inertial frame S'. Denoting by

$$\mathbf{v}_{\mathrm{f}} = \left(\frac{\mathrm{d}\mathbf{r}'}{\mathrm{d}t}\right)_{\mathrm{f}}, \qquad \mathbf{v}_{\mathrm{m}} = \left(\frac{\mathrm{d}\mathbf{r}}{\mathrm{d}t}\right)_{\mathrm{m}} \tag{6.18}$$

the velocity of the particle with respect to the fixed and moving frames, we can rewrite Eq. (6.16) in the more compact form

$$\boxed{\mathbf{v}_{\mathrm{f}} = \mathbf{v}_{\mathrm{m}} + \omega \times \mathbf{r} + \mathbf{V}.} \tag{6.19}$$

Differentiating again this relation in the inertial frame we obtain

$$\begin{aligned}\left(\frac{\mathrm{d}\mathbf{v}_{\mathrm{f}}}{\mathrm{d}t}\right)_{\mathrm{f}} &= \left(\frac{\mathrm{d}\mathbf{v}_{\mathrm{m}}}{\mathrm{d}t}\right)_{\mathrm{m}} + \omega \times \mathbf{v}_{\mathrm{m}} + \dot{\omega} \times \mathbf{r} + \omega \times (\mathbf{v}_{\mathrm{m}} + \omega \times \mathbf{r}) + \left(\frac{\mathrm{d}\mathbf{V}}{\mathrm{d}t}\right)_{\mathrm{f}} \\ &= \left(\frac{\mathrm{d}\mathbf{V}}{\mathrm{d}t}\right)_{\mathrm{f}} + \left(\frac{\mathrm{d}\mathbf{v}_{\mathrm{m}}}{\mathrm{d}t}\right)_{\mathrm{m}} + 2\omega \times \mathbf{v}_{\mathrm{m}} + \omega \times (\omega \times \mathbf{r}) + \dot{\omega} \times \mathbf{r}.\end{aligned}$$

Taking into account that

$$\left(\frac{\mathrm{d}\mathbf{v}_{\mathrm{f}}}{\mathrm{d}t}\right)_{\mathrm{f}} = \left(\frac{\mathrm{d}^2\mathbf{r}'}{\mathrm{d}t^2}\right)_{\mathrm{f}} = \mathbf{a}_{\mathrm{f}}, \qquad \left(\frac{\mathrm{d}\mathbf{v}_{\mathrm{m}}}{\mathrm{d}t}\right)_{\mathrm{m}} = \left(\frac{\mathrm{d}^2\mathbf{r}}{\mathrm{d}t^2}\right)_{\mathrm{m}} = \mathbf{a}_{\mathrm{m}}$$

are the particle's *accelerations* in the fixed and moving frames, and denoting by

$$\mathbf{A} := \left(\frac{\mathrm{d}\mathbf{V}}{\mathrm{d}t}\right)_{\mathrm{f}} = \left(\frac{\mathrm{d}^2\mathbf{R}}{\mathrm{d}t^2}\right)_{\mathrm{f}}$$

the acceleration of the origin of the moving frame S with respect to the fixed one S', we finally obtain the fundamental relation

$$\boxed{\mathbf{a}_{\mathrm{f}} = \mathbf{a}_{\mathrm{m}} + \mathbf{A} + 2\omega \times \mathbf{v}_{\mathrm{m}} + \omega \times (\omega \times \mathbf{r}) + \dot{\omega} \times \mathbf{r}.} \tag{6.20}$$

From the previous equation it follows that if the particle is acted upon by a force **F**, as measured in the *inertial* frame S', its equation of motion in the *moving* frame S will be

$$m\mathbf{a}_{\mathrm{m}} = \mathbf{F} - m\mathbf{A} - 2m\boldsymbol{\omega} \times \mathbf{v}_{\mathrm{m}} - m\boldsymbol{\omega} \times (\boldsymbol{\omega} \times \mathbf{r}) - m\dot{\boldsymbol{\omega}} \times \mathbf{r} =: \mathbf{F} + \mathbf{F}_{\mathrm{in}}\,. \tag{6.21}$$

Hence in the moving frame Newton's second law must be modified by adding to the *real force* **F** the *fictitious force*

$$\mathbf{F}_{\mathrm{in}} = -m\mathbf{A} - 2m\boldsymbol{\omega} \times \mathbf{v}_{\mathrm{m}} - m\boldsymbol{\omega} \times (\boldsymbol{\omega} \times \mathbf{r}) - m\dot{\boldsymbol{\omega}} \times \mathbf{r}\,. \tag{6.22}$$

It is important to note that this fictitious force is an *inertial* force, since it is proportional to the particle's *mass* m.

The first term in the fictitious force $\mathbf{F}_{\mathrm{in}}$ is simply due to the *acceleration* of the origin of the non-inertial frame S with respect to the inertial one S', and thus vanishes if this point moves with *constant velocity* relative to S'. The remaining terms in $\mathbf{F}_{\mathrm{i}}$ are due to the *rotation of the axes* of the moving reference frame. While the last term vanishes if the angular velocity $\boldsymbol{\omega}$ is *constant*, the second and the third terms are in general nonzero even when $\boldsymbol{\omega}$ is constant. The term $-m\boldsymbol{\omega} \times (\boldsymbol{\omega} \times \mathbf{r})$ is the so-called **centrifugal force**, since it is a vector in the plane determined by $\boldsymbol{\omega}$ and $\mathbf{r}$, perpendicular to $\boldsymbol{\omega}$ and pointing *away* from the axis determined by this vector. The term $-2m\boldsymbol{\omega} \times \mathbf{v}_{\mathrm{m}}$, which depends on the particle's velocity, is known as the **Coriolis force**. Note that the centrifugal force is of *second order* in $\boldsymbol{\omega}$, while the Coriolis force is of *first order*. It is therefore to be expected that the former force should be negligible compared to the latter for small angular velocities $|\boldsymbol{\omega}|$.

From the previous remarks it is also clear that the fictitious force $\mathbf{F}_{\mathrm{in}}$ vanishes identically if and only if

$$\left(\frac{\mathrm{d}^2\mathbf{R}}{\mathrm{d}t^2}\right)_{\mathrm{f}} = \boldsymbol{\omega} = 0$$

for all t. By Eq. (6.7), the vanishing of $\boldsymbol{\omega}(t)$ for all t is equivalent to the condition that the rotation matrix $O(t)$ be *constant*. From the discussion of Section 2.3.4 on Galileo's relativity principle, this is the same as saying that S is also an *inertial* frame. In other words, *inertial forces are absent only in inertial reference frames.*

6.4 MOTION OF A PARTICLE RELATIVE TO THE ROTATING EARTH

We shall apply in this section the equation of motion (6.21) obtained above to study the dynamics of a particle moving near Earth's surface. We shall neglect Earth's motion around the Sun, and assume that Earth rotates around its south-north axis in

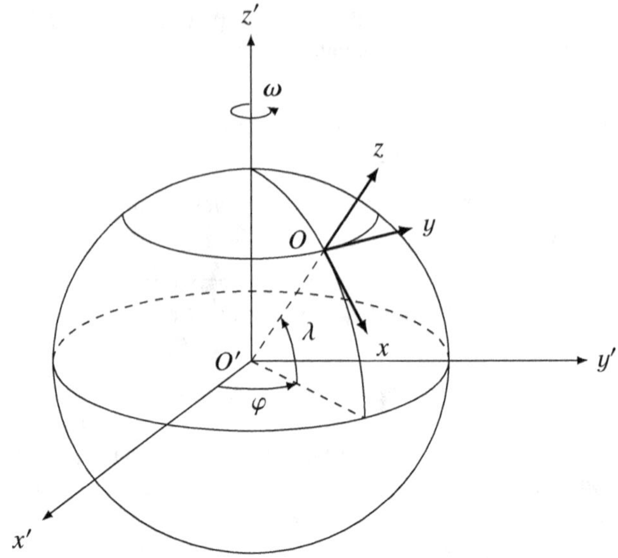

Figure 6.1. Terrestrial axes at a point O on Earth's surface.

the west-east direction with constant angular velocity of magnitude[1]

$$\omega = \frac{2\pi \text{ rad}}{1 \text{ sidereal day}} \simeq \frac{2\pi \text{ rad}}{86164.1 \text{ s}} \simeq 7.29212 \cdot 10^{-5} \text{ rad s}^{-1}\,.$$

Let us choose a (moving) frame of **terrestrial axes** in the manner indicated in Fig. 6.1. More precisely, the origin O of the terrestrial frame S is a point on Earth's surface with *latitude* λ and *longitude* φ, the vector $\mathbf{e}_3$ (z axis) is directed along the vector $\mathbf{R}$ joining Earth's center with the point O, the vector $\mathbf{e}_1$ (x axis) is tangent to the *meridian* passing through O (in a southerly direction), and the vector $\mathbf{e}_2$ (y axis) is then tangent to the *parallel* passing through O (in an easterly direction). In other words:

> The x axis is directed toward the *south*, the y axis toward the *east*, and the z axis along the *vertical*.

As fixed axes we shall take a frame with origin O' located at Earth's center such that the vector $\mathbf{e}'_3$ is directed along the South-North Pole axis. Hence Earth's angular velocity is given by

$$\boldsymbol{\omega} = \omega \mathbf{e}'_3\,.$$

[1]By definition, a *sidereal day* is the time taken by Earth to perform a complete rotation around its axis, while a *solar day* (equal to 24 hours) is the interval between two consecutive transits of the Sun across the meridian of any point on Earth's surface. Due to Earth's rotation around the Sun, the sidereal day is about 4 minutes shorter than the solar day.

Note that the vectors $\mathbf{e}_i$ of the moving (terrestrial) frame are respectively the unit vectors $\mathbf{e}_\theta$, $\mathbf{e}_\varphi$, and $\mathbf{e}_r$ of the spherical coordinate system at the point $\mathbf{r}$, with $\theta = \frac{\pi}{2} - \lambda$. Using Eqs. (2.7) we thus obtain

$$\boxed{\begin{aligned}\mathbf{e}_1 &= \sin\lambda\cos\varphi\mathbf{e}'_1 + \sin\lambda\sin\varphi\mathbf{e}'_2 - \cos\lambda\mathbf{e}'_3\,,\\ \mathbf{e}_2 &= -\sin\varphi\mathbf{e}'_1 + \cos\varphi\mathbf{e}'_2\,,\\ \mathbf{e}_3 &= \cos\lambda\cos\varphi\mathbf{e}'_1 + \cos\lambda\sin\varphi\mathbf{e}'_2 + \sin\lambda\mathbf{e}'_3\,.\end{aligned}}$$

From the previous equations (or simply from Fig. 6.1) it follows that in the terrestrial frame Earth's angular velocity is given by

$$\boxed{\boldsymbol{\omega} = \omega(-\cos\lambda\mathbf{e}_1 + \sin\lambda\mathbf{e}_3)\,.} \tag{6.23}$$

Let us next write down the equation of motion (6.21) in the terrestrial frame for a particle of mass m moving in the vicinity of the point O. We shall assume, for the time being, that the only force acting on the particle is Earth's gravitational attraction $m\mathbf{g}_0$, where

$$\mathbf{g}_0 = -\frac{GM}{r'^3}\mathbf{r}'\,,$$

M is Earth's mass and $\mathbf{r}' = \mathbf{R} + \mathbf{r}$ is the particle's position vector relative to the fixed frame. If the particle remains close enough to the point O on Earth's surface, we can replace the vector $\mathbf{r}'$ by $\mathbf{R}$, and thus take

$$\mathbf{g}_0 = -\frac{GM}{R^3}\mathbf{R} = -\frac{GM}{R^2}\mathbf{e}_3 = -g_0\mathbf{e}_3\,,$$

where

$$\boxed{g_0 := \frac{GM}{R^2} \simeq 9.80665\ \mathrm{m\,s^{-2}}}$$

is the *acceleration due to gravity* (also called **standard gravity**) on Earth's surface. In the rest of this chapter we shall write, for simplicity,

$$\boxed{\mathbf{v}_\mathrm{m} = \dot{\mathbf{r}}\,, \qquad \mathbf{a}_\mathrm{m} = \ddot{\mathbf{r}}\,.}$$

The particle's equation of motion in the terrestrial frame is therefore

$$\ddot{\mathbf{r}} = \mathbf{g}_0 - \mathbf{A} - 2\boldsymbol{\omega}\times\dot{\mathbf{r}} - \boldsymbol{\omega}\times(\boldsymbol{\omega}\times\mathbf{r})\,.$$

This expression can be simplified taking into account that in this case

$$\mathbf{V} = \left(\frac{\mathrm{d}\mathbf{R}}{\mathrm{d}t}\right)_\mathrm{f} = \left(\frac{\mathrm{d}\mathbf{R}}{\mathrm{d}t}\right)_\mathrm{m} + \boldsymbol{\omega}\times\mathbf{R} = \boldsymbol{\omega}\times\mathbf{R}\,,$$

since $\mathbf{R} = R\mathbf{e}_3$ is constant in the terrestrial frame. Differentiating with respect to t (and taking into account that $\boldsymbol{\omega}$ is constant) we obtain

$$\mathbf{A} = \left(\frac{\mathrm{d}\mathbf{V}}{\mathrm{d}t}\right)_\mathrm{f} = \boldsymbol{\omega}\times\mathbf{V} = \boldsymbol{\omega}\times(\boldsymbol{\omega}\times\mathbf{R})\,.$$

Thus the particle's equation of motion reduces to

$$\boxed{\ddot{\mathbf{r}} = \mathbf{g} - 2\boldsymbol{\omega} \times \dot{\mathbf{r}} - \boldsymbol{\omega} \times (\boldsymbol{\omega} \times \mathbf{r})\,,} \tag{6.24a}$$

where the vector

$$\boxed{\mathbf{g} := \mathbf{g}_0 - \boldsymbol{\omega} \times (\boldsymbol{\omega} \times \mathbf{R})\,,} \tag{6.24b}$$

which is *constant* in the terrestrial frame, is known as the **effective gravity** at the point O, i.e., the acceleration relative to the terrestrial frame experienced by a particle instantaneously at rest at the point O on Earth's surface. Obviously, if apart from gravity an additional force $\mathbf{F}$ is exerted on the particle, its equation of motion is

$$\boxed{\ddot{\mathbf{r}} = \frac{\mathbf{F}}{m} + \mathbf{g} - 2\boldsymbol{\omega} \times \dot{\mathbf{r}} - \boldsymbol{\omega} \times (\boldsymbol{\omega} \times \mathbf{r})\,.} \tag{6.25}$$

Exercise 6.5. Show that at a point of latitude λ the plumb line deviates from the vertical by an angle $\delta(\lambda)$ given by

$$\boxed{\tan\delta(\lambda) = \frac{\omega^2 R \sin\lambda\cos\lambda}{g_0 - \omega^2 R\cos^2\lambda}\,.} \tag{6.26}$$

Find the latitude λ for which $\delta(\lambda)$ is maximum and the maximum value of $\delta(\lambda)$.

Solution. By definition, the plumb line is the direction determined by a string from which a mass hangs at rest, i.e., the direction opposite to the string's tension $\mathbf{T}$ at equilibrium. To find this direction, it is enough to note that the equation of motion of the mass is obtained substituting $\mathbf{F} = \mathbf{T}$ in Eq. (6.25), that is

$$\ddot{\mathbf{r}} = \mathbf{g} - 2\boldsymbol{\omega} \times \dot{\mathbf{r}} - \boldsymbol{\omega} \times (\boldsymbol{\omega} \times \mathbf{r}) + \frac{\mathbf{T}}{m}\,.$$

Since the mass is at rest $\dot{\mathbf{r}} = \ddot{\mathbf{r}} = 0$, and thus

$$\frac{\mathbf{T}}{m} = -\mathbf{g} + \boldsymbol{\omega} \times (\boldsymbol{\omega} \times \mathbf{r}) \simeq -\mathbf{g}\,, \tag{6.27}$$

where we have neglected the term $\boldsymbol{\omega} \times (\boldsymbol{\omega} \times \mathbf{r})$ taking into account that

$$|\boldsymbol{\omega} \times (\boldsymbol{\omega} \times \mathbf{r})| \leqslant \omega^2 r \ll g_0 \simeq |\mathbf{g}|\,.$$

From Eq. (6.27) it follows that the direction of the plumb line is approximately that of the effective gravity $\mathbf{g}$. Taking into account that $\mathbf{g}_0 = -g_0\mathbf{e}_3$ and

$$\begin{aligned}\boldsymbol{\omega} \times (\boldsymbol{\omega} \times \mathbf{R}) &= (\boldsymbol{\omega} \cdot \mathbf{R})\boldsymbol{\omega} - \omega^2\mathbf{R} = \omega^2 R\sin\lambda(-\cos\lambda\mathbf{e}_1 + \sin\lambda\mathbf{e}_3) - \omega^2 R\mathbf{e}_3\\ &= -\omega^2 R\cos\lambda(\sin\lambda\mathbf{e}_1 + \cos\lambda\mathbf{e}_3)\end{aligned}$$

we obtain

$$\mathbf{g} = g_0 \Big(\gamma \sin\lambda \cos\lambda \mathbf{e}_1 - (1 - \gamma \cos^2\lambda) \mathbf{e}_3 \Big) , \qquad \text{with} \quad \gamma := \frac{\omega^2 R}{g_0} \simeq 3.455 \cdot 10^{-3} .$$

The vector $\mathbf{g}$ has a component

$$g_1 = \gamma g_0 \sin\lambda \cos\lambda = \frac{\omega^2 R}{2} \sin 2\lambda$$

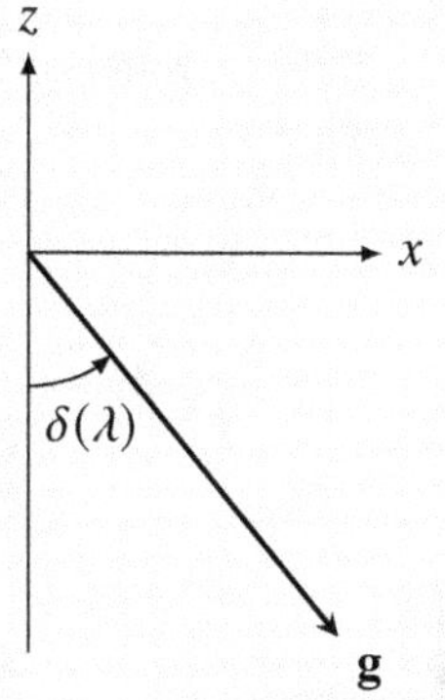

Figure 6.2. Effective gravity $\mathbf{g}$ (in the Northern Hemisphere).

in the direction of the basis vector $\mathbf{e}_1$ (cf. Fig. 6.2). Therefore $\mathbf{g}$ deviates from the vertical (i.e., the direction of the z axis) to the south in the Northern Hemisphere ($\lambda > 0$) and to the north in the Southern one ($\lambda < 0$). The tangent of the angle $\delta(\lambda)$ between the vector $\mathbf{g}$ and the vertical is given by

$$\tan\delta(\lambda) = \frac{g_1}{|g_3|} = \frac{\gamma \sin\lambda \cos\lambda}{1 - \gamma \cos^2\lambda} = \boxed{\frac{\gamma \sin 2\lambda}{2 - \gamma - \gamma \cos 2\lambda}} ,$$

which coincides with Eq. (6.26) by the definition of γ. Differentiating the previous expression we obtain

$$\frac{1}{2\gamma} \frac{\mathrm{d}}{\mathrm{d}\lambda} \tan\delta(\lambda) = \frac{(2 - \gamma - \gamma \cos 2\lambda) \cos 2\lambda - \gamma \sin^2 2\lambda}{(2 - \gamma - \gamma \cos 2\lambda)^2} = \frac{(2 - \gamma) \cos 2\lambda - \gamma}{(2 - \gamma - \gamma \cos 2\lambda)^2} .$$

Thus the angle $\delta(\lambda)$ will be maximum (in absolute value) when

$$\boxed{\cos 2\lambda = \frac{\gamma}{2 - \gamma} ,} \tag{6.28}$$

and its maximum value $\delta_{\max}$ verifies

$$\tan\delta_{\max} = \operatorname{sgn}\lambda\,\frac{\gamma\sqrt{1-\frac{\gamma^2}{(2-\gamma)^2}}}{2-\gamma-\frac{\gamma^2}{2-\gamma}} = \frac{\gamma\operatorname{sgn}\lambda}{\sqrt{(2-\gamma)^2-\gamma^2}} = \boxed{\frac{\gamma\operatorname{sgn}\lambda}{2\sqrt{1-\gamma}}}$$
$$\simeq 1.73\cdot 10^{-3}\operatorname{sgn}\lambda\,,$$

or equivalently

$$\delta_{\max} \simeq 5.68'\ \operatorname{sgn}\lambda\,.$$

Since γ is of the order of 10^{-3}, it follows from Eq. (6.28) that the latitude $\lambda_{\max}$ for which $\delta(\lambda)$ is maximum can be expressed as $\lambda_{\max} = \pm(\frac{\pi}{4}-\varepsilon)$, with $\varepsilon > 0$ small. The value of ε can be approximately computed expanding $\cos 2\lambda_{\max}$ to first order in ε:

$$\cos 2\lambda_{\max} = \cos\left(\frac{\pi}{2}-2\varepsilon\right) = \sin 2\varepsilon \simeq 2\varepsilon = \frac{\gamma}{2-\gamma} \simeq \frac{\gamma}{2}$$
$$\Longrightarrow \quad \varepsilon \simeq \frac{\gamma}{4} \simeq 8.636\cdot 10^{-4}\ \text{rad} = 2.969'\,.$$

The equation of motion (6.24) is *exact*. In fact, this equation is a system of (inhomogeneous) *linear* second-order ordinary differential equations with *constant coefficients* in the components of the vector $\mathbf{r}$. This type of systems can in principle be exactly solved, for instance, by transforming them into a first-order system in $(\mathbf{r}, \dot{\mathbf{r}})$ and using the matrix exponential. In practice, it is preferable to first simplify Eq. (6.24) taking into account the different orders of magnitude of its terms. More precisely, the second term in (6.24b) and the last term in (6.24a) are at most of order $\gamma \sim 10^{-3}$ and $\gamma r/R$, respectively, relative to g_0. Thus, if $r \ll R$ the equation of motion can be approximated by

$$\boxed{\ddot{\mathbf{r}} = \mathbf{g}_0 - 2\boldsymbol{\omega}\times\dot{\mathbf{r}}\,.} \tag{6.29}$$

Integrating once with respect to t we obtain

$$\dot{\mathbf{r}} = \mathbf{g}_0 t - 2\boldsymbol{\omega}\times\mathbf{r} + \mathbf{c}\,,$$

where $\mathbf{c}$ is a constant vector in the terrestrial frame. Although this system can again be *exactly* solved (it is an inhomogeneous linear system of first-order ordinary differential equations with constant coefficients), it is more convenient in practice to take advantage of the fact that for small speeds $|\dot{\mathbf{r}}|$ the first term of the RHS of Eq. (6.29) is much larger than the second one, since

$$\frac{g_0}{\omega} \simeq 1.34483\cdot 10^5\ \mathrm{m\,s^{-1}}\,.$$

This fact makes it possible to obtain an approximate solution of Eq. (6.29), considered as the first-order equation in the velocity

$$\boxed{\dot{\mathbf{v}} = \mathbf{g}_0 - 2\boldsymbol{\omega}\times\mathbf{v}}\,, \tag{6.30}$$

by expanding $\mathbf{v}$ in powers of ω:

$$\boxed{\mathbf{v}(t) = \mathbf{v}_1(t) + \omega\mathbf{v}_2(t) + O(\omega^2)\,,}$$

with $\mathbf{v}_{1,2}$ independent of ω and

$$\mathbf{v}(0) := \mathbf{v}_0 \quad \Longrightarrow \quad \mathbf{v}_1(0) = \mathbf{v}_0\,, \quad \mathbf{v}_2(0) = 0\,.$$

Substituting into Eq. (6.30) we have

$$\dot{\mathbf{v}}_1 + \omega\dot{\mathbf{v}}_2 = \mathbf{g}_0 - 2\boldsymbol{\omega} \times \mathbf{v}_1 + O(\omega^2)\,,$$

whence, equating to zero the terms $O(1)$ and $O(\omega)$ in both sides of the previous expression, we obtain

$$\dot{\mathbf{v}}_1 = \mathbf{g}_0\,, \qquad \omega\dot{\mathbf{v}}_2 = -2\boldsymbol{\omega} \times \mathbf{v}_1\,.$$

Solving for $\mathbf{v}_1$ in the first of these equations and substituting the result into the second one we have

$$\mathbf{v}_1 = \mathbf{g}_0 t + \mathbf{v}_0\,, \qquad \omega\dot{\mathbf{v}}_2 = -2\boldsymbol{\omega} \times \mathbf{v}_0 - 2t\boldsymbol{\omega} \times \mathbf{g}_0 \overset{\mathbf{v}_2(0)=0}{\Longrightarrow} \omega\mathbf{v}_2 = -2t\boldsymbol{\omega} \times \mathbf{v}_0 - t^2\boldsymbol{\omega} \times \mathbf{g}_0\,.$$

Hence

$$\mathbf{v} \simeq \mathbf{v}_1 + \omega\mathbf{v}_2 = \mathbf{v}_0 + \mathbf{g}_0 t - 2t\boldsymbol{\omega} \times \mathbf{v}_0 - t^2\boldsymbol{\omega} \times \mathbf{g}_0\,,$$

and integrating with respect of t we finally obtain

$$\boxed{\mathbf{r} \simeq \mathbf{r}_0 + \mathbf{v}_0 t + \mathbf{g}_0\frac{t^2}{2} - t^2\boldsymbol{\omega} \times \mathbf{v}_0 - \frac{t^3}{3}\,\boldsymbol{\omega} \times \mathbf{g}_0\,.} \tag{6.31}$$

Exercise 6.6. A particle is thrown vertically from a point on Earth's surface with latitude λ until it reaches a height h. Show that the particle lands at a point $(4/3)\sqrt{8h^3/g_0}\,\omega\cos\lambda$ west of the starting point. (Consider only small heights h and neglect air resistance.)

Solution. We have

$$\mathbf{r}_0 = 0\,, \quad \mathbf{v}_0 = v_0\mathbf{e}_3 \implies -\boldsymbol{\omega} \times \mathbf{v}_0 = \omega v_0(\cos\lambda\mathbf{e}_1 - \sin\lambda\mathbf{e}_3) \times \mathbf{e}_3 = -\omega v_0 \cos\lambda\,\mathbf{e}_2$$

and

$$-\boldsymbol{\omega} \times \mathbf{g}_0 = \omega g_0 \cos\lambda\,\mathbf{e}_2\,.$$

From Eq. (6.31) with the previous values of $\mathbf{r}_0$ and $\mathbf{v}_0$ we obtain

$$\mathbf{r} \simeq v_0 t\,\mathbf{e}_3 - \frac{g_0}{2}\,t^2\,\mathbf{e}_3 - \left(\omega v_0 t^2 \cos\lambda - \frac{\omega g_0}{3}\,t^3 \cos\lambda\right)\mathbf{e}_2\,.$$

Hence the law of motion is approximately

$$x = 0\,, \qquad y = \omega v_0 \cos\lambda\, t^2 \left(\frac{g_0 t}{3v_0} - 1\right), \qquad z = v_0 t - \frac{g_0}{2}\, t^2\,.$$

The particle lands at the time $t_0 > 0$ for which $z = 0$, namely

$$t_0 = \frac{2v_0}{g_0}\,.$$

The value of the y coordinate at this time is thus

$$y(t_0) = -\frac{4}{3}\,\frac{\omega v_0^3}{g_0^2}\,\cos\lambda \leqslant 0\,.$$

We thus see that the particle deviates to the *west*, both in the northern and in the southern hemisphere, with maximum deviation at the equator ($\lambda = 0$). In order to express the deviation $y(t_0)$ in terms of the maximum height h, it suffices to note that this height is reached when $\dot{z} = 0$:

$$\dot{z} = v_0 - g_0 t = 0 \quad \Longrightarrow \quad t = \frac{v_0}{g_0} \quad \Longrightarrow \quad z = h = \frac{v_0^2}{2g_0}\,.$$

Hence

$$y(t_0) = -\frac{4\omega}{3g_0^2}\,(2g_0 h)^{3/2}\cos\lambda = \boxed{-\frac{4}{3}\,\omega\cos\lambda\,\sqrt{\frac{8h^3}{g_0}}}\,.$$

For instance,

$$h = 100\,\text{m}\,, \quad \lambda = 40^\circ\, 25' \ \text{(Madrid's latitude)} \quad \Longrightarrow \quad y(t_0) = -6.686\ \text{cm}\,.$$

Exercise 6.7. Redo the previous problem assuming that the particle is dropped from a height h over the vertical.

Solution. In this case

$$\mathbf{r}_0 = h\mathbf{e}_3\,, \qquad \mathbf{v}_0 = 0\,,$$

and substituting into Eq. (6.31) we obtain

$$x = 0\,, \qquad y = \frac{\omega g_0}{3}\, t^3 \cos\lambda\,, \qquad z = h - \frac{g_0}{2}\, t^2\,.$$

The particle lands when

$$t_0 = \sqrt{\frac{2h}{g_0}}\,,$$

and its deviation in the y direction is thus

$$y(t_0) = \frac{\omega}{3}\cos\lambda\sqrt{\frac{8h^3}{g_0}}.$$

Since $y(t_0) \geqslant 0$, the particle deviates *eastward* in both hemispheres.

6.5 FOUCAULT'S PENDULUM

In 1851 J.B.L. Foucault experimentally demonstrated Earth's rotation using the pendulum that nowadays bears his name, schematically represented in Fig. 6.3.

Since we are only interested in studying the small oscillations, we shall assume that the pendulum's length l is very large compared to the coordinates x, y, z of its bob. The total force acting on the pendulum's bob is thus

$$\mathbf{F} = m\mathbf{g}_0 + \mathbf{T},$$

where the tension $\mathbf{T}$ of the pendulum's string is given by

$$\mathbf{T} = T\,\frac{l\mathbf{e}_3 - \mathbf{r}}{|l\mathbf{e}_3 - \mathbf{r}|} \simeq T\left(\mathbf{e}_3 - \frac{\mathbf{r}}{l}\right).$$

With this approximation the equation of motion reads

$$\ddot{\mathbf{r}} = \mathbf{g}_0 + \frac{T}{m}\left(\mathbf{e}_3 - \frac{\mathbf{r}}{l}\right) - 2\boldsymbol{\omega}\times\dot{\mathbf{r}},$$

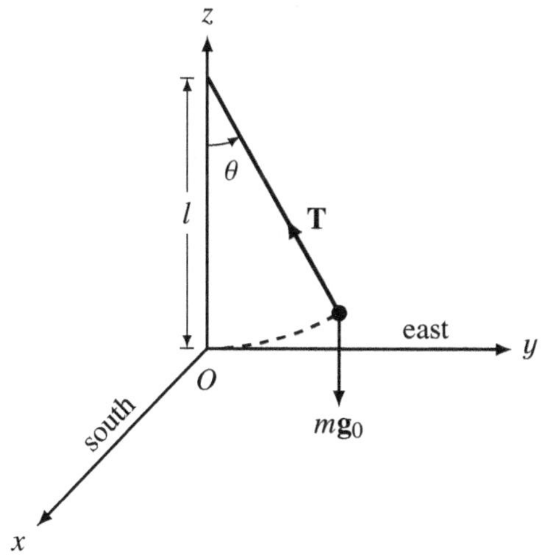

Figure 6.3. Foucault's pendulum.

where

$$\omega \times \dot{\mathbf{r}} = \omega \begin{vmatrix} \mathbf{e}_1 & \mathbf{e}_2 & \mathbf{e}_3 \\ -\cos\lambda & 0 & \sin\lambda \\ \dot{x} & \dot{y} & \dot{z} \end{vmatrix} = \omega\big[-\dot{y}\sin\lambda\,\mathbf{e}_1 + (\dot{x}\sin\lambda + \dot{z}\cos\lambda)\mathbf{e}_2 - \dot{y}\cos\lambda\,\mathbf{e}_3\big],$$

and thus

$$\boxed{\begin{aligned} \ddot{x} &= -\frac{Tx}{lm} + 2\omega\dot{y}\sin\lambda \\ \ddot{y} &= -\frac{Ty}{lm} - 2\omega(\dot{x}\sin\lambda + \dot{z}\cos\lambda) \\ \ddot{z} &= -g_0 + \frac{T}{m}\left(1 - \frac{z}{l}\right) + 2\omega\dot{y}\cos\lambda\,. \end{aligned}} \tag{6.32}$$

Note that $z \ll \sqrt{x^2 + y^2}$, since calling θ the angle between the pendulum and the vertical we have

$$\sqrt{x^2 + y^2} = l\sin\theta \simeq l\theta\,, \qquad z = l(1 - \cos\theta) \simeq \frac{l\theta^2}{2}\,.$$

We can thus neglect in Eqs. (6.32) the quantities z, $\dot{z}$, and $\ddot{z}$ compared to x, y, and their derivatives. In particular, from the last equation we obtain

$$\frac{T}{m} \simeq g_0 - 2\omega\dot{y}\cos\lambda \simeq g_0\,,$$

since g_0/ω is of the order of $10^5\ \mathrm{m\,s^{-1}}$. Substituting this approximation into the first two equations (6.32) and dropping the term proportional to $\dot{z}$ we finally obtain the following system for the coordinates (x, y):

$$\boxed{\begin{aligned} \ddot{x} + \alpha^2 x &= 2\omega\dot{y}\sin\lambda \\ \ddot{y} + \alpha^2 y &= -2\omega\dot{x}\sin\lambda\,, \end{aligned}} \tag{6.33}$$

where

$$\boxed{\alpha := \sqrt{\frac{g_0}{l}}}$$

is the pendulum's natural frequency.

The previous equations are easily solved introducing the complex variable

$$u = x + \mathrm{i}y\,,$$

in terms of which they adopt the simple form

$$\boxed{\ddot{u} + 2\mathrm{i}\Omega\,\dot{u} + \alpha^2 u = 0\,, \qquad \text{with } \Omega := \omega\sin\lambda\,.} \tag{6.34}$$

This is a linear homogeneous second-order ordinary differential equation with constant coefficients, whose characteristic polynomial

$$p(s) = s^2 + 2\mathrm{i}\Omega\, s + \alpha^2$$

possesses the two pure imaginary roots

$$s_\pm = -\mathrm{i}\Omega \pm \mathrm{i}\sqrt{\Omega^2+\alpha^2} = -\mathrm{i}\Omega \pm \mathrm{i}\alpha\sqrt{1+\frac{\Omega^2}{\alpha^2}} = \mathrm{i}\alpha\left(\pm 1 - \frac{\Omega}{\alpha} + O\left(\frac{\Omega^2}{\alpha^2}\right)\right)$$

For all practical purposes, we can neglect the term of order Ω^2/α^2 in the previous expression, since

$$\frac{\Omega^2}{\alpha^2} \leqslant \frac{\omega^2 l}{g_0} = \frac{l}{1.84422\cdot 10^9\,\mathrm{m}}\,.$$

We thus have

$$s_\pm \simeq -\mathrm{i}\Omega \pm \mathrm{i}\,\alpha\,,$$

and the general solution of Eq. (6.34) is therefore given by

$$\boxed{u = \mathrm{e}^{-\mathrm{i}\Omega t}\left(c_1\mathrm{e}^{\mathrm{i}\alpha t} + c_2\mathrm{e}^{-\mathrm{i}\alpha t}\right),} \tag{6.35}$$

where the constants c_1, c_2 are in general *complex*.

Let us find, for instance, the solution of Eqs. (6.33) with the initial conditions

$$x(0) = x_0 > 0\,,\quad y(0) = 0\,,\quad \dot{x}(0) = \dot{y}(0) = 0\,, \tag{6.36}$$

i.e., when the pendulum's bob is initially at rest in the Oxz plane at a distance x_0 from the vertical. If Earth did not rotate around its north-south axis, i.e., if $\omega = 0$, the solution of the equations of motion (6.33) with the initial conditions (6.36) would be

$$x = x_0\cos(\alpha t)\,,\qquad y = 0\,.$$

In other words, the pendulum would oscillate with frequency α and amplitude x_0 about the vertical in the Oxz plane. On the other hand, when $\omega > 0$ the solution (6.35) verifying the initial conditions (6.36) is easily found taking into account that

$$u(0) = x(0) + \mathrm{i}y(0) = x_0\,,\qquad \dot{u}(0) = \dot{x}(0) + \mathrm{i}\dot{y}(0) = 0\,. \tag{6.37}$$

We thus have

$$\begin{cases} c_1 + c_2 = x_0\,,\\ \mathrm{i}\alpha(c_1 - c_2) - \mathrm{i}\Omega(c_1+c_2) = \mathrm{i}\alpha(c_1-c_2) - \mathrm{i}\Omega x_0 = 0\\ \implies\ c_1 - c_2 = \dfrac{\Omega}{\alpha}\,x_0 \simeq 0\,,\end{cases}$$

whose approximate solution is

$$c_1 = c_2 = \frac{1}{2}\,x_0\,.$$

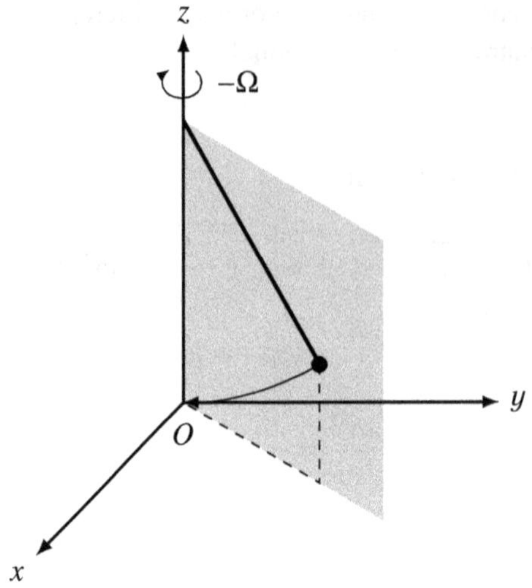

Figure 6.4. Rotation of the plane of Foucault's pendulum (shaded gray).

Thus the sought-for solution of Eq. (6.34) is approximately

$$\boxed{u = x_0 \mathrm{e}^{-\mathrm{i}\Omega t} \cos(\alpha t)}\,.$$

The complex number $\mathrm{e}^{-\mathrm{i}\Omega t}$ is the point on the unit circle making an angle $-\Omega t$ with the real (x) axis, obtained by rotating the unit coordinate vector $\mathbf{e}_1$ by an angle $-\Omega t$ about the z axis. Since $x_0 \cos(\alpha t)$ is real, the previous equation can be rewritten in real terms as

$$(x, y) = x_0 \cos(\alpha t)\, \mathbf{n}(t)\,, \qquad \text{with} \quad \boxed{\mathbf{n}(t) := R_3(-\Omega t)\mathbf{e}_1\,.}$$

From this equation it follows that at each instant t *the pendulum's plane,* determined by the vectors $\mathbf{e}_3$ and $\mathbf{n}(t)$, *makes an angle* $-\Omega t$ *with the* Oxz *plane*. The pendulum's motion can thus be viewed as the composition of two periodic motions, namely a "fast" oscillation with period $2\pi/\alpha$ in the plane determined by the vectors $\mathbf{e}_3$ and $\mathbf{n}(t)$ and a "slow" rotation of this plane about the z axis with period $2\pi/|\Omega| \gg 2\pi/\alpha$ (cf. Fig. 6.4). In particular:

> In the Northern Hemisphere the pendulum's plane rotates *clockwise*, i.e., in the *east-south* direction (since $\dot{\varphi} = -\Omega = -\omega \sin\lambda < 0$), *with angular velocity* $\Omega = \omega \sin\lambda$. In the Southern Hemisphere the rotation of the pendulum's plane is *counterclockwise* (since $\sin\lambda < 0$), and in the equator ($\lambda = 0$) no such rotation occurs.

Note that in each period $2\pi/\alpha$ of the pendulum (time in which $\cos(\alpha t)$ performs a complete oscillation) the angle between the pendulum's plane and the Oxz plane

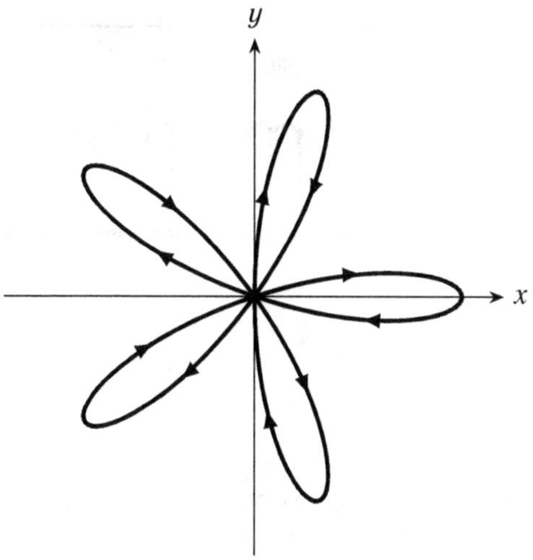

Figure 6.5. Projection onto the Oxy plane of the trajectory of the pendulum's bob for $\alpha/\Omega = 5$.

increases by $-2\pi\Omega/\alpha$, which as noted before is a very small quantity. The *period* of the rotation of the pendulum's plane is given by

$$\tau = \frac{2\pi}{|\Omega|} = \frac{2\pi}{\omega}|\csc\lambda| = |\csc\lambda| \text{ sidereal days}.$$

For instance, at a latitude of 30° the period is 2 sidereal days, while in Madrid ($\lambda = 40°\,25'$) it is 1.5424 sidereal days. Note, finally, that (with the approximations made) the motion of the pendulum's bob is not exactly periodic unless the ratio α/Ω is a rational number (cf. Fig. 6.5).

CHAPTER 7

Rigid body motion

IN this chapter we provide a concise overview of the dynamics of a rigid body. Due to the constraints imposed by rigidity, a generic rigid body only has six degrees of freedom, making it a comparatively simple yet fundamental mechanical system. Building on the description of rotations in three-dimensional space developed in the previous chapter, we show how the energy and angular momentum of a rigid body can be decomposed into two components: one arising from the motion of its center of mass, and the other from the motion of the body's particles relative to the center of mass. We next express the body's internal angular momentum in terms of its inertia tensor, and analyze the main properties of this tensor. Using the inertia tensor, we derive Euler's equations of motion and apply them to study the inertial motion of a symmetric rigid body. We then introduce Euler's angles and show how to parametrize an arbitrary rotation in terms of these angles. The Euler angles also provide a framework for determining the orientation of the body's axes from a solution of Euler's equations. Finally, we combine these concepts with the general Lagrangian formalism to analyze in detail the motion of a symmetric top on a rough surface (Lagrange's top).

7.1 DEGREES OF FREEDOM

A **rigid body** is a system of N particles of mass m_α ($\alpha = 1, \ldots, N$) in which the *distance* $|\mathbf{r}_\alpha - \mathbf{r}_\beta|$ between any two particles is *constant*. In other words, a rigid body is a mechanical system of N particles subject to the $N(N-1)/2$ time-independent holonomic constraints (not all of them independent!)

$$\boxed{(\mathbf{r}_\alpha - \mathbf{r}_\beta)^2 = l_{\alpha\beta}^2 = \text{const.}, \qquad 1 \leqslant \alpha < \beta \leqslant N.} \tag{7.1}$$

- We shall assume in what follows that Newton's third law holds in its *strongest version*, i.e., that the constraint force $\mathbf{F}_{\alpha\beta}$ exerted by particle β on particle α satisfies

$$\boxed{\mathbf{F}_{\alpha\beta} = -\mathbf{F}_{\beta\alpha} \parallel \mathbf{r}_\alpha - \mathbf{r}_\beta\,.}$$

DOI: 10.1201/9781003600633-7

It is easy to see that if this is the case the constraints (7.1) are *ideal*, i.e., that the *principle of virtual work* holds. Indeed, the work done by the constraint forces in an infinitesimal displacement[1] $\mathbf{dr}_\alpha$ ($\alpha = 1, \dots, N$) of the system's particles is given by[2]

$$\sum_{\alpha\neq\beta} \mathbf{F}_{\alpha\beta}\cdot \mathbf{dr}_\alpha = \frac{1}{2}\sum_{\alpha\neq\beta}(\mathbf{F}_{\alpha\beta}\cdot\mathbf{dr}_\alpha + \mathbf{F}_{\beta\alpha}\cdot\mathbf{dr}_\beta) = \frac{1}{2}\sum_{\alpha\neq\beta}\mathbf{F}_{\alpha\beta}\cdot(\mathbf{dr}_\alpha - \mathbf{dr}_\beta)\,. \quad (7.2)$$

On the other hand, differentiating the constraint equation we obtain

$$(\mathbf{r}_\alpha - \mathbf{r}_\beta)\cdot(\mathbf{dr}_\alpha - \mathbf{dr}_\beta) = 0\,, \qquad 1 \leqslant \alpha < \beta \leqslant N\,,$$

whence it follows (since the vectors $\mathbf{F}_{\alpha\beta}$ and $\mathbf{r}_\alpha - \mathbf{r}_\beta$ are parallel) that

$$\mathbf{F}_{\alpha\beta}\cdot(\mathbf{dr}_\alpha - \mathbf{dr}_\beta) = 0\,, \qquad 1 \leqslant \alpha < \beta \leqslant N\,.$$

We thus see that all the terms in the last sum in Eq. (7.2) vanish identically, and as a consequence the total work done by the constraint forces is indeed zero.

- We shall say that a rigid body is *generic* if it contains three non-collinear particles[3].

> In a generic rigid body it is always possible to construct a set of moving axes, the so-called **body axes**, with respect to which all of the body's particles are *fixed*, i.e., *at rest.*

In other words, the position vectors $\mathbf{r}_\alpha$ ($1 \leqslant \alpha \leqslant N$) of *all* the particles in the body are *constant* in the frame of body axes.

Proof. Let P, Q, R be three non-collinear points in the rigid body. A set of body axes is obtained, for example, taking as origin the point P, the x_1 axis in the direction of the vector $\overrightarrow{PQ}$, the x_2 axis in the direction of the line in the plane PQR perpendicular to $\overrightarrow{PQ}$, oriented so that the x_2 coordinate of the point R is positive, and the x_3 axis in the direction of $\overrightarrow{PQ}\times\overrightarrow{PR}$ (cf. Fig. 7.1). Indeed, in this frame the points P, Q, and R are fixed by construction (see next exercise). We shall next show that the coordinates (x, y, z) relative to this frame of any other point S in the body are constant. To this end, let $(a, 0, 0)$ and $(b, c, 0)$ denote the coordinates of Q and R in the frame just defined (where $a, c > 0$ by construction). If r_1, r_2, and r_3 are the (fixed) distances of S to the points P, Q, R we have

$$x^2 + y^2 + z^2 = r_1^2, \quad (x-a)^2 + y^2 + z^2 = r_2^2, \quad (x-b)^2 + (y-c)^2 + z^2 = r_3^2\,. \quad (7.3)$$

[1]Since the rigid body constraints (7.1) are time-independent, virtual displacements coincide with real ones.

[2]In what follow, sums over *Greek* indices $\alpha, \beta, \gamma, \dots$ will implicitly run from 1 to N, while *Latin* indices $i, j, k, \dots$ will take the values 1, 2, 3.

[3]It is easy to show that if three points of a rigid body are collinear at some instant they must remain collinear at any other time. Indeed, three non-collinear points $\mathbf{r}_\alpha$, $\mathbf{r}_\beta$, and $\mathbf{r}_\gamma$ in the body determine a non-degenerate triangle whose sides have constant lengths $l_{\alpha\beta}$, $l_{\alpha\gamma}$, and $l_{\beta\gamma}$. By a well-known result in plane geometry, the three angles of this triangle are also fixed. In particular, the triangle determined by the three points $\mathbf{r}_\alpha$, $\mathbf{r}_\beta$, and $\mathbf{r}_\gamma$ remains non-degenerate at all times, so that these points are never collinear.

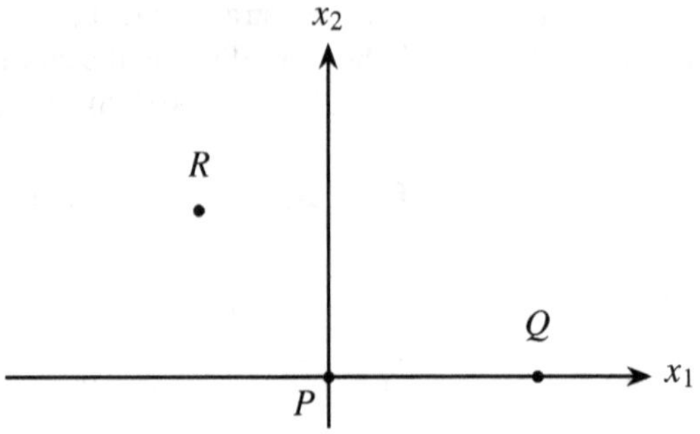

Figure 7.1. Moving reference frame axes (the x_3 axis points out of the page toward the viewer).

Subtracting the second equation from the first we obtain

$$2ax = r_1^2 - r_2^2 + a^2 \quad \Longrightarrow \quad x = \frac{a}{2} + \frac{r_1^2 - r_2^2}{2a},$$

so that the x coordinate is constant. Likewise, subtracting the third equation from the second we deduce that

$$y = \frac{c}{2} + \frac{1}{2c}\left[2(a-b)x + r_2^2 - r_3^2 + b^2 - a^2)\right]$$

is also constant. Finally, from the above formulas for x and y, along with any of the three equations (7.3), it follows that z^2 is constant. By continuity, this implies that z is also constant. ■

Exercise 7.1. Show that the points P, Q, and R have constant coordinates in the frame constructed in the previous proof.

Solution. Obviously $P = (0,0,0)$ and $Q = (a,0,0)$ (with $a = |\overrightarrow{PQ}|$) have constant coordinates in this frame. If $(x,y,0)$ are the coordinates of the point R, calling $l_1 = |\overrightarrow{PR}|$ and $l_2 = |\overrightarrow{QR}|$ we have

$$x^2 + y^2 = l_1^2, \qquad (x-a)^2 + y^2 = l_2^2.$$

Subtracting the second equation from the first we obtain

$$x = \frac{a}{2} + \frac{1}{2a}(l_1^2 - l_2^2).$$

Substituting into the first equation yields y^2, which in turn determines y since $y > 0$ by construction.

• Obviously, there is an infinite number of body axes, obtained from the frame we have just constructed by translating its origin to any point fixed in the body and applying a *constant* rotation to its axes.

More precisely, we shall say that a point is **fixed in the body** if its coordinates are constant (i.e., time-independent) in a frame of body axes. Such a point is, for instance, the body's *center of mass*. Indeed, let O and O' respectively denote the origin of the frame of body axes and of the inertial frame, and let $\mathbf{a} = \overrightarrow{OO'}$. The position vector of the body's CM in the inertial frame is by definition the vector

$$\overrightarrow{O'C} = \frac{1}{M} \sum_{\alpha} m_\alpha \mathbf{r}'_\alpha ,$$

where $M = \sum_\alpha m_\alpha$ is the body's total mass and $\mathbf{r}'_\alpha$ is the position vector of the α-th particle relative to the inertial frame. Moreover, if $\mathbf{r}$ and $\mathbf{r}'$ respectively denote the position vectors of a point in space in the frame of body axes and in the inertial frame we have

$$\mathbf{r} = \mathbf{a} + \mathbf{r}' .$$

In particular, the position vector of the body's CM in the frame of body axes is given by

$$\overrightarrow{OC} = \mathbf{a} + \frac{1}{M} \sum_{\alpha} m_\alpha \mathbf{r}'_\alpha = \mathbf{a} + \frac{1}{M} \sum_{\alpha} m_\alpha (\mathbf{r}_\alpha - \mathbf{a}) = \mathbf{a} + \frac{1}{M} \sum_{\alpha} m_\alpha \mathbf{r}_\alpha - \frac{\mathbf{a}}{M} \sum_{\alpha} m_\alpha$$
$$= \frac{1}{M} \sum_{\alpha} m_\alpha \mathbf{r}_\alpha ,$$

where $\mathbf{r}_\alpha$ denotes the position vector of the α-th particle in this frame. Since the components of the position vectors $\mathbf{r}_\alpha$ are constant by construction in the frame of body axes, so are the coordinates of the CM in this frame. Note, finally, that from the previous argument it also follows that the position vector of the CM in the frame of body axes is still given by the usual formula $(1/M) \sum_\alpha m_\alpha \mathbf{r}_\alpha$.

A *generic* rigid body has 6 degrees of freedom.

To see this, note that in order to determine the coordinates of any particle in the rigid body at an arbitrary time t, given its coordinates at the initial instant t_0, it suffices to know the position of the origin P of a set of body axes $\{\mathbf{e}_i\}_{i=1}^3$ together with the rotation matrix $O(t)$ relating these axes to those of an inertial (fixed) reference frame $\{\mathbf{e}'_i\}_{i=1}^3$ (cf. Eq. (6.1)). Indeed, all the vectors $\mathbf{r}_\alpha$ are known at $t = t_0$, and since they are *constant* the coordinates $\mathbf{r}'_\alpha(t)$ of the position vector of particle α in the fixed frame at an arbitrary time t can be computed through the formula[4]

$$\mathbf{r}'_\alpha(t) = \overrightarrow{O'P} + O(t)\mathbf{r}_\alpha .$$

[4]In the formula that follows $\mathbf{r}_\alpha$ denotes the vector whose components are the coordinates of the position vector of particle α in the body frame. As explained in Section 2.3.3, to obtain the coordinates of the *same* vector in the fixed frame we have to multiply $\mathbf{r}_\alpha$ by the rotation matrix $O(t)$.

The vector $\overrightarrow{O'P}$ (i.e., the position vector of the point P relative to the fixed frame) is determined by three parameters (for instance, its Cartesian coordinates), while the matrix $O(t) \in \mathrm{SO}(3)$ can be specified by another three independent parameters (for example, the two spherical coordinates of the rotation axis $\mathbf{n}$ and the rotation angle $\theta \in [0, \pi]$). In practice, the most widespread way of determining the rotation matrix $O(t)$ is through three angles, the so-called *Euler angles*, that we shall define in Section 7.6.

Exercise 7.2. How many degrees of freedom has a *linear rigid body* (i.e., a rigid body all of whose particles are collinear)?

Solution. The coordinates of every particle in the linear body can be determined at any time if we know the position of one of its particles and a unit vector in the body's direction (i.e., the line joining this particle with a second particle in the body). The position of the particle is determined by its three Cartesian coordinates, while the unit vector is specified by (for instance) its polar and azimuthal angles (θ, φ) relative to a fixed line. Thus a linear body has only $3 + 2 = 5$ degrees of freedom.

The above considerations also apply to the *continuous version* of a rigid body, which consists of a continuous mass distribution of density $\rho(\mathbf{r})$ over a volume $\Omega \subset \mathbb{R}^3$ whose *shape* does not change with time. In other words, the location and shape of the body at any instant t is obtained by applying a *rigid motion* (an overall translation followed by a rotation, or vice versa) to the set Ω. More precisely, at any given instant t the mass distribution is concentrated on the set $\Omega(t) \subset \mathbb{R}^3$ given by

$$\Omega(t) = O(t)\Omega + \mathbf{X}(t),$$

for some vector $\mathbf{X}(t) \in \mathbb{R}^3$ and rotation matrix $O(t) \in \mathrm{SO}(3)$. The state of the system is thus completely determined by the three components of the vector $\mathbf{X}(t)$ together with the three parameters needed to specify the rotation matrix $O(t)$. As in the discrete case, this implies that *a continuous rigid body has* 6 *degrees of freedom.* Of course, we can also have continuous bodies whose mass density is concentrated on a surface, or even a curve, in $\mathbb{R}^3$. Note, finally, that the *center of mass* of a continuous rigid body is naturally defined by

$$\boxed{\mathbf{R} = \frac{1}{M}\int_\Omega \rho(\mathbf{r})\mathbf{r}\,\mathrm{d}^3\mathbf{r},}$$

where

$$\boxed{M = \int_\Omega \rho(\mathbf{r})\,\mathrm{d}^3\mathbf{r}}$$

is the body's total mass. In particular, if the mass density ρ is constant then

$$\boxed{\mathbf{R} = \frac{1}{V}\int_\Omega \mathbf{r}\,\mathrm{d}^3\mathbf{r},}$$

where V is the body's volume. Similar considerations apply to a continuous rigid body whose mass is distributed over a surface or on a curve in $\mathbb{R}^3$, replacing the volume element by the surface or line element and the volume density by the surface or line density.

7.2 ANGULAR MOMENTUM AND KINETIC ENERGY

We shall next compute the angular momentum of the rigid body with respect to an *inertial frame*, which we shall often call for short (as in the previous chapter) the *fixed* (or *space*) *frame*.

> From now on, unless otherwise stated we shall take the rigid body's *center of mass* as the origin of the set of body axes.

Let us denote, as usual, by $\mathbf{R}(t)$ the vector joining the origin O' of the fixed frame with the center of mass O (i.e., the origin of the set of body axes). If $\mathbf{r}_\alpha$ and $\mathbf{r}'_\alpha$ are respectively the position vectors of the α-th particle with respect to the body axes (which plays the role of the *moving frame* in the last chapter) and the fixed ones we have

$$\boxed{\mathbf{r}'_\alpha = \mathbf{R} + \mathbf{r}_\alpha\,.}$$

In this case $\dot{\mathbf{r}}_\alpha = 0$, since the particles which make up the rigid body are at rest with respect to the frame of body axes. Hence Eq. (6.19) reduces to

$$\boxed{\mathbf{v}'_\alpha = \mathbf{V} + \boldsymbol{\omega} \times \mathbf{r}_\alpha\,,} \tag{7.4}$$

where $\boldsymbol{\omega}$ denotes the instantaneous angular velocity of the set of body axes with respect to the fixed ones and

$$\boxed{\mathbf{v}'_\alpha := \left(\frac{\mathrm{d}\mathbf{r}'_\alpha}{\mathrm{d}t}\right)_{\mathrm{f}}\,, \qquad \mathbf{V} := \left(\frac{\mathrm{d}\mathbf{R}}{\mathrm{d}t}\right)_{\mathrm{f}}\,.}$$

- By Eq. (7.4), the infinitesimal change of the position vector of particle α from a time t to a time $t + \mathrm{d}t$ is given by

$$\mathrm{d}\mathbf{r}'_\alpha = \mathbf{v}'_\alpha\,\mathrm{d}t = \mathbf{V}\,\mathrm{d}t + \boldsymbol{\omega}\,\mathrm{d}t \times \mathbf{r}_\alpha = \mathrm{d}\mathbf{R} + \boldsymbol{\omega}\,\mathrm{d}t \times \mathbf{r}_\alpha\,.$$

Hence:

> The instantaneous motion of the body, as seen from the fixed reference frame, can be viewed as an infinitesimal translation followed by an infinitesimal rotation about the axis parallel to $\boldsymbol{\omega}$ passing through the CM by an angle $\omega(t)\,\mathrm{d}t$.

The previous assertion can also be proved directly, taking into account that the rate of change of the position vector with respect to the CM of any particle α, as measured in the fixed frame, is given by

$$\left(\frac{\mathrm{d}\mathbf{r}_\alpha}{\mathrm{d}t}\right)_{\mathrm{f}} = \dot{\mathbf{r}}_\alpha + \omega(t) \times \mathbf{r}_\alpha = \omega(t) \times \mathbf{r}_\alpha \,. \tag{7.5}$$

Hence from the point of view of the fixed frame the position vectors of all particles in the rigid body relative to the CM rotate *instantaneously* with the *same* angular velocity $\omega(t)$ about an axis parallel to the vector $\omega(t)/\omega(t)$ passing through the CM.

The rigid body's linear momentum with respect to the inertial frame is given by

$$\boxed{\mathbf{P} = \sum_\alpha m_\alpha \mathbf{v}'_\alpha = \sum_\alpha m_\alpha \mathbf{V} + \omega \times \sum_\alpha m_\alpha \mathbf{r}_\alpha = M\mathbf{V} \,,} \tag{7.6}$$

where

$$M = \sum_\alpha m_\alpha$$

is the body's total mass and we have used the identity

$$\sum_\alpha m_\alpha \mathbf{r}_\alpha = 0 \tag{7.7}$$

(since the LHS is proportional to the position vector of the CM with respect to the CM itself). As expected, Eq. (7.6) coincides with Eq. (2.82) in Section 2.6.2.

Let us next find the rigid body's angular momentum with respect to the origin O' of the set of fixed axes (as measured in this frame), defined by

$$\boxed{\mathbf{L} = \sum_\alpha m_\alpha \mathbf{r}'_\alpha \times \mathbf{v}'_\alpha \,.}$$

Using Eq. (7.4) for $\mathbf{v}'_\alpha$ and the identity (7.7) we easily obtain

$$\begin{aligned} \mathbf{L} &= \sum_\alpha m_\alpha (\mathbf{R} + \mathbf{r}_\alpha) \times (\mathbf{V} + \omega \times \mathbf{r}_\alpha) \\ &= M\mathbf{R} \times \mathbf{V} + \mathbf{R} \times \Big(\omega \times \sum_\alpha m_\alpha \mathbf{r}_\alpha\Big) + \Big(\sum_\alpha m_\alpha \mathbf{r}_\alpha\Big) \times \mathbf{V} + \sum_\alpha m_\alpha \mathbf{r}_\alpha \times (\omega \times \mathbf{r}_\alpha) \\ &= \boxed{M\mathbf{R} \times \mathbf{V} + \sum_\alpha m_\alpha \mathbf{r}_\alpha \times (\omega \times \mathbf{r}_\alpha)} \,. \end{aligned} \tag{7.8}$$

Note that this expression coincides with Eq. (2.87) from Section 2.6.2, by virtue of Eq. (7.5). The first term in Eq. (7.8) is simply the angular momentum of a particle located at the CM with mass equal to the body's total mass. To interpret the second term, note first that the angular momentum of the rigid body with respect to any point P, as measured in the fixed frame S', is by definition

$$\mathbf{L}_P := \sum_\alpha m_\alpha \overrightarrow{PP_\alpha} \times \mathbf{v}'_\alpha,$$

where P_α is the position of the particle α. In particular, taking P as the CM we have

$$\mathbf{L}_{\mathrm{CM}} = \sum_\alpha m_\alpha \mathbf{r}_\alpha \times \mathbf{v}'_\alpha = \sum_\alpha m_\alpha \mathbf{r}_\alpha \times (\omega \times \mathbf{r}_\alpha)\,, \tag{7.9}$$

where we have used again Eq. (7.4) for $\mathbf{v}'_\alpha$ and the identity (7.7). By Eq. (7.8) we then have

$$\mathbf{L} = M\mathbf{R} \times \mathbf{V} + \mathbf{L}_{\mathrm{CM}}\,. \tag{7.10}$$

It is important to note that, although $\mathbf{L}_{\mathrm{CM}}$ is the rigid body's angular momentum with respect to the CM, *it is computed in the space frame* S', since the particle's velocities $\mathbf{v}'_\alpha$ in Eq. (7.9) are measured in this frame.

Proceeding in the same way, we can simplify the body's kinetic energy (with respect to the inertial frame)

$$T = \frac{1}{2}\sum_\alpha m_\alpha \mathbf{v}'^2_\alpha\,.$$

Indeed, using again Eq. (7.4) for $\mathbf{v}'_\alpha$ and the identity (7.7) we obtain the expression

$$T = \frac{1}{2}\sum_\alpha m_\alpha \left(\mathbf{V} + \omega \times \mathbf{r}_\alpha\right)^2 = \frac{1}{2} M\mathbf{V}^2 + \frac{1}{2}\sum_\alpha m_\alpha \left(\omega \times \mathbf{r}_\alpha\right)^2, \tag{7.11}$$

which again coincides with Eq. (2.91) of Section 2.6.2 on account of Eq. (7.5). We can thus write

$$T = \frac{1}{2} M\mathbf{V}^2 + T_{\mathrm{rot}}\,, \qquad T_{\mathrm{rot}} = \frac{1}{2}\sum_\alpha m_\alpha \left(\omega \times \mathbf{r}_\alpha\right)^2, \tag{7.12}$$

where the first term in T is the CM's translational energy while the second one is the body's **rotational energy** about its CM, since

$$\frac{1}{2}\sum_\alpha m_\alpha \left(\omega \times \mathbf{r}_\alpha\right)^2 = \frac{1}{2}\sum_\alpha m_\alpha \left(\frac{\mathrm{d}\mathbf{r}_\alpha}{\mathrm{d}t}\right)^2_{\mathrm{f}}\,.$$

Using the identities

$$\mathbf{a} \times (\mathbf{b} \times \mathbf{c}) = (\mathbf{a}\cdot\mathbf{c})\mathbf{b} - (\mathbf{a}\cdot\mathbf{b})\mathbf{c}\,, \qquad (\mathbf{a} \times \mathbf{b})^2 = \mathbf{a}^2\mathbf{b}^2 - (\mathbf{a}\cdot\mathbf{b})^2$$

Eqs. (7.9) and (7.12) can be recast in the alternative form

$$\mathbf{L}_{\mathrm{CM}} = \sum_\alpha m_\alpha \big[r_\alpha^2 \,\omega - (\omega \cdot \mathbf{r}_\alpha)\mathbf{r}_\alpha \big] , \quad T_{\mathrm{rot}} = \frac{1}{2} \sum_\alpha m_\alpha \big[\omega^2 r_\alpha^2 - (\omega \cdot \mathbf{r}_\alpha)^2 \big] , \tag{7.13}$$

which yields the important identity

$$T_{\mathrm{rot}} = \frac{1}{2}\, \omega \cdot \mathbf{L}_{\mathrm{CM}} . \tag{7.14}$$

Note that in all of the previous formulas the vectors $\mathbf{L}_{\mathrm{CM}}$ and ω, and hence the rotational energy T_{rot}, are in general functions of time.

7.3 INERTIA TENSOR

7.3.1 Definition and elementary properties

The expressions obtained in the previous section for the angular momentum with respect to the CM and the rotational energy of a rigid body can be greatly simplified with the help of the so-called *inertia tensor*. Since the rotational energy is expressed in terms of $\mathbf{L}_{\mathrm{CM}}$ through Eq. (7.14), we can restrict ourselves to the angular momentum. The key observation is that Eq. (7.13) clearly indicates that, although in general $\mathbf{L}_{\mathrm{CM}}$ is *not* parallel to the angular velocity ω, it is a *linear function* thereof. In other words, we can write

$$\mathbf{L}_{\mathrm{CM}} = I\omega , \tag{7.15}$$

where $I : \mathbb{R}^3 \to \mathbb{R}^3$ is a linear map, which can be represented by a 3×3 matrix whose entries we shall now compute. To this end it suffices to note that, if $x_{\alpha i}$ and ω_i (with $i = 1, 2, 3$) respectively denote the i-th components of the vectors $\mathbf{r}_\alpha$ and ω in *any* basis, the i-th component of $\mathbf{L}_{\mathrm{CM}}$ (in the same basis) is given by

$$\begin{aligned}
L_{\mathrm{CM},i} &= \omega_i \sum_\alpha m_\alpha r_\alpha^2 - \sum_\alpha m_\alpha x_{\alpha i} \sum_j \omega_j x_{\alpha j} \\
&= \omega_i \sum_\alpha m_\alpha r_\alpha^2 - \sum_j \omega_j \sum_\alpha m_\alpha x_{\alpha i} x_{\alpha j} \\
&= \sum_j \omega_j \delta_{ij} \sum_\alpha m_\alpha r_\alpha^2 - \sum_j \omega_j \sum_\alpha m_\alpha x_{\alpha i} x_{\alpha j} \\
&= \sum_j \omega_j \sum_\alpha m_\alpha \big(\delta_{ij} r_\alpha^2 - x_{\alpha i} x_{\alpha j} \big) .
\end{aligned}$$

We thus have

$$L_{\mathrm{CM},i} = \sum_j I_{ij}\omega_j\,, \tag{7.16a}$$

where the matrix element I_{ij} is given by

$$I_{ij} = \sum_\alpha m_\alpha\big(\delta_{ij} r_\alpha^2 - x_{\alpha i} x_{\alpha j}\big)\,. \tag{7.16b}$$

The linear map I with matrix elements given by Eq. (7.16b) is known as the rigid body's **inertia tensor**[5]. It is important to note that, although both $\mathbf{L}_{\mathrm{CM}}$ and ω in general depend on t, the matrix elements (7.16b) of the inertia tensor are *constant in the body frame*, since the Cartesian coordinates $x_{\alpha i}$ ($i = 1, 2, 3$) of the body's particles in a frame of body axes do not depend on time. In other words:

In a frame of body axes, the inertia tensor is a *constant matrix* characteristic of the rigid body.

Unless otherwise stated, we shall always suppose that the matrix elements of the inertia tensor are computed in a frame of body axes in what follows.

- From Eq. (7.16b) it immediately follows that the inertia tensor is *symmetric*:

$$I_{ij} = I_{ji}\,, \qquad i, j = 1, 2, 3\,.$$

The diagonal matrix elements of the inertia tensor are given by

$$I_{ii} = \sum_\alpha m_\alpha\big(x_{\alpha j}^2 + x_{\alpha k}^2\big)\,, \qquad i = 1, 2, 3\,,$$

with (i, j, k) different from each other. In other words,

$$I_{ii} = \sum_\alpha m_\alpha d_{\alpha i}^2\,, \tag{7.17}$$

where $d_{\alpha i}$ is the distance of the α-th particle to the i-th axis. Hence the matrix element I_{ii} is the so-called **moment of inertia** of the body with respect to the axis $\mathbf{e}_i$. Likewise, the off-diagonal matrix elements of I

$$I_{ij} = -\sum_\alpha m_\alpha x_{\alpha i} x_{\alpha j}\,, \qquad 1 \leqslant i \neq j \leqslant 3\,,$$

[5]The name "tensor" is due to the fact that in general a linear map is a tensor with one covariant and one contravariant indices. Note, however, that in *orthogonal* Cartesian coordinates there is no distinction between covariant and contravariant indices.

are the negatives of the body's **products of inertia**. For a continuous rigid body Ω with mass density $\rho(\mathbf{r})$, the previous expressions must be replaced by their obvious continuous analogues

$$I_{ij} = \int_\Omega \rho(\mathbf{r})\left(\delta_{ij}r^2 - x_i x_j\right) \mathrm{d}^3\mathbf{r}\,, \qquad i,j = 1,2,3\,, \tag{7.18}$$

or, in more detail,

$$I_{ii} = \int_\Omega \rho(\mathbf{r})\left(x_j^2 + x_k^2\right) \mathrm{d}^3\mathbf{r}\,, \qquad i = 1,2,3\,,$$

(with (i,j,k) different from each other) and

$$I_{ij} = -\int_\Omega \rho(\mathbf{r}) x_i x_j \,\mathrm{d}^3\mathbf{r}\,, \qquad 1 \leqslant i \neq j \leqslant 3\,.$$

Analogous expressions are obtained for a continuous body whose mass is distributed on a surface or along a curve, replacing the volume element $\mathrm{d}^3\mathbf{r}$ with the surface element $\mathrm{d}S$ or the line element $\mathrm{d}s$.

• From the identity (7.14) it follows that the body's rotational energy can be expressed in terms of its angular velocity ω and the inertia tensor I through the formula

$$T_{\mathrm{rot}} = \frac{1}{2}\,\omega \cdot (I\omega)\,. \tag{7.19}$$

Note also that the previous expression can be written using matrix notation as

$$T_{\mathrm{rot}} = \frac{1}{2}\,\omega^{\mathsf{T}}\mathsf{I}\,\omega\,, \tag{7.20}$$

if we interpret ω as the *column* vector $\begin{pmatrix}\omega_1\\ \omega_2\\ \omega_3\end{pmatrix}$ and I (sans serif I) denotes the real 3×3 matrix with elements I_{ij}. In other words, T_{rot} is a *quadratic form* in the components of ω, whose matrix elements are the matrix elements (7.16b) of the inertia tensor. Since $T_{\mathrm{rot}} \geqslant 0$ for all ω, this quadratic form—or, equivalently, the inertia tensor I—is *positive semidefinite*. In fact:

The inertia tensor is *positive definite* if and only if the body is generic.

Indeed, if I were not positive definite, by Eq. (7.12) there would exist a nonzero vector ω such that

$$2T_{\mathrm{rot}} = \sum_\alpha m_\alpha (\omega \times \mathbf{r}_\alpha)^2 = 0\,.$$

Since all the terms in the sum are nonnegative, the last equality is only possible if $\boldsymbol{\omega} \times \mathbf{r}_\alpha$ vanishes for all $\alpha = 1, \dots, N$, i.e., if all the particles lie on the line parallel to $\boldsymbol{\omega}$ passing through the CM.

- From Eq. (7.19) it follows that a rigid body's rotational energy can also be expressed as

$$T_{\text{rot}} = \frac{1}{2}\,\omega^2\,\mathbf{n}\cdot I\mathbf{n}\,,$$

where $\mathbf{n} = \boldsymbol{\omega}/\omega$ is the direction of the instantaneous axis of rotation of the body axes. Since $I_{ii} = \mathbf{e}_i \cdot I\mathbf{e}_i$, by Eq. (7.17) we can also write

$$\mathbf{n}\cdot I\mathbf{n} = \sum_\alpha m_\alpha d_\alpha(\mathbf{n})^2 =: I_{\mathbf{n}},$$

where $d_\alpha(\mathbf{n})$ and $I_{\mathbf{n}}$ respectively denote the distance of particle α to the line through the CM parallel to the vector $\mathbf{n}$ and the body's moment of inertia with respect to this axis. It follows that the rotational energy T_{rot} can be expressed in terms of $I_{\mathbf{n}}$ as

$$\boxed{T_{\text{rot}} = \frac{1}{2}\,I_{\mathbf{n}}\,\omega^2\,.}$$

7.3.2 Steiner's theorem

We shall next determine how the inertia tensor changes when we compute it with respect to a point P *fixed in the body* that does not necessarily coincide with the center of mass C. If we denote by $\widetilde{\mathbf{r}}_\alpha$ the position vector of the α-th particle with respect to the point P, the inertia tensor I_P with respect to P is defined by

$$\boxed{(I_P)_{ij} = \sum_\alpha m_\alpha\big(\delta_{ij}\widetilde{\mathbf{r}}_\alpha^2 - \widetilde{x}_{\alpha i}\widetilde{x}_{\alpha j}\big)\,.} \tag{7.21}$$

Taking into account that

$$\widetilde{\mathbf{r}}_\alpha = \mathbf{r}_\alpha - \mathbf{a}\,, \qquad \mathbf{a} := \overrightarrow{CP}\,,$$

we obtain

$$\begin{aligned}(I_P)_{ij} &= \sum_\alpha m_\alpha\big[\delta_{ij}(\mathbf{r}_\alpha - \mathbf{a})^2 - (x_{\alpha i} - a_i)(x_{\alpha j} - a_j)\big] = I_{ij} + M(\mathbf{a}^2\delta_{ij} - a_i a_j)\\ &\quad - 2\delta_{ij}\mathbf{a}\cdot\sum_\alpha m_\alpha\mathbf{r}_\alpha + a_i\sum_\alpha m_\alpha x_{\alpha j} + a_j\sum_\alpha m_\alpha x_{\alpha i}\,.\end{aligned}$$

The last three terms vanish on account of the identity (7.7), so that we finally have

$$\boxed{(I_P)_{ij} = I_{ij} + M(\mathbf{a}^2\delta_{ij} - a_i a_j)\,.} \tag{7.22}$$

The previous formula is known as *Steiner's theorem*.

• The last term in Eq. (7.22) is nothing but the inertia tensor of a particle of mass M located at the point P. Note also that this equation is invariant under $\mathbf{a} \mapsto -\mathbf{a}$, so it remains valid if we set $\mathbf{a} = \overrightarrow{PC}$.

It often happens that there is a point P *fixed in the body* which is *also fixed in some inertial frame*; for example, if the body is rotating around a *fixed* axis we can choose as P any point on the axis of rotation. When this is the case, it is possible—and, in fact, usually advantageous—to *take P as the origin O' of the inertial frame.* It then follows that the vector $\mathbf{R} = \overrightarrow{O'O} = \overrightarrow{PC}$ is *constant in the frame of body axes*, since its endpoints are both fixed in the body. Thus in this case $\dot{\mathbf{R}} = 0$, and by Eq. (7.4) the velocity of the CM in the inertial frame can be simply expressed as

$$\mathbf{V} = \omega \times \mathbf{R},$$

so that

$$\mathbf{v}'_\alpha = \omega \times \mathbf{R} + \omega \times \mathbf{r}_\alpha = \omega \times (\mathbf{r}_\alpha + \mathbf{R}) = \omega \times \mathbf{r}'_\alpha .$$

The angular momentum $\mathbf{L}$ with respect to $O' = P$ is then given by

$$\mathbf{L} = \sum_\alpha m_\alpha \mathbf{r}'_\alpha \times \mathbf{v}'_\alpha = \sum_\alpha m_\alpha \mathbf{r}'_\alpha \times (\omega \times \mathbf{r}'_\alpha),$$

i.e., is obtained replacing $\mathbf{r}_\alpha$ by $\mathbf{r}'_\alpha$ in Eq. (7.9) for $\mathbf{L}_{\mathrm{CM}}$. In other words, in this case the body's *total* angular momentum is given by

$$\boxed{\mathbf{L} = I_P \omega .} \tag{7.23}$$

Note that I_P is still a *constant* matrix characteristic of the rigid body considered (and of the orientation of the axes in the body frame), since in this case $\mathbf{r}'_\alpha = \mathbf{R} + \mathbf{r}_\alpha$ is still a constant vector in the body frame. Likewise,

$$\boxed{T = \frac{1}{2}\sum_\alpha m_\alpha \mathbf{v}'^2_\alpha = \frac{1}{2}\sum_\alpha m_\alpha (\omega \times \mathbf{r}'_\alpha)^2 = \frac{1}{2}\omega \cdot I_P \omega .} \tag{7.24}$$

The inertia tensor I_P can be computed from I applying Steiner's theorem (7.22), taking into account that in this case $\mathbf{a} = \overrightarrow{CP} = -\overrightarrow{PC} = -\overrightarrow{O'O} = -\mathbf{R}$:

$$\boxed{(I_P)_{ij} = I_{ij} + M(\mathbf{R}^2 \delta_{ij} - X_i X_j),}$$

where X_i $(i = 1, 2, 3)$ are the components of the vector $\mathbf{R}$ in the body frame. Note that the last term in Eq. (7.22) is the inertia tensor with respect to O' of a particle of mass M located at the CM.

Note. From now on we shall usually *omit the subindex* when the point with respect to which the inertia tensor is computed is clear from the context.

7.3.3 Principal axes of inertia

Let us next see how the components of the inertia tensor (7.16b) with respect to the CM change when we perform a *constant* rotation of the frame of body axes. More precisely, let

$$\boxed{\widetilde{\mathbf{e}}_j = \sum_j a_{ij}\mathbf{e}_i\,, \qquad i = 1,2,3\,,} \tag{7.25}$$

be a second positively oriented frame fixed in the body. Then the *change of basis matrix*

$$\boxed{A := (a_{ij})_{1\leqslant i,j\leqslant 3}}$$

is a *constant* proper orthogonal matrix (i.e., $A \in \mathrm{SO}(3)$ is time-independent). As is well known, the coordinates (or, in general, the components of any vector) in both frames are related by the dual equation

$$\boxed{x_i = \sum_j a_{ij}\widetilde{x}_j\,;}$$

indeed,

$$\sum_j \widetilde{x}_j\widetilde{\mathbf{e}}_j = \sum_j \widetilde{x}_j \sum_i a_{ij}\mathbf{e}_i = \sum_i \left(\sum_j a_{ij}\widetilde{x}_j\right)\mathbf{e}_i \implies x_i = \sum_j a_{ij}\widetilde{x}_j\,.$$

Denoting by x (sans serif x) the column vector whose components are the coordinates x_i, and similarly $\widetilde{\mathsf{x}} = (\widetilde{x}_1\,\widetilde{x}_2\,\widetilde{x}_3)^{\mathsf{T}}$, we can rewrite the previous relation in matrix form as

$$\boxed{\mathsf{x} = A\,\widetilde{\mathsf{x}}\,.}$$

Likewise, if $\omega = (\omega_1\,\omega_2\,\omega_3)^{\mathsf{T}}$, $\mathsf{L}_{\mathrm{CM}} = (L_{\mathrm{CM},1}\,L_{\mathrm{CM},2}\,L_{\mathrm{CM},3})^{\mathsf{T}}$, with similar definitions for $\widetilde{\omega}$ and $\widetilde{\mathsf{L}}_{\mathrm{CM}}$, we have

$$\omega = A\,\widetilde{\omega}, \qquad \mathsf{L}_{\mathrm{CM}} = A\,\widetilde{\mathsf{L}}_{\mathrm{CM}}.$$

Using this notation, and denoting by I the matrix of the inertia tensor with respect to the original set of axes $\{\mathbf{e}_i : i = 1,2,3\}$, we then obtain

$$\mathsf{L}_{\mathrm{CM}} = \mathsf{I}\,\omega = \mathsf{I}\,A\,\widetilde{\omega} = A\,\widetilde{\mathsf{L}}_{\mathrm{CM}} \implies \widetilde{\mathsf{L}}_{\mathrm{CM}} = A^{-1}\mathsf{I}\,A\,\widetilde{\omega} =: \widetilde{\mathsf{I}}\,\widetilde{\omega}\,.$$

Hence the matrix of the inertia tensor in the new set of body axes is given by

$$\boxed{\widetilde{\mathsf{I}} = A^{-1}\mathsf{I}\,A = A^{\mathsf{T}}\mathsf{I}\,A\,,}$$

as A is orthogonal. Note that in the new set of body axes the body's rotational energy can be expressed as

$$T_{\text{rot}} = \frac{1}{2}\,\omega^{\mathsf{T}}\mathsf{I}\,\omega = \frac{1}{2}\,\widetilde{\omega}^{\mathsf{T}}A^{\mathsf{T}}\mathsf{I}\,A\widetilde{\omega} = \frac{1}{2}\,\widetilde{\omega}^{\mathsf{T}}\widetilde{\mathsf{I}}\,\widetilde{\omega}\,,$$

which agrees with the expression just derived for the matrix $\widetilde{\mathsf{I}}$.

It is well known that *a real symmetric matrix can be diagonalized by means of a proper orthogonal transformation*[6]. In other words, it is always possible to find a matrix $A \in \mathrm{SO}(3)$ such that in the new set of body axes (7.25) we have

$$\widetilde{I}_{ij} = \delta_{ij} I_i\,, \qquad 1 \leqslant i, j \leqslant 3\,,$$

where I_1, I_2, and I_3 are the three *eigenvalues* of the inertia tensor I. Note that A is a *constant* (i.e., time-independent) matrix, since the matrix elements of the inertia tensor are also constant. If the vectors $\widetilde{\mathbf{e}}_j$ are defined by (7.25), where $A \in \mathrm{SO}(3)$ is the proper orthogonal matrix diagonalizing I, then

$$\boxed{I\widetilde{\mathbf{e}}_i = I_i\widetilde{\mathbf{e}}_i\,, \qquad i = 1, 2, 3\,.} \tag{7.26}$$

In other words, the vector $\widetilde{\mathbf{e}}_i$, whose components with respect to the basis $\{\mathbf{e}_1, \mathbf{e}_2, \mathbf{e}_3\}$ are the i-th column of the change of basis matrix A, is an *eigenvector* of the inertia tensor I with eigenvalue I_i. Since A is a proper orthogonal matrix, the vectors $\widetilde{\mathbf{e}}_i$ $(i = 1, 2, 3)$ are a positively oriented orthonormal basis of $\mathbb{R}^3$. These vectors are *fixed in the body*, since their components with respect to the original set of body axes $\{\mathbf{e}_1, \mathbf{e}_2, \mathbf{e}_3\}$ are the matrix elements of the *constant* matrix A. Hence:

It is always possible to find a set of body axes whose unit vectors $\widetilde{\mathbf{e}}_i$ are all *eigenvectors* of the linear map I. In this set of body axes the inertia tensor is represented by the diagonal matrix

$$I = \begin{pmatrix} I_1 & 0 & 0 \\ 0 & I_2 & 0 \\ 0 & 0 & I_3 \end{pmatrix},$$

where I_i is the eigenvalue of the inertia tensor I corresponding to the eigenvector $\widetilde{\mathbf{e}}_i$.

The vectors $\widetilde{\mathbf{e}}_i$ satisfying the relations (7.26) are known as the body's **principal axes of inertia**, and their corresponding eigenvalues I_i are called its **principal moments of inertia**. As is well known, the eigenvalues of the matrix (I_{ij}), i.e., the principal moments of inertia, are the roots of the **secular equation**

$$\boxed{\det\left(I_{ij} - \lambda\delta_{ij}\right) = 0\,.}$$

[6]This is essentially due to the following elementary facts: i) every real symmetric matrix is diagonalizable; ii) its eigenvalues are all real, and iii) two eigenvectors of a real symmetric matrix corresponding to different eigenvalues are orthogonal. From these three facts it easily follows that there exists a (positively oriented) orthonormal basis of eigenvectors of any real orthogonal matrix.

Note that the principal axes of inertia, i.e., the directions of the eigenvectors of the matrix (I_{ij}), are not uniquely determined (up to a sign) unless all the eigenvalues of the inertia tensor are distinct (i.e., they are *simple* roots of the secular equation).

• If the set of body axes $\{\mathbf{e}_1, \mathbf{e}_2, \mathbf{e}_3\}$ is a set of principal axes of inertia, Eqs. (7.15) and (7.19) reduce to

$$\boxed{\mathbf{L}_{\mathrm{CM}} = \sum_i I_i \omega_i \mathbf{e}_i\,, \qquad T_{\mathrm{rot}} = \frac{1}{2}\sum_i I_i \omega_i^2\,.} \tag{7.27}$$

In particular, if the body rotates around its i-th principal axis of inertia we have

$$\boxed{\mathbf{L}_{\mathrm{CM}} = I_i \boldsymbol{\omega}\,, \qquad T_{\mathrm{rot}} = \frac{1}{2} I_i \omega^2\,.}$$

> If the origin of the fixed frame is also a point P fixed in the body, expressions analogous to the previous ones are valid for $\mathbf{L}$ and T replacing I by the inertia tensor I_P with respect to the point P.

• Rigid bodies can be classified into the following three categories, depending on the multiplicity of the eigenvalues of their inertia tensor:

1) *Asymmetric tops:* $I_i \neq I_j$ for all $i \neq j$
2) *Axially symmetric tops:* $I_i = I_j \neq I_k$ (with (i, j, k) distinct)
3) *Spherically symmetric tops*: $I_1 = I_2 = I_3$.

7.3.4 Symmetries

We shall next examine how the *symmetries* of a rigid body Ω of mass density ρ result in simplifications of its inertia tensor.

1) *If Ω and ρ are invariant under the reflection $x_i \mapsto -x_i$, then*

$$\boxed{I_{ij} = 0\,, \qquad \forall j \neq i\,.}$$

Indeed, suppose that (for instance) Ω is invariant under reflection of the x_1 coordinate and $\rho(-x_1, x_2, x_3) = \rho(x_1, x_2, x_3)$. Performing the change of variables

$$x_1 = -x_1'\,, \quad x_2 = x_2'\,, \quad x_3 = x_3'$$

in the integral for I_{1j} (with $j \neq 1$), which by hypothesis maps Ω to itself, we obtain

$$\begin{aligned}
-I_{1j} &= \int_\Omega \rho(\mathbf{r}) x_1 x_j \, \mathrm{d}^3\mathbf{r} = -\int_{\Omega'} \rho(-x_1', x_2', x_3') x_1' x_j' \, \mathrm{d}^3\mathbf{r}' \\
&= -\int_\Omega \rho(x_1', x_2', x_3') x_1' x_j' \, \mathrm{d}^3\mathbf{r}' = I_{1j} \quad \Longrightarrow \quad I_{1j} = 0\,, \qquad j \neq 1\,.
\end{aligned}$$

2) *If* Ω *and* ρ *are invariant under the exchange* $x_i \mapsto x_j$, *then*

$$\boxed{I_{ii} = I_{jj}\,, \qquad I_{ik} = I_{jk} \quad (k \neq i, j)\,.}$$

Indeed, if (for instance) Ω is invariant under $x_1 \mapsto x_2$ and $\rho(x_2, x_1, x_3) = \rho(x_1, x_2, x_3)$, performing the change of variable

$$x_1 = x_2'\,, \quad x_2 = x_1'\,, \quad x_3 = x_3'$$

in the integral for I_{11} we obtain

$$\begin{aligned} I_{11} &= \int_\Omega \rho(\mathbf{r})(x_2^2 + x_3^2)\,\mathrm{d}^3\mathbf{r} = \int_{\Omega'} \rho(x_2', x_1', x_3')(x_1'^2 + x_3'^2)\,\mathrm{d}^3\mathbf{r}' \\ &= \int_\Omega \rho(x_1', x_2', x_3')(x_1'^2 + x_3'^2)\,\mathrm{d}^3\mathbf{r}' = I_{22}\,. \end{aligned}$$

Likewise,

$$\begin{aligned} -I_{13} &= \int_\Omega \rho(\mathbf{r}) x_1 x_3\,\mathrm{d}^3\mathbf{r} = \int_{\Omega'} \rho(x_2', x_1', x_3') x_2' x_3'\,\mathrm{d}^3\mathbf{r}' = \int_\Omega \rho(x_1', x_2', x_3') x_2' x_3'\,\mathrm{d}^3\mathbf{r}' \\ &= -I_{23}\,. \end{aligned}$$

Analogous results hold for the coordinates of the body's *center of mass*. For instance, if Ω and ρ are invariant under the reflection $x_i \mapsto -x_i$ then the i-th coordinate of the CM vanishes, since (taking, for definiteness, $i = 1$)

$$\begin{aligned} MX_1 &= \int_\Omega \rho(\mathbf{r}) x_1\,\mathrm{d}^3\mathbf{r} = -\int_{\Omega'} \rho(-x_1', x_2', x_3') x_1'\,\mathrm{d}^3\mathbf{r}' = -\int_\Omega \rho(x_1', x_2', x_3') x_1'\,\mathrm{d}^3\mathbf{r}' \\ &= -MX_1 \quad \Longrightarrow \quad X_1 = 0\,. \end{aligned}$$

Similarly, if Ω and ρ are invariant under the permutation $x_i \mapsto x_j$ then $X_i = X_j$.

Example 7.1. *Inertia tensor of a homogeneous solid of revolution*
Consider a *homogeneous* (i.e., with $\rho = \text{const.}$) rigid body Ω in the shape of a *solid of revolution* about a certain axis. Taking the z axis of the set of body axes in the direction of the body's axis of revolution, in cylindrical coordinates[a] (r, φ, z) the body is described by an equation of the form

$$0 \leqslant r \leqslant f(z)\,, \quad z_1 \leqslant z \leqslant z_2\,, \quad 0 \leqslant \varphi \leqslant 2\pi\,.$$

The symmetry under rotations about the z axis implies the invariance of the body under the transformations

$$x_1 \mapsto -x_1\,, \qquad x_2 \mapsto -x_2\,, \qquad x_1 \mapsto x_2\,.$$

Hence the body's center of mass is a point on the z axis, which we shall take as the origin of coordinates, since we are interested in computing the inertia tensor with respect to the CM. By the above symmetries, the components of the inertia tensor satisfy

$$I_{11} = I_{22}, \qquad I_{ij} = 0 \quad (i \neq j).$$

Hence in this case the inertia tensor is *diagonal*, with principal moments of inertia $I_i = I_{ii}$ given by

$$I_1 = I_2 = \rho \int_{z_1}^{z_2} \mathrm{d}z \int_0^{f(z)} \mathrm{d}r \int_0^{2\pi} r\,\mathrm{d}\varphi \cdot (z^2 + r^2 \sin^2 \varphi)$$

$$= \boxed{\pi\rho \int_{z_1}^{z_2} z^2 f^2(z)\,\mathrm{d}z + \frac{\pi\rho}{4} \int_{z_1}^{z_2} f^4(z)\,\mathrm{d}z},$$

$$I_3 = \rho \int_{z_1}^{z_2} \mathrm{d}z \int_0^{f(z)} \mathrm{d}r \int_0^{2\pi} r\,\mathrm{d}\varphi \cdot r^2 = \boxed{\frac{\pi\rho}{2} \int_{z_1}^{z_2} f^4(z)\,\mathrm{d}z}.$$

The mass density ρ can be expressed in terms of the body's total mass M through the formula

$$\rho = \frac{M}{V} = \frac{M}{\pi \int_{z_1}^{z_2} f^2(z)\,\mathrm{d}z}.$$

Note that, in general, $I_1 = I_2 \neq I_3$; more precisely, we have

$$I_1 = I_2 = \frac{1}{2} I_3 + \pi\rho \int_{z_1}^{z_2} z^2 f^2(z)\,\mathrm{d}z.$$

Thus a solid of revolution is in general an axially symmetric top. Note also that in this case the axis of revolution is a principal axis of inertia, as is any axis perpendicular to it.

For instance, in the case of a *cylinder* of radius a and height h we can take $f(z) = a$, $z_1 = -h/2$ and $z_2 = h/2$, since by symmetry the CM of a cylinder is equidistant from its bases. Hence

$$\boxed{\begin{aligned} I_3 &= \frac{\pi}{2}\rho a^4 h = \frac{1}{2} M a^2, \\ I_1 = I_2 &= \frac{1}{2} I_3 + \pi\rho a^2 \int_{-h/2}^{h/2} z^2\,\mathrm{d}z = \frac{1}{4} M a^2 + 2\pi\rho a^2 \int_0^{h/2} z^2\,\mathrm{d}z \\ &= \frac{1}{4} M a^2 + \frac{1}{12}\pi\rho a^2 h^3 = \frac{1}{4} M\left(a^2 + \frac{h^2}{3}\right). \end{aligned}}$$

In particular, a cylinder is a spherically symmetric top if and only if $h = \sqrt{3}\,a$.

[a] We are temporarily denoting the distance to the z axis as r instead of the usual notation ρ to avoid confusion with the mass density.

7.4 EQUATIONS OF MOTION OF A RIGID BODY

7.4.1 Equations of motion in an inertial frame

Since a rigid body has (in general) 6 degrees of freedom, it should be expected that its motion be determined by 6 differential equations. The first three of these equations are obviously the equations of motion of the body's CM, which, as we saw in Section 2, read

$$\boxed{M\left(\frac{\mathrm{d}^2\mathbf{R}}{\mathrm{d}t^2}\right)_{\mathrm{f}} = \mathbf{F}\,.} \tag{7.28}$$

In the RHS

$$\mathbf{F} = \sum_{\alpha} \mathbf{F}_{\alpha}$$

denotes the sum of the *external* forces acting on the particles making up the body (recall that, by Newton's third law, the sum of the internal forces vanishes). The remaining three differential equations can be taken as the equations of motion of the body's angular momentum with respect to the origin O' of the fixed frame, namely

$$\boxed{\left(\frac{\mathrm{d}\mathbf{L}}{\mathrm{d}t}\right)_{\mathrm{f}} = \mathbf{N}\,.} \tag{7.29}$$

Here

$$\mathbf{N} = \sum_{\alpha} \mathbf{r}'_{\alpha} \times \mathbf{F}_{\alpha}$$

denotes the total torque (with respect to O') of the *external* forces acting on the body. Indeed, as we saw in Section 2, if we assume that Newton's third law holds in its stronger sense the torque of the internal forces vanishes.

In equation (7.29) both the angular momentum and the total torque of the external forces are computed with respect to the origin O' of the fixed frame. In fact, *Eq.* (7.29) *still holds if we replace* $\mathbf{L}$ *by* $\mathbf{L}_{\mathrm{CM}}$ *and* $\mathbf{N}$ *by the total torque of the external forces with respect to the CM*, given by

$$\boxed{\mathbf{N}_{\mathrm{CM}} = \sum_{\alpha} \mathbf{r}_{\alpha} \times \mathbf{F}_{\alpha}\,.} \tag{7.30}$$

Indeed, from the relation $\mathbf{L} = M\mathbf{R} \times \mathbf{V} + \mathbf{L}_{\mathrm{CM}}$ and the CM's equation of motion it follows that

$$\left(\frac{\mathrm{d}\mathbf{L}}{\mathrm{d}t}\right)_{\mathrm{f}} = \left(\frac{\mathrm{d}\mathbf{L}_{\mathrm{CM}}}{\mathrm{d}t}\right)_{\mathrm{f}} + \mathbf{R} \times \mathbf{F} = \mathbf{N} = \sum_{\alpha} (\mathbf{R} + \mathbf{r}_{\alpha}) \times \mathbf{F}_{\alpha} = \mathbf{R} \times \mathbf{F} + \mathbf{N}_{\mathrm{CM}}\,,$$

and thus

$$\boxed{\left(\frac{\mathrm{d}\mathbf{L}_{\mathrm{CM}}}{\mathrm{d}t}\right)_{\mathrm{f}} = \mathbf{N}_{\mathrm{CM}}\,.} \tag{7.31}$$

In general, if the total force acting on the body *vanishes* the torque $\mathbf{N}$ is independent of the point with respect to which it is computed.

Indeed, if $\mathbf{F} := \sum_\alpha \mathbf{F}_\alpha = 0$ and $\mathbf{a}$ is a fixed vector we have

$$\sum_\alpha (\mathbf{r}'_\alpha + \mathbf{a}) \times \mathbf{F}_\alpha = \sum_\alpha \mathbf{r}'_\alpha \times \mathbf{F}_\alpha + \mathbf{a} \times \mathbf{F} = \sum_\alpha \mathbf{r}'_\alpha \times \mathbf{F}_\alpha .$$

Exercise 7.3. Show that a necessary condition for a generic rigid body to be at equilibrium in an inertial frame is that $\mathbf{F} = \mathbf{N} = 0$. Is this condition sufficient?

Solution. By definition, the body is at equilibrium in an inertial frame if $\mathbf{v}'_\alpha = 0$ for all t and for all α, i.e., if

$$\mathbf{v}'_\alpha = \mathbf{V} + \omega \times \mathbf{r}_\alpha = 0, \qquad \forall t,\ \forall \alpha .$$

The previous condition implies that

$$\omega \times (\mathbf{r}_\alpha - \mathbf{r}_\beta) = 0\,, \ \ \forall \alpha \neq \beta .$$

If $\omega \neq 0$ then $\mathbf{r}_\alpha - \mathbf{r}_\beta$ would be parallel to ω for fixed β and all $\alpha \neq \beta$, and hence the body would lie on the straight line parallel to ω passing through $\mathbf{r}_\beta$. Since the body is generic by hypothesis, we must have $\omega = 0$, which implies that $\mathbf{V} = 0$. Conversely, if $\mathbf{V} = \omega = 0$ it is clear that $\mathbf{v}'_\alpha = 0$ for all α. Thus the necessary and sufficient condition for a generic body to be at equilibrium is that $\mathbf{V} = \omega = 0$.

Suppose, to begin with, that the body is at equilibrium. Then $\mathbf{V} = \omega = 0$, and therefore $\mathbf{L} = M\mathbf{R} \times \mathbf{V} + I\omega = 0$. Substituting into the equations of motion (7.28)–(7.29) we immediately obtain $\mathbf{F} = \mathbf{N} = 0$.

Conversely, suppose that $\mathbf{F} = \mathbf{N} = 0$. From the vanishing of $\mathbf{F}$ and the equation of motion for $\mathbf{R}$ we deduce that $\mathbf{V}$ is constant in all inertial frames, and therefore it vanishes in some inertial frame that we shall take as the *fixed frame*. Moreover, by the last framed remark $\mathbf{F} = \mathbf{N} = 0$ implies that $\mathbf{N}_{\mathrm{CM}} = \mathbf{N} = 0$, and therefore

$$\left(\frac{\mathrm{d}\mathbf{L}_{\mathrm{CM}}}{\mathrm{d}t}\right)_{\mathrm{f}} = \mathbf{N}_{\mathrm{CM}} = 0 \quad \Longrightarrow \quad \mathbf{L}_{\mathrm{CM}} = I\omega(t) = \mathbf{L}_{\mathrm{CM}}(0) = I\omega(0) \ \ \forall t .$$

If the body is generic its inertia tensor I is invertible, and hence

$$\omega(t) = \omega(0) \quad \forall t .$$

We have thus shown that if $\mathbf{F} = \mathbf{N} = 0$ then $\mathbf{V} = 0$ and ω is *constant* (although not necessarily 0!) in some inertial frame. To show that the body is at equilibrium, we need to assume that $\omega(0) = 0$ in this frame, so that $\mathbf{V} = \omega = 0$. In other words, the vanishing of $\mathbf{F}$ and $\mathbf{N}$ guarantees that the body is at equilibrium if and only if *the body is initially at rest in some inertial frame.*

Remark 7.1. The condition $\omega(0) = 0$ is *essential* to guarantee that the body is at equilibrium when $\mathbf{F} = \mathbf{N} = 0$. Indeed, we shall see in Section 7.5 that when $\mathbf{F}$ and $\mathbf{N}$ vanish the body can still rotate with constant angular velocity about a principal axis of inertia passing through the CM. ■

Exercise 7.4. Show that the condition for equilibrium of a rigid body found above is equivalent to $\mathbf{F} = \mathbf{N}_{\mathrm{CM}} = 0$.

Solution. If $\mathbf{F} = 0$, the torque of the external forces does not depend on the point with respect to which it is taken, so that in this case $\mathbf{N} = \mathbf{N}_{\mathrm{CM}}$.

7.4.2 Motion in a constant external field

A particular case which often occurs in practice arises when the external forces $\mathbf{F}_\alpha$ acting on the rigid body are due to a *constant external field* $\mathbf{f}$, to which the particles couple through a "charge" $\lambda \in \mathbb{R}$. In this case

$$\mathbf{F}_\alpha = \lambda_\alpha \mathbf{f}, \qquad \alpha = 1, \dots, N, \tag{7.32}$$

where $\mathbf{f}$ is independent of α and λ_α is the charge of particle α. For instance, *Earth's gravitational force* is of this form if the body's extension is small compared to its distance to Earth's center, so that Earth's gravitational field is approximately uniform inside the body (in this case $\lambda_\alpha = m_\alpha$, $\mathbf{f} = \mathbf{g}$). The same is true for the electric force due to a *uniform electric field* $\mathbf{E}$ (in this case $\lambda_\alpha = e_\alpha$ is the electric charge and $\mathbf{f} = \mathbf{E}$).

If the external forces are of the form (7.32) we have

$$\mathbf{F} = \sum_\alpha \mathbf{F}_\alpha = \mathbf{f} \sum_\alpha \lambda_\alpha = \Lambda \mathbf{f}, \qquad \mathbf{N} = \Big(\sum_\alpha \lambda_\alpha \mathbf{r}'_\alpha\Big) \times \mathbf{f},$$

$\Lambda := \sum_\alpha \lambda_\alpha$ being the body's total charge. If (as is the case with the gravitational force) $\Lambda \neq 0$, we define the body's *center of charge* by the equation

$$\mathbf{X} = \frac{1}{\Lambda} \sum_\alpha \lambda_\alpha \mathbf{r}'_\alpha = \frac{1}{\Lambda} \sum_\alpha \lambda_\alpha (\mathbf{R} + \mathbf{r}_\alpha) = \mathbf{R} + \frac{1}{\Lambda} \sum_\alpha \lambda_\alpha \mathbf{r}_\alpha . \tag{7.33}$$

Note that the point $\mathbf{X}$ is *fixed* in the body, since its position vector with respect to the CM (also fixed in the body)

$$\mathbf{X} - \mathbf{R} = \frac{1}{\Lambda} \sum_\alpha \lambda_\alpha \mathbf{r}_\alpha =: \mathbf{X}_{\mathrm{CM}}$$

is a constant vector in the frame of body axes. In particular, for the gravitational field $\mathbf{X} = \mathbf{R}$ and $\mathbf{X}_{\mathrm{CM}} = 0$. In terms of the vector $\mathbf{X}$, the torque $\mathbf{N}$ can be concisely expressed as

$$\boxed{\mathbf{N} = \mathbf{X} \times \mathbf{F}.} \tag{7.34}$$

In other words:

> If the total charge doesn't vanish, the total torque of the external forces coincides with the torque of the total external force applied at the *center of charge*. In particular, when computing the torque of the gravitational forces acting on a rigid body we can always assume that they are applied at its *center of mass*, i.e., that
>
> $$\boxed{\mathbf{N} = \mathbf{R} \times \mathbf{F}.}$$

Likewise, if $\Lambda \neq 0$ the torque of the external forces with respect to the CM is given by

$$\mathbf{N}_{\mathrm{CM}} = \left(\sum_{\alpha} \lambda_\alpha \mathbf{r}_\alpha\right) \times \mathbf{f} = \mathbf{X}_{\mathrm{CM}} \times \mathbf{F}.$$

In the case of the gravitational force $\mathbf{X}_{\mathrm{CM}} = 0$, and hence $\mathbf{N}_{\mathrm{CM}} = 0$. In other words:

> The torque with respect to the CM of the gravitational forces $\mathbf{F}_\alpha = m_\alpha \mathbf{g}$ *vanishes* (assuming again that the body's size is negligible compared to its distance to Earth's center).

Note, finally, that the forces (7.32) are clearly *conservative*, with potential

$$\boxed{U = -\mathbf{f} \cdot \sum_{\alpha} \lambda_\alpha \mathbf{r}'_\alpha = -\Lambda \mathbf{f} \cdot \mathbf{X} = -\mathbf{F} \cdot \mathbf{X},}$$

where the last two equalities are valid only when $\Lambda \neq 0$. Thus (assuming again that $\Lambda \neq 0$) *when computing the potential energy we can assume in this case that the total constant external force* $\mathbf{F}$ *is applied at the point* $\mathbf{X}$. In particular, the potential energy of a rigid body due to Earth's gravitational field is simply

$$\boxed{U = -M\mathbf{g} \cdot \mathbf{R}.}$$

Note. If the total charge Λ vanishes the center of charge cannot be defined by Eq. (7.33). In this case the total torque of the external forces is clearly independent of the point with respect to which it is taken, since the total external force $\mathbf{F} = \Lambda \mathbf{f}$ vanishes.

7.4.3 Euler's equations

Since the relation between angular momentum and angular velocity is particularly simple in a frame of principal axes of inertia fixed in the body, it is convenient to formulate the equation of motion of the angular momentum in such a frame. To this end, we shall assume that the point P with respect to which $\mathbf{L}$, I, and $\mathbf{N}$ are computed is either the CM or (when it exists) a point simultaneously *fixed* in the body and in an inertial frame, that we shall take as the origin O' of the latter frame. To cover both situations we shall when necessary use the more descriptive notation $\mathbf{L}_P$, $\mathbf{N}_P$, and

I_P to respectively denote the angular momentum, torque, and inertia tensor taking as origin the point P (in our old notation, $\mathbf{L}_C \equiv \mathbf{L}_{\mathrm{CM}}$, $\mathbf{L}_{O'} \equiv \mathbf{L}$, $\mathbf{N}_C \equiv \mathbf{N}_{\mathrm{CM}}$, $\mathbf{N}_{O'} \equiv \mathbf{N}$, and $I_C \equiv I$). With this notation we can write

$$\left(\frac{\mathrm{d}\mathbf{L}_P}{\mathrm{d}t}\right)_{\mathrm{f}} = \mathbf{N}_P\,, \qquad \mathbf{L}_P = I_P\boldsymbol{\omega}\,, \qquad E := \frac{1}{2}\,\boldsymbol{\omega}\cdot\mathbf{L}_P = \begin{cases} T_{\mathrm{rot}}\,, & P = C \\ T\,, & P = O'\,, \end{cases}$$

where I_P is *constant* in the frame of body axes (since P is by hypothesis fixed in the body). *We shall usually drop the subindex, and simply write*

$$\left(\frac{\mathrm{d}\mathbf{L}}{\mathrm{d}t}\right)_{\mathrm{f}} = \mathbf{N}\,, \qquad \mathbf{L} = I\boldsymbol{\omega}, \qquad \frac{1}{2}\,\boldsymbol{\omega}\cdot\mathbf{L} = E \tag{7.35}$$

to deal with both cases at the same time. From Eqs. (6.14) and (7.29) we then obtain the relation

$$\dot{\mathbf{L}} + \boldsymbol{\omega}\times\mathbf{L} = \mathbf{N}\,,$$

where as usual the dot denotes time derivative with respect to the frame of body axes. Using the relation between $\mathbf{L}$ and $\boldsymbol{\omega}$, and taking into account that I_{ij} is constant in a frame of body axes, we immediately obtain

$$I\dot{\boldsymbol{\omega}} + \boldsymbol{\omega}\times(I\boldsymbol{\omega}) = \mathbf{N}\,. \tag{7.36}$$

If the body axes are principal axes of inertia, the i-th component of this vector equation is simply

$$I_i\dot{\omega}_i + \omega_j(I_k\omega_k) - \omega_k(I_j\omega_j) = N_i\,, \qquad i = 1,2,3\,,$$

or equivalently

$$I_i\dot{\omega}_i - (I_j - I_k)\omega_j\omega_k = N_i\,, \qquad i = 1,2,3\,, \tag{7.37a}$$

where

$$(i,j,k) = \textit{cyclic}\ \text{permutation of}\ (1,2,3)\,. \tag{7.37b}$$

Equations (7.37), i.e., the system

$$\begin{aligned} I_1\dot{\omega}_1 - (I_2 - I_3)\omega_2\omega_3 &= N_1\,,\\ I_2\dot{\omega}_2 - (I_3 - I_1)\omega_1\omega_3 &= N_2\,,\\ I_3\dot{\omega}_3 - (I_1 - I_2)\omega_1\omega_2 &= N_3\,, \end{aligned} \tag{7.38}$$

are known as **Euler's equations**. We emphasize that these equations are valid if both $\mathbf{N}$ and I are computed either with respect to the CM or (when possible) to a point simultaneously fixed in the body and in the inertial frame. Moreover, the quantities ω_i and N_i appearing in Euler's equations are the components of the vectors $\boldsymbol{\omega}$ and $\mathbf{N}$ *in a frame of principal axes of inertia* (in general *not* inertial!).

If the total torque $\mathbf{N}$ of the external forces vanishes and the origin O' of the inertial frame is a point fixed in the body, $\mathbf{L}_{O'} \equiv \mathbf{L}$ and T are conserved. Similarly, if $\mathbf{N}_{\text{CM}}$ vanishes then $\mathbf{L}_{\text{CM}}$ and the rotational energy T_{rot} are conserved.

Proof. Using the notation in Eq. (7.35), we only need to show that

$$\mathbf{N} = 0 \quad \Longrightarrow \quad \left(\frac{\mathrm{d}\mathbf{L}}{\mathrm{d}t}\right)_{\mathrm{f}} = 0, \quad \dot{E} = 0 .$$

(Note that E is a *scalar*, so that it is not necessary to specify the frame with respect to which the time derivative is taken.) The conservation of $\mathbf{L}$ (in the fixed frame) follows immediately from its equation of motion (7.35), while the conservation of E is obtained differentiating the third relation (7.35) *in the frame of body axes* (which is correct, since E is a *scalar*). Indeed, in this frame we have

$$\frac{\mathrm{d}}{\mathrm{d}t}(\omega \cdot I\omega) = \frac{\mathrm{d}}{\mathrm{d}t}\sum_i I_i\omega_i^2 = 2\sum_i I_i\omega_i\dot{\omega}_i = 2\omega \cdot I\dot{\omega}$$

From Euler's equations in their vector form (7.36) with $\mathbf{N} = 0$ it then follows that

$$\dot{E} = \omega \cdot I\dot{\omega} = -\omega \cdot \big(\omega \times (I\omega)\big) = 0,$$

as $\omega \times (I\omega)$ is perpendicular to ω. ■

Remark 7.2. Since the magnitude of the angular momentum $L = |\mathbf{L}|$ and the energy E are *scalars*, when $\mathbf{N} = 0$ both L and E are also constant (i.e., time-independent) *in the body frame.* ■

7.5 INERTIAL MOTION OF A SYMMETRIC TOP

We shall study in this section the rotational motion of an *axially symmetric* top when the total torque of the external forces with respect to either the CM, or a point simultaneously fixed in the body and in an inertial frame (when such a point exists), vanishes. This will obviously happen (in both cases) if the body is *free*, that is, in the absence of external forces. More generally, as we saw at the end of Section 7.4.2, the torque $\mathbf{N}_{\text{CM}}$ will vanish provided that the only external force acting on the body is Earth's gravity (assuming the body's size to be negligible compared to its distance to Earth's center). Before starting our analysis, it is convenient to prove the following fundamental fact regarding the angular velocity vector ω:

The instantaneous angular velocity ω of a set of axes with respect to another is *additive.*

In other words, let S_0, S_1, and S_2 be three sets of axes, and suppose that at a certain time t the axes of S_1 have an angular velocity ω_1 relative to those of S_0, and the axes of S_2 have in turn an angular velocity ω_2 with respect to those of S_1. Then the angular velocity of the axes of S_2 relative to those of S_0 is

$$\boxed{\omega = \omega_1 + \omega_2 \,.} \tag{7.39}$$

Indeed, let $\{\mathbf{e}'_i\}_{1\leqslant i\leqslant 3}$, $\{\mathbf{e}_i\}_{1\leqslant i\leqslant 3}$, and $\{\mathbf{e}''_i\}_{1\leqslant i\leqslant 3}$ respectively denote the axes of the frames S_0, S_1, and S_2. By definition of angular velocity,

$$\left(\frac{\mathrm{d}\mathbf{e}_i}{\mathrm{d}t}\right)_0 = \omega_1 \times \mathbf{e}_i \,, \qquad \left(\frac{\mathrm{d}\mathbf{e}''_i}{\mathrm{d}t}\right)_1 = \omega_2 \times \mathbf{e}''_i \,, \qquad \left(\frac{\mathrm{d}\mathbf{e}''_i}{\mathrm{d}t}\right)_0 = \omega \times \mathbf{e}''_i \,.$$

But then

$$\left(\frac{\mathrm{d}\mathbf{e}''_i}{\mathrm{d}t}\right)_0 = \left(\frac{\mathrm{d}\mathbf{e}''_i}{\mathrm{d}t}\right)_1 + \omega_1 \times \mathbf{e}''_i = \omega_2 \times \mathbf{e}''_i + \omega_1 \times \mathbf{e}''_i = (\omega_1 + \omega_2) \times \mathbf{e}''_i \,,$$

whence Eq. (7.39) follows.

7.5.1 Motion relative to the body frame

Recall that we are denoting by $\mathbf{L}$ and I the angular momentum and the inertia tensor with respect to either O' or the CM, depending on whether $\mathbf{N} = 0$ or $\mathbf{N}_{\mathrm{CM}} = 0$ (in the first case, it is assumed that O' is simultaneously fixed in the body and the inertial system).

By definition, in an axially symmetric top two principal moments of inertia, which we shall take as I_1 and I_2, coincide, while the third one (i.e., I_3) differs from the other two. In other words, we have $I_1 = I_2 \neq I_3$. In this case the $\mathbf{e}_3$ axis is a principal axis of inertia (with moment of inertia I_3), as is any axis perpendicular to it (with moment of inertia $I_1 = I_2$). In particular, from the discussion in Example 7.1 it follows that a solid of revolution about the x_3 axis is an axially symmetric top with symmetry axis along the vector $\mathbf{e}_3$. This is not, however, the most general example; for instance, a homogeneous rectangular parallelepiped with sides $l_1 = l_2 \neq l_3$ is also an axially symmetric top with $I_1 = I_2 \neq I_3$ (exercise).

Substituting $\mathbf{N} = 0$ and $I_1 = I_2$ in Euler's equations (7.38) we obtain the simpler system

$$\boxed{\begin{aligned} I_1\dot{\omega}_1 - (I_1 - I_3)\omega_2\omega_3 &= 0 \,, \\ I_1\dot{\omega}_2 - (I_3 - I_1)\omega_1\omega_3 &= 0 \,, \\ I_3\dot{\omega}_3 &= 0 \,, \end{aligned}} \tag{7.40}$$

whence it immediately follows (assuming that $I_3 \neq 0$, i.e., that the body is not collinear) that

$$\boxed{\omega_3 = \text{const.}}$$

Calling

$$\boxed{\Omega := \frac{I_3 - I_1}{I_1}\,\omega_3\,,} \tag{7.41}$$

the first two equations read

$$\dot{\omega}_1 = -\Omega\,\omega_2\,, \qquad \dot{\omega}_2 = \Omega\,\omega_1,$$

or, in complex notation,

$$\boxed{\dot{\omega}_1 + \mathrm{i}\dot{\omega}_2 = \mathrm{i}\Omega(\omega_1 + \mathrm{i}\omega_2)\,.}$$

The solution of this linear first-order differential equation is

$$\omega_1 + \mathrm{i}\omega_2 = \big(\omega_1(0) + \mathrm{i}\omega_2(0)\big)\mathrm{e}^{\mathrm{i}\Omega t}\,. \tag{7.42}$$

From this equation it follows that

$$\omega_1^2 + \omega_2^2 = \left|\omega_1 + \mathrm{i}\omega_2\right|^2 = \omega_1(0)^2 + \omega_2(0)^2 =: \omega_0^2$$

is constant, and so are $\omega = \sqrt{\omega_0^2 + \omega_3^2}$ and the angle $\alpha = \arctan(\omega_0/\omega_3)$ between the vectors $\boldsymbol{\omega}$ and $\mathbf{e}_3$. We have thus shown the following:

> The magnitude of the projection of $\boldsymbol{\omega}$ onto the plane perpendicular to $\mathbf{e}_3$, ω_3, ω, and the angle α between the vectors $\boldsymbol{\omega}$ and $\mathbf{e}_3$ are all constant.

In real terms, Eq. (7.42) can be written as

$$\omega_1\mathbf{e}_1 + \omega_2\mathbf{e}_2 = R_3(\Omega t)\cdot\big(\omega_1\mathbf{e}_1(0) + \omega_2\mathbf{e}_2(0)\big)\,,$$

where $R_3(\varphi)$ is a rotation about the $\mathbf{e}_3$ axis by an angle φ. On the other hand,

$$\omega_3\mathbf{e}_3 = R_3(\Omega t)\cdot(\omega_3\mathbf{e}_3) = R_3(\Omega t)\cdot\big(\omega_3(0)\mathbf{e}_3\big)\,,$$

since ω_3 is constant. Adding both equations we finally obtain

$$\boxed{\boldsymbol{\omega} = R_3(\Omega t)\cdot\boldsymbol{\omega}(0)\,.}$$

In other words:

> In the *frame of body axes*, the vector $\boldsymbol{\omega}$ rotates about the $\mathbf{e}_3$ axis with constant angular velocity Ω (cf. Fig. 7.2).

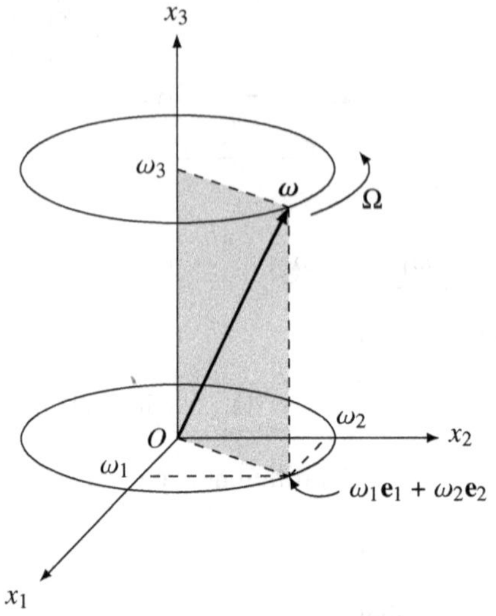

Figure 7.2. Procession of the vector $\boldsymbol{\omega}$ about the $\mathbf{e}_3$ body axis.

• The previous result could have been deduced directly from Euler's equations, since

$$\dot{\boldsymbol{\omega}} = \dot{\omega}_1 \mathbf{e}_1 + \dot{\omega}_2 \mathbf{e}_2 = \Omega(-\omega_2 \mathbf{e}_1 + \omega_1 \mathbf{e}_2) = \Omega \mathbf{e}_3 \times \boldsymbol{\omega}\,.$$

• Note also that if $\omega_3 > 0$ the angular velocity Ω is positive for $I_3 > I_1$ ("flat" body), whereas it is negative for $I_3 < I_1$ ("tall" body); indeed,

$$I_3 - I_1 = \int_\Omega \rho(\mathbf{r})(x_1^2 - x_3^2)\,\mathrm{d}^3\mathbf{r}.$$

• A particular solution of Euler's equations consists of a *rotation about a principal axis of inertia with constant angular velocity.* Indeed, in this case $\dot{\boldsymbol{\omega}} = 0$ and $I\boldsymbol{\omega} \parallel \boldsymbol{\omega}$, so that $I\dot{\boldsymbol{\omega}} + \boldsymbol{\omega} \times (I\boldsymbol{\omega}) = 0$. (Note that the axis of rotation $\boldsymbol{\omega}/\omega$ is also fixed in space, since $\dot{\boldsymbol{\omega}} = 0$ holds both in the body and in the fixed frame). *We shall in what follows disregard these* (trivial) *solutions.*

• By the previous observation, *we can assume that ω_3 and α do not vanish.* Indeed, if $\alpha = 0$ then $\boldsymbol{\omega} = \omega_3 \mathbf{e}_3$ is constant and hence the body rotates with constant angular velocity about its symmetry axis, which is a principal axis of inertia. Likewise, if $\omega_3 = 0$ then $\Omega = 0$, $\boldsymbol{\omega} = \omega_1 \mathbf{e}_1 + \omega_2 \mathbf{e}_2$ is constant, and thus the body rotates again around a principal axis of inertia perpendicular to the axis of symmetry with constant angular velocity. ■

Since we are assuming that $\omega_3 \neq 0$, we can choose the direction of the $\mathbf{e}_3$ axis so that $\omega_3 > 0$, and hence $\alpha \in (0, \pi/2)$. In terms of the angle α we can write

$$\boxed{\omega_3 = \omega \cos\alpha\,, \qquad \omega_0 = \omega \sin\alpha\,, \qquad \omega_1 + \mathrm{i}\omega_2 = \omega \sin\alpha\, \mathrm{e}^{\mathrm{i}(\Omega t + \beta)}\,,} \tag{7.43}$$

where β is the angle between the vectors $\omega_1(0)\mathbf{e}_1 + \omega_2(0)\mathbf{e}_2$ and $\mathbf{e}_1$ (which could be taken as zero choosing appropriately the initial time). The previous results can be expressed in a more geometric language as follows:

Relative to the *frame of body axes,* the vector $\boldsymbol{\omega}$ moves tracing out a cone with axis $\mathbf{e}_3$ and half-angle α, with constant angular velocity Ω. This cone is called the **body cone** or, more correctly, the *cone fixed in the body.*

The motion of the angular momentum $\mathbf{L}$ *relative to the body axes* is easily determined from the second equation (7.35), which can be expressed in complex notation as

$$\boxed{\begin{aligned} &L_3 = I_3\omega_3 = I_3\omega\cos\alpha = \text{const.}\,, \\ &L_1 + \mathrm{i}L_2 = I_1(\omega_1 + \mathrm{i}\omega_2) = I_1\omega\sin\alpha\,\mathrm{e}^{\mathrm{i}(\Omega t+\beta)}\,. \end{aligned}} \tag{7.44}$$

In other words, $\mathbf{L}$ lies on the plane determined by the vectors $\mathbf{e}_3$ and $\boldsymbol{\omega}$, with L_3, $L_1^2 + L_2^2$, L, and the angle θ between $\mathbf{L}$ and $\mathbf{e}_3$ all constant with respect to the frame of body axes. Note that the fact that L is constant in this frame follows also from the fact that it is constant in the fixed frame (since $\mathbf{N} = 0$). Similarly, that the vectors $\mathbf{e}_3$, $\boldsymbol{\omega}$, and $\mathbf{L}$ are coplanar can be proved directly remarking that

$$\mathbf{e}_3 \cdot (\boldsymbol{\omega} \times \mathbf{L}) = (I_2 - I_1)\omega_1\omega_2 = 0\,.$$

From the previous discussion it follows that:

Relative to the *frame of body axes,* the angular momentum $\mathbf{L}$ rotates with constant angular velocity Ω about the $\mathbf{e}_3$ axis.

The angle θ between $\mathbf{L}$ and $\mathbf{e}_3$ is easily computed noting that

$$\boxed{\tan\theta = \frac{\sqrt{L_1^2 + L_2^2}}{L_3} = \frac{I_1\omega_0}{I_3\omega_3} = \frac{I_1}{I_3}\tan\alpha\,;} \tag{7.45}$$

in particular, $\theta > \alpha$ for a "tall" body. Obviously, the angle between $\mathbf{L}$ and $\boldsymbol{\omega}$ is $|\theta - \alpha|$.

7.5.2 Motion relative to the fixed frame

We shall next describe the motion of the body in the *fixed frame*. Relative to this frame, the vector $\mathbf{L}$ is *constant*, since by hypothesis the torque of the external forces vanishes. The direction of this vector, which is therefore constant in the fixed frame, is known as the **invariant direction** and is usually taken as the $\mathbf{e}'_3$ axis:

$$\boxed{\mathbf{e}'_3 = \frac{\mathbf{L}}{L}\,.}$$

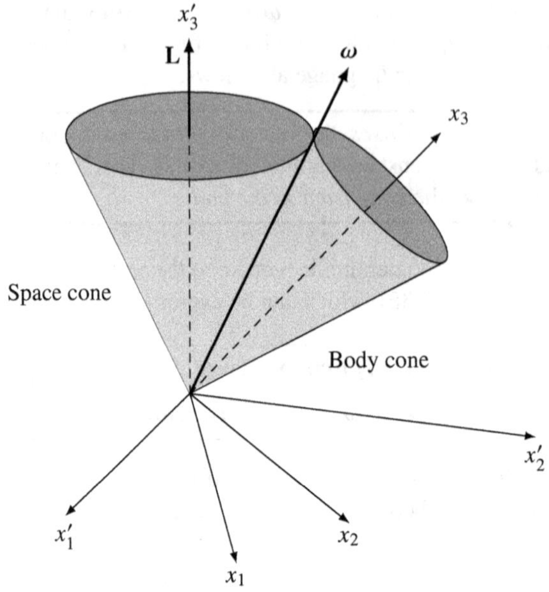

Figure 7.3. Space and body cones (in the case $I_1 > I_3$).

The vectors $\boldsymbol{\omega}$ and $\mathbf{e}_3$ both rotate around $\mathbf{L}$ since, as we have just seen, the angle between these vectors and the angular momentum, as well as their magnitude, are constant. Moreover, $\boldsymbol{\omega}$ and $\mathbf{e}_3$ rotate with the *same angular velocity* Ω_p, since they are coplanar with $\mathbf{L}$. In other words:

Relative to the *fixed frame*, the vector $\boldsymbol{\omega}$ moves tracing out a cone of axis $\mathbf{L}$ and half-angle $|\theta - \alpha|$ with angular velocity Ω_p. This cone is known as the **space cone** (more precisely, the *cone fixed in space*).

Note that the body and space cones are tangent at all times along their common generatrix parallel to the vector $\boldsymbol{\omega}$ (cf. Fig. 7.3).

To compute the angular velocity Ω_p, consider a third system of axes $\mathbf{e}_i''$ $(i = 1, 2, 3)$ with

$$\mathbf{e}_3'' = \mathbf{e}_3\,, \qquad \mathbf{e}_1'' = \frac{\mathbf{L}}{L}\,, \qquad \mathbf{e}_2'' = \mathbf{e}_3'' \times \mathbf{e}_1'' = \frac{\mathbf{e}_3 \times \mathbf{L}}{|\mathbf{e}_3 \times \mathbf{L}|} = \frac{\mathbf{e}_3 \times \mathbf{L}}{\sqrt{L_1^2 + L_2^2}}\,.$$

Note that by construction the vectors $\mathbf{e}_1''$ and $\mathbf{e}_3'' = \mathbf{e}_3$ span the same plane as $\mathbf{L}$ and $\mathbf{e}_3$, i.e.,

$$\operatorname{lin}\{\mathbf{e}_1'', \mathbf{e}_3''\} = \operatorname{lin}\{\mathbf{L}, \mathbf{e}_3\}\,.$$

Thus the angular velocity $\boldsymbol{\omega}'$ of the axes $\{\mathbf{e}_i''\}_{1\leqslant i\leqslant 3}$ with respect to the fixed frame $\{\mathbf{e}_i'\}_{1\leqslant i\leqslant 3}$ is equal to the angular velocity with which the plane spanned by $\mathbf{e}_3$ and $\mathbf{L}$ rotates around $\mathbf{e}_3' = \mathbf{L}/L$, which coincides with the angular velocity $\Omega_p\mathbf{e}_3' = \Omega_p\mathbf{L}/L$

of the rotation of the vector $\mathbf{e}_3$ around the invariant direction $\mathbf{e}'_3$. On the other hand, the angular velocity ω'' of the body axes $\{\mathbf{e}_i\}_{1\leqslant i\leqslant 3}$ relative to the frame $\{\mathbf{e}''_i\}_{1\leqslant i\leqslant 3}$ is equal to $-\Omega\mathbf{e}_3$, since the plane $\operatorname{lin}\{\mathbf{L}, \mathbf{e}_3\} = \operatorname{lin}\{\mathbf{e}''_1, \mathbf{e}''_3\}$ rotates with angular velocity Ω around $\mathbf{e}_3$ relative to the body axes $\{\mathbf{e}_i\}_{1\leqslant i\leqslant 3}$. By the additivity of angular velocities, the angular velocity ω of the body axes $\{\mathbf{e}_i\}_{1\leqslant i\leqslant 3}$ with respect to the fixed frame $\{\mathbf{e}'_i\}_{1\leqslant i\leqslant 3}$, is given by

$$\boxed{\omega = \omega' + \omega'' = \Omega_p \frac{\mathbf{L}}{L} - \Omega\mathbf{e}_3\,.} \tag{7.46}$$

The previous equation, together with the argument leading to its proof, shows that in the fixed frame the body's motion can be described as the composition of a *precession* of its symmetry axis $\mathbf{e}_3$ about the invariant direction (i.e., $\mathbf{L}/|\mathbf{L}|$) with angular velocity Ω_p and a *rotation* around its symmetry axis with angular velocity $-\Omega$.

Let us, finally, compute the angular velocity of precession Ω_p. To this end, note first of all that from Fig. 7.4 and Eq. (7.43) it follows that

$$\omega_0 = \omega \sin\alpha = \Omega_p \sin\theta\,,$$

and therefore

$$\Omega_p = \omega\frac{\sin\alpha}{\sin\theta} = \omega\sin\alpha\frac{L}{\sqrt{L_1^2 + L_2^2}} = \frac{L}{I_1}\,. \tag{7.47}$$

By Eqs. (7.44) we then have

$$\boxed{\Omega_p = \frac{\omega}{I_1}\sqrt{I_1^2\sin^2\alpha + I_3^2\cos^2\alpha} = \omega\sqrt{1 + \frac{I_3^2 - I_1^2}{I_1^2}\cos^2\alpha}\,.}$$

In particular, $\Omega_p < \omega$ for a "tall" body, whereas $\Omega_p > \omega$ for a "flat" one.

Exercise 7.5. Deduce Eq. (7.47) directly from Eq. (7.46).

Solution. Taking the scalar product of Eq. (7.46) with $\mathbf{e}_3$ we obtain

$$\Omega_p\frac{L_3}{L} = \Omega_p\frac{\omega_3 I_3}{L} = \omega_3 + \Omega = \omega_3 + \frac{I_3 - I_1}{I_1}\omega_3 = \frac{I_3\omega_3}{I_1} \quad\Longrightarrow\quad \Omega_p = \frac{L}{I_1}\,,$$

as before.

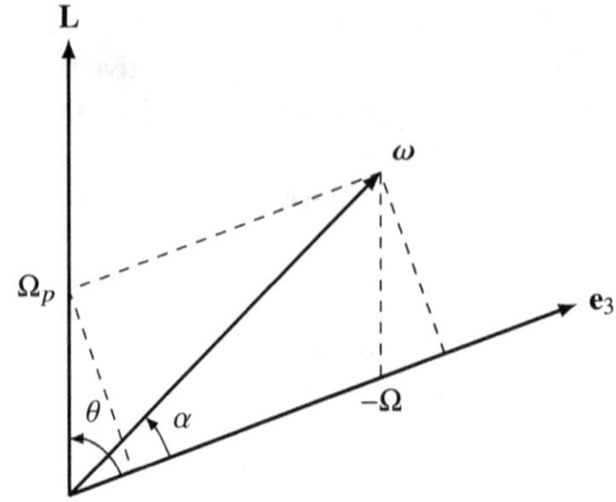

Figure 7.4. Vectors $\mathbf{L}$, ω, and $\mathbf{e}_3$ in the inertial motion of a rigid body symmetric about the $\mathbf{e}_3$ axis (in the case $I_1 > I_3$).

Exercise 7.6. Study the stability of the rotation of an asymmetric top (for which $I_i \neq I_j$ for $i \neq j$) about one of its principal axes of inertia in the case of inertial motion.

Solution. Suppose, for instance, that the body is rotating around its principal axis of inertia $\mathbf{e}_3$ with angular velocity $\boldsymbol{\omega} = \omega_3 \mathbf{e}_3$. Note, first of all, that ω_3 must be *constant*, as is easily deduced from Euler's equations with $\mathbf{N} = 0$ and $\omega_1 = \omega_2 = 0$. Let us next see what happens if we slightly modify the initial conditions

$$\omega_1(0) = \omega_2(0) = 0\,, \qquad \omega_3(0) = \omega_3$$

leading to the previous solution. To first order in the small quantities ω_1 and ω_2, the product $\omega_1\omega_2$ can be taken as zero, so that the third Euler equation implies that ω_3 remains approximately constant. Differentiating with respect to time the first two Euler equations (with ω_3 constant) we easily obtain

$$\ddot{\omega}_i + \frac{\omega_3^2}{I_1 I_2}(I_1 - I_3)(I_2 - I_3)\omega_i = 0\,, \qquad i = 1, 2\,.$$

The solution $\omega_1 = \omega_2 = 0$, $\omega_3 = \text{const.}$ will be stable provided that the solutions of the previous equations are *oscillatory*, and unstable otherwise. Thus the stability condition is that the product $(I_1 - I_3)(I_2 - I_3)$ be *positive*, namely that either $I_3 < I_{1,2}$ or $I_3 > I_{1,2}$. In other words:

> The inertial rotation about a principal axis of inertia of an asymmetric top is *stable* if and only if the corresponding principal moment of inertia is either *maximum* or *minimum*.

For example, for a rectangular parallelepiped of sides $l_1 < l_2 < l_3$ and mass M the inertia tensor with respect to the CM is given by

$$I = \frac{M}{12}\begin{pmatrix} l_2^2 + l_3^2 & 0 & 0 \\ 0 & l_1^2 + l_3^2 & 0 \\ 0 & 0 & l_1^2 + l_2^2 \end{pmatrix}.$$

Therefore

$$I_3 < I_2 < I_1,$$

so that according to the previous result only the rotations about the x_1 and x_3 axes (corresponding to the shortest and longest sides) are stable.

Exercise 7.7. Study the inertial motion of an *asymmetric* rigid body with $I_1 > I_2 > I_3$ in the case $L^2 = 2I_2E$.

Solution. In a frame of principal axes of inertia the conservation of the magnitude of the angular momentum and the (rotational) kinetic energy read

$$\boxed{\sum_i I_i^2\omega_i^2 = L^2\,, \qquad \sum_i I_i\omega_i^2 = 2E\,.}$$

Combining these equations we obtain

$$\begin{aligned} I_2(I_2 - I_1)\omega_2^2 + I_3(I_3 - I_1)\omega_3^2 &= L^2 - 2I_1E\,, \\ I_1(I_1 - I_3)\omega_1^2 + I_2(I_2 - I_3)\omega_2^2 &= L^2 - 2I_3E\,, \end{aligned}$$

and hence

$$\begin{aligned} \omega_1^2 &= \frac{L^2 - 2I_3E - I_2(I_2 - I_3)\omega_2^2}{I_1(I_1 - I_3)} = \frac{I_2 - I_3}{I_1(I_1 - I_3)}\left(2E - I_2\omega_2^2\right), \\ \omega_3^2 &= \frac{L^2 - 2I_1E - I_2(I_2 - I_1)\omega_2^2}{I_3(I_3 - I_1)} = \frac{I_1 - I_2}{I_3(I_1 - I_3)}\left(2E - I_2\omega_2^2\right) = \frac{I_1(I_1 - I_2)}{I_3(I_2 - I_3)}\,\omega_1^2\,. \end{aligned}$$

The Euler equation for ω_2 is thus

$$\dot\omega_2 = \pm\sqrt{\frac{(I_2 - I_3)(I_1 - I_2)}{I_1 I_3}}\left(\frac{L^2}{I_2^2} - \omega_2^2\right),$$

whose general solution is

$$\pm\nu(t - t_0) = \operatorname{arctanh}(I_2\omega_2/L) \quad\Longleftrightarrow\quad \omega_2 = \pm\frac{L}{I_2}\tanh\big(\nu(t - t_0)\big)$$

with

$$\nu := \frac{L}{I_2}\sqrt{\frac{(I_2 - I_3)(I_1 - I_2)}{I_1 I_3}}\,.$$

Choosing appropriately the origin of t and the direction of the x_2 principal axis, we can simply write

$$\boxed{\omega_2 = \frac{L}{I_2}\,\tanh(\nu t)\,.}$$

From the previous equations for ω_1 and ω_3 we then obtain (taking into account that, by the Euler equation for ω_2, ω_1 and ω_3 must have opposite signs, and changing the direction of the x_1 principal axis if necessary)

$$\boxed{\omega_1 = L\sqrt{\frac{I_2 - I_3}{I_1 I_2 (I_1 - I_3)}}\;\mathrm{sech}(\nu t)\,, \qquad \omega_3 = -L\sqrt{\frac{I_1 - I_2}{I_2 I_3 (I_1 - I_3)}}\;\mathrm{sech}(\nu t)\,.}$$

Thus when $t \to \infty$ we have

$$\omega_2 \to \frac{L}{I_2}\,, \qquad \omega_{1,3} \to 0\,;$$

in other words, in the limit $t \to \infty$ the body rotates around the x_2 principal axis of inertia with constant angular velocity.

Remark 7.3. In the generic case $I_i \neq I_j$ for $i \neq j$, *the solution of Euler's equations with* $\mathbf{N} = 0$ *can be found by quadratures* using the conservation of E and L^2. Indeed, let us suppose, as in the previous exercise, that $I_1 > I_2 > I_3$. Proceeding as before we can solve for $\omega_{1,3}$ in terms of L and E to find

$$\omega_1^2 = \frac{L^2 - 2I_3E - I_2(I_2 - I_3)\omega_2^2}{I_1(I_1 - I_3)}\,, \qquad \omega_3^2 = \frac{2I_1E - L^2 - I_2(I_1 - I_2)\omega_2^2}{I_3(I_1 - I_3)}\,. \tag{7.48}$$

From these equations it follows that the motion is only possible if $2EI_3 \leqslant L^2 \leqslant 2EI_1$. Note that $L^2 = 2EI_3$ implies that

$$\omega_1 = \omega_2 = 0, \qquad \omega_3^2 = \frac{2I_1E - L^2}{I_3(I_1 - I_3)} = \frac{L^2}{I_3^2}\,,$$

so that the body is rotating around its third principal axis of inertia with constant angular velocity, and similarly if $L^2 = 2EI_1$. To exclude these trivial solutions, we shall suppose in what follows that $2EI_3 < L^2 < 2EI_1$. From the Euler equation for ω_2 we then obtain

$$\int \frac{\mathrm{d}\omega_2}{\sqrt{\left[L^2 - 2I_3E - I_2(I_2 - I_3)\omega_2^2\right]\left[2I_1E - L^2 - I_2(I_1 - I_2)\omega_2^2\right]}} = \pm\frac{t - t_0}{I_2\sqrt{I_1 I_3}}\,. \tag{7.49}$$

Once $\omega_2(t)$ is found from this equation (which can be explicitly done in terms of Jacobian elliptic functions), $\omega_1(t)$ and $\omega_3(t)$ can be immediately determined using Eq. (7.48). ■

Exercise 7.8. Show that ω_2 in Eq. (7.49) behaves as the x coordinate of a particle moving in a certain effective one-dimensional potential. Find the particle's effective energy and analyze the qualitative features of the motion of ω_2.

Solution. From Eq. (7.49) it immediately follows that

$$\frac{1}{2}\,\dot{\omega}_2^2 + U(\omega_2) = 0\,,$$

with

$$U(\omega_2) = -(2I_1I_2^2I_3)^{-1}\Big(L^2 - 2I_3E - I_2(I_2 - I_3)\omega_2^2\Big)\Big(2I_1E - L^2 - I_2(I_1 - I_2)\omega_2^2\Big)\,.$$

Thus ω_2 behaves as the x coordinate of a particle of unit mass and zero energy subject to the one-dimensional potential $U(x)$. Since the coefficients

$$L^2 - 2I_3E =: a_1^2, \quad I_2(I_2 - I_3) =: b_1^2, \quad 2I_1E - L^2 := a_2^2, \quad I_2(I_1 - I_2) =: b_2^2,$$

are all positive, the effective potential behaves as shown in Fig. 7.5 with

$$r_1 = \min\left\{\frac{a_1}{b_1}, \frac{a_2}{b_2}\right\}, \qquad r_2 = \max\left\{\frac{a_1}{b_1}, \frac{a_2}{b_2}\right\}$$

(where $a_i, b_i > 0$). From Eq. (7.48) we have

$$a_1^2 - b_1^2\omega_2^2 = I_1(I_1 - I_3)\omega_1^2 \geqslant 0\,, \qquad a_2^2 - b_2^2\omega_2^2 = I_3(I_1 - I_3)\omega_1^2 \geqslant 0,$$

and therefore $|\omega_2(t)| \leqslant \min\left\{\frac{a_1}{b_1}, \frac{a_2}{b_2}\right\} = r_1$. Hence the motion of ω_2 (and, as a consequence, ω_1 and ω_3) is *always bounded*. Moreover, an elementary calculation shows that the absolute maximum $U_{\max}$ of $U(\omega_2)$ is reached for

$$\omega_2 = \pm\left(\frac{r_1^2 + r_2^2}{2}\right)^{1/2},$$

and that

$$U_{\max} = \frac{b_1^2 b_2^2}{8I_1I_2^2I_3}\,(r_2^2 - r_1^2)^2 \geqslant 0\,.$$

Since the effective energy is zero, it follows from Fig. (7.5) that $\omega_2(t)$—and, hence, $\omega_1(t)$ and $\omega_3(t)$—*is a periodic function of t if $U_{\max} > 0$*, whereas *when $U_{\max} = 0$*

the motion of $\omega_2(t)$ is bounded but not periodic (and similarly for $\omega_1(t)$ and $\omega_3(t)$). Note, finally, that

$$\begin{aligned} U_{\max} = 0 \iff r_1^2 = r_2^2 &\iff a_1^2 b_2^2 = a_2^2 b_1^2 \\ &\iff (L^2 - 2I_3E)(I_1 - I_2) = (2I_1E - L^2)(I_2 - I_3) \\ &\iff L^2 = 2EI_2, \end{aligned}$$

which is the case dealt with in Exercise 7.7.

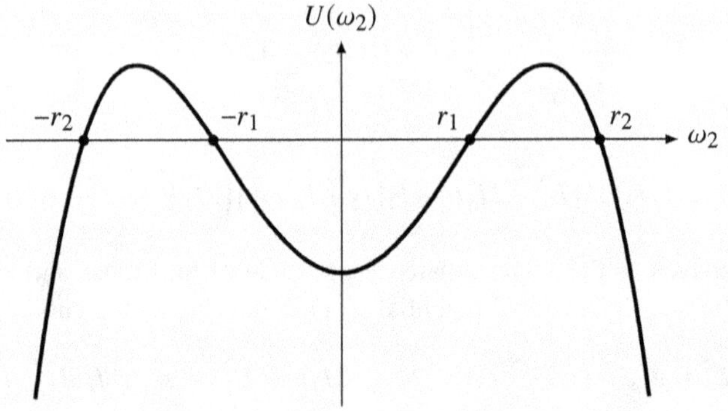

Figure 7.5. Effective potential $U(\omega_2)$.

7.6 EULER ANGLES

Euler's equations, derived in Section 7.4.3, have two major drawbacks, namely: i) they are formulated in terms of the components of the angular velocity $\omega(t)$ *in the frame of body axes*, whose motion with respect to a fixed inertial frame is not known a priori, and ii) at best, they determine the angular velocity, but do not directly yield the orientation of the frame of body axes with respect to the inertial one as a function of time. To sidestep both drawbacks, in this section we shall describe a convenient way of determining the orientation of the frame of body axes (or, in general, the relative orientation of two frames) in terms of three appropriate angles. Once this is done, we shall express the angular velocity $\omega(t)$ in terms of these angles, which, by virtue of Euler's equations, will lead to a second-order system of differential equations for them. Solving this system one can in principle find the orientation of the body axes at all times, which is equivalent to determining the body's rotational motion.

We shall start by showing how to bring the fixed axes $\{\mathbf{e}_i'\}_{1\leqslant i\leqslant 3}$ to the body axes $\{\mathbf{e}_i\}_{1\leqslant i\leqslant 3}$ by means of three successive rotations. We begin by performing a rotation about the x_3' axis by a suitable angle $\phi \in [0, 2\pi)$, so that the rotated x_1' axis points along the direction of the vector $\mathbf{e}_3' \times \mathbf{e}_3$ (cf. Fig. 7.6 (a)). Calling $\{\mathbf{e}_i''\}_{1\leqslant i\leqslant 3}$ the new frame thus obtained we have

$$\mathbf{e}_i'' = R_{\mathbf{e}_3'}(\phi)\, \mathbf{e}_i',$$

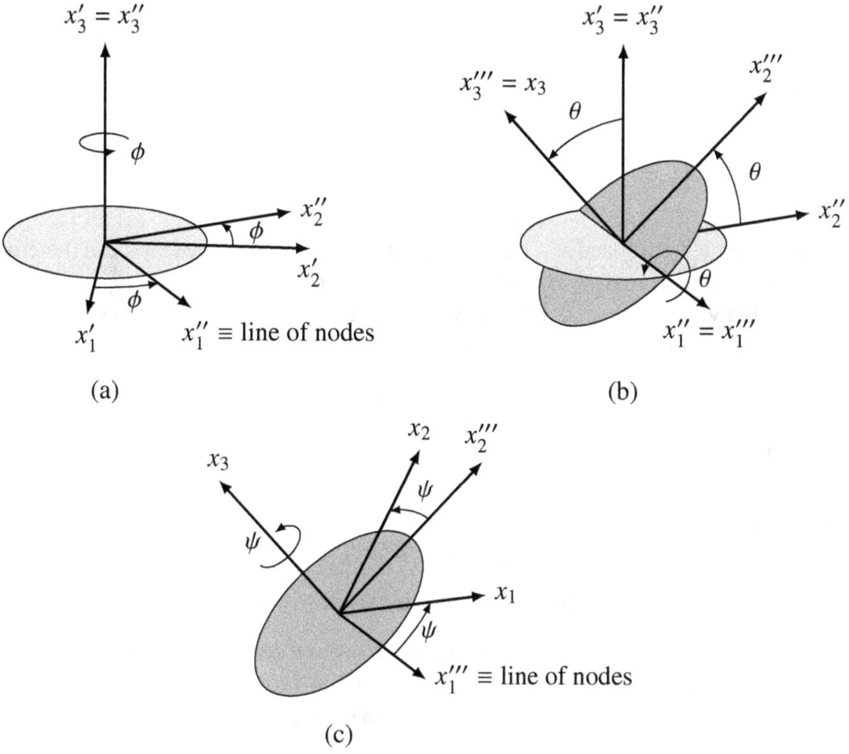

Figure 7.6. Euler angles.

and, in particular,

$$\mathbf{e}_3'' = \mathbf{e}_3' \,.$$

In matrix notation, if $\mathbf{x}' = (x_1', x_2', x_3')$ and $\mathbf{x}'' = (x_1'', x_2'', x_3'')$ respectively denote the coordinates of the *same* vector with respect to the $\{\mathbf{e}_i'\}_{1\leqslant i\leqslant 3}$ and $\{\mathbf{e}_i''\}_{1\leqslant i\leqslant 3}$ axes, regarded as column vectors, then

$$\mathbf{x}' = R_3(\phi)\,\mathbf{x}'' \,.$$

Since the new x_1'' axis, called the **line of nodes**, is perpendicular to both the x_3 and x_3' axes, we can bring the x_3' axis to the x_3 axis through a rotation by an appropriate angle $\theta \in [0, \pi]$ around the x_1'' axis (cf. Fig. 7.6 (b)). Denoting by $\{\mathbf{e}_i'''\}_{1\leqslant i\leqslant 3}$ the new axes thus obtained we can write

$$\mathbf{e}_i''' = R_{\mathbf{e}_1''}(\theta)\,\mathbf{e}_i'' \,,$$

and, in particular,

$$\mathbf{e}_1''' = \mathbf{e}_1'' , \qquad \mathbf{e}_3''' = \mathbf{e}_3 .$$

In terms of coordinates,

$$\mathbf{x}'' = R_1(\theta)\,\mathbf{x}''' , \qquad R_1(\theta) := \begin{pmatrix} 1 & 0 & 0 \\ 0 & \cos\theta & -\sin\theta \\ 0 & \sin\theta & \cos\theta \end{pmatrix} .$$

The new axes $x_1''' = x_1''$ and x_2''' lie on the plane perpendicular to $\mathbf{e}_3$, and can therefore be brought to the x_1 and x_2 axes of the body frame through a suitable rotation about the x_3 axis by an angle $\psi \in [0, 2\pi)$, without changing the orientation of this axis (cf. Fig. 7.6 (c)). We thus have

$$\mathbf{e}_i = R_{\mathbf{e}_3}(\psi)\,\mathbf{e}_i''' = R_{\mathbf{e}_3'''}(\psi)\,\mathbf{e}_i''' ,$$

or in terms of coordinates

$$\mathbf{x}''' = R_3(\psi)\,\mathbf{x} .$$

We thus have

$$\mathbf{x}' = R(\phi, \theta, \psi)\,\mathbf{x} , \tag{7.50}$$

where $R(\phi, \theta, \psi) = R_3(\phi) R_1(\theta) R_3(\psi)$ is the matrix with elements

$$\begin{pmatrix} \cos\psi\cos\phi - \cos\theta\sin\psi\sin\phi & -\sin\psi\cos\phi - \cos\theta\cos\psi\sin\phi & \sin\theta\sin\phi \\ \cos\psi\sin\phi + \cos\theta\sin\psi\cos\phi & -\sin\psi\sin\phi + \cos\theta\cos\psi\cos\phi & -\sin\theta\cos\phi \\ \sin\theta\sin\psi & \sin\theta\cos\psi & \cos\theta \end{pmatrix} .$$

The inverse transformation $\mathbf{x}' \mapsto \mathbf{x}$ can be easily found as follows:

$$\mathbf{x} = R_3(\psi)^{-1} R_1(\theta)^{-1} R_3(\phi)^{-1} \mathbf{x}' = R_3(-\psi) R_1(-\theta) R_3(-\phi)\,\mathbf{x}' = R(-\psi, -\theta, -\phi)\,\mathbf{x}' .$$

Exercise 7.9. Show that the angles $(\phi, \theta, \psi) \in [0, 2\pi) \times [0, \pi] \times [0, 2\pi)$ are *uniquely determined* by the rotation matrix $R(\phi, \theta, \psi)$. In other words,

$$R(\phi, \theta, \psi) = R(\phi', \theta', \psi') \quad \Longrightarrow \quad (\phi', \theta', \psi') = (\phi, \theta, \psi).$$

Solution. Indeed, equating the last rows of $R(\phi, \theta, \psi)$ and $R(\phi', \theta', \psi')$ we obtain

$$(\sin\theta\sin\psi, \sin\theta\cos\psi, \cos\theta) = (\sin\theta'\sin\psi', \sin\theta'\cos\psi', \cos\theta'),$$

and therefore (since both (θ, ψ) and (θ', ψ') belong to the range $[0, \pi] \times [0, 2\pi)$ of the spherical coordinate system)

$$(\theta', \psi') = (\theta, \psi).$$

Equating now the last column of $R(\phi, \theta, \psi)$ to that of $R(\phi', \theta', \psi')$ we arrive at the equation

$$\sin\theta(\sin\phi, -\cos\phi) = \sin\theta(\sin\phi', -\cos\phi').$$

If $\sin\theta \neq 0$ (i.e., $\theta \neq 0, \pi$) we have

$$(\sin\phi, -\cos\phi) = (\sin\phi', -\cos\phi') \quad \Longrightarrow \quad \phi = \phi',$$

since both ϕ and ϕ' belong to the interval $[0, 2\pi)$. On the other hand, if $\sin\theta = 0$ then $\cos\theta = \pm 1$, and equating the first column of $R(\phi, \theta, \psi)$ to that of $R(\phi', \theta', \psi')$ we obtain

$$R(\pm\psi)\begin{pmatrix}\cos\phi\\ \sin\phi\end{pmatrix} = R(\pm\psi)\begin{pmatrix}\cos\phi'\\ \sin\phi'\end{pmatrix}, \qquad \text{with} \quad R(\alpha) := \begin{pmatrix}\cos\alpha & -\sin\alpha\\ \sin\alpha & \cos\alpha\end{pmatrix}.$$

Since the two-dimensional rotation matrix $R(\alpha)$ is invertible, we deduce that

$$(\cos\phi, \sin\phi) = (\cos\phi', \sin\phi').$$

This again implies that $\phi = \phi'$, as both ϕ and ϕ' belong to the interval $[0, 2\pi)$.

The angles

$$\phi \in [0, 2\pi), \qquad \theta \in [0, \pi], \qquad \psi \in [0, 2\pi), \tag{7.51}$$

which, as we have seen, fully and uniquely determine the orientation of the frame of body axes with respect to the fixed inertial frame, are called **Euler angles**.

Note that the Euler angles have the following geometric interpretation (cf. Fig. 7.6):

$$\boxed{\begin{aligned}&\phi = \text{angle between the } x_1' \text{ axis and the line of nodes}\\ &\theta = \text{angle between the } x_3' \text{ and } x_3 \text{ axes}\\ &\psi = \text{angle between the line of nodes and the } x_1 \text{ axis.}\end{aligned}} \tag{7.52}$$

By the additivity of angular velocities proved in Section 7.5, the angular velocity $\omega(t)$ can be expressed in terms of the Euler angles by the formula

$$\boxed{\omega = \dot\phi\, \mathbf{e}_3' + \dot\theta \mathbf{e}_1'' + \dot\psi\, \mathbf{e}_3\,.} \tag{7.53}$$

It is also of interest to express the angular velocity in the frame of body axes $\{\mathbf{e}_i\}_{1\leqslant i\leqslant 3}$, since Euler's equations are written in terms of the components of ω in this frame. To this end, note first of all that

$$\mathbf{e}_1'' = \mathbf{e}_1''' = R_{\mathbf{e}_3}(\psi)^{-1}\mathbf{e}_1 = R_{\mathbf{e}_3}(-\psi)\mathbf{e}_1 = \cos\psi\mathbf{e}_1 - \sin\psi\mathbf{e}_2\,,$$

while

$$\mathbf{e}_3' = R(\phi, \theta, \psi)^{-1}\mathbf{e}_3 = R(\phi, \theta, \psi)^{\mathsf{T}}\mathbf{e}_3 = \sin\theta\sin\psi\mathbf{e}_1 + \sin\theta\cos\psi\mathbf{e}_2 + \cos\theta\mathbf{e}_3. \tag{7.54}$$

Using these formulas in Eq. (7.53) we finally obtain the following expression for the components ω_i of ω in the frame of body axes:

$$\begin{aligned} \omega_1 &= \dot{\phi}\sin\theta\sin\psi + \dot{\theta}\cos\psi\,, \\ \omega_2 &= \dot{\phi}\sin\theta\cos\psi - \dot{\theta}\sin\psi\,, \\ \omega_3 &= \dot{\psi} + \dot{\phi}\cos\theta\,. \end{aligned} \tag{7.55}$$

Substituting these equations into Euler's equations (7.38) we obtain a system of second-order nonlinear differential equations for the three Euler angles (ϕ, θ, ψ), whose solution with appropriate initial conditions determines the orientation of the body axes relative to the inertial frame at all times. In practice, if the components N_i are explicitly known in terms of the angular velocity components ω_i it is often preferable to first solve Euler's equations (7.38) for $\omega(t)$ and then compute the Euler angles using Eqs. (7.55).

To see how the Euler angles can be expressed in terms of the components ω_i of the angular velocity *in the body frame*, let us start by choosing the $\mathbf{e}'_3$ axis of the fixed frame to coincide with the invariant direction $\mathbf{L}/L$, so that $\mathbf{L} = L\mathbf{e}'_3$. Using Eq. (7.54) we then obtain

$$L_1 = I_1\omega_1 = L\sin\theta\sin\psi\,, \quad L_2 = I_2\omega_2 = L\sin\theta\cos\psi\,, \quad L_3 = I_3\omega_3 = L\cos\theta, \tag{7.56}$$

from which we easily deduce that

$$\cos\theta = \frac{I_3\omega_3}{L}\,, \qquad \tan\psi = \frac{I_1\omega_1}{I_2\omega_2}\,. \tag{7.57}$$

These equations determine[7] the angles θ and ψ. From Eqs. (7.55)–(7.56) it then follows that $\dot{\phi}$ can be expressed purely in terms of ω_3:

$$\dot{\phi} = \frac{\omega_1\sin\psi + \omega_2\cos\psi}{\sin\theta} = \frac{I_1\omega_1^2 + I_2\omega_2^2}{L\sin^2\theta} = \frac{I_1\omega_1^2 + I_2\omega_2^2}{L\left(1 - \frac{I_3^2\omega_3^2}{L^2}\right)} = \frac{L(2E - I_3\omega_3^2)}{L^2 - I_3^2\omega_3^2}\,. \tag{7.58}$$

Integrating the previous equation with respect to t yields the angle ϕ.

The procedure just outlined can be applied, for instance, to determining the inertial motion of a generic rigid body with $I_1 > I_2 > I_3$. Indeed, in this case $\mathbf{N} = 0$, and $\omega_i(t)$ can be computed by quadratures as explained in Remark 7.3. The body's orientation at all times can then be found by quadratures from Eqs. (7.57)–(7.58).

[7]More precisely, in order to fully determine the angle ψ we need to use the relation $\operatorname{sgn}(\sin\psi) = \operatorname{sgn}(\omega_1)$, which follows from the first equation (7.56).

Example 7.2. As an example, let us determine the inertial motion of a symmetric top using the above method. Taking the x_3 axis along the body's symmetry axis, the solution of Euler's equations is given by Eq. (7.43), namely

$$\omega_1 = \omega \sin\alpha \cos(\Omega t + \beta), \quad \omega_2 = \omega \sin\alpha \sin(\Omega t + \beta), \quad \omega_3 = \omega \cos\alpha,$$

with $\omega = |\boldsymbol{\omega}|$ and α integration constants and Ω given by Eq. (7.41). Since in this case $I_1 = I_2$, from the second Eq. (7.57) we obtain

$$\tan\psi = \frac{\omega_1}{\omega_2} = \cot(\Omega t + \beta) = \tan\left(\tfrac{\pi}{2} - \Omega t - \beta\right) \implies \psi = \pm\frac{\pi}{2} - \Omega t - \beta.$$

Assuming, as in Section 7.5, that $0 < \alpha < \pi/2$, we must take the "+" sign in the previous equation, since

$$\sin\psi = \sin\left(\pm\tfrac{\pi}{2} - \Omega t - \beta\right) = \pm\cos\left(\Omega t + \beta\right)$$

must have the same sign as $\omega_1 = \omega \sin\alpha \cos(\Omega t + \beta)$. We thus have

$$\psi = -\Omega t + \frac{\pi}{2} - \beta \implies \dot\psi = -\Omega.$$

Thus the body rotates about its symmetry axis (i.e., the x_3 axis) with constant angular velocity $-\Omega$, as we saw in Section 7.5. Similarly, the first Eq. (7.57) yields

$$\cos\theta = \frac{I_3}{L}\,\omega\cos\alpha. \tag{7.59}$$

This is the same result obtained in Section 7.5, as

$$\tan^2\theta = \sec^2\theta - 1 = \frac{L^2}{I_3^2\omega_3^2} - 1 = \frac{I_1^2(\omega_1^2 + \omega_2^2)}{I_3^2\omega_3^2} = \frac{I_1^2}{I_3^2}\tan^2\alpha,$$

which is equivalent to Eq. (7.45) (indeed, by Eq. (7.59) $\tan\theta$ must be positive). Finally, by Eqs. (7.57) and (7.58) we have

$$\dot\phi = \frac{I_1(\omega_1^2 + \omega_2^2)}{L\left(1 - \frac{I_3^2\omega_3^2}{L^2}\right)} = \frac{L I_1(\omega_1^2 + \omega_2^2)}{I_1^2(\omega_1^2 + \omega_2^2)} = \frac{L}{I_1}.$$

As expected, $\dot\phi$ is equal to the angular velocity Ω_p of the precession of the body's symmetry (x_3) axis around the invariant direction (i.e., the x_3' axis). Note also that, since θ is constant, by Eq. (7.53) the body's angular velocity is given by

$$\boldsymbol{\omega} = \dot\phi\,\mathbf{e}_3' + \dot\psi\,\mathbf{e}_3 = \frac{L}{I_1}\,\mathbf{e}_3' - \Omega\,\mathbf{e}_3 = \Omega_p\frac{\mathbf{L}}{L} - \Omega\,\mathbf{e}_3,$$

again in agreement with Eq. (7.46).

Exercise 7.10. Determine the Euler angles in the inertial motion of an asymmetric top with $I_2 > I_3 > I_1$ in the special case where $L^2 = 2I_3E$. Using the result obtained, describe the body's motion as seen from the fixed frame.

Solution. We can rewrite the solution of Exercise 7.7 as

$$\omega_1 = aL\,\text{sech}(\nu t), \qquad \omega_2 = bL\,\text{sech}(\nu t), \qquad \omega_3 = \frac{L}{I_3}\tanh(\nu t),$$

with

$$a = -\sqrt{\frac{I_2 - I_3}{I_1 I_3 (I_2 - I_1)}}, \qquad b = \sqrt{\frac{I_3 - I_1}{I_2 I_3 (I_2 - I_1)}}, \qquad \nu = \frac{L}{I_3}\sqrt{\frac{(I_3 - I_1)(I_2 - I_3)}{I_1 I_2}}.$$

From Eqs. (7.57) we obtain

$$\theta = \arccos\big(\tanh(\nu t)\big), \qquad \tan\psi = \frac{I_1 a}{I_2 b} = -\sqrt{\frac{I_1(I_2 - I_3)}{I_2(I_3 - I_1)}} \qquad \Longrightarrow \qquad \dot{\psi} = 0.$$

Finally, since

$$2E - I_3\omega_3^2 = \frac{L^2}{I_3} - I_3\omega_3^2 = \frac{1}{I_3}\left(L^2 - I_3^2\omega_3^2\right),$$

by Eq. (7.58) we have

$$\dot{\phi} = \frac{L}{I_3} = \text{const.}$$

Since $\dot{\psi} = 0$, the motion of the body consists of a precession of its x_3 principal axis—i.e., the principal axis corresponding to the intermediate moment of inertia—about the invariant (x_3') direction with constant angular velocity $\dot{\phi}$, combined with a steady nutation (wobbling) of the x_3 principal axis relative to the invariant direction. More precisely, the angle θ between this principal axis and the invariant direction decreases from $\theta = \pi$ for $t \to -\infty$ to $\theta = 0$ for $t \to \infty$. Note that for very large t this motion becomes practically indistinguishable from a pure rotation around the invariant direction with constant angular velocity $\dot{\phi}$.

7.7 LAGRANGE'S TOP

In this section we shall study the motion of an axially symmetric rigid body with a fixed point subject only to the force of Earth's gravity, usually called **Lagrange's top** (cf. Fig. 7.7). Taking the top's fixed point as the origin of the inertial (fixed) frame, its kinetic energy is given by

$$T = \frac{1}{2}\omega \cdot (I\omega) = \frac{1}{2}I_1(\omega_1^2 + \omega_2^2) + \frac{1}{2}I_3\omega_3^2,$$

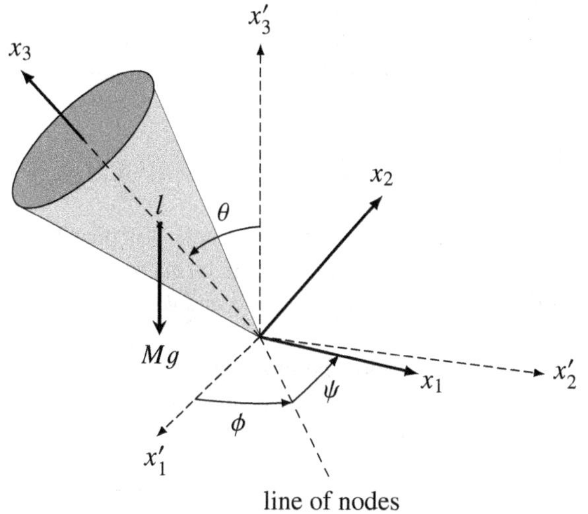

Figure 7.7. Lagrange's top.

where I_i $(i = 1, 2, 3)$ is the top's i-th principal moment of inertia *with respect to the origin* and its symmetry axis has been chosen as the x_3 axis. Using Eqs. (7.55) for the components of ω we easily obtain

$$T = \frac{1}{2} I_1(\dot{\theta}^2 + \dot{\phi}^2 \sin^2\theta) + \frac{1}{2} I_3(\dot{\psi} + \dot{\phi}\cos\theta)^2 .$$

The top's potential energy is the potential of the total external force $\mathbf{F} = -Mg\mathbf{e}_3'$ applied at the CM, i.e.,

$$V = MgX_3' .$$

By symmetry, the top's CM lies on the symmetry axis x_3, at a distance l from the origin. We thus have

$$V = Mgl\cos\theta ,$$

and the system's Lagrangian is therefore given by

$$\boxed{L = T - V = \frac{1}{2} I_1(\dot{\theta}^2 + \dot{\phi}^2 \sin^2\theta) + \frac{1}{2} I_3(\dot{\psi} + \dot{\phi}\cos\theta)^2 - Mgl\cos\theta .} \tag{7.60}$$

The canonical momenta associated to the generalized coordinates (θ, ϕ, ψ) are

$$\boxed{\begin{aligned} p_\theta &= \frac{\partial L}{\partial \dot{\theta}} = I_1\dot{\theta} , \\ p_\phi &= \frac{\partial L}{\partial \dot{\phi}} = I_1 \sin^2\theta\,\dot{\phi} + I_3\cos\theta(\dot{\psi} + \dot{\phi}\cos\theta) , \\ p_\psi &= \frac{\partial L}{\partial \dot{\psi}} = I_3(\dot{\psi} + \dot{\phi}\cos\theta) . \end{aligned}} \tag{7.61}$$

Since L is independent of the angles ϕ and ψ, the canonical momenta p_ϕ and p_ψ are conserved. Moreover, since L is also independent of the time t, the energy integral h, which coincides with the total energy $E = T + V$ as L is quadratic in the generalized velocities, is also conserved:

$$h = T + V = \frac{1}{2} I_1(\dot{\theta}^2 + \dot{\phi}^2 \sin^2\theta) + \frac{1}{2} I_3(\dot{\psi} + \dot{\phi}\cos\theta)^2 + Mgl\cos\theta = E\,. \qquad (7.62)$$

We shall see that using these three conservation laws it is possible to determine the motion of the θ coordinate by quadratures (i.e., in terms of an integral), which in turn determines the motion of the other two angles ϕ and ψ using the conservation of p_ϕ and p_ψ.

By the third Eq. (7.55), the canonical momentum p_ψ is equal to $I_3\omega_3$, i.e., to the component of the angular momentum $\mathbf{L}$ in the direction of the $\mathbf{e}_3$ axis, which we shall denote by L_3. Similarly, from Eqs. (7.54) and (7.55) it follows that

$$\begin{aligned} L'_3 = \mathbf{L}\cdot\mathbf{e}'_3 &= \sin\theta(L_1\sin\psi + L_2\cos\psi) + \cos\theta L_3 \\ &= I_1\sin\theta(\omega_1\sin\psi + \omega_2\cos\psi) + I_3\omega_3\cos\theta \\ &= I_1\sin^2\theta\dot{\phi} + I_3\cos\theta(\dot{\psi} + \dot{\phi}\cos\theta) = p_\phi\,. \end{aligned}$$

On the other hand, the torque of the gravitational force

$$\mathbf{N} = -Mgl\mathbf{e}_3 \times \mathbf{e}'_3$$

is perpendicular to the vectors $\mathbf{e}_3$ and $\mathbf{e}'_3$ (i.e., is directed along the line of nodes), and hence $N_3 = \mathbf{n}\cdot\mathbf{e}_3$ and $N'_3 = \mathbf{N}\cdot\mathbf{e}'_3$ both vanish. Applying the equation of motion of the angular momentum in the *fixed* frame with $N'_3 = \mathbf{N}\cdot\mathbf{e}'_3 = 0$ we immediately deduce that L'_3 is conserved. The conservation of $L_3 = I_3\omega_3$ can also be directly proved from elementary considerations by noting that

$$\begin{aligned} \frac{\mathrm{d}}{\mathrm{d}t}(\mathbf{L}\cdot\mathbf{e}_3) = \mathbf{e}_3\cdot\left(\frac{\mathrm{d}\mathbf{L}}{\mathrm{d}t}\right)_{\mathrm{f}} + \mathbf{L}\cdot\left(\frac{\mathrm{d}\mathbf{e}_3}{\mathrm{d}t}\right)_{\mathrm{f}} &= \mathbf{e}_3\cdot\mathbf{N} + \mathbf{L}\cdot(\boldsymbol{\omega}\times\mathbf{e}_3) = N_3 + \mathbf{e}_3\cdot(\mathbf{L}\times\boldsymbol{\omega}) \\ &= \mathbf{e}_3\cdot(\mathbf{L}\times\boldsymbol{\omega}) = (I_1 - I_2)\omega_1\omega_2 = 0\,. \end{aligned}$$

Alternatively, from the third Euler equation (7.38) with $N_3 = 0$ it follows that ω_3, and therefore $L_3 = I_3\omega_3$, is constant in the body frame. Since L_3 is a scalar, it must also be constant in the inertial frame.

The motion of the angle θ is determined substituting the relations

$$\dot{\psi} + \dot{\phi}\cos\theta = \frac{p_\psi}{I_3}\,, \qquad I_1\sin^2\theta\,\dot{\phi} = p_\phi - p_\psi\cos\theta \qquad (7.63)$$

into Eq. (7.62), which yields a first-order differential equation in θ with separable variables:

$$\boxed{\frac{1}{2}I_1\dot{\theta}^2 + \frac{(p_\phi - p_\psi\cos\theta)^2}{2I_1\sin^2\theta} + Mgl\cos\theta = E - \frac{p_\psi^2}{2I_3}\,.} \qquad (7.64)$$

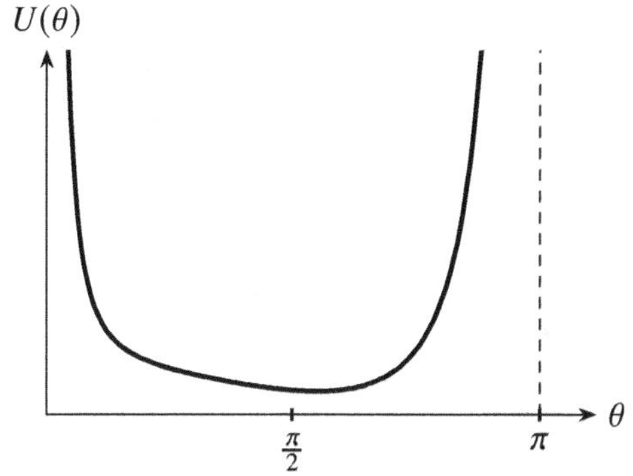

Figure 7.8. Effective potential $U(\theta)$ in Eq. (7.66).

This is formally the equation of motion of a particle of unit mass and energy

$$\varepsilon = \frac{1}{I_1}\left(E - \frac{p_\psi^2}{2I_3}\right) \tag{7.65}$$

in the effective one-dimensional potential[8]

$$\boxed{U(\theta) = \frac{(a - b\cos\theta)^2}{2\sin^2\theta} + c\cos\theta\,,} \qquad 0 \leqslant \theta \leqslant \pi\,, \tag{7.66a}$$

with

$$\boxed{a = \frac{p_\phi}{I_1}\,, \qquad b = \frac{p_\psi}{I_1}\,, \qquad c = \frac{Mgl}{I_1} > 0} \tag{7.66b}$$

(cf. Fig. 7.8). The above potential is singular for $\theta = 0$ or $\theta = \pi$. However, from the conservation of p_ϕ and p_ψ it follows that if at some instant $\theta = 0$ (respectively $\theta = \pi$), then $a = b$ (resp. $a = -b$). We shall show below that in such a case U is, in fact, regular at $\theta = 0$ (resp. at $\theta = \pi$), and that moreover $U'(0) = 0$ (resp. $U'(\pi) = 0$).

We shall assume, for the time being, that

$$a \neq \pm b\,,$$

and hence $\theta \neq 0, \pi$. In terms of the variable

$$\boxed{u = \cos\theta \in (-1, 1)}$$

the conservation of energy equation (7.64) can be written as

$$\boxed{\dot{u}^2 = f(u)\,,} \tag{7.67a}$$

[8]For the angle θ to range from $\pi/2$ to π the top should actually be a gyroscope.

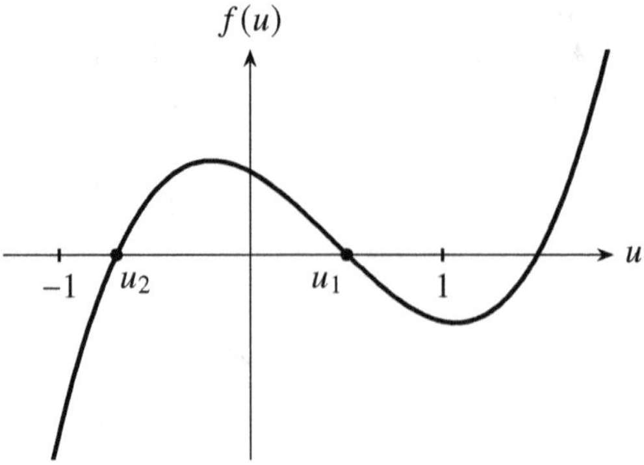

Figure 7.9. Polynomial $f(u)$ in Eq. (7.67b) (in the generic case $a \neq b$).

where

$$\boxed{f(u) = 2(\varepsilon - cu)(1 - u^2) - (a - bu)^2} \tag{7.67b}$$

is a third-degree polynomial. Formally, the previous equation can be used to solve for the motion of the θ coordinate through the formula

$$t - t_0 = \pm \int^{\cos\theta} \frac{du}{\sqrt{f(u)}} = \pm \int^{\cos\theta} \frac{du}{\sqrt{2(\varepsilon - cu)(1 - u^2) - (a - bu)^2}} .$$

This expression is of little use in practice, since the integral can be evaluated only in terms of elliptic functions. On the other hand, Eqs. (7.67) provide a simple qualitative description of the motion of the top that we shall develop next.

First of all, since $a \neq \pm b$ the points $u = \pm 1$ are *not* roots of f. Secondly, as $c > 0$ the polynomial $f(u)$ is positive for $u \to \infty$ and negative for $u \to -\infty$. On the other hand, this polynomial must have either one or three real roots (counting multiplicities). At least one of these roots should belong to the physical interval $(-1, 1)$, since otherwise f would not change sign in this interval and therefore would be negative there, as $f(\pm 1) < 0$. As a matter of fact, in the interval $(-1, 1)$ there must be *exactly two* roots of f (counting multiplicities), since if there were an odd number of roots $f(-1)$ and $f(1)$ would have opposite signs. This implies that f necessarily has three real roots, two of which (counting multiplicities) belong to the interval $(-1, 1)$ while the third one lies on the half-line $(1, \infty)$. Hence the graph of f looks qualitatively as shown in Fig. 7.9.

Denoting by $u_2 < u_1$ the two roots of f in $(-1, 1)$, the motion takes place in the region $u_2 \leqslant u \leqslant u_1$, or equivalently $\theta_1 \leqslant \theta \leqslant \theta_2$, where $\theta_i = \arccos u_i$. In other words, the top's symmetry axis oscillates between the angles θ_1 and θ_2, a wobbling motion that is called **nutation** (from the Latin "nutatio," nodding). If

$$\sin^2\theta\,\dot{\phi} = a - b\cos\theta = a - bu$$

does not change sign when $\cos\theta$ ranges between u_2 and u_1, the path traced out by the axis of the top on the unit sphere is similar to that shown in Fig. 7.10 (left), while if

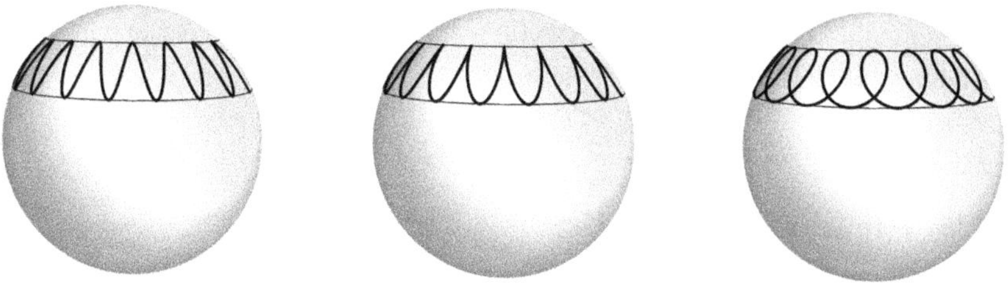

Figure 7.10. Path traced out on the unit sphere by the symmetry (x_3) axis of Lagrange's top (the thin horizontal lines are the curves $\theta = \theta_1$ and $\theta = \theta_2$).

$a - bu$ changes sign in the interval (u_2, u_1) it resembles that represented in Fig. 7.10 (right).

Let us analyze next the limiting case in which $a - bu$ vanishes at one of the *endpoints* of the interval $[u_2, u_1]$, which for the moment we shall denote by u_0, and has therefore constant sign for other values of $u \in [u_2, u_1]$. Since u_0 is by hypothesis one of the turning points $u_{1,2}$ of the potential $f(u)$, setting $u_0 = \cos\theta_0$ we have

$$a = bu_0 \implies f(u) = 2(\varepsilon - cu)(1-u^2) - b^2(u-u_0)^2, \quad f(u_0) = 0 \implies \varepsilon = cu_0,$$

and therefore

$$f(u) = (u - u_0)\left[2c(u^2 - 1) - b^2(u - u_0)\right].$$

Since $u_0 = \cos\theta_0 < 1$ and $c > 0$, and hence

$$f'(u_0) = 2c(u_0^2 - 1) < 0,$$

u_0 must then be equal to the *largest* root u_1 (cf. Fig. 7.9), and therefore $\theta_0 = \theta_1$. The precession velocity

$$\dot{\phi} = \frac{a - bu}{1 - u^2} = \frac{b(u_1 - u)}{1 - u^2}$$

thus verifies $b\dot{\phi} > 0$ for $\theta_1 < \theta \leqslant \theta_2$ and $\dot{\phi} = 0$ for $\theta = \theta_1$. In other words, the path of the axis of the top on the unit sphere has a *cusp* at $\theta = \theta_1$ (cf. Fig. 7.10, center). Note, finally, that the case in which $\dot{\phi}$ vanishes for $\theta = \theta_1$ occurs precisely when the top is thrown in the usual way, i.e., when initially $\dot{\theta}(0) = \dot{\phi}(0) = 0$, since (denoting by $u_0 = \cos\theta_0 \neq \pm 1$ the initial value of u)

$$\dot{\phi}(0) = 0 \implies a - bu_0 = 0, \quad \dot{\theta}(0) = 0 \implies \dot{u}(0) = f(u_0) = -\sin\theta_0\,\dot{\theta}(0) = 0.$$

Let us now study under what conditions it is possible for the top to precess but not nutate, i.e., when do the equations of motion have a constant solution $\theta = \theta_0$. (For the moment, we shall continue to assume that $a \neq \pm b$, and therefore $0 < \theta_0 < \pi$.) In this case the equation of motion for the variable $u = \cos\theta$,

$$\ddot{u} = \frac{1}{2} f'(u),$$

obtained by differentiating Eq. (7.67a) with respect to time, must have an equilibrium solution $u = u_0$. For this to be the case we must have

$$f(u_0) = f'(u_0) = 0 \,. \tag{7.68}$$

Note that if θ is constant so is the precession velocity

$$\Omega_p := \dot{\phi} = \frac{a - bu_0}{1 - u_0^2} \,,$$

(recall that $\theta_0 \neq 0, \pi$, and hence $u_0 \neq \pm 1$). Using conditions (7.68) we obtain

$$2(\varepsilon - cu_0) = \frac{(a - bu_0)^2}{1 - u_0^2} = (1 - u_0^2)\Omega_p^2 \,,$$

and therefore

$$\frac{1}{2} f'(u_0) = -2u_0(\varepsilon - cu_0) - c(1 - u_0^2) + b(a - bu_0) = -(1 - u_0^2)(u_0\Omega_p^2 - b\Omega_p + c) \,.$$

The precession velocity Ω_p must therefore verify the quadratic equation

$$u_0\Omega_p^2 - b\Omega_p + c = 0 \,.$$

This equation possesses real roots if and only if its discriminant is non-negative, i.e., if

$$b^2 - 4cu_0 \geqslant 0 \,,$$

or equivalently

$$\boxed{p_\psi^2 = I_3^2\omega_3^2 \geqslant 4MglI_1 \cos\theta_0 \,.} \tag{7.69}$$

Note that this condition automatically holds if $\theta_0 \geqslant \pi/2$. If Eq. (7.69) is satisfied, there will be in general *two* precession frequencies given by[9]

$$\begin{aligned} \Omega_{p,\pm} &= \frac{1}{2u_0}\left(b \pm \sqrt{b^2 - 4cu_0}\right) = \frac{b}{2u_0}\left(1 \pm \sqrt{1 - \frac{4cu_0}{b^2}}\right) \\ &= \frac{I_3\omega_3}{2I_1 \cos\theta_0}\left(1 \pm \sqrt{1 - \frac{4MglI_1 \cos\theta_0}{I_3^2\omega_3^2}}\right). \end{aligned}$$

If the frequency ω_3 is much larger than the term $(MglI_1|\cos\theta_0|)^{1/2}/I_3$, the radical in the previous equation can be approximated by

$$1 - \frac{2MglI_1 \cos\theta_0}{I_3^2\omega_3^2} \,,$$

[9]For $\theta_0 = \pi/2$, or equivalently $u_0 = 0$, the precession velocity is $\Omega_p = c/b = Mgl/(I_3\omega_3)$.

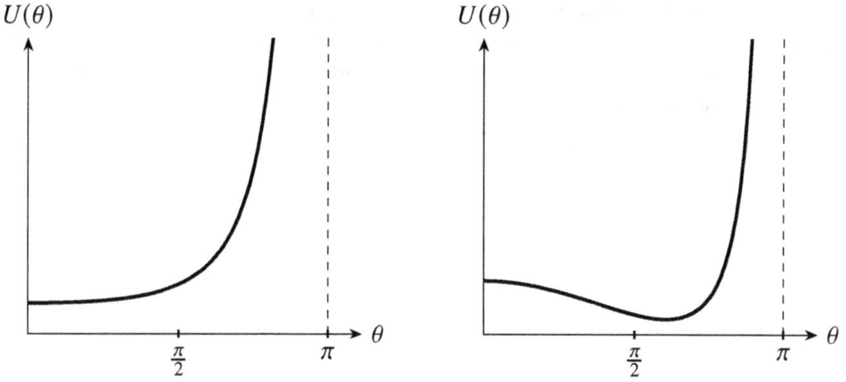

Figure 7.11. Effective potential $U(\theta)$ for $a = b \geqslant 2\sqrt{c}$ (left) and $a = b < 2\sqrt{c}$ (right).

and hence

$$\Omega_{p,+} \simeq \frac{I_3\omega_3}{I_1\cos\theta_0}, \qquad \Omega_{p,-} \simeq \frac{Mgl}{I_3\omega_3}.$$

In practice, friction causes ω_3 to slowly decrease until the condition (7.69) stops being fulfilled, at which point the top begins to wobble.

Let us finally determine the top's motion in the case $a = b$ (the case $a = -b$ is dealt with in a similar fashion). The effective potential $U(\theta)$ reduces now to

$$U(\theta) = \frac{a^2}{2}\frac{(1-\cos\theta)^2}{\sin^2\theta} + c\cos\theta = \frac{a^2}{2}\tan^2(\theta/2) + c\cos\theta. \tag{7.70}$$

Differentiating with respect to θ we obtain

$$\begin{aligned} U'(\theta) &= \frac{a^2}{2}\tan(\theta/2)\sec^2(\theta/2) - 2c\sin(\theta/2)\cos(\theta/2) \\ &= 2c\tan(\theta/2)\sec^2(\theta/2)\left(\frac{a^2}{4c} - \cos^4(\theta/2)\right), \qquad 0 < \theta < \pi. \end{aligned}$$

Thus when $a^2 \geqslant 4c$, or equivalently

$$p_\psi^2 = I_3^2\omega_3^2 \geqslant 4MglI_1, \tag{7.71}$$

the derivative of U vanishes for $\theta = 0$ and is positive for $0 < \theta < \pi$. Therefore the effective potential behaves as shown in Fig. 7.11 (left). In particular, in this case $\theta = 0$ is a *stable equilibrium*, and there are no other equilibrium solutions $\theta =$ const. On the other hand, if $a^2 < 4c$, i.e.,

$$I_3^2\omega_3^2 < 4MglI_1, \tag{7.72}$$

then $\theta = 0$ is an *unstable equilibrium* (a relative maximum), since $U'(\theta) < 0$ for sufficiently small $\theta > 0$. More precisely, in this case, the derivative of U is negative for

$$0 < \theta < \theta_0 = 2\arccos\left(\left(a^2/(4c)\right)^{1/4}\right) = 2\arccos\left(\left(I_3^2\omega_3^2/(4MglI_1)\right)^{1/4}\right),$$

vanishes for $\theta = \theta_0$, and is positive for $\theta_0 < \theta < \pi$. It follows that $\theta = \theta_0$ is actually a *stable equilibrium,* and the potential behaves as shown in Fig. 7.11 (right). In practice, friction makes the frequency ω_3 diminish until condition (7.72) is verified, after which the top starts wobbling ("sleeping top").

CHAPTER 8

Introduction to relativistic mechanics

Newtonian mechanics is built upon the existence of a universal time, shared by all inertial observers. This fundamental tenet is also the basis of Galileo's relativity principle introduced in Chapter 2. However, by the late 19th century it became evident that this principle is incompatible with Maxwell's equations, which govern all electromagnetic phenomena, including light. In 1905 Einstein formulated a new relativity principle, which posits that *all* laws of physics—encompassing both mechanics and electromagnetism—must be the same for all inertial observers. A key consequence of this principle is the invariance of the speed of light, a notion that defies everyday intuition but has been consistently confirmed, beginning with the celebrated Michelson–Morley experiment.

In this chapter we develop the physical consequences of Einstein's relativity principle, and study how the laws of classical mechanics must be modified to comply with it. We begin by deriving the Lorentz transformations, which relate the space-time coordinates of events in different inertial frames. These transformations have important physical implications such as the relativity of simultaneity, time dilation and the Lorentz–Fitzgerald contraction of lengths, all of which ultimately arise from the non-universal character of time. We then develop the framework of relativistic dynamics, introducing fundamental concepts like four-velocity and four-momentum. We explore the conservation of four-momentum and its profound consequence, the equivalence of mass and energy, epitomized in Einstein's iconic relation $E = mc^2$.

We also discuss the modifications of Newton's laws of motion necessary to render them consistent with the principle of relativity, introducing the concepts of four-force and relativistic force and deriving the relativistic Lagrangian. As a simple application of these notions, we analyze the motion of a particle subject to a constant force, known as hyperbolic motion. We then examine the Lorentz force as a fundamental example of relativistic force, introducing the electromagnetic field tensor and discussing its transformation properties. We conclude by applying these concepts to derive

DOI: 10.1201/9781003600633-8

the Liénard–Wiechert electromagnetic potentials of a moving charge using basic relativistic principles.

8.1 THE PRINCIPLES OF SPECIAL RELATIVITY

As we saw in Chapter 2, *the laws of mechanics have the same form in* all *inertial frames*. More precisely, consider (for the sake of simplicity) the Galilean boost

$$\boxed{t' = t\,, \qquad x_1' = x_1 - vt\,, \qquad x_2' = x_2\,, \qquad x_3' = x_3\,,} \tag{8.1}$$

relating the space-time coordinates (t, x_1, x_2, x_3) of an event in an inertial reference frame S with their counterparts (t', x_1', x_2', x_3') in another inertial frame S', whose origin O' moves with *constant* velocity $v\mathbf{e}_1$ with respect to S and whose axes are parallel to those of S. Newton's second law in the S frame

$$m\mathbf{a} = \mathbf{F}(t, \mathbf{r}, \dot{\mathbf{r}})$$

then becomes in S'

$$m\mathbf{a}' = \mathbf{F}'(t', \mathbf{r}', \dot{\mathbf{r}}')\,,$$

where

$$\mathbf{F}'(t', \mathbf{r}', \dot{\mathbf{r}}') = \mathbf{F}(t, \mathbf{r}, \dot{\mathbf{r}}) \quad (\text{with } \mathbf{r}' = \mathbf{r} - vt\mathbf{e}_1, \;\; \dot{\mathbf{r}}' = \dot{\mathbf{r}} - v\mathbf{e}_1).$$

In other words, in both frames the particle's acceleration is the quotient between the force acting on it and its mass, the force being the *same* in both frames but expressed in terms of the particle's coordinates and velocities in each of them. An equivalent way of stating this principle, known as **Galileo's relativity principle**, is the following:

> No *mechanical* experiment can discriminate between two inertial frames.

Indeed, mechanical experiments are ultimately based on Newton's second law, which determines the *acceleration* of particles, and this acceleration is the same in S as in S':

$$\boxed{\mathbf{a}' = \ddot{\mathbf{r}}' = \frac{\mathrm{d}^2}{\mathrm{d}t^2}(\mathbf{r} - vt\mathbf{e}_1) = \ddot{\mathbf{r}} = \mathbf{a}\,.}$$

In other words, the following (Galilean) *relativity principle* holds:

> All inertial frames are *equivalent* from the point of view of Newtonian mechanics.

At the end of the 19th century, the question arose whether Galileo's relativity principle also applied to Maxwell's equations, which govern electromagnetic phenomena—in particular, the propagation of *electromagnetic waves*, including *light*. Stated differently: is it possible to distinguish between two inertial frames by some kind of electromagnetic (in particular, optical) phenomenon?

To answer the previous question, recall that in empty space the electromagnetic potentials $A_0 := \Phi/c$ and $\mathbf{A} = (A_1, A_2, A_3)$ obey the wave equation[1]

$$\frac{1}{c^2}\frac{\partial^2 A_\mu}{\partial t^2} - \sum_{i=1}^{3}\frac{\partial^2 A_\mu}{\partial x_i^2} = 0\,, \qquad \mu = 0, \ldots, 3\,, \tag{8.2}$$

where

$$c = \frac{1}{\sqrt{\varepsilon_0 \mu_0}}$$

is a *universal constant* depending on the *constant* parameters ε_0, μ_0 appearing in Maxwell's equations. We must therefore examine how Eq. (8.2) transforms under the Galilean boost (8.1). To this end, note first of all that

$$A'_\mu(t', \mathbf{r}') = A_\mu(t, \mathbf{r})\,,$$

since $A_0 = \Phi/c$ is a scalar and, although $\mathbf{A}$ is a vector, the axes of S and S' are assumed to be parallel. Hence the transformed potentials $A'_\mu(t', \mathbf{r}')$ also satisfy Eq. (8.2), that is

$$\frac{1}{c^2}\frac{\partial^2 A'_\mu}{\partial t^2} - \sum_{i=1}^{3}\frac{\partial^2 A'_\mu}{\partial x_i^2} = 0\,, \qquad \mu = 0, \ldots, 3\,.$$

Taking into account that

$$\frac{\partial}{\partial t} = \frac{\partial}{\partial t'} - v\frac{\partial}{\partial x'_1}\,, \qquad \frac{\partial}{\partial x_i} = \frac{\partial}{\partial x'_i}\,, \qquad i = 1, 2, 3\,,$$

we immediately obtain the transformed equation

$$\frac{1}{c^2}\frac{\partial^2 A'_\mu}{\partial t'^2} - \left(1 - \frac{v^2}{c^2}\right)\frac{\partial^2 A'_\mu}{\partial x'^2_1} - \frac{\partial^2 A'_\mu}{\partial x'^2_2} - \frac{\partial^2 A'_\mu}{\partial x'^2_3} - \frac{2v}{c^2}\frac{\partial^2 A'_\mu}{\partial t' \partial x'_1} = 0\,, \qquad \mu = 0, \ldots, 3\,,$$

which is *not* a wave equation[2] in the space-time coordinates (t', x'_1, x'_2, x'_3) for any value of $v \neq 0$.

The non-invariance of the wave equation (8.2) (or, equivalently, Maxwell's equations) under Galilean transformations raises the following three possibilities, which can only be decided by experiment:

1) There exists a privileged reference frame in which Eqs. (8.2) (or, equivalently, Maxwell's equations) are valid, and electromagnetic waves propagate with speed $c = 1/\sqrt{\varepsilon_0 \mu_0}$. Consequently, the relativity principle—namely, *the equivalence of all inertial frames*—holds only for mechanics but not for electromagnetism.
2) The relativity principle holds both for mechanics and electromagnetism, but Maxwell's equations are incorrect.

[1]We shall suppose in this chapter that the electromagnetic potentials verify the Lorenz gauge (2.54).
[2]Note, however, that for $v \ll c$ the wave equation is approximately invariant under a Galilean boost.

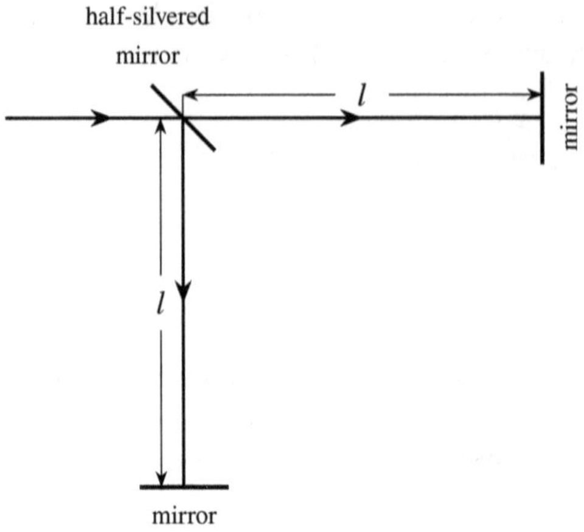

Figure 8.1. The Michelson–Morley experiment.

3) The relativity principle holds both for mechanics and electromagnetism, but Eqs. (8.1)—which follow from a fundamental tenet of Newtonian mechanics, namely the absolute character of time—are not the correct formulas relating the space-time coordinates of the same event in two inertial frames.

At the end of the 19th century, it was generally thought that the correct hypothesis was the first one. The theoretical basis for this opinion was the belief that electromagnetic waves propagated in a material medium filling all space called the *ether*, and therefore that Eqs. (8.2)—or, equivalently, Maxwell's equations—only held in an inertial frame at rest with respect to the ether. It was also believed that this privileged inertial frame coincided with that of distant stars, usually identified with Newton's "absolute space." If this hypothesis were true, it would be possible in principle to experimentally detect the motion of an inertial frame relative to the ether ("absolute motion") by measuring the velocity of electromagnetic waves in it.

In 1887, Michelson and Morley conducted a highly sensitive interferometric experiment to detect Earth's motion with respect to the ether. The experiment was based on studying the trajectory of a light beam that is divided by a half-silvered mirror into two perpendicular rays (cf. Fig. 8.1), so that the time taken by each of these rays to return to the mirror is different if the device is moving with respect to the ether. Even if this effect is very small (of the order of $v^2/c^2 \sim 10^{-8}$, where v is Earth's speed relative to the ether, believed to be approximately equal to its speed with respect to the Sun), it is possible to observe it by studying the interference fringes produced when both rays recombine. Although the experiment was repeated numerous times, *a negative result was always obtained*, i.e., no relative speed of Earth with respect to the ether was detected. This result was totally unexpected and certainly surprising, since, even admitting that at some point in its orbit Earth's speed relative to the ether vanishes, Earth's velocity varies along its orbit as well as throughout the day (due to Earth's rotation around its axis).

For almost two decades, the Michelson–Morley experiment remained without an explanation consistent with other known phenomena (like, e.g., the aberration of light or the speed of light in moving material media) which discarded the theory of ether drag. Finally, in 1905 Einstein observed that the negative result of this experiment (as well as all of the above mentioned phenomena) can be explained on the basis of the following two fundamental assumptions:

1) The laws of *physics* are the same in *all* inertial frames (**relativity principle**).
2) The speed of electromagnetic waves *in vacuo* is the *universal constant* $c = 1/\sqrt{\varepsilon_0\mu_0}$.

These two postulates are the foundations of the **special theory of relativity**[3] (SR). The first postulate is evidently an extension of Galileo's relativity principle to *all* laws of physics (including electromagnetism), not just mechanics. Combining this postulate with the second one we immediately reach the following conclusion:

The speed of electromagnetic waves *in vacuo* is equal to c in *all* inertial frames.

Of course, this principle satisfactorily explains the negative result of the Michelson–Morley experiment, since it implies that the two light rays in this experiment travel with the same speed c. It is, however, profoundly anti-intuitive from the point of view of Newtonian mechanics, since it violates the familiar *law of addition of velocities*

$$\dot{\mathbf{r}} = \dot{\mathbf{r}}' + v\mathbf{e}_1$$

which follows immediately differentiating Eq. (8.1). As a consequence, *the Galilean transformation* (8.1) cannot *be correct*. The same conclusion is reached by noting that the wave equation (8.2) for the electromagnetic potentials—which is equivalent to Maxwell's equations—is not invariant under the Galilean boost (8.1), in contradiction with the two postulates of special relativity stated above.

8.2 LORENTZ TRANSFORMATIONS

Note. Throughout this chapter, v shall usually denote the x_1 *component* of the velocity of the origin of a frame S' relative to another frame S, which can thus be positive or negative. To avoid confusion, the *magnitude* of the velocity shall be denoted by $|\mathbf{v}| = |v|$.

8.2.1 Deduction of the equations of the transformation

As we have just remarked, the Galilean boost (8.1) is *not* compatible with the postulates of the special theory of relativity. We shall apply in this section these postulates,

[3]The general theory extends the relativity principle to non-inertial frames, thereby developing a theory of gravitation (based on space-time geometry) compatible with the postulates of special relativity.

together with the *homogeneity* of space-time and the *isotropy* of space[4], to deduce the correct equations of the transformation relating the space-time coordinates $(t, \mathbf{r})$ and $(t', \mathbf{r}')$ of the same event in two different inertial frames S and S'. We shall assume, as in the previous section, that the axes of both frames are parallel, their origins coincide at some instant and the velocity of the origin O' of S' relative to S is $\mathbf{v} = v\mathbf{e}_1$. Choosing suitably the origin of time in S and S', we can always arrange for O and O' to coincide at $t = t' = 0$. Therefore

$$x_\mu = 0\,, \quad \mu = 0, \dots, 3 \quad \Longrightarrow \quad x'_\mu = 0\,, \quad \forall \mu = 0, \dots, 3\,, \tag{8.3}$$

where we have introduced the notation

$$\boxed{x_0 := ct\,,}$$

and similarly for x'_0. From now on we shall tacitly assume that condition (8.3) is satisfied, unless otherwise stated.

i) To begin with, using the *homogeneity* of space-time it can be shown that the transformation relating the coordinates x'_μ and x_μ is *linear*, i.e., that

$$x'_\mu = \sum_{\nu=0}^{3} \Lambda_{\mu\nu}(v)\, x_\nu\,, \qquad \mu = 0, \dots, 3\,,$$

where the coefficients $\Lambda_{\mu\nu}(v)$ depend only on the relative velocity between both frames.

Indeed, consider a clock moving with *constant* velocity with respect to S, and hence (by the first postulate of SR) to S'. If $x_i(t)$ and $x'_i(t')$ $(i = 1, 2, 3)$ are the clock's spatial coordinates in the frames S and S', we then have

$$\frac{\mathrm{d}^2 x_i}{\mathrm{d}t^2} = \frac{\mathrm{d}^2 x'_i}{\mathrm{d}t'^2} = 0\,, \qquad i = 1, 2, 3\,.$$

On the other hand, by the homogeneity of space-time the time τ measured by a moving clock must satisfy

$$\frac{\mathrm{d}\tau}{\mathrm{d}t} = \text{const.}\,, \qquad \frac{\mathrm{d}\tau}{\mathrm{d}t'} = \text{const.}$$

From these equations it easily follows that

$$\frac{\mathrm{d}^2 x_i}{\mathrm{d}\tau^2} = \frac{\mathrm{d}^2 x'_i}{\mathrm{d}\tau^2} = 0\,, \qquad i = 1, 2, 3\,,$$

and hence

$$\frac{\mathrm{d}^2 x_\mu}{\mathrm{d}\tau^2} = \frac{\mathrm{d}^2 x'_\mu}{\mathrm{d}\tau^2} = 0\,, \qquad i = 0, \dots, 3\,.$$

[4]Space-time must be *homogeneous*, i.e., all its points must be equivalent. Likewise, space should be *isotropic*, by which is meant that all spatial directions should be equivalent.

The previous equations imply that

$$\frac{\mathrm{d}x'_\mu}{\mathrm{d}\tau} = \sum_{\nu=0}^{3} \frac{\partial x'_\mu}{\partial x_\nu} \frac{\mathrm{d}x_\nu}{\mathrm{d}\tau}, \qquad \frac{\mathrm{d}^2 x'_\mu}{\mathrm{d}\tau^2} = \sum_{\nu,\sigma=0}^{3} \frac{\partial^2 x'_\mu}{\partial x_\nu \partial x_\sigma} \frac{\mathrm{d}x_\nu}{\mathrm{d}\tau} \frac{\mathrm{d}x_\sigma}{\mathrm{d}\tau} = 0. \tag{8.4}$$

Denoting by u_i the components of the clock's velocity with respect to the frame S we have

$$\frac{\mathrm{d}x_i}{\mathrm{d}\tau} = \frac{u_i}{c} \frac{\mathrm{d}x_0}{\mathrm{d}\tau},$$

so that Eq. (8.4) becomes

$$c^2 \frac{\partial^2 x'_\mu}{\partial x_0^2} + 2c \sum_{i=1}^{3} \frac{\partial^2 x'_\mu}{\partial x_0 \partial x_i} u_i + \sum_{i,j=1}^{3} \frac{\partial^2 x'_\mu}{\partial x_i \partial x_j} u_i u_j = 0.$$

Since u_k $(k = 1, \ldots, 3)$ is arbitrary, it follows that

$$\frac{\partial^2 x'_\mu}{\partial x_\nu \partial x_\sigma} = 0, \qquad \forall \mu, \nu, \sigma = 0, \ldots, 3,$$

as claimed.

ii) Secondly, it is easy to check that the spatial coordinates transversal to the velocity $\mathbf{v}$ must be equal in both frames, i.e., that

$$\boxed{x'_2 = x_2\,, \qquad x'_3 = x_3\,.}$$

Indeed, since the transformation $x_\mu \mapsto x'_\mu$ is linear we must have

$$x'_2 = \sum_\mu a_\mu(v) x_\mu\,,$$

where the summation index ranges from 0 to 3. (In general, from now on *Greek* indices will always range from 0 to 3, while *Latin* ones will run from 1 to 3). Since $x_2 = 0$ implies that $x'_2 = 0$ (recall that the axes of S and S' are parallel), all the coefficients a_μ vanish except for a_2, and therefore

$$x'_2 = a_2(v) x_2\,.$$

By the *isotropy of space*, the coefficient a_2 can only depend on $|v|$, i.e.,

$$a_2(v) = a_2(-v).$$

By the relativity principle, the velocity of O relative to O' must be $-v\mathbf{e}_1$, and hence

$$x_2 = a_2(-v) x'_2 = a_2(-v) a_2(v) x_2 = a_2^2(v) x_2 \quad \Longrightarrow \quad a_2(v) = \pm 1\,.$$

By continuity (since $a_2(0) = 1$) we must have $a_2(v) = 1$, and thus $x_2 = x'_2$ as claimed. Obviously, a similar argument applies to x_3 and x'_3.

iii) Since the origin of S' moves with velocity $v\mathbf{e}_1$ relative to S, the coordinate x_1' must vanish when $x_1 - vt = 0$, and thus (since the relation between x_μ' and x_ν is *linear*)

$$x_1' = \gamma(v)(x_1 - vt), \tag{8.5}$$

where γ is an *even* function of v by the isotropy of space. The relativity principle implies the analogous relation

$$x_1 = \gamma(v)(x_1' + vt'). \tag{8.6}$$

Solving for t' in the previous equation and using the value of x_1' from Eq. (8.5) we obtain

$$x_1 = \gamma^2(v)(x_1 - vt) + \gamma(v)vt' \quad \Longrightarrow \quad t' = \gamma(v)\left[t + \left(\gamma(v)^{-2} - 1\right)\frac{x_1}{v}\right]. \tag{8.7}$$

Equations (8.5) and (8.7) determine the transformation $(t, x_1) \mapsto (t', x_1')$ in terms of the unknown coefficient $\gamma(v)$.

So far we have only applied the relativity principle (the first postulate of SR) and the homogeneity and isotropy of space-time. If we assumed at this point that $t' = t$ (i.e., that time is *absolute*), from Eq. (8.7) we would immediately conclude that $\gamma(v) = 1$, which yields the Galilean boost (8.1). We know, however, that this transformation is incorrect, so that necessarily $t' \neq t$. This contradicts one of the fundamental assumptions of Newtonian mechanics, namely the absolute character of time. In fact, in order to find the correct relation between t and t' we must apply Einstein's second postulate, which so far had played no role in our argument. More precisely, according to this postulate the equation $x_1 - ct = 0$ (which describes the propagation of a plane electromagnetic wave in the x_1 direction emitted at $t = t' = 0$ from the origin of both frames) must imply $x_1' - ct' = 0$. Substituting these relations into Eqs. (8.5) and (8.6) we obtain

$$ct' = \gamma(v)(c - v)t, \qquad ct = \gamma(v)(c + v)t'.$$

Multiplying both equations and canceling the common factor tt' we easily arrive at the relation

$$c^2 = \gamma^2(v)(c^2 - v^2) \quad \Longrightarrow \quad \gamma(v) = \pm\frac{1}{\sqrt{1 - \frac{v^2}{c^2}}}.$$

We must again take, by continuity, the "+" sign (since when $v = 0$ we must have $t = t'$, and hence $\gamma(0) = 1$), so that

$$\boxed{\gamma(v) = \frac{1}{\sqrt{1 - \frac{v^2}{c^2}}}.} \tag{8.8}$$

Substituting into Eqs. (8.5) and (8.7) we finally arrive at the equations relating the coordinates x_μ and x'_μ in both inertial frames:

$$t' = \gamma(v)\left(t - \frac{vx_1}{c^2}\right), \quad x'_1 = \gamma(v)(x_1 - vt), \qquad x'_k = x_k \quad (k = 2, 3). \tag{8.9}$$

The transformation (8.8)–(8.9) between the coordinates x_μ and x'_μ, which replaces the Galilean boost (8.1), is known as a **Lorentz boost** (in the x_1 direction). Note that in terms of the coordinate $x_0 = ct$ (which has dimensions of length), and using the dimensionless parameter $\beta := v/c$, the previous equations adopt the more symmetric form

$$x'_0 = \gamma(v)(x_0 - \beta x_1), \quad x'_1 = \gamma(v)(x_1 - \beta x_0), \quad x'_k = x_k \qquad (k = 2, 3). \tag{8.10}$$

The following facts are a direct consequence of Eq. (8.9) for a Lorentz boost:

i) From Eq. (8.8) for the function $\gamma(v)$ it follows that *the relative speed between two inertial frames must be* strictly less *than the speed c of electromagnetic waves* in vacuo.
ii) In particular, *the speed of all material particles* (i.e., with non-vanishing mass) *is necessarily less than c*, since a set of such particles can be used to construct a reference frame.
iii) In fact, it is easy to show that *the propagation speed of* any *physical signal cannot exceed c*, where by physical "signal" is meant the exchange of *information* between two observers.

To prove the last assertion, suppose that a signal is sent from a point P to a second point Q with speed $u > c$ measured in an inertial frame S. Let us choose the axes of S in such a way that P and Q both lie on the x_1 axis with a spatial separation $\Delta x_1 > 0$, and let $\Delta t > 0$ be the time taken by the signal to reach Q according to S (cf. Fig. 8.2). By Eq. (8.9), the corresponding time measured in the second inertial frame S' is

$$\Delta t' = \gamma(v)\left(\Delta t - \frac{v\Delta x_1}{c^2}\right) = \gamma(v)\Delta t\left(1 - \frac{uv}{c^2}\right).$$

If the speed v of the origin of S' relative to S satisfies

$$\frac{c^2}{u} < v < c,$$

which is possible since $u > c$ by hypothesis, we will have $\Delta t' < 0$. In other words, according to S' the signal is received by Q *before* it was emitted by P, which violates the *causality principle* (a cause must always precede its effect).

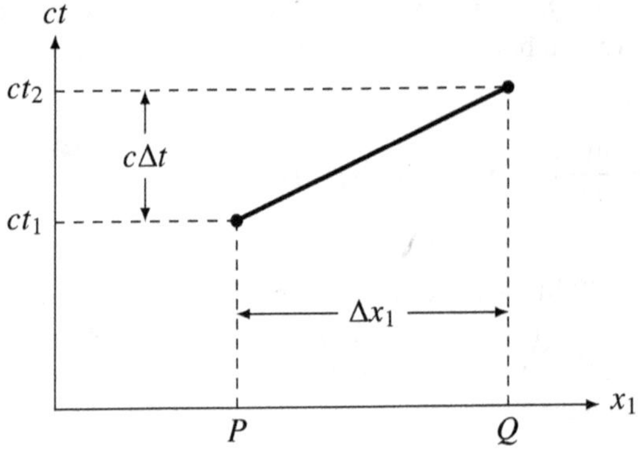

Figure 8.2. Transmission of a signal from P to Q with sped $u > c$ relative to an inertial frame S.

Exercise 8.1. Show that in general the Lorentz transformation between two inertial frames S and S' with parallel axes is given by

$$t' = \gamma(v)\left(t - \frac{\mathbf{v}\cdot\mathbf{x}}{c^2}\right), \qquad \mathbf{x}' = \mathbf{x} - \gamma(v)\mathbf{v}t + \big(\gamma(v) - 1\big)\frac{\mathbf{v}\cdot\mathbf{x}}{v^2}\,\mathbf{v}, \tag{8.11}$$

where $\mathbf{v}$ is the velocity of O' relative to S.

Solution. Setting $\mathbf{n} = \mathbf{e}_1 = \mathbf{v}/v$ we have

$$x_1 = \mathbf{n}\cdot\mathbf{x} = \frac{\mathbf{x}\cdot\mathbf{v}}{v}, \qquad x_2\mathbf{e}_2 + x_3\mathbf{e}_3 = \mathbf{x} - x_1\mathbf{n} = \mathbf{x} - \frac{\mathbf{x}\cdot\mathbf{v}}{v^2}\,\mathbf{v}$$

and hence

$$t' = \gamma(v)\left(t - \frac{\mathbf{v}\cdot\mathbf{x}}{c^2}\right), \qquad \mathbf{x}' = \gamma(v)\left(\frac{\mathbf{x}\cdot\mathbf{v}}{v} - vt\right)\frac{\mathbf{v}}{v} + \mathbf{x} - \frac{\mathbf{x}\cdot\mathbf{v}}{v^2}\,\mathbf{v},$$

which upon simplification yields Eq. (8.11).

8.2.2 Relativistic addition of velocities

Although we have just shown that the two postulates of SR lead to the equations (8.8)–(8.9) of a Lorentz boost, we must still check that this transformation is actually consistent with these postulates.

As to the first postulate, using the Lorentz transformation equations and setting

$$u_i := \frac{\mathrm{d}x_i}{\mathrm{d}t}, \quad u_i' := \frac{\mathrm{d}x_i'}{\mathrm{d}t'} = \frac{\mathrm{d}x_i'}{\mathrm{d}t}\Big/\frac{\mathrm{d}t'}{\mathrm{d}t}, \qquad i = 1, 2, 3,$$

we immediately obtain

$$u_1' = \frac{dx_1'}{dt'} = \frac{u_1 - v}{1 - \dfrac{u_1 v}{c^2}}, \qquad u_k' = \frac{dx_k'}{dt'} = \frac{u_k}{\gamma(v)\left(1 - \dfrac{u_1 v}{c^2}\right)} \qquad (k = 2, 3). \tag{8.12}$$

Thus if a particle moves with constant velocity $\mathbf{u}$ with respect to S it also moves with constant velocity $\mathbf{u}'$ relative to S'. This is consistent with *Newton's first law* (i.e., if the *law of inertia* applies in the inertial frame S it will also apply in S').

The expression of $\mathbf{u}$ as a function of $\mathbf{u}'$ can be obtained by solving for u_i in terms of u_i' from the previous equations, or more easily (by the relativity principle) replacing v by $-v$ and u_i' by u_i:

$$u_1 = \frac{u_1' + v}{1 + \dfrac{u_1' v}{c^2}}, \qquad u_k = \frac{u_k'}{\gamma(v)\left(1 + \dfrac{u_1' v}{c^2}\right)} \qquad (k = 2, 3). \tag{8.13}$$

This is the **relativistic law for the addition of velocities**, which replaces its Galilean analogue $\mathbf{u} = \mathbf{u}' + v\mathbf{e}_1$. From Eq. (8.13) it follows that

$$\left(1 + \frac{u_1' v}{c^2}\right)^2 (\mathbf{u}^2 - c^2) = u_1'^2 + \frac{1}{\gamma^2(v)}(u_2'^2 + u_3'^2) + v^2 - c^2 - \frac{v^2 u_1'^2}{c^2} = \frac{\mathbf{u}'^2 - c^2}{\gamma^2(v)}.$$

Since $|u_1'| \leqslant c$ and $|v| < c$ the first term in parentheses in the LHS is always positive, and therefore $\mathbf{u}^2 - c^2$ and $\mathbf{u}'^2 - c^2$ have the same sign. In particular, if $|\mathbf{u}'| = c$ then $|\mathbf{u}| = c$, which is consistent with the second postulate of special relativity. Moreover, from the previous equation it also follows that if $|\mathbf{u}'| < c$ then $|\mathbf{u}| < c$. In other words:

The addition of two speeds smaller than the speed of light produces a speed which is also smaller than c.

Exercise 8.2. Find the relation between the velocities $\mathbf{u}$ and $\mathbf{u}'$ when the relative velocity $\mathbf{v}$ between the inertial frames S and S' with parallel axes is not necessarily directed along the $\mathbf{e}_1 = \mathbf{e}_1'$ direction.

Solution. When $\mathbf{v} = v\mathbf{e}_1$ we have

$$u_1' = \frac{\mathbf{u}' \cdot \mathbf{v}}{v} \quad \Longrightarrow \quad \frac{u_1' v}{c^2} = \frac{\mathbf{u}' \cdot \mathbf{v}}{c^2}, \quad (u_1' + v)\mathbf{e}_1 = (u_1' + v)\frac{\mathbf{v}}{v} = \left(1 + \frac{\mathbf{u}' \cdot \mathbf{v}}{v^2}\right)\mathbf{v}.$$

Using equation (8.13) we thus obtain

$$\mathbf{u} = \frac{\mathbf{u}' - u_1'\mathbf{e}_1}{\gamma(v)\left(1 + \frac{\mathbf{u}'\cdot\mathbf{v}}{c^2}\right)} + \frac{u_1'\mathbf{e}_1 + \mathbf{v}}{1 + \frac{\mathbf{u}'\cdot\mathbf{v}}{c^2}} = \frac{\mathbf{u}' + \left[\gamma(v) + (\gamma(v) - 1)\frac{\mathbf{u}'\cdot\mathbf{v}}{v^2}\right]\mathbf{v}}{\gamma(v)\left(1 + \frac{\mathbf{u}'\cdot\mathbf{v}}{c^2}\right)},$$

which is the relation sought.

Exercise 8.3. If

$$u_1' = \frac{u_1 - v}{1 - \frac{u_1 v}{c^2}},$$

prove the identity

$$\gamma(u_1') = \left(1 - \frac{u_1 v}{c^2}\right)\gamma(u)\gamma(v).$$

Solution. Indeed,

$$\gamma(u_1')^{-2} = 1 - \frac{(u_1 - v)/c^2}{(1 - \frac{u_1 v}{c^2})^2}$$

$$\implies \gamma(u_1')^{-2}\left(1 - \frac{u_1 v}{c^2}\right)^2 = \left(1 - \frac{u_1 v}{c^2}\right)^2 - \frac{(u_1 - v)^2}{c^2}$$

$$= 1 + \frac{u_1^2 v^2}{c^4} - \frac{u_1^2}{c^2} - \frac{v^2}{c^2} = 1 - \frac{v^2}{c^2} - \frac{u_1^2}{c^2}\left(1 - \frac{v^2}{c^2}\right) = \gamma(u)^{-2}\gamma(v)^{-2}.$$

This is equivalent to the proposed identity, since

$$1 - \frac{u_1 v}{c^2} \geqslant 0.$$

8.2.3 Interval

Consider the propagation of a light signal (in general, an electromagnetic pulse) emitted from the origin of S at time $t = 0$, governed by the equation

$$c^2 t^2 - \mathbf{x}^2 = 0\,, \qquad \mathbf{x} := (x_1, x_2, x_3)$$

in the inertial frame S. By the second postulate, the equation of the wave front in another frame S' whose origin O' coincides with O at $t = t' = 0$ should be

$$c^2 t'^2 - \mathbf{x}'^2 = 0\,.$$

Hence $c^2t^2 - \mathbf{x}^2 = 0$ must imply $c^2t'^2 - \mathbf{x}'^2 = 0$. In fact, using the Lorentz transformation equations (8.8)–(8.9) in the expression $c^2t'^2 - \mathbf{x}'^2$ we readily obtain

$$c^2t'^2 - \mathbf{x}'^2 = \gamma^2(v)\left(ct - \frac{vx_1}{c}\right)^2 - \gamma^2(v)(x_1 - vt)^2 - x_2^2 - x_3^2$$
$$= \gamma^2(v)(c^2 - v^2)t^2 - \gamma^2(v)\left(1 - \frac{v^2}{c^2}\right)x_1^2 - x_2^2 - x_3^2 = c^2t^2 - \mathbf{x}^2 .$$

We have thus shown the following fundamental property of Lorentz transformations:

The quadratic form $c^2t^2 - \mathbf{x}^2 \equiv x_0^2 - \mathbf{x}^2$ is *invariant* under the Lorentz transformation (8.8)–(8.9).

In general, the **interval** between two events with space-time coordinates x_μ and $x_\mu + \Delta x_\mu$ (with $x_0 = ct$) is defined by

$$\Delta s^2 := c^2\Delta t^2 - \sum_{i=1}^{3} \Delta x_i^2 = \Delta x_0^2 - \sum_{i=1}^{3} \Delta x_i^2 = \Delta x_0^2 - \Delta\mathbf{x}^2 . \tag{8.14}$$

Note that, in spite of what the notation might suggest, the interval Δs^2 *may be negative.* Since the Lorentz transformation (8.8)–(8.9) is *linear*, the differences Δx_μ transform in the same way as the coordinates x_μ, and hence:

The interval between two events is invariant under the Lorentz transformation (8.8)–(8.9):

$$\Delta s^2 = \Delta x_0^2 - \Delta\mathbf{x}^2 = \Delta x_0'^2 - \Delta\mathbf{x}'^2 . \tag{8.15}$$

Thus the interval between two events is an *intrinsic* property of their mutual relation, independent of the reference frame used to describe them.

By definition, the interval between two events is **time-like** if $\Delta s^2 > 0$, **light-like** if $\Delta s^2 = 0$, and **space-like** if $\Delta s^2 < 0$ (cf. Fig. (8.3)).

Note that

$$\Delta s^2 > 0 \iff \Delta x_0 \neq 0\,, \quad \left|\frac{\Delta\mathbf{x}}{\Delta x_0}\right| < 1\,,$$

$$\Delta s^2 < 0 \iff |\Delta\mathbf{x}| \neq 0\,, \quad \frac{|\Delta x_0|}{|\Delta\mathbf{x}|} < 1\,,$$

$$\iff \Delta x_0 = 0, \quad |\Delta\mathbf{x}| > 0 \quad \text{or} \quad \Delta x_0 \neq 0\,, \quad \left|\frac{\Delta\mathbf{x}}{\Delta x_0}\right| > 1\,,$$

$$\Delta s^2 = 0 \iff \Delta x_0 = |\Delta\mathbf{x}| = 0 \quad \text{or} \quad \Delta x_0 \neq 0\,, \quad \left|\frac{\Delta\mathbf{x}}{\Delta x_0}\right| = 1\,.$$

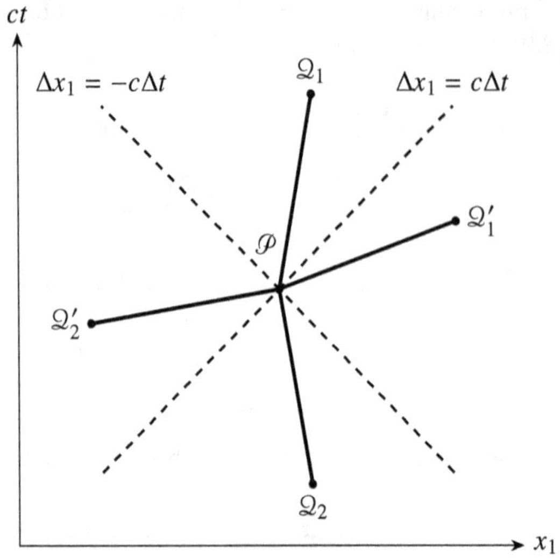

Figure 8.3. Events $\mathcal{P}$, $\mathcal{Q}_i$, $\mathcal{Q}'_i$ ($i = 1, 2$). For the sake of simplicity, we have taken $x_2 = x_3 = 0$. The intervals $\mathcal{Q}_i - \mathcal{P}$ are time-like, whereas the remaining intervals $\mathcal{Q}'_i - \mathcal{P}$ are space-like. The event $\mathcal{Q}_1$ is in the future of $\mathcal{P}$ ($t(\mathcal{Q}_1) - t(\mathcal{P}) > 0$), while $\mathcal{Q}_2$ is in its past ($t(\mathcal{Q}_2) - t(\mathcal{P}) < 0$). The event $\mathcal{Q}'_2$ cannot have influenced $\mathcal{P}$, nor $\mathcal{Q}'_1$ have been influenced by $\mathcal{P}$.

It follows that *two events separated by a time-like (or light-like) interval can influence each other* (in particular, one can be the cause of the other), since it is possible to transmit a signal from one to the other at a speed $|\Delta\mathbf{x}|/|\Delta t|$ not exceeding the speed of light. On the contrary, *two events separated by a space-like interval cannot influence each other,* since a hypothetical signal transmitted from one to the other would travel at a speed greater than c.

If the interval between two events is *time-like*, there exists an inertial reference frame relative to which they occur at the *same point* in space.

Indeed, let us choose the axes of the original inertial frame S so that

$$\Delta x_2 = \Delta x_3 = 0\,,$$

and consider a second inertial frame S' moving with velocity $v\mathbf{e}_1$ relative to S. Since

$$\Delta x'_2 = \Delta x'_3 = 0\,, \qquad \Delta x'_1 = \gamma(\Delta x_1 - v\Delta t)\,,$$

in order to guarantee that $\Delta\mathbf{x}' = 0$ it suffices to take

$$v = \frac{\Delta x_1}{\Delta t}\,.$$

This is certainly possible, since $\Delta s^2 > 0$ implies that

$$\frac{|v|}{c} = \left|\frac{\Delta x_1}{\Delta x_0}\right| < 1\,.$$

The time lapse $\Delta t'$ between both events, measured in the frame S' relative to which they take place at the same point in space, is known as the **proper time lapse** and is usually denoted by $\Delta\tau$. It follows from the invariance of the interval that in any other inertial frame S we have

$$\Delta s^2 = c^2\Delta t^2 - \Delta\mathbf{x}^2 = c^2\Delta t'^2 = c^2\Delta\tau^2 \quad \Longrightarrow \quad \boxed{\Delta\tau = \Delta t\sqrt{1 - \frac{\Delta\mathbf{x}^2}{\Delta x_0^2}}}\,,$$

where we have taken into account that $\Delta\tau = \Delta t'$ and Δt have the same sign[5] (cf. Eq. (8.16) below). Thus *the coordinate time lapse* Δt *is always* greater than or equal to *the proper time lapse*, and in fact only coincides with the latter in the inertial frame with respect to which both events take place at the same point in space.

> If the interval between two events is *space-like*, it is possible to find an inertial frame relative to which they appear to be *simultaneous*.

Indeed (supposing, again, that $\Delta x_2 = \Delta x_3 = 0$), since

$$\Delta t' = \gamma\left(\Delta t - \frac{v\Delta x_1}{c^2}\right)$$

$\Delta t'$ will vanish provided that

$$v = \frac{c^2\Delta t}{\Delta x_1} = \frac{c\Delta x_0}{\Delta x_1}\,.$$

This is again possible, since $\Delta s^2 < 0$ implies that

$$\frac{|v|}{c} = \left|\frac{\Delta x_0}{\Delta x_1}\right| < 1\,.$$

Note also that in this case

$$\sqrt{-\Delta s^2} = |\Delta x_1'|$$

coincides with the distance between both events in the reference frame relative to which they are simultaneous, known as the events' **proper distance**. Since

$$|\Delta x_1'| = \sqrt{-\Delta s^2} = \sqrt{\Delta\mathbf{x}^2 - \Delta x_0^2} \leqslant |\Delta\mathbf{x}|\,,$$

[5]Since $\mathcal{P}$ and $\mathcal{Q}$ are separated by a time-like interval, it is possible to transmit a signal from $\mathcal{P}$ to $\mathcal{Q}$ or vice versa. If Δt and $\Delta t'$ had opposite signs the effect would precede the cause in either S or S', which would of course violate the causality principle.

the proper distance is always *less than or equal to* the spatial distance $|\Delta\mathbf{x}|$ in any other inertial frame, and only coincides with the latter in a frame in which both events are simultaneous.

Finally, if two events are separated by a light-like interval, then

$$\Delta s^2 = c^2\Delta t^2 - \Delta\mathbf{x}^2 = 0\,,$$

and hence both events lie along the path of a light ray.

• Consider two events separated by a space-like interval, like $\mathscr{P}$ and $\mathscr{Q}_1'$ in Fig. 8.3. As we have just seen, although in the inertial frame S the event $\mathscr{P}$ precedes $\mathscr{Q}_1'$, there is an inertial frame relative to which $\mathscr{P}$ and $\mathscr{Q}_1'$ are simultaneous. In fact, it is easy to show that there are inertial frames S'' in which $\mathscr{Q}_1'$ *precedes* $\mathscr{P}$ (exercise). In other words:

> The concept of *simultaneity* is not absolute, but depends on the inertial reference frame used.

This fact is known as the **relativity of simultaneity**, and it represents one of the most radical departures from the Newtonian notion of time in special relativity.

• It is important to realize that *the relativity of simultaneity does not violate the causality principle*, since it applies to events separated by a *space-like* interval, between which there can be no transfer of information (indeed, a hypothetical signal connecting both events would travel with a speed $|\Delta\mathbf{x}|/|\Delta t|$ greater than c). Thus two events separated by a space-like interval cannot influence one another.

• On the other hand, if the interval between two events $\mathscr{P} \neq \mathscr{Q}$ is *time-like* or *light-like*, and $\mathscr{P}$ precedes $\mathscr{Q}$ in an inertial frame S, the same must be true in *any* other inertial frame S' related to S by a Lorentz transformation (8.8)–(8.9), since otherwise the causality principle would be violated. Indeed, the coordinate time lapses Δt and $\Delta t'$ between both events satisfy[6]

$$\Delta t' = \gamma(v)\Delta t\left(1 - \frac{v}{c}\frac{\Delta x_1}{\Delta x_0}\right), \tag{8.16}$$

where the term in parentheses is always positive if $\Delta s^2 > 0$ (since $|v|/c < 1$ and $|\Delta x_1|/|\Delta x_0| \leqslant 1$).

8.2.4 Minkowski product

If $x = (x_0, \mathbf{x})$ and $y = (y_0, \mathbf{y})$ denote the space-time coordinates of two events in a certain inertial frame S, from the invariance of the interval and of the quadratic form

[6] Recall that if two different events are separated by a time-like or light-like interval the coordinate time lapse Δt cannot vanish in any inertial frame.

$x_0^2 - \mathbf{x}^2$ it follows that

$$
\begin{aligned}
(y_0 - x_0)^2 - (\mathbf{y} - \mathbf{x})^2 &= y_0^2 - \mathbf{y}^2 + x_0^2 - \mathbf{x}^2 - 2(x_0 y_0 - \mathbf{x}\mathbf{y}) = (y_0' - x_0')^2 - (\mathbf{y}' - \mathbf{x}')^2 \\
&= y_0'^2 - \mathbf{y}'^2 + x_0'^2 - \mathbf{x}'^2 - 2(x_0' y_0' - \mathbf{x}'\mathbf{y}') \\
&= y_0^2 - \mathbf{y}^2 + x_0^2 - \mathbf{x}^2 - 2(x_0' y_0' - \mathbf{x}'\mathbf{y}'),
\end{aligned}
$$

and thus

$$x_0 y_0 - \mathbf{x}\mathbf{y} = x_0' y_0' - \mathbf{x}'\mathbf{y}'. \tag{8.17}$$

In other words, *the bilinear form*

$$\boxed{x \cdot y := x_0 y_0 - \mathbf{x}\mathbf{y},} \tag{8.18}$$

known as the **Minkowski product**, *is also invariant under Lorentz transformations* (8.8)–(8.9). Note that, according to this definition,

$$\boxed{x^2 := x \cdot x = x_0^2 - \mathbf{x}^2, \qquad \Delta s^2 = (\Delta x)^2.} \tag{8.19}$$

The vector space $\mathbb{R}^4$ whose elements are the space-time coordinates of events (or **space-time**, for short), endowed with the Minkowski product (8.18), is usually known as **Minkowski space**. Note that, since the quadratic form (8.19) associated with the Minkowski product (essentially, the interval) is *not* positive definite, Eq. (8.18) does *not* define a true scalar product in Minkowski space. We can, however, use the Minkowski product to endow Minkowski space with a geometric structure which is of great help in uncovering the properties of space-time in special relativity.

8.2.5 Lorentz group

Let S and S' be two inertial frames whose origins coincide for $t = t' = 0$, and denote by $\mathbf{v}$ the velocity of the origin of S' relative to S. In order to find the relation between the space-time coordinates x and x' of a certain event respectively in S and S', we can proceed as follows. First of all, consider an inertial frame S'' at rest relative to S, whose x_1'' axis is in the direction of the relative velocity $\mathbf{v}$. We then have

$$x'' = R_1 x,$$

where R_1 is a rotation of the spatial coordinates:

$$x_0'' = x_0, \quad \mathbf{x}'' = \mathcal{R}\mathbf{x},$$

with $\mathcal{R} \in \mathrm{SO}(3)$. Secondly, let S''' be a new inertial frame moving with velocity $\mathbf{v} = v\mathbf{e}_1''$ (with $v = |\mathbf{v}| > 0$) relative to S'', with axes parallel to those of S'' and whose origin coincides with that of S'' for $t'' = t''' = 0$. Hence the coordinates in S'' and S''' are related by

$$x''' = L(v)x'',$$

where $L(v)$ is the Lorentz transformation (8.8)–(8.9) with x replaced by x'' and x' by x'''. Finally, since S' and S''' move with the same velocity $\mathbf{v}$ with respect to S, and their origins initially coincide, the space-time coordinates x' and x''' are related simply by a spatial rotation R_2, i.e.,

$$x' = R_2 x''' .$$

Combining these equations we finally obtain

$$\boxed{x' = R_2 L(v) R_1 x =: \Lambda x .} \tag{8.20}$$

The transformation Λ, which is known as a **general Lorentz transformation**, is the most general transformation relating the coordinates of the same event in two inertial frames whose space-time origins coincide (i.e., $t = x_i = 0 \iff t' = x'_i = 0$). Obviously, if we do not make this assumption we obtain the **Poincaré transformation**

$$\boxed{x' = \Lambda x + a ,}$$

with $a \in \mathbb{R}^4$ constant.

The Lorentz boost (8.8)–(8.9) can be written in matrix form as

$$x' = L(v) x , \tag{8.21}$$

where $L(v)$ is the 4×4 matrix

$$L(v) = \begin{pmatrix} \gamma(v) & -\beta(v)\gamma(v) & 0 & 0 \\ -\beta(v)\gamma(v) & \gamma(v) & 0 & 0 \\ 0 & 0 & 1 & 0 \\ 0 & 0 & 0 & 1 \end{pmatrix}, \qquad \beta(v) := \frac{v}{c} . \tag{8.22}$$

Using matrix notation, the Minkowski product of two **four-vectors** $x, y \in \mathbb{R}^4$ can be expressed as

$$\boxed{x \cdot y = x^{\mathsf{T}} \eta y ,}$$

where x, y in the RHS are regarded as column vectors and η is the diagonal matrix

$$\eta = \begin{pmatrix} 1 & 0 & 0 & 0 \\ 0 & -1 & 0 & 0 \\ 0 & 0 & -1 & 0 \\ 0 & 0 & 0 & -1 \end{pmatrix} . \tag{8.23}$$

The invariance of the Minkowski product under the transformation (8.9) can be written in matrix form as

$$x' \cdot y' = \big(L(v)x\big)^{\mathsf{T}} \eta \big(L(v)y\big) = x^{\mathsf{T}} \big(L(v)^{\mathsf{T}} \eta L(v)\big) y = x \cdot y = x^{\mathsf{T}} \eta y , \qquad \forall x, y \in \mathbb{R}^4 ,$$

or equivalently

$$L(v)^{\mathsf{T}} \eta L(v) = \eta . \tag{8.24}$$

On the other hand, the Minkowski product is also invariant under *rotations*, since they do not affect time and leave invariant the scalar product of the spatial components of two four-vectors:

$$R^{\mathsf{T}}\eta R = \eta \tag{8.25}$$

for any rotation R. If $\Lambda = R_2 L(v) R_1$ is a general Lorentz transformation, from Eqs. (8.24)–(8.25) it immediately follows that

$$\boxed{\Lambda^{\mathsf{T}}\eta\Lambda = \eta\,.} \tag{8.26}$$

In other words:

> The Minkowski product, and hence the interval, are *invariant* under general Lorentz transformations.

From the mathematical point of view, the set of matrices satisfying Eq. (8.26) make up a group usually denoted by $O(1,3)$ and known as the **Lorentz group**, of fundamental importance in physics. It can be shown that general Lorentz transformations (8.20) are a *subgroup* of the Lorentz group, denoted by $SO_+(1,3)$ and known as *proper orthochronous*, defined by Eq. (8.26) and the additional conditions $\det\Lambda = 1$ and $\Lambda_{00} > 0$.

• Consider, again, the Lorentz boost in the x_1 direction (8.8)–(8.9). Since $\beta(v) \in (-1,1)$, there is a unique $\phi \in \mathbb{R}$ such that

$$\beta(v) = \tanh\phi\,.$$

In terms of this parameter, usually called *rapidity*, $\gamma(v)$ is given by

$$\gamma(v) = \frac{1}{\sqrt{1-\beta(v)^2}} = \frac{1}{\sqrt{1-\tanh^2\phi}} = \cosh\phi\,,$$

and thus the matrix $L(v)$ adopts the simple form

$$L(v) = \begin{pmatrix} \cosh\phi & -\sinh\phi & 0 & 0 \\ -\sinh\phi & \cosh\phi & 0 & 0 \\ 0 & 0 & 1 & 0 \\ 0 & 0 & 0 & 1 \end{pmatrix}. \tag{8.27}$$

Suppose that we successively perform two Lorentz boosts with velocities $v_1 = c\tanh\phi_1$ and $v_2 = c\tanh\phi_2$. Using the addition formulas satisfied by cosh and sinh it is easy to show that the resulting transformation is another Lorentz boost, with rapidity $\phi_1 + \phi_2$. The velocity of this boost is thus

$$v = c\tanh(\phi_1+\phi_2) = c\,\frac{\tanh\phi_1 + \tanh\phi_2}{1+\tanh\phi_1\tanh\phi_2} = \frac{v_1+v_2}{1+\dfrac{v_1 v_2}{c^2}}\,.$$

We obtain in this way the relativistic law for the addition of two parallel velocities $v_1\mathbf{e}_1$ and $v_2\mathbf{e}_2$, which is a particular case of Eq. (8.13).

8.3 PHYSICAL CONSEQUENCES OF LORENTZ TRANSFORMATIONS

Equations (8.8)–(8.9) have important physical consequences that we shall briefly review in this section.

8.3.1 Time dilation

Let, again, S and S' be two inertial frames with parallel axes[7] moving with relative velocity $v\mathbf{e}_1$, and consider a clock fixed at the origin of S'. According to S', $\mathbf{x}' = 0$ for all t' along the clock's trajectory. Hence when the clock records a time t' the corresponding time t measured in S is given by

$$t = \gamma(v)\left(t' + \frac{vx_1'}{c^2}\right) = \gamma(v)t' = \frac{t'}{\sqrt{1-\frac{v^2}{c^2}}} > t' . \tag{8.28}$$

In other words, the clock fixed at the origin of S' appears to run *slow* relative to the clocks in S. For small velocities v compared to the speed of light c the difference $t - t'$ is very small, since

$$t = t'\left(1 + \frac{v^2}{2c^2} + O(v^4/c^4)\right).$$

However, for velocities comparable to c this difference can be arbitrarily large, since it tends to infinity as $v \to c$. For instance, if $v = 3c/5$ we have $t = 5t'/4$. It is important to bear in mind the following considerations:

- The effect just described, known as **time dilation**, is *symmetric between both inertial frames*, in accordance with Einstein's first postulate. In other words, the relation between the time t recorded by a clock fixed at the origin of S and the corresponding time t' measured by the clocks in S' is

$$t' = \gamma(v)t , \tag{8.29}$$

  since now $\mathbf{x} = 0$ for all t.

- The apparent discrepancy between Eqs. (8.28) and (8.29) is resolved taking into account that in these equations both t and t' denote *different* times. The point is that in both cases there is a clear *asymmetry* between the **proper time** measured by a *single* clock at rest in a certain inertial frame and the **coordinate time** recorded by the *clocks* in another frame relative to which the clock is moving—necessarily *more than one*, since the "ticks" of a clock at rest in an inertial frame occur in *different* positions as seen from another inertial frame. It would be incorrect to say

[7]From now on, we shall tacitly assume that the origins of S and S' coincide at $t = t' = 0$ unless otherwise stated.

that time flows more slowly in S than in S', or vice versa, since *all inertial frames are equivalent*, and there is no *absolute* motion (or rest). It is however true that *the proper time of a clock runs more slowly than the* coordinate *time measured by the clocks in any inertial frame in motion with respect to the clock.*

- Time dilation is constantly being verified in experiments measuring the *half-life* of elementary particles. By definition, the half-life $t_{1/2}$ of a certain particle is the lapse of time after the particle is produced for which the probability that the particle decays reaches $1/2$. In other words, given a large sample of such particles produced at $t = 0$ about half of the sample will have decayed at $t = t_{1/2}$. If the half-life of a particle is Δt_0 in an inertial frame relative to which the particle is at rest—i.e., in the particle's *rest frame*—its half-life in the laboratory frame will be

$$\Delta t = \gamma(v)\Delta t_0 \,, \tag{8.30}$$

v being the particle's velocity with respect to the latter frame. The half-life Δt_0 can often be computed using quantum field theory techniques, which makes it possible to check the validity of Eq. (8.30) by measuring v and Δt. All the (extremely numerous) experiments performed to date have confirmed the validity of Eq. (8.30). For instance, muons present in cosmic rays can reach a speed

$$v = 0.999\, c$$

when they enter Earth's atmosphere. For this value of v, the muon's half-life measured in Earth's frame (approximately inertial) is given by

$$\Delta t = \frac{\Delta t_0}{\sqrt{1 - (1 - 10^{-3})^2}} \simeq 22.3663\, \Delta t_0 \,.$$

In the case of muons,

$$\Delta t_0 \simeq 1.5 \cdot 10^{-6}\,\mathrm{s} \quad \Longrightarrow \quad \Delta t \simeq 3.35 \cdot 10^{-5}\,\mathrm{s}\,.$$

Note that the distance traveled by the muon in the time Δt is

$$v\Delta t \sim 10\,\mathrm{Km}\,,$$

while the distance traveled at that speed during the time Δt_0 is merely

$$v\Delta t_0 \sim 450\,\mathrm{m}\,.$$

Thus, if it weren't for time dilation, muons in cosmic rays would decay long before reaching Earth's surface.

Example 8.1. *Twins paradox.*
Suppose that a traveler departs from the origin O of an inertial frame S with velocity $v\mathbf{e}_1$ and after a certain time $\Delta t/2$ (measured in S) reverts its velocity, arriving back to O at a time Δt (cf. Fig. 8.4). What is the time elapsed according to the traveler? In the first part of the trip (until reaching the event denoted by $\mathcal{P}$ in Fig. 8.4 left), the traveler's reference frame is an inertial frame S' moving with constant velocity

$v\mathbf{e}_1$ with respect to S. Thus the time assigned to the event $\mathscr{P}$ by the traveler is

$$\Delta t_1' = \frac{\Delta t}{2\gamma(v)}$$

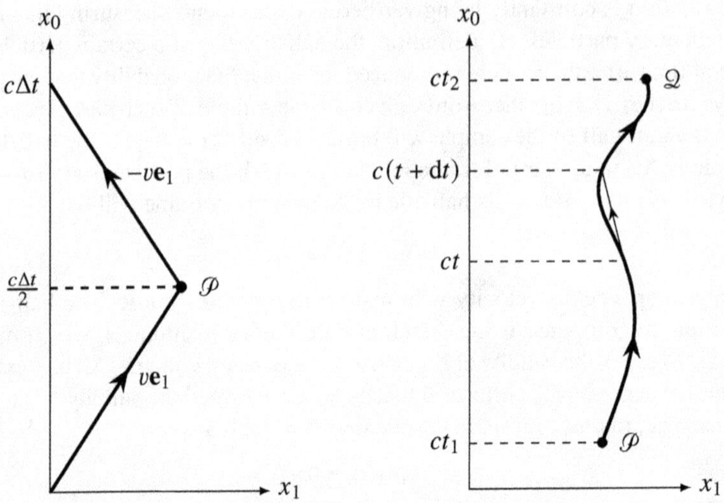

Figure 8.4. World line of the traveler in the twins paradox (left) and of the material particle used in the definition of proper time (right).

(cf. Eq. (8.28)). In the second part of the trip (from $\mathscr{P}$ on), the traveler's reference frame is *another* inertial frame S'' whose velocity with relative to S is $-v\mathbf{e}_1$. The travel time according to S for this part of the trip is again (by symmetry) $\Delta t/2$, while for the traveler the corresponding time lapse will be

$$\Delta t_2' = \frac{\Delta t}{2\gamma(-v)} = \frac{\Delta t}{2\gamma(v)} .$$

Thus the trip's total duration according to the traveler is

$$\Delta t' = \Delta t_1' + \Delta t_2' = \frac{\Delta t}{\gamma(v)} = \sqrt{1 - \frac{v^2}{c^2}}\, \Delta t ,$$

which can be considerably less than Δt if v/c is close to 1. This result may seem paradoxical, since one might think that from the point of view of the traveler it is the observer at O who has moved with speed $\mp v\mathbf{e}_1$, and hence the duration of the trip measured by the traveler should be $\gamma(v)\Delta t > \Delta t$.

The fallacy consists in assuming that the relation between the observer at O and the traveler is *symmetric*, which is far from being the case. Indeed, while O is at rest in an *inertial* reference frame at *all* times, the traveler is at rest with respect to *no* inertial frame during the *whole* trip, due to the change in the direction of his or her velocity at $\mathcal{P}$. In other words, while the observer at O has not been subject to any acceleration, the traveler has felt an (infinite) acceleration when changing course. It is clear that this will happen *regardless of the trajectory* described by the traveler. Indeed, since this trajectory begins and ends at the origin of the inertial frame S, the traveler must necessarily feel an acceleration at some point (otherwise he or she would move away from the observer at constant speed).

More generally, suppose that a material particle follows a trajectory C with equation

$$\mathbf{x} = \mathbf{x}(t), \qquad t_1 \leqslant t \leqslant t_2,$$

relative to an inertial frame S. We shall define the particle's **proper time** lapse as the time elapsed between the two events $\mathcal{P} = (ct_1, \mathbf{x}(t_1))$ and $\mathcal{Q} = (ct_2, \mathbf{x}(t_2))$ according to a clock (i.e., an observer) traveling with the particle, i.e., for which the particle is at rest at all times. Since such an observer does not define an inertial frame unless its velocity $\dot{\mathbf{x}}(t)$ is constant, in order to compute the proper time we subdivide the particle's trajectory in Minkowski space, known as its **worldline**, in small, approximately straight arcs. In each of these arcs the coordinate time of S varies between t and $t + \mathrm{d}t$, and the particle's velocity is approximately constant and equal to $\dot{\mathbf{x}}(t)$ (cf. Fig. 8.4 right). Hence the proper time $\mathrm{d}\tau$ taken by the particle to trace out this infinitesimal arc is equal to the proper time lapse measured by an inertial frame S' moving with speed $\dot{\mathbf{x}}(t)$ relative to S, namely

$$\boxed{\mathrm{d}\tau = \sqrt{1 - \frac{\dot{\mathbf{x}}^2(t)}{c^2}}\,\mathrm{d}t\,.} \tag{8.31}$$

"Adding up" all these infinitesimal proper times $\mathrm{d}\tau$, i.e., integrating with respect to t, we obtain the following expression for the total lapse of proper time $\Delta\tau(C)$ as the particle travels from $\mathcal{P}$ to $\mathcal{Q}$ along C:

$$\boxed{\Delta\tau(C) = \int_{t_1}^{t_2} \sqrt{1 - \frac{\dot{\mathbf{x}}^2(t)}{c^2}}\,\mathrm{d}t\,.} \tag{8.32}$$

Note that $\Delta\tau(C)$ is invariant under Lorentz transformations by its very definition. This can also be checked directly, since by Eq. (8.31) we have

$$\mathrm{d}\tau^2 = \frac{1}{c^2}(c^2\,\mathrm{d}t^2 - \mathrm{d}\mathbf{x}^2) = \frac{\mathrm{d}s^2}{c^2}\,.$$

Obviously $\Delta\tau(C)$ is always *less than or equal to* the coordinate time lapse $\Delta t = t_2 - t_1$, and $\Delta\tau(C) = \Delta t$ if and only if $\dot{\mathbf{x}}(t) = 0$ for all $t \in [t_1, t_2]$, i.e., if the particle is *at rest* relative to S. It is also important to realize that *the proper time $\Delta\tau$ depends in general on the trajectory followed by the particle,* and not just on the initial and final events $(t_i, \mathbf{x}(t_i))$, $i = 1, 2$ (cf. Fig. 8.5).

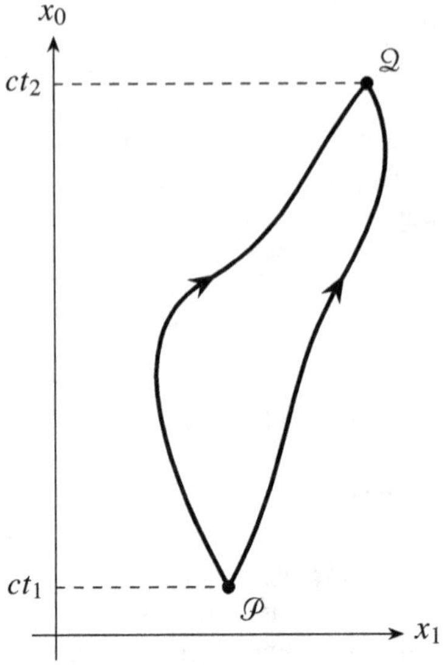

Figure 8.5. World lines connecting two events $\mathcal{P}$ and $\mathcal{Q}$.

Exercise 8.4. Let $\mathcal{P}$ and $\mathcal{Q}$ be two events separated by a *time-like* interval. Prove that the worldline with endpoints $\mathcal{P}$ and $\mathcal{Q}$ along which the elapsed proper time is *maximum* is a straight line (corresponding to rectilinear motion with constant speed). What happens if the two points are separated instead by a *space-like* or *light-like* interval?

Solution. Suppose first that the events $\mathcal{P}$ and $\mathcal{Q}$ are separated by a time-like interval, and let C denote the straight worldline from $\mathcal{P}$ to $\mathcal{Q}$. Since $\Delta\tau$ is Lorentz invariant, we can compute the proper time $\Delta\tau(C)$ along the worldline C in any inertial frame. In particular, choosing the frame S in which $\mathcal{P}$ and $\mathcal{Q}$ take place at the same point in space (i.e., the *proper frame* for these events) we have

$$\Delta\tau(C) = t_2 - t_1 \geqslant \int_{t_1}^{t_2} \sqrt{1 - \frac{\dot{\mathbf{x}}^2(t)}{c^2}}\, dt\,, \tag{8.33}$$

where the last expression is the proper time elapsed along an arbitrary path $\mathbf{x} = \mathbf{x}(t)$. This shows that the proper time $\Delta\tau$ is maximum along C, as claimed. Moreover, to have equality in Eq. (8.33) we must have $\dot{\mathbf{x}}(t) = 0$ for all $t \in [t_1, t_2]$, i.e., the particle must travel along the straight worldline (which is a vertical line in the proper frame of the two events).

On the other hand, if $\mathcal{P}$ and $\mathcal{Q}$ are joined by a space- or light-like interval, *no* curve joining $\mathcal{P}$ and $\mathcal{Q}$ can be the worldline of a material particle (i.e., no material particle can travel from $\mathcal{P}$ to $\mathcal{Q}$). Indeed, along the worldline of a material particle we must have

$$|\Delta \mathbf{x}| = \left| \int_{t_1}^{t_2} \dot{\mathbf{x}}(t)\, dt \right| \leqslant \int_{t_1}^{t_2} |\dot{\mathbf{x}}(t)|\, dt < c\Delta t \quad \Longrightarrow \quad \Delta s^2 > 0 .$$

In fact, if the interval separating $\mathcal{P}$ and $\mathcal{Q}$ is space-like then both events are *simultaneous* in an appropriate inertial frame, so that not even light can travel from $\mathcal{P}$ to $\mathcal{Q}$.

Finally, if $\mathcal{P}$ and $\mathcal{Q}$ are separated by a *light-like* interval, the only possible worldline joining both events is that of a light ray. Indeed, from the above argument it follows that in this case $|\dot{\mathbf{x}}(t)| = c$ for all t. If $l(C)$ denotes the (spatial) length of the trajectory C (in any frame), we then have

$$\Delta t = \frac{l(C)}{c} = \frac{|\Delta \mathbf{x}|}{c} \quad \Longrightarrow \quad l(C) = |\Delta \mathbf{x}| ,$$

so that the path is indeed a straight line traced out with constant speed c.

8.3.2 Lorentz–Fitzgerald contraction

Let, again, S and S' be two reference frames with parallel axes moving with relative speed $v\mathbf{e}_1$. Consider a ruler *at rest* in S', which we can assume to be determined by two marks at the points x_1' and $x_1' + l_0$ (with $l_0 > 0$) on the x_1' axis. The distance l_0, i.e., the ruler's length in its **proper frame** S', is known as the ruler's **rest length**. To determine the ruler's length l in the frame S, it is necessary to measure the coordinates x_1 and $x_1 + l$ of its endpoints *at the same time* t. Using Eqs. (8.8)–(8.9) of the Lorentz transformation relating S to S' we obtain

$$\Delta x_1' = l_0 = \gamma(v)(\Delta x_1 - v\Delta t) = \gamma(v)\Delta x_1 = \gamma(v) l$$
$$\Longrightarrow \quad l = \frac{l_0}{\gamma(v)} = l_0 \sqrt{1 - \frac{v^2}{c^2}} < l_0 .$$

Thus in the frame S the ruler appears to be *contracted* by a factor $1/\gamma(v) = \left(1 - \frac{v^2}{c^2}\right)^{1/2}$, a phenomenon known as the **Lorentz–Fitzgerald contraction**.

• Note that *this contraction only occurs in the direction of the velocity of the inertial frame* S' (relative to which the ruler is *at rest*) *with respect to* S, since in the transversal directions $x_k = x_k'$ $(k = 2, 3)$.

• Again, it should be stressed that this phenomenon is *symmetric with respect to both reference frames*. In other words, rulers at rest in S also appear to be contracted in S' (along the x_1' direction) by the same factor $1/\gamma(v)$.

• The *asymmetry* is again between the inertial reference frame in which the ruler is *at rest* and any other inertial frame. Indeed, in the ruler's proper frame its length can be determined *directly* (comparing it, for example, with a calibrated ruler), without the need of measuring *simultaneously* the spatial coordinates of its two endpoints.

More precisely, in the ruler's proper frame the worldlines of its endpoints are the vertical lines

$$(t', 0, 0, 0), \qquad (t', l_0, 0, 0),$$

where for simplicity's sake we have taken $x_1' = 0$. In another inertial frame S these worldlines become the lines

$$\Big(\gamma(v)\,t', \gamma(v)vt', 0, 0\Big), \qquad \Big(\gamma(v)\big(t' + \tfrac{vl_0}{c^2}\big), \gamma(v)(l_0 + vt'), 0, 0\Big).$$

According to S, the observer in the ruler's proper frame measures the distance between its endpoints at two *different* instants $t = \gamma(v)t'$ and $t + \Delta t$, separated by a time difference

$$\Delta t = \frac{\gamma(v)vl_0}{c^2}.$$

In this time interval Δt the right endpoint has moved, according to the observer in S, by

$$v\Delta t = \frac{v^2}{c^2}\,\gamma(v)l_0.$$

Thus from this observer's point of view at the time $t = \gamma(v)t'$ the endpoints of the ruler are located at the points

$$x_1 = \gamma(v)vt', \qquad x_1 + \Delta x_1 = \gamma(v)(l_0 + vt') - \frac{v^2}{c^2}\,\gamma(v)l_0,$$

and the ruler's length measured in S is therefore

$$l = \Delta x_1 = \gamma(v)l_0 - \frac{v^2}{c^2}\,\gamma(v)l_0 = \gamma(v)l_0\left(1 - \frac{v^2}{c^2}\right) = \frac{l_0}{\gamma(v)}.$$

We see, in particular, that the Lorentz–Fitzgerald contraction is closely related to the *relativity of simultaneity*.

Exercise 8.5. A space probe is launched toward a star T light years away from Earth. The probe reaches the star after τ years have elapsed according to its on-board clock. Find the speed of the probe (assumed to be uniform).

Solution. If v is the probe's velocity with respect to Earth and d the distance from Earth to the star measured by a terrestrial observer, by definition of light year we have

$$\frac{d}{c} = T \text{ years}.$$

The time t taken by the probe to reach the star, as measured by a terrestrial observer, is

$$t = \frac{d}{v} = \frac{d}{c\beta} = \frac{T}{\beta} \text{ years}.$$

By the principle of *time dilation,* this time can also be expressed as

$$t = \frac{\tau}{\sqrt{1-\beta^2}} \text{ years},$$

where τ is the time lapse measured by the probe's clock (i.e., the probe's *proper time*). Equating both expressions for t we arrive at the equation

$$\frac{T}{\beta} = \frac{\tau}{\sqrt{1-\beta^2}} \iff (1-\beta^2)T^2 = \beta^2\tau^2 \implies \boxed{\beta = \frac{T}{\sqrt{T^2+\tau^2}}}.$$

For instance, if $T = 40$ years and $\tau = 30$ years then $v = 4/5c$. Note that τ can be less than T, i.e., the proper time for the probe's journey to the star can be shorter than the time taken by light to cover the same distance *as measured by an observer on Earth,* due to time dilation. On the other hand, the Earth-star distance (in light years) *according to the probe's inertial frame* is

$$d' = \frac{v\tau}{c} = \beta\tau < \tau,$$

since in this frame the star is moving toward the probe with the velocity $-v\mathbf{e}_1$.

This problem could also have been solved using the relativistic *length contraction.* Indeed, the Earth-star distance T (in light years) measured by a terrestrial observer is a *proper length,* since the star is stationary with respect to Earth. By the formula for the Lorentz–Fitzgerald contraction, the same distance d' measured in the probe's frame is equal to $\sqrt{1-\beta^2}\,T$. Equating this value of d' to the one found above we arrive at the relation

$$\sqrt{1-\beta^2}\,T = \beta\tau,$$

which coincides with the equation obtained using time dilation.

8.4 FOUR-VELOCITY, FOUR-MOMENTUM, AND RELATIVISTIC KINETIC ENERGY

In Newtonian mechanics, the velocity and momentum of a particle of mass m are related by

$$\mathbf{p} = m\mathbf{v}, \tag{8.34}$$

and the particle's equation of motion is Newton's second law

$$\frac{\mathrm{d}\mathbf{p}}{\mathrm{d}t} = \mathbf{F}. \tag{8.35}$$

The previous relations are *incompatible with the postulates of special relativity.* For instance, if m and $\mathbf{F}$ are constant these equations imply that

$$\mathbf{v}(t) = \mathbf{v}(0) + \frac{\mathbf{F}}{m}t,$$

so the particle's speed will become greater than c for $|t|$ large enough. It is clear, therefore, that Eqs. (8.34)–(8.35) cannot be valid (at least for speeds comparable to c), and thus the question arises of what are the correct equations that should replace them. A fundamental guiding principle in this endeavor is the *principle of relativity*, according to which the correct equations should have the *same form* in all inertial reference frames. In other words, they must be *Lorentz covariant*, i.e., they should maintain their form when we apply to them *any* Lorentz (or more generally, Poincaré) transformation. In general, the easiest way of obtaining Lorentz covariant equations is writing down a relation between two scalars (such as the Minkowski product $x \cdot y$, the interval $x^2 = x \cdot x$, etc.), vectors (such as space-time coordinates x) or, in general, *tensors*, under Lorentz transformations. The problem here is that $\mathbf{v}$, $\mathbf{p}$ and $\mathbf{F}$ are vectors in $\mathbb{R}^3$, covariant only under *rotations*. An even more serious issue is that, while in Newtonian mechanics the time t is a *scalar* (essentially invariant under Galilean transformations), according to the theory of special relativity t actually *depends on the reference frame.*

The simplest generalization of the Newtonian definition of velocity

$$\mathbf{v} = \frac{\mathrm{d}\mathbf{x}}{\mathrm{d}t}, \qquad \mathbf{x} = (x_1, x_2, x_3),$$

which is manifestly covariant under Lorentz transformations is the **four-velocity**

$$\boxed{u = \frac{\mathrm{d}x}{\mathrm{d}\tau}.} \tag{8.36}$$

In this equation τ is the particle's proper time, which as we know is related to the coordinate time t in *any* inertial reference frame by

$$\mathrm{d}\tau = \sqrt{1 - \frac{\mathbf{v}^2}{c^2}}\,\mathrm{d}t = \frac{\mathrm{d}t}{\gamma(v)}. \tag{8.37}$$

To show that u transforms as a vector under a general Lorentz transformation $x' = \Lambda x$, it suffices to note that

$$\mathrm{d}x' = \Lambda\,\mathrm{d}x,$$

whereas $\mathrm{d}\tau$ is a Lorentz *scalar* ($\mathrm{d}\tau = \mathrm{d}\tau'$), and therefore

$$u' = \frac{\mathrm{d}x'}{\mathrm{d}\tau'} = \frac{\mathrm{d}x'}{\mathrm{d}\tau} = \Lambda\frac{\mathrm{d}x}{\mathrm{d}\tau},$$

i.e.,

$$\boxed{u' = \Lambda u.}$$

This shows that $u \in \mathbb{R}^4$ is indeed a *vector under Lorentz transformations*, since it transforms in the same way as the coordinates x of a space-time event. In fact, u is actually a vector under *Poincaré transformations* $x' = \Lambda x + a$, since differentiating this equation it still follows that $\mathrm{d}x' = \Lambda\,\mathrm{d}x$. Let us write

$$\boxed{u =: (u_0, \mathbf{u})\,, \qquad \text{with } \mathbf{u} = (u_1, u_2, u_3) \in \mathbb{R}^3\,.}$$

The spatial coordinates of the four-velocity in an arbitrary inertial frame are then given by

$$\mathbf{u} = \frac{\mathrm{d}\mathbf{x}}{\mathrm{d}\tau} = \frac{\mathrm{d}\mathbf{x}}{\mathrm{d}t}\frac{\mathrm{d}t}{\mathrm{d}\tau} = \gamma(v)\mathbf{v}\,. \tag{8.38}$$

In particular, if the particle's velocity is much smaller than c then $\gamma(v) \simeq 1$ and $\mathbf{u} \simeq \mathbf{v}$. As to the time-like coordinate u_0,

$$u_0 = \frac{\mathrm{d}x_0}{\mathrm{d}\tau} = c\,\frac{\mathrm{d}t}{\mathrm{d}\tau} = c\gamma(v)\,, \tag{8.39}$$

and hence

$$\boxed{u = \gamma(v)(c, \mathbf{v})\,.} \tag{8.40}$$

From the previous equation it immediately follows the important relation

$$\boxed{u^2 = c^2\,.} \tag{8.41}$$

This identity can also be deduced directly from the definition of u, since

$$\mathrm{d}x^2 = c^2\,\mathrm{d}t^2 - \mathrm{d}\mathbf{x}^2 = c^2\,\mathrm{d}\tau^2\,.$$

Note. The vector $\mathbf{u}$ is *not* the particle's velocity in any inertial frame. For instance, since

$$\mathbf{u}^2 = \gamma^2(v)v^2 = \frac{v^2}{1 - \frac{v^2}{c^2}}\,,$$

$|\mathbf{u}| > c$ if $v > c/\sqrt{2}$, and in fact $|\mathbf{u}| \to \infty$ for $v \to c$. ■

In view of the definition of the four-velocity, it is natural to define the **four-momentum** p by

$$\boxed{p = mu\,,} \tag{8.42}$$

where $m > 0$ is the particle's mass. By Eqs. (8.40)–(8.41), the components of the four-momentum are

$$\boxed{p = m\gamma(v)(c, \mathbf{v})\,,} \tag{8.43}$$

and its (Minkowski) square is given by

$$\boxed{p^2 = m^2c^2\,.} \tag{8.44}$$

In particular,

$$p_i = m\gamma(v)v_i\,, \qquad i = 1, 2, 3\,, \tag{8.45}$$

so that for small velocities compared to c we have

$$p_i \simeq mv_i \qquad (v \ll c)\,.$$

From now on, *we shall denote by* **p** *the vector*

$$\mathbf{p} = (p_1, p_2, p_3) = m\gamma(v)\mathbf{v}\,, \tag{8.46}$$

which coincides with the non-relativistic momentum $m\mathbf{v}$ only in the limit $v \to 0$. We shall refer to **p** as the *relativistic three-momentum*, and to $m\mathbf{v}$ as the *non-relativistic (or classical) momentum.*

On the other hand, the time-like component p_0 of p is given by

$$\boxed{p_0 = mc\gamma(v) \geqslant mc > 0\,.}$$

Using the identity (8.44), written as

$$p_0^2 = \mathbf{p}^2 + m^2c^2\,, \tag{8.47}$$

and taking into account that $p_0 > 0$, we obtain

$$\boxed{p_0 = \sqrt{\mathbf{p}^2 + m^2c^2}\,.} \tag{8.48}$$

Since u and p are proportional we have

$$\frac{\mathbf{u}}{u_0} = \frac{\mathbf{v}}{c} = \frac{\mathbf{p}}{p_0} \quad \Longrightarrow \quad \boxed{\mathbf{v} = \frac{c\mathbf{p}}{p_0}\,,} \tag{8.49}$$

and hence, by Eq. (8.48),

$$\boxed{\mathbf{v} = \frac{c\mathbf{p}}{\sqrt{\mathbf{p}^2 + m^2c^2}} = \frac{\mathbf{p}/m}{\sqrt{1 + \frac{\mathbf{p}^2}{m^2c^2}}}\,.} \tag{8.50}$$

Note that the previous equation implies that the velocity of a material particle (with non-vanishing mass) must be less than c, in accordance with the principles of special relativity. We can also use Eq. (8.48) to solve for $\gamma(v)$ in terms of **p**:

$$\boxed{\gamma(v) = \frac{p_0}{mc} = \frac{1}{mc}\sqrt{\mathbf{p}^2 + m^2c^2}\,.} \tag{8.51}$$

If $v \ll c$, expanding cp_0 in powers of v/c and keeping only the first non-constant term we obtain

$$cp_0 = mc^2\left(1 - \frac{v^2}{c^2}\right)^{-1/2} = mc^2\left(1 + \frac{v^2}{2c^2} + O(v^4/c^4)\right) = mc^2 + \frac{1}{2}mv^2 + O(v^4/c^2), \tag{8.52}$$

which, apart from the constant mc^2, coincides to first order in v^2/c^2 with the non-relativistic kinetic energy. The previous equation suggests defining the **relativistic kinetic energy** T by

$$\boxed{T = cp_0 - mc^2 = mc^2\big(\gamma(v) - 1\big),} \tag{8.53}$$

in terms of which

$$\boxed{p_0 = \frac{1}{c}\,(mc^2 + T)\,.} \tag{8.54}$$

8.5 FOUR-MOMENTUM CONSERVATION AND RELATIVISTIC ENERGY

Newton's first law establishes the conservation of the non-relativistic momentum $\mathbf{p} = m\mathbf{v}$ of a particle subject to no external forces. The most natural *Lorentz covariant* generalization of this principle is the *conservation of four-momentum* for a relativistic particle moving in the absence of external forces, namely

$$\boxed{p = \text{const.}\,,}$$

or equivalently

$$cp_0 = mc^2 + T = \text{const.}\,, \qquad p_i = m\gamma(v)v_i = \text{const.}$$

These equations reduce to the conservation of non-relativistic kinetic energy and momentum in the limit $v \ll c$. As in the Newtonian case, by Eq. (8.49) both conservation laws are equivalent to the constancy of the components v_i of the ordinary velocity (in any inertial frame).

Consider next the collision of N particles of mass m_n $(n = 1, \ldots, N)$ on which no external forces act. The **total four-momentum** P is then naturally defined by

$$\boxed{P = \sum_{n=1}^{N} p_n =: (P_0, \mathbf{P})\,,} \tag{8.55}$$

where p_n is the four-momentum of the n-th particle. Hence

$$P_0 = \sum_{n=1}^{N} p_{n,0} = c \sum_{n=1}^{N} m_n \gamma(v_n) , \qquad \mathbf{P} = \sum_{n=1}^{N} \mathbf{p}_n = \sum_{n=1}^{N} m_n \gamma(v_n) \mathbf{v}_n . \tag{8.56}$$

According to Newtonian mechanics, even if the collision is not elastic the system's total linear momentum should be conserved. Moreover, this non-relativistic momentum tends to $\mathbf{P}$ in the limit in which the speeds v_n of all the particles are small compared to c. This fact makes it plausible to postulate the conservation of $\mathbf{P}$ also in relativistic mechanics, i.e.,

$$\mathbf{P}_i = \mathbf{P}_f , \tag{8.57}$$

where P_i and P_f denote the total four-momentum respectively before and after the collision. This equation is not *Lorentz covariant*, since only involves the *spatial* components of a four-vector. However, if Eq. (8.57) holds in *all* inertial frames then the *full* four-momentum P is also necessarily conserved, i.e., we must have

$$P_i = P_f . \tag{8.58}$$

Indeed, suppose that $\mathbf{P}_i = \mathbf{P}_f$ holds in some inertial frame S, and consider a second inertial frame S' moving with velocity $w\mathbf{e}_1$ relative to S. Since by hypothesis $\mathbf{P}'_i = \mathbf{P}'_f$ should also hold in S', it follows that

$$P'_{i,1} = \gamma(w) \left(P_{i,1} - \frac{w}{c} P_{i,0} \right) = P'_{f,1} = \gamma(w) \left(P_{f,1} - \frac{w}{c} P_{f,0} \right) .$$

From $P_{i,1} = P_{f,1}$ and the previous equation it follows that $P_{i,0} = P_{f,0}$, and hence $P_i = P_f$. Actually, the relativistic **law of four-momentum conservation** (8.58) has been (and is being) experimentally verified in multiple situations for speeds arbitrarily close to c, for example in the analysis of collisions taking place in particle accelerators.

The conservation of the time-like component of the total four-momentum can be expressed as

$$\sum_n (m_n c^2 + T_n)_i = \sum_n (m_n c^2 + T_n)_f ,$$

or equivalently

$$\left(Mc^2 + T \right)_i = \left(Mc^2 + T \right)_f ,$$

where

$$M = \sum_n m_n , \qquad T = \sum_n T_n$$

respectively denote the system's *total mass* and *total kinetic energy*. It is important to note at this point that in relativistic mechanics *the number of particles before and after a collision need not be the same*, since, as we shall see below, particles can be created or destroyed under the appropriate conditions. For this reason, from now on it shall be understood that the sums over n appearing in expressions like the previous ones are tacitly extended to *all* particles in the system before or after the collision, without explicitly specifying their number N_i (before the collision) or N_f (after the collision).

In Newtonian mechanics the total mass M is conserved[8], and therefore the conservation of P_0 is equivalent to that of the system's kinetic energy

$$T_i = T_f \,.$$

According to what we have just seen, however, in relativistic mechanics only the quantity $cP_0 = Mc^2 + T$ need be conserved, not M and T separately. In particular:

There may be processes in which the system's total mass decreases (resp. increases), provided that this decrease (resp. increase) is compensated by a corresponding increase (resp. decrease) in the kinetic energy.

More precisely, denoting by $\Delta M = M_f - M_i$ and $\Delta T = T_f - T_i$ the change in the system's mass and kinetic energy due to the collision, the conservation of P_0 can be expressed as

$$\boxed{\Delta T = -\Delta(Mc^2) \,.} \tag{8.59}$$

In other words:

Kinetic energy can be transformed into mass, and vice versa. The conversion factor between energy and mass is equal to the square of the velocity of EM waves *in vacuo*.

This is one of the most important predictions of the special theory of relativity, which has so far been experimentally corroborated without exception.

By the previous discussion, we are practically forced to interpret the quantity

$$\boxed{cP_0 = \sum_n cp_{n,0} = \sum_n (m_n c^2 + T_n) = Mc^2 + T}$$

as the system's **total relativistic energy** E (in the absence of external forces). We thus have

$$\boxed{E = cP_0 = Mc^2 + T \,,} \tag{8.60}$$

[8]The conservation of the total mass in Newtonian mechanics is a *consequence* of the conservation of total momentum and *Galilean invariance*. Indeed, applying a Galilean boost with velocity $\mathbf{w}$ to the equality $\mathbf{P}_i = \mathbf{P}_f$ we obtain:

$$\mathbf{P}'_i = \sum_n m_n \mathbf{v}'_{n,i} = \sum_n m_n (\mathbf{v}_{n,i} - \mathbf{w}) = \mathbf{P}_i - M_i \mathbf{w} = \mathbf{P}'_f = \mathbf{P}_f - M_f \mathbf{w} \quad \Longrightarrow \quad M_i = M_f \,.$$

and the system's total momentum can be expressed as

$$\boxed{P = (E/c, \mathbf{P})}$$

For a single particle

$$p = (p_0, \mathbf{p}) = (E/c, m\gamma(v)\mathbf{v}),$$

and from Eq. (8.51) it follows that the relativistic energy can be expressed in terms of the particle's velocity by the formula

$$\boxed{E = cp_0 = mc^2\gamma(v).} \tag{8.61}$$

Note that the total relativistic energy E is necessarily positive. In particular, when $v = 0$ the particle possesses a **rest energy**

$$\boxed{E_0 = mc^2.}$$

Note also that from Eq. (8.48) and (8.60) it follows the important relation

$$\boxed{E = c\sqrt{\mathbf{p}^2 + m^2c^2}} \tag{8.62}$$

between relativistic energy and momentum. Writing this relation as

$$E = mc^2\sqrt{1 + \frac{\mathbf{p}^2}{m^2c^2}}$$

and expanding in powers of $\mathbf{p}^2$ we obtain

$$E = mc^2 + \frac{\mathbf{p}^2}{2m} + O\left(|\mathbf{p}|^4/(m^3c^2)\right).$$

Note, finally, that Eqs. (8.49) and (8.61) yield the following relations between the particle's velocity, energy and momentum:

$$\boxed{\mathbf{v} = \frac{c^2\mathbf{p}}{E}.} \tag{8.63}$$

Note. An alternative formulation of the previous results consists in defining a *velocity dependent mass*

$$m(v) := m\gamma(v) = \frac{m}{\sqrt{1 - \frac{v^2}{c^2}}},$$

in terms of which the relativistic momentum and energy are simply given by

$$\mathbf{p} = m(v)\mathbf{v}, \qquad E = m(v)c^2.$$

Note, however, that the previous formula for the kinetic energy

$$T = \big(m(v) - m\big)c^2 ,$$

does not reduce to the classical expression replacing m by $m(v)$. In any case, we shall not use the concept of variable mass in this text.

8.6 MASSLESS PARTICLES

As we have just seen, the four-momentum p of a particle of mass $m > 0$ has components

$$p = \big(E/c, \mathbf{p}\big) , \qquad \text{with} \quad E = c\sqrt{\mathbf{p}^2 + m^2c^2} .$$

These relations also make sense if the particle's mass vanishes. Indeed, if $m = 0$ the last equation reduces to

$$\boxed{E = c|\mathbf{p}| ,} \tag{8.64}$$

and therefore

$$\boxed{p = (|\mathbf{p}|, \mathbf{p}) .} \tag{8.65}$$

Moreover, for a massive particle the velocity and the relativistic three-momentum are related by Eq. (8.63). Taking the limit as $m \to 0$ of this equation, and using Eq. (8.64), we obtain

$$\boxed{\mathbf{v} = c\,\frac{\mathbf{p}}{|\mathbf{p}|} .} \tag{8.66}$$

Thus *the speed of a massless particle must be equal to c.*

The only known massless particle[9] is the **photon**, which is the *quantum of energy* of the electromagnetic field (i.e., the particle carrying the electromagnetic field's quanta of energy-momentum). According to quantum mechanics, the relation between the energy of a photon and the angular frequency ω of its associated electromagnetic wave is given by *Planck's equation*

$$\boxed{E = \hbar\omega = h\nu = \frac{hc}{\lambda} ,} \tag{8.67}$$

where ν is the wave's frequency, λ its *wavelength*, and

$$h = 2\pi\hbar = 6.62606957 \cdot 10^{-34}\,\mathrm{J\,s}$$

[9]The existence of a massless particle mediating strong interactions, called *gluon*, has been experimentally confirmed, although gluons are not directly observable because they are confined inside hadrons. For theoretical reasons, it is believed that a similar massless particle known as *graviton* should also exist for the gravitational field.

is Planck's constant. From the relations[10]

$$\boxed{\omega = c|\mathbf{k}|\,, \qquad \mathbf{v} = c\,\frac{\mathbf{k}}{|\mathbf{k}|}} \tag{8.68}$$

and Eqs. (8.64), (8.66), and (8.67), it follows that the wave vector **k** of the EM wave associated to the photon is given by

$$\mathbf{k} = \frac{\omega}{c}\frac{\mathbf{v}}{c} = \frac{\omega}{c}\frac{\mathbf{p}}{|\mathbf{p}|} = \frac{\omega\mathbf{p}}{E} = \frac{\mathbf{p}}{\hbar} \quad \Longrightarrow \quad \mathbf{p} = \hbar\mathbf{k}\,.$$

This suggests defining a **wave four-vector** $k = (k_0, \mathbf{k})$ by

$$\boxed{k = p/\hbar\,,}$$

with time-like component

$$\boxed{k_0 = \frac{p_0}{\hbar} = \frac{E}{\hbar c} = \frac{\omega}{c} = \frac{2\pi}{\lambda} = |\mathbf{k}|\,.}$$

It is important to note that k is a *vector under Lorentz transformations*, being proportional to the four-momentum p of the wave's photons. In other words, if S' is another inertial system and $x' = \Lambda x$ we have

$$\boxed{k' = \Lambda\, k\,.} \tag{8.69}$$

More generally, if two inertial frames S and S' are related by a *Poincaré transformation*

$$x' = \Lambda x + a\,, \tag{8.70}$$

where Λ is a general Lorentz transformation, we know that $u' = \Lambda u$. For $m > 0$ momentum and velocity are proportional, and thus

$$p' = \Lambda\, p\,.$$

Since this relation is independent of the particle's mass, taking the limit $m \to 0$ we conclude that it must also hold for massless particles. Finally, since k is proportional to p we conclude that the transformation law of the wave four-vector k under the Poincaré transformation (8.70) is still given by Eq. (8.69). In other words, *the wave four-vector is a vector under Poincaré transformations.*

---

[10]Recall that in a plane wave propagating with speed c the angular frequency ω, the period τ, the wave vector **k**, the wavelength λ, and the propagation velocity **v** are related by

$$\omega = \frac{2\pi}{\tau} = 2\pi\nu\,, \qquad |\mathbf{k}| = \frac{2\pi}{\lambda}\,, \qquad \mathbf{v} = \frac{c\mathbf{k}}{|\mathbf{k}|}\,, \qquad c = \frac{\lambda}{\tau} = \frac{\lambda\omega}{2\pi} = \frac{\omega}{|\mathbf{k}|}\,.$$

All of these relations easily follow from the fact that in a plane wave the wave fronts are moving planes with equation $\omega t - \mathbf{k}\cdot\mathbf{x} = \text{const.}$

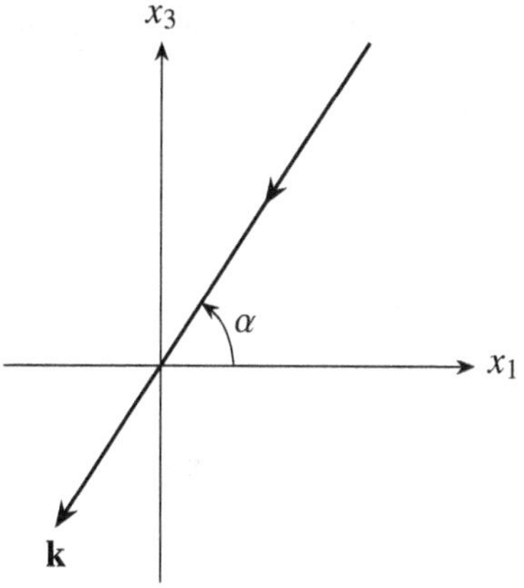

Figure 8.6. Geometry of the relativistic Doppler effect.

8.6.1 Relativistic Doppler effect

The Lorentz covariance of the wave four-vector k makes it easy to deduce the equations of the **relativistic Doppler effect**. Indeed, let S' be an inertial frame traveling with speed $v\mathbf{e}_1$ relative to the laboratory inertial frame S, with axes parallel to those of S and origin not necessarily coinciding with that of S at $t = 0$. Suppose that an electromagnetic wave with frequency ω_0 and wavelength $\lambda_0 = 2\pi c/\omega_0$ is emitted from S' (ω_0 and λ_0 are respectively called the wave's *proper frequency* and *proper wavelength*). Let us choose the axes so that the wave's propagation direction lies in the plane $x_2 = 0$ and makes an angle $\pi + \alpha$ with the x_1 axis according to the observer at S (cf. Fig 8.6), so that

$$\mathbf{k} = -|\mathbf{k}|(\cos\alpha, 0, \sin\alpha)\,.$$

(Note that we can assume without loss of generality that $0 \leqslant \alpha \leqslant \pi/2$, changing the orientation of the x_1 axis if necessary.) By the remark at the end of the previous section, we can find the wave four-vector k' in the frame S' in which the wave was emitted by applying to the wave four-vector k a Lorentz transformation $L(v)$ with velocity $v\mathbf{e}_1$, namely

$$k' = L(v)k\,.$$

Since $k'_2 = k_2 = 0$, the spatial components of k' are also of the form

$$\mathbf{k}' = -|\mathbf{k}'|(\cos\alpha', 0, \sin\alpha')\,.$$

On the other hand, the time-like component k'_0 is given by

$$k'_0 = \frac{\omega_0}{c} = \gamma\Big(k_0 - \beta k_1\Big) = \gamma\Big(\frac{\omega}{c} + \beta|\mathbf{k}|\cos\alpha\Big) = \frac{\gamma\omega}{c}\,(1 + \beta\cos\alpha)\,,$$

and therefore

$$\boxed{\omega = \frac{\omega_0}{\gamma(1+\beta\cos\alpha)} \qquad \Longrightarrow \qquad \lambda = \gamma(1+\beta\cos\alpha)\,\lambda_0\,.} \tag{8.71}$$

The relation between the angles α and α' is also easily computed from the equations

$$k'_1 = \gamma(k_1 - \beta k_0) = \gamma(k_1 - \beta|\mathbf{k}|) = -\gamma|\mathbf{k}|(\cos\alpha + \beta)\,, \qquad k'_3 = k_3 = -|\mathbf{k}|\sin\alpha\,,$$

whence

$$\boxed{\tan\alpha' = \frac{k'_3}{k'_1} = \frac{\sin\alpha}{\gamma(\beta + \cos\alpha)}\,.} \tag{8.72}$$

A particularly important case is the so-called *longitudinal Doppler effect*, in which $\alpha' = 0$, i.e., the electromagnetic wave propagates in the direction of the relative motion between the observer S and the source S'. From the above formulas it follows that $\alpha = 0$, and therefore

$$\boxed{\lambda = \gamma(1 + \beta)\,\lambda_0 = \sqrt{\frac{1+\beta}{1-\beta}}\,\lambda_0\,.} \tag{8.73}$$

Thus if the source S' moves *away* from the observer S (i.e., if $\beta > 0$) then $\lambda > \lambda_0$, so that the observer perceives a *shift toward the red* in the wavelength of the electromagnetic wave emitted by S'. On the contrary, if the source moves *toward* the observer then $\beta < 0$, and hence $\lambda < \lambda_0$. Thus in this case the wavelength of the electromagnetic wave emitted by S' appears *shifted toward the blue* to the observer in S.

On the other hand, if $\alpha = \pi/2$, i.e., when according to the observer in S the wavefront is *perpendicular* to the direction of the emitter's velocity, from Eqs. (8.71)–(8.72) we obtain

$$\boxed{\tan\alpha' = \frac{1}{\gamma\beta}\,, \qquad \lambda = \gamma\lambda_0 > \lambda_0\,.}$$

Hence in this case the observer perceives a *shift toward the red* regardless of the sign of v. This is the so-called *transversal Doppler effect*, which does not have a classical analogue.

8.6.2 Compton effect

We shall consider next the so-called **Compton effect**, which occurs when a photon is scattered by an electron. In the inertial frame in which the electron is at rest (which usually coincides with the laboratory frame), the initial momenta of the photon and the electron can be taken as

$$p_\gamma = \left(\frac{E}{c}, |\mathbf{p}|, 0, 0\right) = \frac{E}{c}\,(1, 1, 0, 0)\,, \qquad p_e = (mc, 0, 0, 0)\,,$$

m being the electron's mass. Let us choose the axes of the frame S so that the collision takes place in the $x_3 = 0$ plane, and denote by θ the angle between the three-momentum of the scattered photon and the x_1 axis. The photon's momentum after the collision is then given by

$$p'_\gamma = \frac{E'}{c}\,(1, \cos\theta, \sin\theta, 0)\,.$$

By the law of four-momentum conservation, we must then have

$$p_\gamma + p_e = p'_\gamma + p'_e\,,$$

or equivalently

$$p_e + (p_\gamma - p'_\gamma) = p'_e\,.$$

Squaring and taking into account that

$$p_\gamma^2 = p_\gamma'^2 = 0\,, \qquad p_e^2 = p_e'^2 = m^2c^2$$

we arrive at the relation

$$p_e(p_\gamma - p'_\gamma) = p_\gamma p'_\gamma\,,$$

in which we have eliminated the momentum p'_e of the scattered electron. Substituting the previous expressions for p_γ, p'_γ, and p_e we obtain

$$m(E - E') = \frac{EE'}{c^2}(1 - \cos\theta) \quad \Longrightarrow \quad mc^2\left(\frac{1}{E'} - \frac{1}{E}\right) = 1 - \cos\theta\,, \tag{8.74}$$

and taking into account Eq. (8.67) we finally arrive at the relation

$$\boxed{\lambda' - \lambda = \frac{h}{mc}\,(1 - \cos\theta)} \tag{8.75}$$

known as **Compton's equation**. We thus see that the wavelength of the scattered photon is always *greater than or equal* to the wavelength of the incoming one.

Exercise 8.6. Show that the angle $-\theta_e$ between the velocity of the scattered electron and the x_1 axis and its kinetic energy T_e are determined by the equations

$$\cot\theta_e = \left(1 + \frac{E}{mc^2}\right)\tan(\theta/2)\,, \qquad T_e = \frac{E}{1 + \frac{mc^2}{2E}\,\csc^2(\theta/2)}\,.$$

Solution. By energy-momentum conservation, we have

$$\frac{E}{c} + mc = \frac{E'}{c} + mc\gamma(v), \tag{8.76a}$$

$$\frac{E}{c} = \frac{E'}{c}\cos\theta + mv\gamma(v)\cos\theta_e, \tag{8.76b}$$

$$0 = \frac{E'}{c}\sin\theta - mv\gamma(v)\sin\theta_e. \tag{8.76c}$$

From the first equation and the identity $T_e = mc^2(\gamma(v) - 1)$ we obtain

$$T_e = E - E'.$$

On the other hand, by Eq. (8.74) we have

$$\frac{1}{E'} = \frac{1}{E} + \frac{2}{mc^2}\sin^2(\theta/2) \quad \Longrightarrow \quad E' = \frac{E}{1 + \frac{2E}{mc^2}\sin^2(\theta/2)},$$

and therefore

$$T_e = E - \frac{E}{1 + \frac{2E}{mc^2}\sin^2(\theta/2)} = \frac{\frac{2E^2}{mc^2}\sin^2(\theta/2)}{1 + \frac{2E}{mc^2}\sin^2(\theta/2)} = \frac{E}{1 + \frac{mc^2}{2E}\csc^2(\theta/2)}.$$

The angle θ_e is easily determined eliminating $v\gamma(v)$ from Eqs. (8.76b)–(8.76c):

$$\cot\theta_e = \frac{E - E'\cos\theta}{E'\sin\theta} = \frac{E}{E'}\csc\theta - \cot\theta = \left(1 + \frac{2E}{mc^2}\sin^2(\theta/2)\right)\csc\theta - \cot\theta.$$

Using the trigonometric identities

$$\csc\theta - \cot\theta = \frac{1 - \cos\theta}{\sin\theta} = \frac{2\sin^2(\theta/2)}{2\sin(\theta/2)\cos(\theta/2)} = \tan(\theta/2),$$

$$2\sin^2(\theta/2)\csc\theta = \frac{2\sin^2(\theta/2)}{2\sin(\theta/2)\cos(\theta/2)} = \tan(\theta/2)$$

we easily arrive at the proposed equation for $\cot\theta_e$.

8.7 RELATIVISTIC COLLISIONS

The conservation of the (four-)momentum of a system of particles on which no external forces act is of fundamental importance in the study of *collisions* in the framework of the special theory relativity. Indeed, as we saw in the previous sections, in the absence of external forces the system's total momentum P is conserved, so in particular the momentum P_i immediately before a collision must coincide with the momentum P_f after it (cf. Eq. (8.58)). This conservation law is equivalent to the *conservation of relativistic energy*

$$\boxed{P_0 = \sum_n p_{n,0} = \sum_n \gamma(v_n)m_n c} \tag{8.77}$$

along with the *conservation of three-momentum*

$$\boxed{\mathbf{P} = \sum_n \mathbf{p}_n = \sum_n \gamma(v_n)m_n\mathbf{v}_n\,.} \tag{8.78}$$

8.7.1 Center of momentum frame

The relation (8.58) is valid in any inertial reference frame. In the analysis of the collisions of a system of ultra-relativistic particles (moving at speeds comparable to c) there is, however, a particularly useful inertial frame known as the **center of momentum** (CM) **frame** (also known as the *zero momentum* frame). This is a frame, analogous to the center of mass frame in Newtonian mechanics, in which the spatial components of the system's total momentum vanish, i.e., in which the equality

$$\mathbf{P} = 0$$

holds. In order to establish the existence of such a frame, it suffices to show that the total momentum P of a system of particles is a *time-like* four-vector, i.e., that $P^2 > 0$ (cf. the framed remark on p. 306). This fact is a direct consequence of the following general result:

> The sum $P = \sum_n p_n$ of any number of *future time-like* four-vectors p_n (for which $p_n^2 > 0$ and $p_{n,0} > 0$ for all n) is also a future time-like four-vector.

Proof. Indeed, since p_n is a future time-like vector we have

$$p_n^2 = p_{n,0}^2 - \mathbf{p}_n^2 > 0 \quad \Longrightarrow \quad |p_{n,0}| = p_{n,0} > |\mathbf{p}_n| \,.$$

Thus, if p_m is another such vector then

$$\mathbf{p}_n \cdot \mathbf{p}_m \leqslant |\mathbf{p}_n||\mathbf{p}_m| < p_{n,0}p_{m,0}$$

and therefore

$$p_n \cdot p_m = p_{n,0}p_{m,0} - \mathbf{p}_n \cdot \mathbf{p}_m > 0 \,.$$

Hence

$$P^2 = \Big(\sum_n p_n\Big)^2 = \sum_{n,m} p_n \cdot p_m = \sum_n p_n^2 + \sum_{n \neq m} p_n \cdot p_m > 0 \,,$$

and of course (since $p_{n,0} > 0$ for all n)

$$P_0 = \sum_n p_{n,0} > 0 \,. \qquad \blacksquare$$

• It is easy to see that the previous result extends to the case in which some of the four-vectors (but *not all*) are light-like, i.e., it is valid as long as $p_n^2 \geqslant 0$ for all n and $p_k^2 > 0$ for some k (with, as before, $p_{n,0} > 0$ for all n).

8.7.2 Threshold energy

Consider a process like

$$a + b \to a + b + c \,,$$

in which two particles a and b collide producing a third particle c as a result of the collision. In the laboratory frame one of the particles (for instance, b) is the target (i.e., $\mathbf{p}_b = 0$), while the other one (the projectile) has a three-momentum $\mathbf{p}_a \neq 0$. What is the **threshold energy** of particle a, that is, the minimum energy that this particle must have so that the creation of the c particle is possible?

Obviously, the conservation of relativistic energy requires that

$$\frac{E_a}{c^2} + m_b = m_a\gamma(v_a') + m_b\gamma(v_b') + m_c\gamma(v_c'),$$

where the primes indicate the speeds after the collision in the laboratory frame. Since $\gamma(v_i') \geqslant 1$, from this relation it follows that

$$E_a \geqslant (m_a + m_c)c^2 .$$

However, in order to achieve equality in the previous inequality it is necessary that $\gamma(v_a') = \gamma(v_b') = \gamma(v_c') = 1$, i.e., $v_a' = v_b' = v_c' = 0$. This is, however, *impossible*, since by momentum conservation $\mathbf{p}_a' + \mathbf{p}_b' + \mathbf{p}_c' = \mathbf{p}_a \neq 0$, so the speeds of all three particles cannot vanish after the collision. Hence the threshold energy for the process is *strictly greater* than $(m_a + m_c)c^2$.

Let us next compute the threshold energy E_{min} in the more general process

$$a + b \longrightarrow c_1 + \cdots + c_N , \tag{8.79}$$

in which the production of an arbitrary number of particles c_i of mass $m_i > 0$ is allowed. To this end, we analyze the collision in the center of momentum (CM) frame, in which the total momentum (before or after the collision) is given by

$$P_{\text{CM}} = \frac{E_{\text{CM}}}{c}(1,0,0,0) .$$

Computing the CM energy E_{CM} after the collision we obtain

$$E_{\text{CM}} = \sum_i m_i\gamma(v_i)c^2 \geqslant \sum_i m_i c^2 = Mc^2 .$$

Note that in this case equality can be achieved if all the particles are at rest in the CM frame—i.e., if all of them move with the same speed $\mathbf{v}$ in the laboratory frame—which is of course possible since none of them has zero mass. Therefore the minimum value of the energy in the CM frame is simply Mc^2:

$$E_{\text{CM}} \geqslant Mc^2 .$$

In order to find the threshold energy of particle a in the laboratory frame, it suffices to apply the law of momentum conservation and the invariance of the Minkowski product, which yield the relation

$$P_{\text{CM}}^2 = \frac{E_{\text{CM}}^2}{c^2} = P_L^2 = (p_a + p_b)^2 = c^2(m_a^2 + m_b^2) + 2p_a \cdot p_b . \tag{8.80}$$

Here P_L is the initial momentum in the laboratory frame, and p_a and p_b the momenta of particles a and b before the collision *in the laboratory frame:*

$$p_a = \left(\frac{E_a}{c}, \mathbf{p}_a\right), \qquad p_b = m_b c(1,0,0,0)\,.$$

Substituting into Eq. (8.80) and operating we obtain

$$\frac{E_{\rm CM}^2}{c^2} = c^2(m_a^2 + m_b^2) + 2E_a m_b\,.$$

Thus the energy of particle a in the laboratory frame is given by

$$E_a = \frac{c^2}{2m_b}\left(\frac{E_{\rm CM}^2}{c^4} - m_a^2 - m_b^2\right).$$

In particular, replacing $E_{\rm CM}$ by its minimum value Mc^2 we obtain the formula

$$\boxed{E_{\rm min} = \frac{c^2}{2m_b}(M^2 - m_a^2 - m_b^2)\,.} \tag{8.81}$$

Note that the previous result is also valid if the a particle (the projectile) is massless.

Exercise 8.7. A proton collides with another proton at rest in the laboratory frame, producing a proton-antiproton pair as a result of the collision ($p+p \longrightarrow p+p+p+\overline{p}$). What is the minimum kinetic energy of the incident proton for this process to be possible?

Solution. Since the mass of a particle is the same as that of its antiparticle, we can apply Eq. (8.81) equation with

$$m_a = m_b \equiv m\,, \qquad M = 4m\,,$$

where $m \simeq 938.27208816\,{\rm MeV}/c^2$ is the proton's mass. We thus obtain

$$E_{\rm min} = \frac{c^2}{2m}(16m^2 - 2m^2) = 7mc^2\,.$$

Hence the minimum kinetic energy of the incident proton is

$$T_{\rm min} = E_{\rm min} - mc^2 = 6mc^2 \simeq 5.63\,{\rm GeV}\,.$$

Exercise 8.8. Show that an *isolated* photon cannot decay into an electron-positron pair ($\gamma \not\rightarrow e^- + e^+$). Prove that, however, the process $\gamma + N \longrightarrow N + e^- + e^+$ (where N is a heavy nucleus) is possible, and that the photon's threshold energy is in this case approximately equal to $2m_e c^2$.

Solution. Let us check, to begin with, that the process $\gamma \to e^- + e^+$ is impossible regardless of the photon's energy. Indeed, in the center of momentum frame of the e^--e^+ pair the final three-momentum $\mathbf{P}$ vanishes, and hence the photon's three-momentum should also vanish in this frame. But this is impossible, since for a massless particle $\mathbf{p} = 0$ implies that $E = c|\mathbf{p}| = 0$, i.e., the particle would have zero energy or momentum. (According to the special theory of relativity the energy of any particle must be strictly positive, even for zero mass.) Let us next consider the process

$$\gamma + N \to N + e^- + e^+$$

mediated by a heavy nucleus N. Using Eq. (8.81) with

$$m_a = 0\,, \qquad m_b = m_N\,, \qquad M = 2m_e + m_N$$

we obtain

$$E_{\min} = \frac{c^2}{2m_N}\left[(2m_e + m_N)^2 - m_N^2\right] = 2m_e c^2\left(1 + \frac{m_e}{m_N}\right) \gtrsim 2m_e c^2\,,$$

since $m_e \ll m_N$.

8.8 RELATIVISTIC DYNAMICS

8.8.1 Four-force and relativistic force

In Newtonian mechanics, the motion of a material particle is governed by Newton's second law

$$\frac{\mathrm{d}\mathbf{p}}{\mathrm{d}t} = \mathbf{F}\,, \tag{8.82}$$

which holds in any inertial frame. From the point of view of the special theory of relativity, the most natural generalization of the previous equation is

$$\boxed{\frac{\mathrm{d}p}{\mathrm{d}\tau} = f\,,} \tag{8.83}$$

where

$$f := (f_0, \mathbf{f}) \in \mathbb{R}^4 \tag{8.84}$$

is a four-vector known as **four-force**, depending in general on the particle's space-time coordinates and velocity. Indeed, this equation is *Lorentz covariant*, since p is a vector under Lorentz transformations and the proper time τ is a scalar. In addition, we shall next see that Eq. (8.83) essentially reduces to Newton's second law for small speeds compared to c.

By analogy with Newtonian mechanics, we *define* the **relativistic force F** so that Newton's second law (8.82) holds if we interpret **p** as the *relativistic* three-momentum. Since

$$\frac{d\mathbf{p}}{dt} = \frac{d\mathbf{p}}{d\tau}\frac{d\tau}{dt} = \frac{1}{\gamma(v)}\frac{d\mathbf{p}}{d\tau} = \frac{\mathbf{f}}{\gamma(v)},$$

where v is the particle's velocity, the four-force and the relativistic force are related by

$$\boxed{\mathbf{F} = \frac{\mathbf{f}}{\gamma(v)}.} \tag{8.85}$$

Note that Eq. (8.82) can be written as

$$\frac{d}{dt}\big(\gamma(v)m\mathbf{v}\big) = \frac{d}{dt}\left(\frac{m\mathbf{v}}{\sqrt{1-\frac{v^2}{c^2}}}\right) = \mathbf{F}. \tag{8.86}$$

Obviously, for a given force **F** (for instance, for constant **F**) the previous equation tends to its classical analogue for particle speeds much smaller than c.

Remark 8.1. The fact that the relativistic force **F** is related to the spatial components **f** of a four-vector f by Eq. (8.85) ensures that, if Eq. (8.82) is valid in some inertial frame, it is valid in *all* of them. Of course, Eq. (8.85) imposes very stringent conditions on relativistic forces; in particular, note that although **F** is a vector under rotations it does *not* transform as the spatial components of a four-vector under Lorentz transformations. ■

Let us next show that the time component of the four-force is determined by the spatial ones. To this end, it suffices to differentiate with respect to τ the identity

$$p^2 = p \cdot p = m^2c^2,$$

which yields

$$\boxed{p \cdot f = 0.} \tag{8.87}$$

In other words, *the four-force and the four-momentum are orthogonal* (with respect to the Minkowski product) *at all times*. From the definition of Minkowski product we thus obtain the relation

$$f_0 = \frac{\mathbf{f}\cdot\mathbf{p}}{p_0} = \frac{\mathbf{f}\cdot\mathbf{v}}{c} = \frac{\gamma(v)}{c}\mathbf{F}\cdot\mathbf{v}, \tag{8.88}$$

where we have taken into account Eq. (8.49). Hence the four-force f can be expressed in terms of the relativistic force **F** by the equation

$$\boxed{f = \gamma(v)\left(\frac{\mathbf{F}\cdot\mathbf{v}}{c}, \mathbf{F}\right).} \tag{8.89}$$

In Newtonian mechanics

$$\mathbf{F}\cdot\mathbf{v} = \frac{\mathrm{d}T}{\mathrm{d}t}, \tag{8.90}$$

where

$$T = \frac{1}{2}m\mathbf{v}^2$$

is the particle's kinetic energy. The relativistic analogue of this equation is obtained from the time-like component of the equation of motion (8.83), namely

$$\frac{\mathrm{d}p_0}{\mathrm{d}\tau} = f_0 \, .$$

Indeed, by Eq. (8.88) we have

$$\frac{\mathrm{d}p_0}{\mathrm{d}\tau} = \frac{\mathrm{d}p_0}{\mathrm{d}t}\frac{\mathrm{d}t}{\mathrm{d}\tau} = \gamma(v)\frac{\mathrm{d}p_0}{\mathrm{d}t} = f_0 = \frac{\gamma(v)}{c}\mathbf{F}\cdot\mathbf{v}\,,$$

which yields the identity

$$\frac{\mathrm{d}}{\mathrm{d}t}(cp_0) = \frac{\mathrm{d}}{\mathrm{d}t}(mc^2+T) = \boxed{\frac{\mathrm{d}T}{\mathrm{d}t} = \mathbf{F}\cdot\mathbf{v}\,.} \tag{8.91}$$

Thus Eq. (8.90) is still valid, if we interpret T as the relativistic kinetic energy and $\mathbf{F}$ as the relativistic force.

Suppose now that, *in a certain inertial frame S*, the relativistic force $\mathbf{F}$ can be obtained from a time-independent scalar potential $V(\mathbf{x})$ through the usual equation

$$\mathbf{F} = -\frac{\partial V(\mathbf{x})}{\partial \mathbf{x}}\,. \tag{8.92}$$

This is the case, for instance, for a constant time-independent force (in some inertial frame), with $V = -\mathbf{F}\cdot\mathbf{x}$ linear in the particle's spatial coordinates. If Eq. (8.92) holds, we have

$$\mathbf{F}\cdot\mathbf{v} = \mathbf{F}\cdot\frac{\mathrm{d}\mathbf{x}}{\mathrm{d}t} = -\frac{\partial V}{\partial \mathbf{x}}\cdot\frac{\mathrm{d}\mathbf{x}}{\mathrm{d}t} = -\frac{\mathrm{d}V}{\mathrm{d}t}\,,$$

and Eq. (8.91) can be written as

$$\frac{\mathrm{d}}{\mathrm{d}t}\big(cp_0 + V(\mathbf{x})\big) = 0\,. \tag{8.93}$$

Thus in this case the **total relativistic energy**

$$E = cp_0 + V(\mathbf{x}) = mc^2 + T + V(\mathbf{x}) = mc^2\gamma(v) + V(\mathbf{x}) \tag{8.94}$$

is conserved.

Exercise 8.9. Find the general solution of the equation of motion of a relativistic particle moving in one dimension under a potential $V(x)$ (in a certain inertial frame).

Solution. By conservation of energy we must have

$$mc^2\gamma(\dot{x}) + V(x) = E\,,$$

where the constant E is the total relativistic energy. Since $\gamma(\dot{x}) \geqslant 1$, the motion is only possible in the region $V(x) \leqslant E - mc^2$, where $E - mc^2$ is the analogue of the non-relativistic energy (indeed, for small velocities $|\dot{x}|$ we have $E - mc^2 = \frac{1}{2}m\dot{x}^2 + V(x) + O(|\dot{x}|^4/c^2)$). Squaring and solving for $\dot{x}$ we obtain

$$1 - \frac{\dot{x}^2}{c^2} = \frac{m^2c^4}{\left(E - V(x)\right)^2} \quad\Longrightarrow\quad \dot{x} = \pm c\sqrt{1 - \frac{m^2c^4}{\left(E - V(x)\right)^2}}$$

and hence

$$t = \pm\frac{1}{c}\int \frac{\mathrm{d}x}{\sqrt{1 - \frac{m^2c^4}{\left(E-V(x)\right)^2}}}\,.$$

Note that the expression under the radical is non-negative, on account of the inequality $V(x) \leqslant E - mc^2$.

Exercise 8.10. A particle is moving along the x_2 direction subject to a conservative relativistic force with potential $V(x_2)$ according to an inertial observer S. What is the force acting on the particle according to a second inertial observer S' moving with uniform velocity $w\mathbf{e}_1$ relative to S?

Solution. The four-force f' measured by the observer S' is obtained from the corresponding four-force measured by S applying a Lorentz boost with velocity $w\mathbf{e}_1$; in particular,

$$f_1' = \gamma(w)\left(f_1 - \frac{w}{c}f_0\right), \qquad f_{2,3}' = f_{2,3}.$$

Since

$$f_{1,3} = \gamma(v)F_{1,3} = 0, \qquad f_2 = \gamma(v)F_2 = -\gamma(v)\frac{\partial V(x_2)}{\partial x_2},$$

$$f_0 = \gamma(v)\,\mathbf{F}\cdot\frac{\mathbf{v}}{c} = -\gamma(v)\frac{v_2}{c}\frac{\partial V(x_2)}{\partial x_2},$$

we have

$$f_1' = -\gamma(w)\frac{w}{c}f_0 = \gamma(v)\gamma(w)\frac{wv_2}{c^2}\frac{\partial V(x_2)}{\partial x_2},$$

$$f_2' = f_2 = -\gamma(v)\frac{\partial V(x_2)}{\partial x_2}, \qquad f_3' = f_3 = 0.$$

Thus the components of the relativistic force $\mathbf{F}'$ measured by S' are given by

$$F_1' = \frac{\gamma(v)\gamma(w)}{\gamma(v')}\frac{wv_2}{c^2}\frac{\partial V(x_2')}{\partial x_2'}, \qquad F_2' = -\frac{\gamma(v)}{\gamma(v')}\frac{\partial V(x_2')}{\partial x_2'}, \qquad F_3' = 0,$$

where we have taken into account that $x_2 = x_2'$. On the other hand, from the formula for relativistic addition of velocities (cf. Eq. (8.13)) we have

$$v_1' = -w, \qquad v_{2,3}' = \frac{v_{2,3}}{\gamma(w)},$$

and therefore (since $v_2 = v$)

$$\gamma(v')^{-2} = 1 - \frac{w^2}{c^2} - \frac{v_2^2}{c^2\gamma^2(w)} = 1 - \frac{w^2}{c^2} - \left(1 - \frac{w^2}{c^2}\right)\frac{v_2^2}{c^2}$$
$$= \left(1 - \frac{v_2^2}{c^2}\right)\left(1 - \frac{w^2}{c^2}\right) = \gamma(v)^{-2}\gamma(w)^{-2} \implies \gamma(v') = \gamma(v)\gamma(w).$$

We thus obtain

$$F_1' = \gamma(w)\frac{wv_2'}{c^2}\frac{\partial V(x_2')}{\partial x_2'}, \qquad F_2' = -\frac{1}{\gamma(w)}\frac{\partial V(x_2')}{\partial x_2'}, \qquad F_3' = 0,$$

or equivalently

$$\mathbf{F}' = \gamma(w)\left(\frac{wv_2'}{c^2}\mathbf{e}_1 - \left(1 - \frac{w^2}{c^2}\right)\mathbf{e}_2\right)\frac{\partial V(x_2')}{\partial x_2'}.$$

Hence the force $\mathbf{F}'$ depends on the particle velocity v_2', and is therefore *not* conservative. Note also that if $v_2' \neq 0$ (or, equivalently, $v_2 \neq 0$), the quotient between the two components of $\mathbf{F}'$, given by

$$\frac{F_1'}{F_2'} = -\frac{wv_2'}{c^2}\gamma(w)^2,$$

tends to $\pm\infty$ as the relative velocity w between the two inertial frames tends to c.

8.8.2 Relativistic Lagrangian

If the relativistic three-force $\mathbf{F}(t, \mathbf{x})$ is irrotational *in some inertial frame* S, i.e., if

$$\mathbf{F}(t, \mathbf{x}) = -\frac{\partial V(t, \mathbf{x})}{\partial \mathbf{x}}$$

in S, the relativistic equations of motion

$$\frac{d\mathbf{p}}{dt} = -\frac{\partial V}{\partial \mathbf{x}} \tag{8.95}$$

are the Lagrange equations of a suitable Lagrangian L in the frame S. To show this, it suffices to find a function $L(t, \mathbf{x}, \dot{\mathbf{x}})$ satisfying the equations

$$\frac{\partial L}{\partial \dot{\mathbf{x}}} = \mathbf{p} = m\gamma(v)\mathbf{v} = \frac{m\dot{\mathbf{x}}}{\sqrt{1 - \frac{\dot{\mathbf{x}}^2}{c^2}}}, \qquad \frac{\partial L}{\partial \mathbf{x}} = -\frac{\partial V}{\partial \mathbf{x}}(t, \mathbf{x}).$$

Integrating the first equation with respect to $\dot{\mathbf{x}}$ we obtain

$$L = -mc^2\sqrt{1 - \frac{\dot{\mathbf{x}}^2}{c^2}} + g(t, \mathbf{x}),$$

and from the second one we deduce that $g(t, \mathbf{x}) = -V(t, \mathbf{x}) + h(t)$. Dropping the total time derivative $h(t)$ we thus arrive at the following formula for the **relativistic Lagrangian**:

$$\boxed{L(t, \mathbf{x}, \dot{\mathbf{x}}) = -mc^2\sqrt{1 - \frac{\dot{\mathbf{x}}^2}{c^2}} - V(t, \mathbf{x}).} \tag{8.96}$$

Note that for small velocities $|\dot{\mathbf{x}}| \ll c$ we have

$$\sqrt{1 - \frac{\dot{\mathbf{x}}^2}{c^2}} = 1 - \frac{\dot{\mathbf{x}}^2}{2c^2} + O(|\dot{\mathbf{x}}|^4/c^4),$$

and therefore

$$L = \frac{1}{2}m\dot{\mathbf{x}}^2 - V(t, \mathbf{x}) - mc^2 + O(|\dot{\mathbf{x}}|^4/c^2).$$

Hence in this limit the relativistic Lagrangian reduces to the non-relativistic one, up to the irrelevant constant $-mc^2$.

Remark 8.2. Unlike what happens in non-relativistic mechanics, the velocity-dependent term in the relativistic Lagrangian L is *not* the relativistic kinetic energy

$$T = -mc^2\big(1 - \gamma(\mathbf{v})\big) = -mc^2\left(1 - \frac{1}{\sqrt{1 - \frac{\dot{\mathbf{x}}^2}{c^2}}}\right).$$

In other words, *in relativistic mechanics the Lagrangian is* not *of the form $T - V$.* ■

It should be noted that Eq. (8.95) only holds in the privileged frame S in which the three-force is irrotational. Indeed, the three-force measured in a different inertial frame S' will in general be velocity-dependent, and thus *not* irrotational (see, e.g., Exercise 8.10). Therefore the relativistic Lagrangian (8.96) is *not* Lorentz covariant. In other words, the equations of motion in an arbitrary inertial frame S' are *not* in general the Lagrange equations of the transformed Lagrangian $L(t', \mathbf{x}', \dot{\mathbf{x}}')$. It follows that the action

$$S = \int_{t_1}^{t_2} dt\, L(t, \mathbf{x}, \dot{\mathbf{x}}) = \int_{t_1}^{t_2} dt \left[-mc^2\sqrt{1 - \frac{\dot{\mathbf{x}}^2}{c^2}} - V(t, \mathbf{x})\right]$$

is *not* Lorentz invariant. More precisely, consider the differential of the action

$$\mathrm{d}S = -\left(mc^2\sqrt{1-\frac{\dot{\mathbf{x}}^2}{c^2}} + V(t,\mathbf{x})\right)\mathrm{d}t = -mc^2\,\mathrm{d}\tau - V(t,\mathbf{x})\,\mathrm{d}t,$$

where τ is the particle's proper time. Although the first term in the right-hand side of the previous equation is Lorentz invariant (it is proportional to the differential of the particle's proper time), the second term is not due to the presence of the differential of the coordinate time t.

When the relativistic Lagrangian is time-independent, i.e., when V depends only on $\mathbf{x}$ and the force is therefore *conservative*, the energy integral $h(\mathbf{x}, \dot{\mathbf{x}})$ is conserved. As expected, in this case

$$h(\mathbf{x},\dot{\mathbf{x}}) = \frac{\partial L}{\partial \dot{\mathbf{x}}}\cdot\dot{\mathbf{x}} - L = \frac{m\dot{\mathbf{x}}^2}{\sqrt{1-\frac{\dot{\mathbf{x}}^2}{c^2}}} + mc^2\sqrt{1-\frac{\dot{\mathbf{x}}^2}{c^2}} + V(\mathbf{x}) = \frac{mc^2}{\sqrt{1-\frac{\dot{\mathbf{x}}^2}{c^2}}} + V(\mathbf{x})$$

is nothing but the total relativistic energy (8.94) introduced above.

From the relativistic Lagrangian (8.96) we can easily derive the Hamiltonian formulation of the equations of motion in the usual way. Indeed, the canonical momenta (in Cartesian coordinates) are

$$p_i = \frac{\partial L}{\partial \dot{x}_i} = \frac{m\dot{x}_i}{\sqrt{1-\frac{\dot{\mathbf{x}}^2}{c^2}}}, \tag{8.97}$$

from which we can easily solve for the velocities $\dot{x}_i$. Indeed, we have

$$\mathbf{p}^2 = \frac{m^2\dot{\mathbf{x}}^2}{1-\frac{\dot{\mathbf{x}}^2}{c^2}} \implies \frac{\dot{\mathbf{x}}^2}{c^2} = \frac{\mathbf{p}^2}{\mathbf{p}^2+m^2c^2} \implies 1-\frac{\dot{\mathbf{x}}^2}{c^2} = \frac{m^2c^2}{\mathbf{p}^2+m^2c^2},$$

which substituted into Eq. (8.97) yields

$$\dot{x}_i = \frac{cp_i}{\sqrt{\mathbf{p}^2+m^2c^2}} \tag{8.98}$$

(cf. Eq. (8.50)). The Hamiltonian is the energy integral $h(\mathbf{x}, \dot{\mathbf{x}})$ expressed in terms of the canonical momenta using the previous formula for the velocity, namely

$$\boxed{H = c\sqrt{\mathbf{p}^2+m^2c^2} + V(\mathbf{x}).}$$

Hamilton's canonical equations are therefore Eqs. (8.98), together with

$$\dot{p}_i = -\frac{\partial H}{\partial x_i} = -\frac{\partial V}{\partial x_i}.$$

8.8.3 Hyperbolic motion

The simplest example of relativistic three-force is that of a constant force[11]

$$\boxed{\mathbf{F} = m\mathbf{a}\,,}$$

with $\mathbf{a} \in \mathbb{R}^3$ a constant vector with dimensions of acceleration. We shall next see that in this case, just as in Newtonian mechanics, the particle's equation of motion can be exactly solved. We shall suppose, for the sake of simplicity, that the particle is initially at rest at the origin of coordinates, i.e.,

$$\mathbf{x}(0) = \mathbf{p}(0) = 0\,.$$

Integrating the equation of motion

$$\frac{\mathrm{d}\mathbf{p}}{\mathrm{d}t} = m\mathbf{a}$$

with the initial condition $\mathbf{p}(0) = 0$ we then have

$$\mathbf{p} = m\mathbf{a}\,t\,.$$

Substituting into Eq. (8.50) or (8.98) we obtain

$$\mathbf{v} = \frac{\mathrm{d}\mathbf{x}}{\mathrm{d}t} = \frac{c\mathbf{p}}{\sqrt{\mathbf{p}^2 + m^2c^2}} = \frac{mc\mathbf{a}\,t}{\sqrt{m^2c^2 + m^2a^2t^2}} = \boxed{\frac{\mathbf{a}\,t}{\sqrt{1 + \frac{a^2t^2}{c^2}}}}\,. \tag{8.99}$$

Note that, regardless of the magnitude of the force $\mathbf{F}$ (i.e., of the constant acceleration $\mathbf{a}$), from the previous equation it follows that $v < c$ for all t. Integrating the last equation with respect to t and taking into account that $\mathbf{x}(0) = 0$ we derive the law of motion:

$$\mathbf{x} = \mathbf{a}\int_0^t \frac{s\,\mathrm{d}s}{\sqrt{1 + \frac{a^2s^2}{c^2}}} = \boxed{\frac{c^2\mathbf{a}}{a^2}\left(\sqrt{1 + \frac{a^2t^2}{c^2}} - 1\right)}\,. \tag{8.100}$$

Note that for $a|t| \ll c$ Eqs. (8.99) and (8.100) approximately reduce to their analogues in Newtonian mechanics

$$\mathbf{v} = \mathbf{a}t\,, \qquad \mathbf{x} = \frac{1}{2}\,\mathbf{a}\,t^2\,. \tag{8.101}$$

On the contrary, for $t \to \pm\infty$ the velocity $\mathbf{v}$ tends to $\pm c\mathbf{a}/a$ and, therefore, the particle's speed tends to c (cf. Fig. 8.7), whereas $\mathbf{x} \sim c|t|\mathbf{a}/a$.

[11]The statement that the force acting on a particle is constant is *not* Lorentz invariant, but depends on the inertial frame considered. In other words, even if $\mathbf{F}$ is constant in a given inertial frame S, it need not be constant in another frame S' in motion relative to S. It can be shown, however, that if $\mathbf{F}$ is constant in an inertial frame S it will remain constant in any other frame S' whose velocity with respect to S is parallel to $\mathbf{F}$, and in this case $\mathbf{F}' = \mathbf{F}$ (see Exercise 8.11).

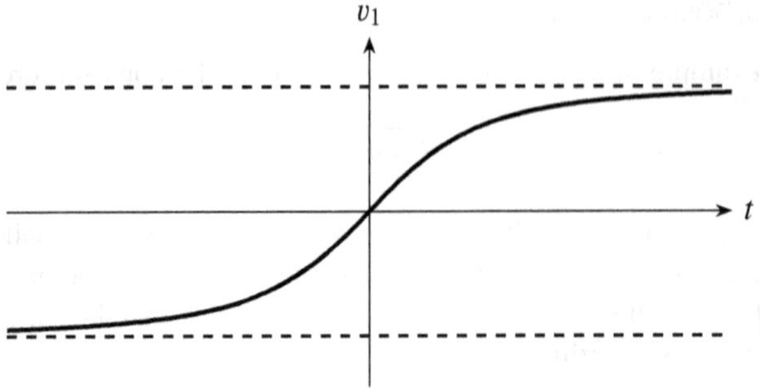

Figure 8.7. Component in the $\mathbf{e}_1$ direction of the velocity of a relativistic particle of mass m subject to a constant force $ma\mathbf{e}_1$ as a function of time (solid curve) and its two asymptotes $v_1 = \pm c$ (dashed lines).

If we choose the axes so that $\mathbf{a} = a\mathbf{e}_1$, the law of motion (8.100) reduces to

$$x_1 = \frac{c^2}{a}\left(\sqrt{1+\frac{a^2t^2}{c^2}} - 1\right) \quad \Longrightarrow \quad \boxed{\left(x_1 + \frac{c^2}{a}\right)^2 - x_0^2 = \frac{c^4}{a^2}, \quad x_1 \geqslant 0}.$$

This is the equation of a (branch of an) *equilateral hyperbola* centered at the point $(0, -c^2/a)$, whose axis is the x_1 axis and having as asymptotes the straight lines

$$x_1 + \frac{c^2}{a} = \pm x_0$$

(cf. Fig. 8.8). Note that in Newtonian mechanics the particle's worldline is the *parabola*

$$x_1 = \frac{a}{2c^2}\, x_0^2$$

(cf. Eq. (8.101)).

From Eq. (8.99) it immediately follows that

$$\frac{\mathrm{d}\tau}{\mathrm{d}t} = \frac{1}{\gamma} = \left(1 + \frac{a^2t^2}{c^2}\right)^{-1/2} \quad \Longrightarrow \quad \tau = \int_0^t \frac{\mathrm{d}s}{\sqrt{1+\frac{a^2s^2}{c^2}}},$$

where for simplicity's sake we have taken $\tau(0) = 0$. Performing the change of variable $as/c = \sinh z$ in the integral we easily obtain

$$\boxed{\tau = \frac{c}{a}\operatorname{arcsinh}(at/c) = \frac{c}{a}\log\left(\frac{at}{c} + \sqrt{1+\frac{a^2t^2}{c^2}}\right).} \tag{8.102}$$

Thus the coordinate time t is related to the proper time τ by

$$\boxed{t = \frac{c}{a}\sinh(a\tau/c).} \tag{8.103}$$

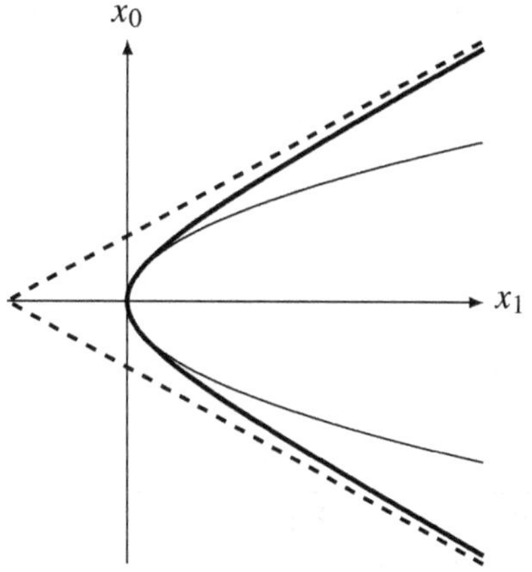

Figure 8.8. World line of a relativistic particle of mass m subject to a constant force ma in the x_1 direction (thick curve), along with its analogue in Newtonian mechanics (thin curve). The dashed lines represent the asymptotes $x_1 = -\frac{c^2}{a} \pm x_0$ of the particle's worldline.

In particular, that for $\tau \gg c/a$ we have

$$t \simeq \frac{c}{2a}\,\mathrm{e}^{a\tau/c} \qquad (\tau \gg c/a)\,,$$

i.e., the coordinate time increases exponentially with the proper time.

It is also of interest to compute $\beta(v)$ and $\gamma(v)$ as functions of the proper time τ. First of all (taking, as before, $\mathbf{a} = a\mathbf{e}_1$), the parameter $\beta(v)$ is easily obtained from Eqs. (8.99) and (8.103):

$$\beta(v) = \frac{v_1}{c} = \frac{\sinh(a\tau/c)}{\sqrt{1+\sinh^2(a\tau/c)}} = \tanh(a\tau/c)\,. \tag{8.104}$$

The γ factor can be derived from the previous expression, or more directly taking into account that it is just the derivative of the coordinate time t with respect to the proper time τ:

$$\gamma(v) = \frac{\mathrm{d}t}{\mathrm{d}\tau} = \cosh(a\tau/c)\,. \tag{8.105}$$

By Eq. (8.53), the particle's kinetic energy is given by

$$T = mc^2\big(\gamma(v)-1\big) = mc^2\big(\cosh(a\tau/c)-1\big) = \boxed{2mc^2\sinh^2\big(a\tau/(2c)\big)}\,. \tag{8.106}$$

This is the energy that must be supplied to the particle to maintain its constant acceleration $\mathbf{a}$ between the *proper* times 0 and τ. By the law of conservation of relativistic energy (8.94), this energy must be equal to the work $\mathbf{F}\cdot\mathbf{x} = m\mathbf{a}\cdot\mathbf{x}$ done by the constant force $\mathbf{F} = m\mathbf{a}$ during this period of time. This fact is also easily verified using Eqs. (8.100) and (8.103):

$$m\mathbf{a} \cdot \mathbf{x} = mc^2 \left(\sqrt{1 + \frac{a^2 t^2}{c^2}} - 1 \right) = mc^2 \big(\cosh(a\tau/c) - 1 \big) . \tag{8.107}$$

Again, for proper times $\tau \gg c/a$ this energy increases exponentially with τ:

$$T \simeq \frac{1}{2} mc^2 \, e^{a\tau/c} \qquad (\tau \gg c/a) .$$

Exercise 8.11. Show that if a force $\mathbf{F}$ is constant in an inertial frame S it is also constant in any other inertial frame S' moving in the direction of $\mathbf{F}$ relative to S, and that moreover $\mathbf{F}' = \mathbf{F}$.

Solution. Let us take the x_1 axis of S in the direction of the force $\mathbf{F}$ and the axes of S' parallel to those of S, and denote by $\mathbf{w} = w\mathbf{e}_1$ the velocity of the origin O' of S' relative to S. In the original frame S the four-force f has components

$$f_0 = \frac{\gamma(v)}{c} \mathbf{F} \cdot \mathbf{v} = \gamma(v) \frac{v_1}{c} F , \qquad f_1 = \gamma(v) F , \qquad f_2 = f_3 = 0 ,$$

and therefore

$$f = F\gamma(v) \left(\frac{v_1}{c}, 1, 0, 0 \right) .$$

The components of the four-force in the frame S' are obtained applying a Lorentz boost of velocity w in the direction of the x_1 axis:

$$f_0' = \gamma(w) \left(f_0 - \frac{w}{c} f_1 \right) = F\gamma(v)\gamma(w) \frac{v_1 - w}{c} ,$$

$$f_1' = \gamma(w) \left(f_1 - \frac{w}{c} f_0 \right) = F\gamma(v)\gamma(w) \left(1 - \frac{v_1 w}{c^2} \right),$$

and of course $f_2' = f_3' = 0$. Taking into account that the x_1 component of the particle's velocity in the frame S' is given by the relativistic law of addition of velocities

$$v_1' = \frac{v_1 - w}{1 - \frac{v_1 w}{c^2}} ,$$

we obtain

$$f' = F\gamma(v)\gamma(w) \left(1 - \frac{v_1 w}{c^2} \right) \left(\frac{v_1'}{c}, 1, 0, 0 \right) .$$

From the identity

$$\gamma(v)\gamma(w) \left(1 - \frac{v_1 w}{c^2} \right) = \gamma(v')$$

(cf. Exercise 8.3) it then follows that

$$f' = F\gamma(v')\left(\frac{v_1'}{c}, 1, 0, 0\right),$$

and in particular

$$\mathbf{F}' = \frac{f_1'}{\gamma(v')}\,\mathbf{e}_1 = F\mathbf{e}_1 = \mathbf{F}\,.$$

Exercise 8.12. The *proper acceleration* of a particle is its instantaneous acceleration relative to its proper inertial frame. i) Express the proper acceleration as a function of the particle's acceleration $\mathbf{a} = \mathrm{d}\mathbf{v}/\mathrm{d}t$ measured in an arbitrary inertial frame. ii) If the particle's velocity relative to a certain inertial frame is always parallel to the vector $\mathbf{e}_1$, show that in that frame the proper acceleration equals $\mathrm{d}\mathbf{u}/\mathrm{d}t$.

Solution.

i) To compute the proper acceleration at a certain time t, let us first determine how the acceleration $\mathbf{a} = \frac{\mathrm{d}\mathbf{v}}{\mathrm{d}t}$ measured in a certain inertial frame S transforms under a Lorentz boost with velocity $w\mathbf{e}_1$. To this end, it suffices to differentiate the law of relativistic addition of velocities (8.12). Setting

$$a_k = \frac{\mathrm{d}v_k}{\mathrm{d}t}, \qquad D = 1 - \frac{v_1 w}{c^2}$$

we thus obtain

$$a_1' = \frac{\mathrm{d}v_1'}{\mathrm{d}t}\bigg/\frac{\mathrm{d}t'}{\mathrm{d}t} = \frac{Da_1 + (v_1 - w)\frac{a_1 w}{c^2}}{\gamma(w)D^3} = \frac{a_1}{\gamma(w)^3 D^3},$$

$$a_k' = \frac{Da_k + v_k\frac{a_1 w}{c^2}}{\gamma^2(w)D^3} \qquad (k = 2, 3)\,.$$

The vector $\mathbf{a}' = (a_1', a_2', a_3')$ is the particle's acceleration measured in an inertial frame S' (with axes parallel to those of S) moving relative to S with velocity $\mathbf{w} = w\mathbf{e}_1$. If we assume that at some instant t the particle is moving in the direction of the $\mathbf{e}_1$ axis, that is, if $\mathbf{v} = v\mathbf{e}_1$ at that instant, taking $v_1 = v = w$ (and therefore $D = \gamma(v)^{-2}$) and $v_2 = v_3 = 0$ in the previous equations yields the particle's acceleration in its proper frame S' at the time t:

$$a_1' = \gamma(v)^3 a_1\,, \qquad a_k' = \gamma(v)^2 a_k \qquad (k = 2, 3)\,.$$

Obviously, in an arbitrary inertial frame (whose $\mathbf{e}_1$ axis need not coincide with the direction of the particle's velocity at the time t) the previous formulas should be replaced by

$$a_{\|}' = \gamma(v)^3 a_{\|}\,, \qquad a_{\perp}' = \gamma(v)^2 a_{\perp}\,,$$

where $a'_{\parallel}$ and $a'_{\perp}$ respectively denote the components of $\mathbf{a}'$ parallel and perpendicular to the particle's velocity at the time t.

ii) If the particle moves at all times in the direction of the x_1 axis with velocity v (not necessarily constant) in a certain inertial frame S, then $x_2 = x_3 = 0$ for all t, and therefore $a_2 = a_3 = 0$. From the previous formulas it then follows that

$$a'_1 = \gamma(v)^3 a_1 = \gamma(v)^3 \frac{dv}{dt}, \qquad a'_2 = a'_3 = 0 .$$

Hence

$$\frac{du_1}{dt} = \frac{d}{dt}\big(\gamma(v)v\big) = \big(\gamma'(v)v + \gamma(v)\big)\frac{dv}{dt} = \left(\gamma(v) + \gamma(v)^3\frac{v^2}{c^2}\right)\frac{dv}{dt} = \gamma(v)^3\frac{dv}{dt} = a'_1 ,$$

as was to be shown. In particular, from the equation of motion under a constant force $\mathbf{F} = F\mathbf{e}_1$ it follows that

$$\frac{\mathbf{F}}{m} = \frac{1}{m}\frac{d\mathbf{p}}{dt} = \frac{d\mathbf{u}}{dt} = \mathbf{a}' .$$

Hence in the hyperbolic motion studied in this section the proper acceleration of the particle is constant and directed along the x_1 axis.

8.9 LORENTZ FORCE

The most important example of a relativistic force is the *Lorentz force*

$$\mathbf{F} = q(\mathbf{E} + \mathbf{v} \times \mathbf{B}), \tag{8.108}$$

where $\mathbf{v}$ denotes the particle's velocity. Indeed, it is an experimental fact that the equation of motion of a particle of charge q in an electric field $\mathbf{E}$ and a magnetic field $\mathbf{B}$ is *exactly* (even at speeds arbitrarily close to c) Eq. (8.82) with the Lorentz force (8.108). We shall check in this section that if the fields $\mathbf{E}$ and $\mathbf{B}$ transform appropriately under Lorentz transformations the equation of motion (8.82), with $\mathbf{F}$ given by Eq. (8.108), is indeed valid in *any* inertial frame. Before proceeding, we shall introduce some convenient notation and briefly review the concepts of (co)vectors and tensors in the context of special relativity.

8.9.1 Vectors, covectors, and tensors in special relativity

As is customary in most textbooks on relativistic mechanics, in this section we shall denote by a^μ the components in a given inertial frame S of a four-vector under Lorentz transformations $a := (a^0, \mathbf{a})$; in particular, $\mathbf{a} = (a^1, a^2, a^3)$. We shall often informally write (a^μ) to denote the four-vector a with components a^μ in the frame S. The components a'^μ of the same vector in another inertial frame S', related to S by a

Poincaré transformation $x' = \Lambda x + b$, can be obtained from the formula

$$a'^{\mu} = \sum_{\nu} \Lambda^{\mu}{}_{\nu} a^{\nu} \tag{8.109}$$

in terms of the matrix elements $(\Lambda^{\mu}{}_{\nu})_{0\leqslant\mu,\nu\leqslant 3} \equiv (\Lambda^{\mu}{}_{\nu})$ of the Lorentz transformation Λ. This relation can be expressed in matrix form as

$$\boxed{a' = \Lambda a,}$$

where $a = (a^0\, a^1\, a^2\, a^3)^{\mathsf{T}}$ is a column vector (and similarly a'). Consider next a linear map $B : \mathbb{R}^4 \to \mathbb{R}$, which in the inertial frame S is defined by an equation of the form

$$B(x) = \sum_{\mu} b_{\mu} x^{\mu}$$

for certain coefficients $b_{\mu} \in \mathbb{R}$. In a different inertial frame S' this equation becomes

$$B(x) = \sum_{\mu} b'_{\mu} x'^{\mu} .$$

To find the relation between the components (b_{μ}) and (b'_{μ}) of the linear form B in the frames S and S', it suffices to note that

$$B(x) = \sum_{\mu} b'_{\mu} x'^{\mu} = \sum_{\mu,\nu} b'_{\mu} \Lambda^{\mu}{}_{\nu} x^{\nu} = \sum_{\nu} x^{\nu} \sum_{\mu} b'_{\mu} \Lambda^{\mu}{}_{\nu} \quad \Longrightarrow \quad b_{\nu} = \sum_{\mu} b'_{\mu} \Lambda^{\mu}{}_{\nu} . \tag{8.110}$$

Note that this relation can be written in matrix form as

$$\boxed{b = b'\Lambda,}$$

where $b = (b_0\, b_1\, b_2\, b_3)$ is a *row* vector (and similarly b'). We can invert this relation using the defining equation (8.26) of Lorentz transformations, which yields[12]

$$\Lambda^{-1} = \eta^{-1}\Lambda^{\mathsf{T}}\eta,$$

and therefore

$$\boxed{b' = b\,\eta^{-1}\Lambda^{\mathsf{T}}\eta\,.}$$

Denoting by $\eta_{\mu\nu}$ and $\eta^{\mu\nu}$ the matrix elements of η and η^{-1}, respectively, we have

$$\boxed{b'_{\mu} = \sum_{\nu,\mu',\nu'} b_{\nu} \eta^{\nu\nu'} \Lambda^{\mu'}{}_{\nu'} \eta_{\mu'\mu} = \sum_{\nu} \Lambda_{\mu}{}^{\nu} b_{\nu} , \quad \text{with } \Lambda_{\mu}{}^{\nu} = \sum_{\mu',\nu'} \eta_{\mu\mu'} \eta^{\nu\nu'} \Lambda^{\mu'}{}_{\nu'} ,} \tag{8.111}$$

[12]Although $\eta^{-1} = \eta$ by Eq. (8.23), we have chosen to distinguish η^{-1} from η since in general relativity η^{-1} need not be equal to η.

where we have made use of the symmetry of η (namely, $\eta_{\mu\mu'} = \eta_{\mu'\mu}$). Note that $\Lambda_\mu{}^\nu$ is formally obtained from $\Lambda^{\mu'}{}_{\nu'}$ by *lowering* the index μ' with $\eta_{\mu\mu'}$ and *raising* the index ν' with $\eta^{\nu\nu'}$.

We thus see that the transformation law (8.109) of the components (a^μ) of a four-vector a is different from the analogous law (8.111) for the components (b_μ) of a linear form B. To emphasize this distinction, linear forms are often called *covectors*, and four-vectors are referred to as *contravariant vectors*. An important example of a covector is the (four-)gradient of a scalar function $\phi(x)$, whose components $\phi_{,\mu}(x)$ are defined by

$$\phi_{,\mu}(x) := \frac{\partial\phi(x)}{\partial x^\mu}.$$

Indeed, it suffices to note that the functions $\phi_{,\mu}(x)$ are the components of the linear form (in $\mathrm{d}x$)

$$\mathrm{d}\phi(x) = \sum_\mu \frac{\partial\phi}{\partial x^\mu}\,\mathrm{d}x^\mu .$$

Given a contravariant vector $a := (a^\mu)$, it is straightforward to check that the quantities

$$a_\mu = \sum_\nu \eta_{\mu\nu}a^\nu$$

transform under Lorentz transformations as the components of a covector. Indeed, since

$$\sum_\mu a_\mu x^\mu = \sum_{\mu,\nu}\eta_{\mu\nu}a^\nu x^\mu = a\cdot x\,,$$

the numbers (a_μ) are the components of the linear form $x \mapsto a\cdot x$. It is thus customary to refer to the numbers (a_μ) as the *covariant components* of the vector a, and in the same vein call (a^μ) its *contravariant components*. Note that from the definition (8.23) of η it follows that

$$\boxed{a_0 = a^0\,, \qquad a_i = -a^i\,.} \tag{8.112}$$

In general, an *r-contravariant and s-covariant tensor* T in Minkowski space can be defined as a set of 4^{r+s} quantities $(T^{\mu_1\dots\mu_r}_{\nu_1\dots\nu_s})$—or, more precisely, an assignment of 4^{r+s} numbers $(T^{\mu_1\dots\mu_r}_{\nu_1\dots\nu_s})$ to each inertial frame S—which transform under a Lorentz transformation Λ as

$$T'^{\mu'_1\dots\mu'_r}_{\nu'_1\dots\nu'_s} = \sum_{\substack{\mu_1,\dots,\mu_r\\ \nu_1,\dots,\nu_s}} \Lambda^{\mu'_1}{}_{\mu_1}\cdots\Lambda^{\mu'_r}{}_{\mu_r}\Lambda_{\nu'_1}{}^{\nu_1}\cdots\Lambda_{\nu'_s}{}^{\nu_s}T^{\mu_1\dots\mu_r}_{\nu_1\dots\nu_s}\,.$$

In other words, the *contravariant indices* μ_i transform as the components of a *contravariant vector* (cf. Eq. (8.109)), while the *covariant indices* ν_j transform as the components of a *covector* (cf. Eq. (8.111)). More formally, T is a mapping from $(V^*)^r \times V^s$ to $\mathbb{R}$, where $V = \mathbb{R}^4$ and V^* is the dual space of V, which is *linear* in each of its $r + s$ arguments. Note, in this respect, that a vector space V is canonically isomorphic to its *bidual* $V^{**} := (V^*)^*$.

8.9.2 Electromagnetic four-force

The condition (8.87), or equivalently $u \cdot f = 0$, is a strong constraint on the form of the covariant four-force f. To begin with, it implies that a *nonzero four-force f must necessarily depend on the particle's velocity,* since even when $\mathbf{f}$ is independent of $\mathbf{v}$ in a certain inertial frame S its time-like component $f^0 = \mathbf{f} \cdot \mathbf{v}/c$ is velocity-dependent. The simplest example of nontrivial covariant four-force is a linear function of the four-velocity u, i.e.,

$$f^\mu = \sum_\nu F^\mu{}_\nu(x) u^\nu = \sum_\nu F^{\mu\nu}(x) u_\nu , \tag{8.113}$$

where

$$F^{\mu\nu}(x) = \sum_\sigma \eta^{\nu\sigma} F^\mu{}_\sigma(x) .$$

Imposing the condition $u \cdot f = 0$ we obtain

$$\sum_\mu f^\mu u_\mu = \sum_{\mu,\nu} F^{\mu\nu}(x) u_\mu u_\nu = \frac{1}{2} \sum_{\mu,\nu} \Big(F^{\mu\nu}(x) + F^{\nu\mu}(x)\Big) u_\mu u_\nu = 0.$$

Since this condition must hold for all values of u_μ, it follows that

$$\boxed{F^{\mu\nu}(x) = -F^{\nu\mu}(x) .}$$

Thus the 4×4 matrix $\big(F^{\mu\nu}(x)\big)_{0\leqslant\mu,\nu\leqslant 3}$ is *antisymmetric*, and has therefore 6 independent components. The relativistic force $\mathbf{F}$ associated with the linear four-force (8.113) has components

$$F^i = \frac{f^i}{\gamma(v)} = \sum_\nu F^{i\nu} \frac{u_\nu}{\gamma(v)} = c F^{i0}(x) - \sum_j F^{ij}(x) v^j ,$$

where we have used Eq. (8.112). This expression is reminiscent of the Lorentz force acting on a charged particle, since it consists of a term independent of the velocity (proportional to the electric field strength) and another one linear in the velocity (associated with the magnetic field). In fact, let us define

$$E^i(x) := c F^{i0}(x) = -c F^{0i}(x) . \tag{8.114}$$

Moreover, since $F^{ij}(x)$ is antisymmetric we can write

$$F^{ij}(x) = -\sum_k \varepsilon_{ijk} B^k(x) , \tag{8.115}$$

where

$$B^k = -\frac{1}{2} \sum_{i,j} \varepsilon_{ijk} F^{ij}(x).$$

and ε_{ijk} is Levi-Civita's completely antisymmetric symbol (cf. Exercise 6.3). We then have

$$F^i = E^i(x) + \sum_{j,k} \varepsilon_{ijk} v^j B^k(x) \iff \mathbf{F} = \mathbf{E}(x) + \mathbf{v} \times \mathbf{B}(x) .$$

This is the correct form of the electromagnetic force for a unit charge. For an arbitrary charge q the electromagnetic four-force is therefore

$$\boxed{f^\mu = q \sum_\mu F^\mu{}_\nu(x) u^\nu = q \sum_\mu F^{\mu\nu}(x) u_\nu ,} \tag{8.116}$$

where the elements of the 4×4 matrix $F^{\mu\nu}(x)$ are related to the electric and magnetic fields $\mathbf{E}(x)$ and $\mathbf{B}(x)$ by Eqs. (8.114)–(8.115). The corresponding relativistic force is the familiar Lorentz force

$$\boxed{\mathbf{F} = q\Big(\mathbf{E}(x) + \mathbf{v} \times \mathbf{B}(x)\Big) .} \tag{8.117}$$

It follows that the relativistic equation of motion of a charge q in an electromagnetic field $(\mathbf{E}(x), \mathbf{B}(x))$ is simply

$$\boxed{\frac{\mathrm{d}\mathbf{p}}{\mathrm{d}t} = q\Big(\mathbf{E}(x) + \mathbf{v} \times \mathbf{B}(x)\Big) ,} \tag{8.118}$$

where $\mathbf{p} = m\gamma(v)\mathbf{v}$ is the relativistic three-momentum. Note that this equation is *exact* (i.e., it holds for particle speeds arbitrarily close to c) and is valid in *all* inertial frames. In fact, by Eq. (8.116) for the Lorentz four-force, the previous equation is equivalent to the manifestly covariant one

$$\boxed{\frac{\mathrm{d}p^\mu}{\mathrm{d}\tau} = q \sum_\nu F^{\mu\nu}(x) u_\nu .} \tag{8.119}$$

8.9.3 Electromagnetic field tensor

From Eq. (8.116), and the fact that f and u are contravariant vectors, it follows that the quantities $(F^{\mu\nu})$ are the components of a twice contravariant antisymmetric tensor under Lorentz transformations (cf. next exercise). By Eqs. (8.114)–(8.115), the components of the tensor $(F^{\mu\nu})$, which is known as the **electromagnetic field tensor**, are related to the fields $\mathbf{E}$ and $\mathbf{B}$ by

$$\boxed{(F^{\mu\nu}) = \begin{pmatrix} 0 & -E^1/c & -E^2/c & -E^3/c \\ E^1/c & 0 & -B^3 & B^2 \\ E^2/c & B^3 & 0 & -B^1 \\ E^3/c & -B^2 & B^1 & 0 \end{pmatrix} .} \tag{8.120}$$

Exercise 8.13. Show that the quantities $(F^{\mu\nu})$ transform under a Lorentz transformation $x' = \Lambda x$ between two inertial frames S and S' as the components of a twice contravariant tensor.

Solution. Indeed, note that, since

$$\frac{u \cdot f}{q} = \sum_{\mu,\nu} F^{\mu\nu} u_\mu u_\nu$$

is a Lorentz scalar, we must have

$$\sum_{\mu',\nu'} F'^{\mu'\nu'} u'_{\mu'} u'_{\nu'} = \sum_{\mu,\nu} F^{\mu\nu} u_\mu u_\nu = \sum_{\mu,\nu,\mu',\nu'} F^{\mu\nu} \Lambda^{\mu'}{}_{\mu} \Lambda^{\nu'}{}_{\nu} u'_{\mu'} u'_{\nu'}$$

for all (time-like) vectors $u \in \mathbb{R}^4$, and hence

$$\boxed{F'^{\mu'\nu'} = \sum_{\mu,\nu} \Lambda^{\mu'}{}_{\mu} \Lambda^{\nu'}{}_{\nu} F^{\mu\nu}\,,} \tag{8.121}$$

as claimed. Note that Eq. (8.121) can be written in matrix form as

$$\boxed{F' = \Lambda F \Lambda^{\mathsf{T}},} \tag{8.122}$$

where

$$F = (F^{\mu\nu}), \qquad F' = (F'^{\mu'\nu'})$$

are 4×4 matrices.

Likewise, $(F_{\mu\nu})$ transforms as a rank 2 (antisymmetric) covariant tensor, which is defined by some authors as the electromagnetic field tensor instead of $(F^{\mu\nu})$. Since

$$F_{\mu\nu} = \sum_{\rho,\sigma} \eta_{\mu\rho} \eta_{\nu\sigma} F^{\rho\sigma} = \eta_{\mu\mu} \eta_{\nu\nu} F^{\mu\nu}\,,$$

by Eq. (8.120), the components of $(F_{\mu\nu})$ are given by

$$(F_{\mu\nu}) = \begin{pmatrix} 0 & E^1/c & E^2/c & E^3/c \\ -E^1/c & 0 & -B^3 & B^2 \\ -E^2/c & B^3 & 0 & -B^1 \\ -E^3/c & -B^2 & B^1 & 0 \end{pmatrix}.$$

Exercise 8.14. Find the transformation law of the fields **E** and **B** under a Lorentz boost in the x^1 direction with velocity v.

Solution. In this case the only nonzero elements of $(\Lambda^{\mu}{}_{\nu})$ are

$$\Lambda^0{}_0 = \Lambda^1{}_1 = \gamma(v)\,, \qquad \Lambda^0{}_1 = \Lambda^1{}_0 = -\beta\gamma(v)\,, \qquad \Lambda^2{}_2 = \Lambda^3{}_3 = 1\,.$$

Using Eq. (8.122), after an elementary calculation we obtain

$$E'^1 = E^1, \qquad E'^2 = \gamma(E^2 - vB^3), \qquad E'^3 = \gamma(E^3 + vB^2), \tag{8.123}$$

$$B'^1 = B^1, \qquad B'^2 = \gamma\left(B^2 + \frac{v}{c^2}E^3\right), \qquad B'^3 = \gamma\left(B^3 - \frac{v}{c^2}E^2\right). \tag{8.124}$$

Denoting respectively by $\mathbf{E}_\parallel$ and $\mathbf{E}_\perp$ the components of $\mathbf{E}$ parallel and perpendicular to the velocity $\mathbf{v}$, and similarly for $\mathbf{B}$, the above equations can be written as

$$\mathbf{E}'_\parallel = \mathbf{E}_\parallel\,, \quad \mathbf{E}'_\perp = \gamma(\mathbf{E}_\perp + \mathbf{v}\times\mathbf{B}_\perp)\,; \qquad \mathbf{B}'_\parallel = \mathbf{B}_\parallel\,, \quad \mathbf{B}'_\perp = \gamma\left(\mathbf{B}_\perp - \frac{\mathbf{v}}{c^2}\times\mathbf{E}_\perp\right).$$

Taking into account that

$$\mathbf{E}_\parallel = \frac{\mathbf{E}\cdot\mathbf{v}}{v^2}\,\mathbf{v}\,, \qquad \mathbf{E}_\perp = \mathbf{E} - \mathbf{E}_\parallel$$

(and similarly for $\mathbf{E}'$, $\mathbf{B}$, $\mathbf{B}'$), and using the identity

$$1 - \gamma = \frac{1-\gamma^2}{1+\gamma} = -\frac{\beta^2\gamma^2}{1+\gamma}\,,$$

after a straightforward calculation we obtain

$$\mathbf{E}' = \gamma(\mathbf{E} + \mathbf{v}\times\mathbf{B}) - \frac{\gamma^2}{1+\gamma}(\mathbf{E}\cdot\boldsymbol{\beta})\,\boldsymbol{\beta}\,, \qquad \mathbf{B}' = \gamma\left(\mathbf{B} - \boldsymbol{\beta}\times\frac{\mathbf{E}}{c}\right) - \frac{\gamma^2}{1+\gamma}(\mathbf{B}\cdot\boldsymbol{\beta})\,\boldsymbol{\beta}\,,$$

where $\boldsymbol{\beta} := \mathbf{v}/c$. Note that, since these equations are written in vector form, they are in fact valid regardless of the direction of the relative velocity $\mathbf{v}$ between the reference frames S and S'.

Exercise 8.15. Compute the electromagnetic field created by a charge q moving with uniform velocity $v\mathbf{e}_1$ in the direction of the x_1 axis with respect to an inertial frame S.

Solution. In the proper frame S' of the charge

$$\mathbf{E}'(t',\mathbf{x}') = \frac{q}{4\pi\varepsilon_0}\frac{\mathbf{x}'}{|\mathbf{x}'|^3}\,, \qquad \mathbf{B}'(t',\mathbf{x}') = 0\,.$$

From Eqs (8.123)–(8.124) with the roles of $(\mathbf{E}, \mathbf{B})$ and $(\mathbf{E}', \mathbf{B}')$ interchanged, and replacing $-\mathbf{v}$ by $\mathbf{v}$, we obtain

$$\begin{aligned}\mathbf{E}(t,\mathbf{x}) &= \frac{q}{4\pi\varepsilon_0|\mathbf{x}'|^3}\left(x'^1\mathbf{e}_1 + \gamma(v)(x'^2\mathbf{e}_2 + x'^3\mathbf{e}_3)\right)\\ &= \frac{q\gamma(v)(\mathbf{x}-\mathbf{v}t)}{4\pi\varepsilon_0\left(\gamma^2(v)(x^1-vt)^2+(x^2)^2+(x^3)^2\right)^{3/2}},\\ \mathbf{B}(t,\mathbf{x}) &= \frac{\gamma(v)}{c^2}\,\mathbf{v}\times\mathbf{E}'_\perp = \frac{\mathbf{v}}{c^2}\times\left(\mathbf{E}'_\parallel + \gamma(v)\mathbf{E}'_\perp\right) = \frac{\mathbf{v}}{c^2}\times\mathbf{E}(t,\mathbf{x}),\end{aligned}$$

where we have assumed that the particle starts from the origin of S at $t = 0$. Note, in particular, that the fields $\mathbf{E}$ and $\mathbf{B}$ are perpendicular at every point in space.

Let us denote by ψ the angle formed by the vector

$$\boldsymbol{r} := \mathbf{x} - \mathbf{v}t$$

with the x^1 axis (i.e., with the vector $\mathbf{v}/v$). We then have

$$x^1 - vt = r\cos\psi\,, \qquad (x^2)^2 + (x^3)^2 = r^2\sin^2\psi,$$

and therefore

$$\begin{aligned}\gamma^2(v)(x^1-vt)^2 + (x^2)^2 + (x^3)^2 &= r^2\left[\gamma^2(v)\cos^2\psi + \sin^2\psi\right]\\ &= r^2\gamma^2(v)\left(1 - \frac{v^2}{c^2}\sin^2\psi\right).\end{aligned} \tag{8.125}$$

This identity makes it possible to express the fields more concisely as

$$\begin{aligned}\mathbf{E}(t,\mathbf{x}) &= \frac{q\boldsymbol{r}}{4\pi\varepsilon_0 r^3}\,\frac{1-\beta^2}{\left(1-\beta^2\sin^2\psi\right)^{3/2}},\\ \mathbf{B}(t,\mathbf{x}) &= \frac{\mu_0 q}{4\pi}\,\frac{\mathbf{v}\times\boldsymbol{r}}{r^3}\,\frac{1-\beta^2}{\left(1-\beta^2\sin^2\psi\right)^{3/2}},\end{aligned}$$

where we have used the relation $\varepsilon_0\mu_0 = 1/c^2$. For a fixed value of r, the intensity of the electric field is minimum in the direction of the velocity of the charge (i.e., for $\psi = 0$), and maximum in the directions perpendicular to it ($\psi = \pi/2$).

In fact, $|\mathbf{E}|_{\min} = q(1-\beta^2)/(4\pi\varepsilon_0 r^2)$ and $|\mathbf{E}|_{\max} = q\gamma(v)/(4\pi\varepsilon_0 r^2)$ respectively tend to 0 and ∞ as $v \to c$. Likewise, since

$$\left|\mathbf{B}(t,\mathbf{x})\right| = \frac{\mu_0 q v(1-\beta^2)}{4\pi r^2}\,\frac{\sin\psi}{\left(1-\beta^2\sin^2\psi\right)^{3/2}},$$

it is straightforward to check that $|\mathbf{B}|_{\min} = 0$ (for $\psi = 0, \pi$) and $|\mathbf{B}|_{\max} = \mu_0 q v\gamma(v)/(4\pi r^2)$ (for $\psi = \pi/2$), with $|\mathbf{B}|_{\max} \to \infty$ if $v \to c$.

8.9.4 Retarded time

The **retarded time** $t_r \equiv t_r(t, \mathbf{x})$ is the time at which an electromagnetic signal must have been emitted from the position occupied by a moving charge so that it reaches at the time t an observer at rest at the point $\mathbf{x}$ in an inertial frame S. In other words, t_r is implicitly defined by

$$\boxed{c(t - t_r) = |\mathbf{x} - \mathbf{x}_0(t_r)|,} \tag{8.126}$$

where $\mathbf{x}_0(t)$ is the trajectory of the moving charge as measured in the inertial frame S. In what follows, a subscript r on a variable will indicate that the variable is evaluated at the *retarded* time t_r.

We shall next show that the electromagnetic field created by a charge q moving with uniform velocity $v\mathbf{e}_1$ can be expressed in terms of the retarded time. Indeed (assuming, as we did in the previous exercise, that the charge starts from the origin at $t = 0$), at the retarded time t_r it was at the point $vt_r\mathbf{e}_1$ according to the inertial frame S. If

$$\boldsymbol{\mathfrak{r}}_r = \mathbf{x} - vt_r\mathbf{e}_1$$

denotes the vector connecting the point $vt_r\mathbf{e}_1$ with the observation point $\mathbf{x}$ at time t, we then have

$$\boldsymbol{\mathfrak{r}}_r = \mathbf{x} - vt_r\mathbf{e}_1 = \boldsymbol{\mathfrak{r}} + v(t - t_r)\mathbf{e}_1 = \boldsymbol{\mathfrak{r}} + \mathfrak{r}_r\boldsymbol{\beta},$$

since

$$\mathfrak{r}_r = c(t - t_r) \tag{8.127}$$

by the definition (8.126) of retarded time. Therefore

$$\boxed{\boldsymbol{\mathfrak{r}} = \mathfrak{r}_r(\mathbf{n}_r - \boldsymbol{\beta}), \quad \text{with} \quad \mathbf{n}_r := \frac{\boldsymbol{\mathfrak{r}}_r}{\mathfrak{r}_r},}$$

and consequently

$$\beta\mathfrak{r}\sin\psi = |\boldsymbol{\mathfrak{r}} \times \boldsymbol{\beta}| = |\boldsymbol{\mathfrak{r}}_r \times \boldsymbol{\beta}|.$$

From the previous equalities it follows that

$$\begin{aligned}\mathfrak{r}^2(1 - \beta^2\sin^2\psi) &= \mathfrak{r}_r^2(\mathbf{n}_r - \boldsymbol{\beta})^2 - |\boldsymbol{\mathfrak{r}}_r \times \boldsymbol{\beta}|^2 = \mathfrak{r}_r^2(\mathbf{n}_r - \boldsymbol{\beta})^2 - \beta^2\mathfrak{r}_r^2 + (\boldsymbol{\mathfrak{r}}_r \cdot \boldsymbol{\beta})^2 \\ &= \mathfrak{r}_r^2 - 2\mathfrak{r}_r^2\mathbf{n}_r \cdot \boldsymbol{\beta} + \mathfrak{r}_r^2(\mathbf{n}_r \cdot \boldsymbol{\beta})^2 = \mathfrak{r}_r^2(1 - \boldsymbol{\beta} \cdot \mathbf{n}_r)^2,\end{aligned} \tag{8.128}$$

and therefore

$$\boxed{\mathbf{E}(t, \mathbf{x}) = \frac{q}{4\pi\varepsilon_0}\frac{(1 - \beta^2)(\mathbf{n}_r - \boldsymbol{\beta})}{\mathfrak{r}_r^2(1 - \boldsymbol{\beta} \cdot \mathbf{n}_r)^3}, \qquad \mathbf{B}(t, \mathbf{x}) = \frac{\mu_0 q}{4\pi}\frac{(1 - \beta^2)\mathbf{v} \times \mathbf{n}_r}{\mathfrak{r}_r^2(1 - \boldsymbol{\beta} \cdot \mathbf{n}_r)^3}.} \tag{8.129}$$

Since these equations are expressed in vector form, they are valid for an arbitrary velocity $\mathbf{v}$.

Note. The retarded time t_r in this case can be easily expressed in closed form as a function of $(t, \mathbf{x})$ by solving Eq. (8.127). Indeed,

$$c^2(t - t_r)^2 = (\mathbf{x} - vt_r\mathbf{e}_1)^2 = \mathbf{x}^2 - 2vt_rx_1 + v^2t_r^2$$
$$\implies (c^2 - v^2)t_r^2 + 2(vx_1 - c^2t)t_r + c^2t^2 - \mathbf{x}^2 = 0\,.$$

The solution of this quadratic equation in t_r is

$$t_r = (c^2 - v^2)^{-1}\Big[c^2t - vx_1 - \sqrt{(c^2t - vx_1)^2 - (c^2t^2 - \mathbf{x}^2)(c^2 - v^2)}\,\Big],$$

where we have taken the "–" sign in front of the square root so that for $v = 0$ we obtain the correct result $t_r = t - |\mathbf{x}|/c \leqslant t$. ■

The above result is important from a theoretical point of view, since it shows that the value of the electromagnetic field at the point $(t, \mathbf{x})$ is entirely determined by the position of the charge at the retarded time t_r, where $t - t_r$ is the time taken by light to travel from the position of the charge at time t_r to the point at which the field is measured. This corroborates the assertion that it is not possible to transmit a signal faster than the speed of light.

8.10 ELECTROMAGNETIC FIELD CREATED BY A MOVING CHARGE

8.10.1 Electromagnetic four-potential

The electromagnetic field tensor can be simply expressed through the **electromagnetic four-potential**

$$A := (A^0, \mathbf{A})\,, \qquad \text{with} \quad A^0 := \frac{\Phi}{c}\,,$$

by means of the formula

$$F^{\mu\nu} = \frac{\partial A^\nu}{\partial x_\mu} - \frac{\partial A^\mu}{\partial x_\nu} = \partial^\mu A^\nu - \partial^\nu A^\mu\,,$$

where

$$(\partial^\mu) = \left(\frac{\partial}{\partial x_\mu}\right) = \left(\frac{\partial}{\partial x^0}, -\nabla\right).$$

Indeed, by Eqs. (2.52), (8.114), and (8.115) we have

$$\frac{\partial A^i}{\partial x_0} - \frac{\partial A^0}{\partial x_i} = \frac{\partial A^i}{\partial x^0} + \frac{\partial A^0}{\partial x^i} = \frac{1}{c}\left(\frac{\partial A^i}{\partial t} + \frac{\partial \Phi}{\partial x^i}\right) = -\frac{E^i}{c} = F^{0i},$$

$$\frac{\partial A^j}{\partial x_i} - \frac{\partial A^i}{\partial x_j} = -\left(\frac{\partial A^j}{\partial x^i} - \frac{\partial A^i}{\partial x^j}\right) = -\sum_{k=1}^{3}\varepsilon_{ijk}B^k = F^{ij}.$$

The operator

$$\boxed{(\partial_\mu) = \left(\frac{\partial}{\partial x^\mu}\right) = \left(\frac{\partial}{\partial x^0}, \nabla\right)}$$

transforms as a covariant four-vector (cf. Eq. (8.110)), since from the Lorentz transformation law $x'^\mu = \sum_\nu \Lambda^\mu{}_\nu x^\nu$ it follows that

$$\partial_\nu \equiv \frac{\partial}{\partial x^\nu} = \sum_\mu \frac{\partial x'^\mu}{\partial x^\nu}\frac{\partial}{\partial x'^\mu} = \sum_\mu \Lambda^\mu{}_\nu \frac{\partial}{\partial x'^\mu} \equiv \sum_\mu \Lambda^\mu{}_\nu \partial'_\mu .$$

From the identity

$$\partial^\mu = \frac{\partial}{\partial x_\mu} = \sum_\nu \frac{\partial x^\nu}{\partial x_\mu}\frac{\partial}{\partial x^\nu} = \sum_\nu \eta^{\mu\nu}\frac{\partial}{\partial x^\nu} \equiv \sum_\nu \eta^{\mu\nu}\partial_\nu$$

we then deduce that (∂^μ) is the *contravariant* four-vector associated to (∂_μ), and therefore

$$\partial'^\mu = \frac{\partial}{\partial x'_\mu} = \sum_\nu \Lambda^\mu{}_\nu \frac{\partial}{\partial x_\nu} \equiv \sum_\nu \Lambda^\mu{}_\nu \partial^\nu .$$

The transformation law (8.121) of the electromagnetic field tensor and the previous equation imply that

$$\begin{aligned} F'^{\mu\nu} &= \partial'^\mu A'^\nu - \partial'^\nu A'^\mu = \sum_\rho \left(\Lambda^\mu{}_\rho \partial^\rho A'^\nu - \Lambda^\nu{}_\rho \partial^\rho A'^\mu\right) = \sum_{\rho,\sigma} \Lambda^\mu{}_\rho \Lambda^\nu{}_\sigma F^{\rho\sigma} \\ &= \sum_{\rho,\sigma} \Lambda^\mu{}_\rho \Lambda^\nu{}_\sigma (\partial^\rho A^\sigma - \partial^\sigma A^\rho). \end{aligned}$$

Exchanging the dummy summation indices ρ and σ in the second term of the last sum we arrive at the equality

$$\sum_\rho \left(\Lambda^\mu{}_\rho \partial^\rho A'^\nu - \Lambda^\nu{}_\rho \partial^\rho A'^\mu\right) = \sum_\rho \left\{\Lambda^\mu{}_\rho \partial^\rho \sum_\sigma \Lambda^\nu{}_\sigma A^\sigma - \Lambda^\nu{}_\rho \partial^\rho \sum_\sigma \Lambda^\mu{}_\sigma A^\sigma\right\},$$

or equivalently

$$\sum_\rho \partial^\rho \left\{\Lambda^\mu_\rho \left(A'^\nu - \sum_\sigma \Lambda^\nu{}_\sigma A^\sigma\right) - \Lambda^\nu_\rho \left(A'^\mu - \sum_\sigma \Lambda^\mu{}_\sigma A^\sigma\right)\right\}.$$

This equation will be identically satisfied by all four-potentials $A(x)$ provided that

$$\boxed{A'^\mu = \sum_\sigma \Lambda^\mu{}_\sigma A^\sigma ,}$$

i.e., that A transforms as a contravariant vector. In other words:

The electromagnetic four-potential $A = (A^\mu)$ is a *contravariant four-vector* under Lorentz transformations.

8.10.2 Covariant relativistic action

Proceeding as in Section 4.2.4, it is straightforward to derive the Lagrangian formulation of the motion of a charged particle in an electromagnetic field with four-potential $A(x) = (\Phi(x)/c, \mathbf{A}(x))$. Indeed, using Eq. (4.44) the relativistic equation of motion (8.118) can be written in terms of the electromagnetic potentials as

$$\frac{\mathrm{d}\mathbf{p}}{\mathrm{d}t} = -q\left(\frac{\partial\Phi}{\partial\mathbf{x}} + \frac{\partial\mathbf{A}}{\partial t}\right) + q\,\dot{\mathbf{x}}\times(\nabla\times\mathbf{A}) = -q\frac{\mathrm{d}\mathbf{A}}{\mathrm{d}t} - q\frac{\partial}{\partial\mathbf{x}}\left(\Phi - \dot{\mathbf{x}}\cdot\mathbf{A}\right),$$

or equivalently

$$\frac{\mathrm{d}}{\mathrm{d}t}\left(\mathbf{p} + q\mathbf{A}\right) + q\frac{\partial}{\partial\mathbf{x}}\left(\Phi - \dot{\mathbf{x}}\cdot\mathbf{A}\right) = 0,$$

where q is the charge of the particle. These are the Euler–Lagrange equations of a Lagrangian $L(t, \mathbf{x}, \dot{\mathbf{x}})$ provided that

$$\frac{\partial L}{\partial\dot{\mathbf{x}}} = \mathbf{p} + q\mathbf{A}(x) = \frac{m\dot{\mathbf{x}}}{\sqrt{1 - \frac{\dot{\mathbf{x}}^2}{c^2}}} + q\mathbf{A}(x), \qquad \frac{\partial L}{\partial\mathbf{x}} = -q\frac{\partial}{\partial\mathbf{x}}\left(\Phi(x) - \dot{\mathbf{x}}\cdot\mathbf{A}(x)\right).$$

Integrating the first equation with respect to $\dot{\mathbf{x}}$ we obtain

$$L = -mc^2\sqrt{1 - \frac{\dot{\mathbf{x}}^2}{c^2}} + q\mathbf{A}\cdot\dot{\mathbf{x}} + g(t, \mathbf{x}),$$

which substituted into the second one yields

$$\frac{\partial g}{\partial\mathbf{x}} + q\,\dot{\mathbf{x}}\cdot\frac{\partial\mathbf{A}}{\partial\mathbf{x}} = -q\frac{\partial}{\partial\mathbf{x}}\left(\Phi(x) - \dot{\mathbf{x}}\cdot\mathbf{A}(x)\right)$$
$$\Longrightarrow \quad \frac{\partial g}{\partial\mathbf{x}} = -q\frac{\partial\Phi(x)}{\partial\mathbf{x}} \quad \Longrightarrow \quad g = -q\Phi(x) + h(t).$$

Discarding the total derivative $h(t)$ we thus obtain the following expression for the relativistic Lagrangian of a particle of charge q in an electromagnetic field with potentials (Φ, A):

$$L = -mc^2\sqrt{1 - \frac{\dot{\mathbf{x}}^2}{c^2}} - q\big(\Phi(x) - \dot{\mathbf{x}}\cdot\mathbf{A}(x)\big). \tag{8.130}$$

Thus the solutions of the equations of motion (8.118) are the stationary points of the **relativistic action** functional

$$S = -\int_{t_1}^{t_2}\mathrm{d}t\left[mc^2\sqrt{1 - \frac{\dot{\mathbf{x}}^2}{c^2}} + q\big(\Phi(x) - \dot{\mathbf{x}}\cdot\mathbf{A}(x)\big)\right]. \tag{8.131}$$

Although the action (8.131) is apparently non-covariant (as it involves the coordinate time t in a certain inertial frame), since Eqs. (8.118) actually hold in *any* inertial frame it should be possible to express this action in a manifestly covariant form independent of the reference frame used. To this end, let us introduce an arbitrary parameter[13] s such that $s(t_1) = s_1$ and $s(t_2) = s_2$, and note that

$$-L\,dt = mc\sqrt{c^2\,dt^2 - d\mathbf{x}^2} + q\big(\Phi(x)\,dt - \mathbf{A}(x)\cdot d\mathbf{x}\big)$$
$$= ds\left[mc\sqrt{c^2\left(\frac{dt}{ds}\right)^2 - \left(\frac{d\mathbf{x}}{ds}\right)^2} + q\left(\Phi(x)\frac{dt}{ds} - \mathbf{A}(x)\cdot\frac{d\mathbf{x}}{ds}\right)\right].$$

Thus the relativistic action (8.131) can be written in the manifestly covariant form

$$S = -\int_{s_1}^{s_2} ds\left[mc\sqrt{\left(\frac{dx}{ds}\right)^2} + qA(x)\cdot\frac{dx}{ds}\right], \tag{8.132}$$

where the dot denotes Minkowski product and

$$\left(\frac{dx}{ds}\right)^2 \equiv \frac{dx}{ds}\cdot\frac{dx}{ds}.$$

As a consequence, *the worldlines $x(s)$ of a particle moving in an electromagnetic field with four-potential $A(x)$ are the* stationary points *of the action functional* (8.132).

Exercise 8.16. Show that the Euler–Lagrange equations of the manifestly covariant relativistic action (8.132) yield the covariant equation of motion

$$\frac{dp_\mu}{d\tau} = q\sum_\nu F_{\mu\nu}(x)u^\nu,$$

obtained from Eq. (8.119) by lowering the index μ.

Solution. The Lagrangian of the covariant relativistic action (8.132) is

$$L(x,\dot{x}) = -mc\sqrt{\dot{x}^2} - qA(x)\cdot\dot{x}, \qquad \text{with} \quad \dot{x} \equiv \frac{dx}{ds}.$$

[13] To be more precise, the coordinate time t should be a monotonically increasing function of the worldline parameter s. Note that this condition is relativistically invariant, i.e., it holds in every inertial frame if it is satisfied in a particular frame. Indeed, since

$$\frac{dt}{ds} = \frac{dt}{d\tau}\frac{d\tau}{ds},$$

and $\frac{dt}{d\tau} > 0$ in any inertial frame, the condition $\frac{dt}{ds} > 0$ is equivalent to the invariant condition $\frac{d\tau}{ds} > 0$.

The canonical momenta are therefore given by

$$\frac{\partial L}{\partial \dot{x}^\mu} = -\frac{mc\dot{x}_\mu}{\sqrt{\dot{x}^2}} - qA_\mu(x),$$

and the Euler–Lagrange equations read

$$-\frac{\mathrm{d}}{\mathrm{d}s}\left(\frac{mc\dot{x}_\mu}{\sqrt{\dot{x}^2}} + qA_\mu(x)\right) = \frac{\partial L}{\partial x^\mu} = -q\frac{\partial A(x)}{\partial x^\mu}\cdot\dot{x} = -q\sum_\nu \dot{x}^\nu \partial_\mu A_\nu(x),$$

or equivalently

$$mc\dot{x}_\mu \frac{\mathrm{d}}{\mathrm{d}s}(\dot{x}^2)^{-1/2} + \frac{mc}{\sqrt{\dot{x}^2}}\ddot{x}_\mu + q\sum_\nu \dot{x}^\nu \partial_\nu A_\mu(x) = q\sum_\nu \dot{x}^\nu \partial_\mu A_\nu(x).$$

These equations hold for *any* choice of the worldline parameter s. Choosing s as the particle's proper time τ we have $\dot{x} = u$, and therefore

$$p = m\dot{x}, \qquad \dot{x}^2 = u^2 = c^2, \qquad \frac{\mathrm{d}}{\mathrm{d}s}(\dot{x}^2)^{-1/2} = \frac{\mathrm{d}}{\mathrm{d}s}c^{-1} = 0.$$

Hence the above equations reduce to

$$m\ddot{x}_\mu = q\sum_\nu \dot{x}^\nu \left(\partial_\mu A_\nu(x) - \partial_\nu A_\mu(x)\right) = q\sum_\nu F_{\mu\nu}(x)\dot{x}^\nu,$$

as claimed.

Remark 8.3. The covariant relativistic Lagrangian

$$L(x,\dot{x}) = -mc\sqrt{\dot{x}^2} - qA(x)\cdot\dot{x}, \qquad \dot{x} \equiv \frac{\mathrm{d}x}{\mathrm{d}s}, \tag{8.133}$$

is a homogeneous function of degree one in the generalized velocities $\dot{x}^\mu$, since clearly

$$L(x,\lambda\dot{x}) = L(x,\dot{x}), \qquad \forall\lambda > 0.$$

Hence the energy integral is in this case identically zero, since by Euler's theorem we have

$$\sum_\mu \dot{x}^\mu \frac{\partial L}{\partial \dot{x}^\mu} = L.$$

As a consequence, it is not possible to develop a covariant Hamiltonian formulation of the equations of motion (8.118) using the Lagrangian (8.133). ■

Exercise 8.17. Consider the covariant Lagrangian

$$\mathcal{L} = \frac{1}{2}m\dot{x}^2 + qA(x)\cdot\dot{x}, \qquad \dot{x} \equiv \frac{\mathrm{d}x}{\mathrm{d}s}, \tag{8.134}$$

where s is an arbitrary parameter such that $\frac{\mathrm{d}s}{\mathrm{d}\tau} > 0$.

i) Check that $\dot{x}^2$ is a constant of the motion.
ii) Show that the worldlines $x = x(\tau)$ of a charge q moving in the presence of the electromagnetic four-potential $A(x)$ are the trajectories of $\mathcal{L}$ satisfying the condition

$$\dot{x}^2 = c^2.$$

Solution.
i) The Lagrange equations of the Lagrangian (8.134) are

$$\frac{\mathrm{d}}{\mathrm{d}s}\left(m\dot{x}_\mu + qA_\mu(x)\right) = m\ddot{x}_\mu + q\sum_\nu \dot{x}^\nu \partial_\nu A_\mu(x) = q\dot{x}\cdot\frac{\partial A(x)}{\partial x^\mu} = q\sum_\nu \dot{x}^\nu \partial_\mu A_\nu(x),$$

or equivalently

$$m\frac{\mathrm{d}^2 x_\mu}{\mathrm{d}s^2} = q\sum_\nu \left(\partial_\mu A_\nu(x) - \partial_\nu A_\mu(x)\right)\frac{\mathrm{d}x^\nu}{\mathrm{d}s}. \tag{8.135}$$

On the other hand, since the Lagrangian (8.134) is independent of s the energy integral

$$h = \sum_\mu \frac{\partial \mathcal{L}}{\partial \dot{x}^\mu}\dot{x}^\mu - \mathcal{L} = \frac{1}{2}m\dot{x}^2 \equiv \frac{m}{2}\left(\frac{\mathrm{d}x}{\mathrm{d}s}\right)^2$$

is conserved. This equation relates the parameter s to the particle's proper time along each trajectory, since

$$\mathrm{d}x^2 = c^2\,\mathrm{d}\tau^2 = \left(\frac{\mathrm{d}x}{\mathrm{d}s}\right)^2 \mathrm{d}s^2 = \frac{2h}{m}\,\mathrm{d}s^2$$

$$\Longrightarrow \quad \frac{\mathrm{d}s}{\mathrm{d}\tau} = \sqrt{\frac{mc^2}{2h}} \quad \Longrightarrow \quad s = \sqrt{\frac{mc^2}{2h}}(\tau - \tau_0), \tag{8.136}$$

where τ_0 is an arbitrary constant and we have taken into account that $\frac{\mathrm{d}s}{\mathrm{d}\tau} > 0$.
ii) Let $x = x(s)$ be a trajectory of the Lagrangian (8.134), i.e., a solution of Eqs. (8.135). Performing the change of parameter (8.136) in Eqs. (8.135) we obtain

$$m\frac{\mathrm{d}^2 x_\mu}{\mathrm{d}\tau^2} = q\sqrt{\frac{mc^2}{2h}}\sum_\nu \left(\partial_\mu A_\nu(x) - \partial_\nu A_\mu(x)\right)\frac{\mathrm{d}x^\nu}{\mathrm{d}\tau}.$$

These are the equations of motion (8.118) of the charge q if and only if

$$h \equiv \frac{1}{2}m\dot{x}^2 = \frac{1}{2}mc^2 \quad \Longrightarrow \quad \dot{x}^2 = c^2,$$

as was to be shown. Finally, when this is the case Eq. (8.136) implies that $s = \tau - \tau_0$, i.e., the parameter s essentially coincides with the proper time of the particle.

8.10.3 The Liénard–Wiechert potentials

Consider a point charge q moving with constant velocity $\mathbf{v}$ relative to an inertial reference frame S. In its proper frame S', the four-potential of the electromagnetic field created by the particle can be taken as

$$A'^0(x') = \frac{q}{4\pi\varepsilon_0 c|\mathbf{x}'|}, \qquad \mathbf{A}'(x') = 0. \tag{8.137}$$

If we assume that the particle starts at the origin of S at $t = 0$, the components of the four-potential in the inertial frame S are:

$$A^0(x) = \gamma(v)A'^0(x') = \frac{q\gamma(v)}{4\pi\varepsilon_0 c|\mathbf{x}'|}, \qquad \mathbf{A}(x) = \boldsymbol{\beta}\gamma(v)A'^0(x') = \boldsymbol{\beta}A^0(x),$$

with

$$\mathbf{x}' = \mathbf{x} + \big(\gamma(v) - 1\big)(\boldsymbol{\beta}\cdot\mathbf{x})\boldsymbol{\beta} - \gamma(v)\mathbf{v}t$$

(cf. equation (8.11)). To simplify these equations, note that by Eqs. (8.125)–(8.128) we have (indicating again with the subscript r that a variable is evaluated at the *retarded time* t_r)

$$\begin{aligned}\frac{|\mathbf{x}'|}{\gamma(v)} &= \imath_r(1 - \boldsymbol{\beta}\cdot\mathbf{n}_r) = \imath_r - \boldsymbol{\beta}\cdot(\mathbf{x} - \mathbf{x}_0(t_r)) \\ &= c(t - t_r) - \boldsymbol{\beta}\cdot\big(\mathbf{x} - \mathbf{x}_0(t_r)\big) = \frac{1}{c\gamma(v)}\,u\cdot\big(x - x_0(t_r)\big),\end{aligned}$$

where

$$x_0(t) = (ct, \mathbf{v}t) = (ct, \mathbf{x}_0(t))$$

is the equation of the worldline of the charge in uniform rectilinear motion. Since $(1, \boldsymbol{\beta}) = u/(c\gamma(v))$, the four-potential $A(x)$ can therefore be expressed in the covariant form

$$A(x) = \frac{q}{4\pi\varepsilon_0 c}\,\frac{u}{u\cdot(x - x_0(t_r))}.$$

Although the previous equation was derived under the assumption that the velocity of the charge q (and therefore its four-velocity u) was *constant*, we shall now show that the correct equation, regardless of the trajectory of the charge, is:

$$\boxed{A(x) = \frac{q}{4\pi\varepsilon_0 c}\,\frac{u(t_r)}{u(t_r)\cdot(x - x_0(t_r))},} \tag{8.138}$$

where

$$u(t) = \frac{\mathrm{d}}{\mathrm{d}\tau}x_0(t) = \gamma(v)(c, \mathbf{v}(t)), \qquad \mathbf{v}(t) = \frac{\mathrm{d}\mathbf{x}_0(t)}{\mathrm{d}t} = \dot{\mathbf{x}}_0(t).$$

Indeed, this equation is written in a covariant form, since the retarded time t_r is also determined by the Lorentz covariant conditions

$$(x - x_0(t_r))^2 = c^2(t - t_r)^2 - |\mathbf{x} - \mathbf{x}_0(t_r)|^2 = 0, \qquad t_r < t.$$

Therefore, it suffices to show that Eq. (8.138) reduces at each instant to Eq. (8.137) in the proper frame of the moving charge (which in general varies with time). In this frame, at the retarded time t_r we have:

$$x_0(t_r) = (ct_r, 0)\,, \quad u(t_r) = (c, 0)$$
$$\Longrightarrow \quad u(t_r)\cdot(x - x_0(t_r)) = c^2(t-t_r) = c|\mathbf{x}-\mathbf{x}_0(t_r)| = c|\mathbf{x}|\,.$$

It follows that Eq. (8.138) reduces to (8.137), as claimed.

We can write down Eq. (8.138) in an arbitrary inertial frame using the identity

$$\frac{1}{c\gamma(v_r)}u(t_r)\cdot(x-x_0(t_r)) = c(t-t_r) - \boldsymbol{\beta}(t_r)\cdot(\mathbf{x}-\mathbf{x}_0(t_r)) = |\mathbf{x}-\mathbf{x}_0(t_r)|(1-\boldsymbol{\beta}\cdot\mathbf{n}), \tag{8.139}$$

where $v_r = v(t_r)$ and for simplicity we have set

$$\boxed{\mathbf{v} = \mathbf{v}(t_r) = \dot{\mathbf{x}}_0(t_r)\,, \qquad \boldsymbol{\beta} := \frac{\mathbf{v}(t_r)}{c}\,, \qquad \mathbf{n} := \frac{\mathbf{x}-\mathbf{x}_0(t_r)}{|\mathbf{x}-\mathbf{x}_0(t_r)|}}$$

omitting the subscript r in $\mathbf{v}$, $\boldsymbol{\beta}$, and $\mathbf{n}$. Using Eq. (8.139), the potentials (8.138) can be more explicitly written

$$\boxed{\Phi(t,\mathbf{x}) = cA^0(x) = \frac{q}{4\pi\varepsilon_0|\mathbf{x}-\mathbf{x}_0(t_r)|}\,\frac{1}{1-\boldsymbol{\beta}\cdot\mathbf{n}}\,, \qquad \mathbf{A}(t,\mathbf{x}) = \frac{\boldsymbol{\beta}}{c}\,\Phi(t,\mathbf{x})\,.} \tag{8.140}$$

These are called the **Liénard–Wiechert potentials.**

In order to facilitate the calculation of the fields $\mathbf{E}(t,\mathbf{x})$ and $\mathbf{B}(t,\mathbf{x})$ from Eq. (8.140), it is convenient to use equation (8.139) to represent the electric potential $\Phi(t,\mathbf{x})$ in the equivalent form:

$$\boxed{\Phi(t,\mathbf{x}) = \frac{q}{4\pi\varepsilon_0 D}\,, \qquad D := c(t-t_r) - \boldsymbol{\beta}(t_r)\cdot(\mathbf{x}-\mathbf{x}_0(t_r))\,.} \tag{8.141}$$

To determine $\mathbf{E}$ and $\mathbf{B}$, we first need to compute the partial derivatives of the retarded time t_r with respect to the spacetime coordinates x^μ. These partial derivatives are calculated by implicitly differentiating the relation

$$c^2(t-t_r)^2 = (\mathbf{x}-\mathbf{x}_0(t_r))^2,$$

that is

$$c^2(t-t_r)\left(\frac{\partial t}{\partial x^\mu} - \frac{\partial t_r}{\partial x^\mu}\right) = c|\mathbf{x}-\mathbf{x}_0(t_r)|\left(\frac{\partial t}{\partial x^\mu} - \frac{\partial t_r}{\partial x^\mu}\right) = (\mathbf{x}-\mathbf{x}_0(t_r))\cdot\left(\frac{\partial \mathbf{x}}{\partial x^\mu} - \mathbf{v}\frac{\partial t_r}{\partial x^\mu}\right)$$
$$\Longrightarrow \quad c\left(\frac{\partial t}{\partial x^\mu} - \frac{\partial t_r}{\partial x^\mu}\right) = \mathbf{n}\cdot\left(\frac{\partial \mathbf{x}}{\partial x^\mu} - \mathbf{v}\frac{\partial t_r}{\partial x^\mu}\right),$$

whence

$$\frac{\partial t_r}{\partial x^\mu} = \frac{\dfrac{\partial t}{\partial x^\mu} - \dfrac{\mathbf{n}}{c}\dfrac{\partial \mathbf{x}}{\partial x^\mu}}{1-\boldsymbol{\beta}\cdot\mathbf{n}} \quad \Longrightarrow \quad \boxed{\frac{\partial t_r}{\partial t} = \frac{1}{1-\boldsymbol{\beta}\cdot\mathbf{n}}, \qquad \frac{\partial t_r}{\partial \mathbf{x}} = -\frac{\mathbf{n}/c}{1-\boldsymbol{\beta}\cdot\mathbf{n}}}. \tag{8.142}$$

The components of the electric field are given by

$$\begin{aligned} E^i &= -\frac{\partial \Phi}{\partial x^i} - \frac{\partial A^i}{\partial t} = -\frac{\partial \Phi}{\partial x^i} - \frac{\beta^i}{c}\frac{\partial \Phi}{\partial t} - \frac{\dot{\beta}^i \Phi}{c}\frac{\partial t_r}{\partial t} \\ &= \frac{q}{4\pi\varepsilon_0 D^2}\left(\frac{\partial D}{\partial x^i} + \frac{\beta^i}{c}\frac{\partial D}{\partial t}\right) - \frac{q}{4\pi\varepsilon_0 c D}\frac{\dot{\beta}^i}{1-\boldsymbol{\beta}\cdot\mathbf{n}}. \end{aligned}$$

On the other hand,

$$\frac{\partial D}{\partial x^i} + \frac{\beta^i}{c}\frac{\partial D}{\partial t} = -\beta^i + \frac{\partial D}{\partial t_r}\frac{\partial t_r}{\partial x_i} + \beta_i + \frac{\beta_i}{c}\frac{\partial D}{\partial t_r}\frac{\partial t_r}{\partial t} = \frac{\partial D}{\partial t_r}\left(\frac{\partial t_r}{\partial x^i} + \frac{\beta^i}{c}\frac{\partial t_r}{\partial t}\right),$$

with

$$\frac{\partial t_r}{\partial x^i} + \frac{\beta^i}{c}\frac{\partial t_r}{\partial t} = \frac{\beta^i - n^i}{c(1-\boldsymbol{\beta}\cdot\mathbf{n})}, \qquad \frac{\partial D}{\partial t_r} = -c(1-\beta^2) - \dot{\boldsymbol{\beta}}\cdot\big(\mathbf{x} - \mathbf{x}_0(t_r)\big)$$

and

$$\dot{\boldsymbol{\beta}} = \frac{\mathrm{d}\boldsymbol{\beta}}{\mathrm{d}t}(t_r) = \frac{1}{c}\ddot{\mathbf{x}}_0(t_r).$$

Using the identities

$$\mathbf{x} - \mathbf{x}_0(t_r) = \mathscr{r}\,\mathbf{n}, \qquad \mathscr{r} = |\mathbf{x} - \mathbf{x}_0(t_r)|, \qquad D = \mathscr{r}(1-\boldsymbol{\beta}\cdot\mathbf{n}),$$

we finally obtain

$$\mathbf{E} = \frac{q}{4\pi\varepsilon_0 \mathscr{r}^2}\frac{(1-\beta^2)(\mathbf{n}-\boldsymbol{\beta})}{(1-\boldsymbol{\beta}\cdot\mathbf{n})^3} + \frac{q}{4\pi\varepsilon_0 c\mathscr{r}}\frac{(\mathbf{n}-\boldsymbol{\beta})(\dot{\boldsymbol{\beta}}\cdot\mathbf{n}) - (1-\boldsymbol{\beta}\cdot\mathbf{n})\dot{\boldsymbol{\beta}}}{(1-\boldsymbol{\beta}\cdot\mathbf{n})^3}, \tag{8.143}$$

or more concisely

$$\boxed{\mathbf{E} = \frac{q}{4\pi\varepsilon_0 \mathscr{r}^2}\frac{(1-\beta^2)(\mathbf{n}-\boldsymbol{\beta})}{(1-\boldsymbol{\beta}\cdot\mathbf{n})^3} + \frac{q}{4\pi\varepsilon_0 c\mathscr{r}}\frac{\mathbf{n}\times\big((\mathbf{n}-\boldsymbol{\beta})\times\dot{\boldsymbol{\beta}}\big)}{(1-\boldsymbol{\beta}\cdot\mathbf{n})^3}.} \tag{8.144}$$

The magnetic field $\mathbf{B} = \nabla\times\mathbf{A}$ is given by

$$\begin{aligned} \mathbf{B} &= \frac{1}{c}\nabla\times(\boldsymbol{\beta}\Phi) = -\frac{\boldsymbol{\beta}}{c}\times\nabla\Phi + \frac{\Phi}{c}\nabla\times\boldsymbol{\beta} = -\frac{\boldsymbol{\beta}}{c}\times\nabla\Phi + \frac{\Phi}{c}\nabla t_r\times\dot{\boldsymbol{\beta}} \\ &= -\frac{\boldsymbol{\beta}}{c}\times\nabla\Phi - \frac{\Phi}{c^2}\frac{\mathbf{n}\times\dot{\boldsymbol{\beta}}}{1-\boldsymbol{\beta}\cdot\mathbf{n}}. \end{aligned}$$

Taking into account that

$$-\nabla\Phi = \mathbf{E}+\frac{\partial \mathbf{A}}{\partial t} = \mathbf{E}+\frac{\partial}{\partial t}\left(\frac{\boldsymbol{\beta}}{c}\Phi\right) = \mathbf{E}+\frac{\boldsymbol{\beta}}{c}\frac{\partial\Phi}{\partial t}+\frac{\Phi}{c}\dot{\boldsymbol{\beta}}\frac{\partial t_r}{\partial t} = \mathbf{E}+\frac{\boldsymbol{\beta}}{c}\frac{\partial\Phi}{\partial t}+\frac{\Phi}{c}\frac{\dot{\boldsymbol{\beta}}}{1-\boldsymbol{\beta}\cdot\mathbf{n}}$$

we obtain

$$\mathbf{B} = \frac{\boldsymbol{\beta}}{c}\times\mathbf{E} - \frac{\Phi}{c^2}\frac{(\mathbf{n}-\boldsymbol{\beta})\times\dot{\boldsymbol{\beta}}}{1-\boldsymbol{\beta}\cdot\mathbf{n}} = \frac{\mathbf{n}}{c}\times\mathbf{E}+\frac{\boldsymbol{\beta}-\mathbf{n}}{c}\times\mathbf{E}-\frac{\Phi}{c^2}\frac{(\mathbf{n}-\boldsymbol{\beta})\times\dot{\boldsymbol{\beta}}}{1-\boldsymbol{\beta}\cdot\mathbf{n}}. \tag{8.145}$$

On the other hand, from equation (8.143) it follows that

$$\frac{\boldsymbol{\beta}-\mathbf{n}}{c}\times\mathbf{E} = \frac{q}{4\pi\varepsilon_0 c^2 \mathscr{r}}\frac{(\mathbf{n}-\boldsymbol{\beta})\times\dot{\boldsymbol{\beta}}}{(1-\boldsymbol{\beta}\cdot\mathbf{n})^2} = \frac{\Phi}{c^2}\frac{(\mathbf{n}-\boldsymbol{\beta})\times\dot{\boldsymbol{\beta}}}{1-\boldsymbol{\beta}\cdot\mathbf{n}},$$

so that the last two terms in the RHS of Eq. (8.145) cancel. We thus arrive at the following explicit formula for the magnetic field **B**:

$$\mathbf{B} = \frac{\mathbf{n}}{c}\times\mathbf{E} = \frac{\mu_0 q}{4\pi\mathscr{r}^2}\frac{(1-\beta^2)(\mathbf{v}\times\mathbf{n})}{(1-\boldsymbol{\beta}\cdot\mathbf{n})^3}+\frac{\mu_0 q}{4\pi\mathscr{r}}\frac{\left[(\mathbf{n}\cdot\dot{\boldsymbol{\beta}})\boldsymbol{\beta}+(1-\boldsymbol{\beta}\cdot\mathbf{n})\dot{\boldsymbol{\beta}}\right]\times\mathbf{n}}{(1-\boldsymbol{\beta}\cdot\mathbf{n})^3}. \tag{8.146}$$

Note. It is important to remember that in equations (8.140), (8.144), and (8.146) the quantities $\mathscr{r}$, $\mathbf{n}$, $\boldsymbol{\beta}$, must be $\dot{\boldsymbol{\beta}}$ are evaluated at the *retarded time* t_r, that is

$$\mathscr{r} = |\mathbf{x}-\mathbf{x}_0(t_r)|, \quad \mathbf{n} = \frac{\mathbf{x}-\mathbf{x}_0(t_r)}{|\mathbf{x}-\mathbf{x}_0(t_r)|}, \quad \boldsymbol{\beta} = \frac{\dot{\mathbf{x}}_0(t_r)}{c}, \quad \dot{\boldsymbol{\beta}} = \frac{\ddot{\mathbf{x}}_0(t_r)}{c}.$$

The fields (8.144) and (8.146) consist of two terms. The first term, inversely proportional to $\mathscr{r}^2 = |\mathbf{x}-\mathbf{x}_0(t_r)|^2$, i.e., the square of the distance from the observation point **x** to the position of the charge creating the field *at the retarded time* t_r, is dominant when this distance is small. This first term exactly matches equation (8.129), to which equations (8.144) and (8.146) reduce when $\dot{\boldsymbol{\beta}}(t_r) = 0$ (i.e., when the instantaneous acceleration of the charge is zero). On the other hand, for large $\mathscr{r}$ the first term in Eqs. (8.144) and (8.146) is negligible compared to the second one, which is inversely proportional to $\mathscr{r}$. This second term is responsible for the fact that *electromagnetic radiation* can propagate over long distances, as the term independent of $\dot{\boldsymbol{\beta}}(t_r)$ decays too quickly as $\mathscr{r}\to\infty$. The **radiation fields** (dominant for $\mathscr{r} \gg 1$) are then

$$\mathbf{E}_{\text{rad}} = \frac{q}{4\pi\varepsilon_0 c\mathscr{r}}\frac{\mathbf{n}\times((\mathbf{n}-\boldsymbol{\beta})\times\dot{\boldsymbol{\beta}})}{(1-\boldsymbol{\beta}\cdot\mathbf{n})^3},$$

$$\mathbf{B}_{\text{rad}} = \frac{\mu_0 q}{4\pi\mathscr{r}}\frac{\left[(\mathbf{n}\cdot\dot{\boldsymbol{\beta}})\boldsymbol{\beta}+(1-\boldsymbol{\beta}\cdot\mathbf{n})\dot{\boldsymbol{\beta}}\right]\times\mathbf{n}}{(1-\boldsymbol{\beta}\cdot\mathbf{n})^3} = \frac{\mathbf{n}}{c}\times\mathbf{E}_{\text{rad}}.$$

Note that these radiation terms are only present if $\dot{\boldsymbol{\beta}}(t_r) \neq 0$, i.e., if the charge creating the field is *accelerated*. In particular, it follows that *only* accelerated *particles can radiate electromagnetic energy.*

Exercise 8.18. Show that the *Poynting vector* $\mathbf{S}_{\rm rad} = \mu_0^{-1}\,\mathbf{E}_{\rm rad} \times \mathbf{B}_{\rm rad}$ of the radiation fields is given by:

$$\mathbf{S}_{\rm rad} = \frac{E_{\rm rad}^2}{\mu_0 c}\,\mathbf{n} = \frac{q^2}{16\pi^2\varepsilon_0 c\mathfrak{r}^2}\,\frac{|\mathbf{n}\times((\mathbf{n}-\boldsymbol{\beta})\times\dot{\boldsymbol{\beta}})|^2}{(1-\boldsymbol{\beta}\cdot\mathbf{n})^6}\,\mathbf{n}\,.$$

Prove that in the non-relativistic limit $\boldsymbol{\beta} \to 0$ we have

$$\mathbf{S}_{\rm rad} \simeq \frac{q^2}{16\pi^2 c^3\varepsilon_0\mathfrak{r}^2}\,|\mathbf{n}\times\mathbf{a}|^2\,\mathbf{n},$$

where $\mathbf{a}$ is the acceleration of the charge creating the field.

Solution. Using the previous formulas for $\mathbf{E}_{\rm rad}$ and $\mathbf{B}_{\rm rad}$ and the identity $c^2 = (\varepsilon_0\mu_0)^{-1}$ we obtain

$$\begin{aligned}\mathbf{S}_{\rm rad} &= \mu_0^{-1}\,\mathbf{E}_{\rm rad}\times\mathbf{B}_{\rm rad} = (\mu_0 c)^{-1}\,\mathbf{E}_{\rm rad}\times(\mathbf{n}\times\mathbf{E}_{\rm rad}) = (\mu_0 c)^{-1}\,\mathbf{E}_{\rm rad}^2\,\mathbf{n}\\ &= \frac{q^2}{16\pi^2\varepsilon_0 c\mathfrak{r}^2}\,\frac{|\mathbf{n}\times((\mathbf{n}-\boldsymbol{\beta})\times\dot{\boldsymbol{\beta}})|^2}{(1-\boldsymbol{\beta}\cdot\mathbf{n})^6}\,\mathbf{n},\end{aligned}$$

where we have taken into account that $\mathbf{n}\cdot\mathbf{E}_{\rm rad} = 0$. In the non-relativistic limit we can set $\boldsymbol{\beta} = 0$ in the previous formula, with the result

$$\mathbf{S}_{\rm rad} \simeq \frac{q^2}{16\pi^2\varepsilon_0 c^3\mathfrak{r}^2}\,\big|\mathbf{n}\times(\mathbf{n}\times\mathbf{a})\big|^2\,\mathbf{n} = \frac{q^2}{16\pi^2\varepsilon_0 c^3\mathfrak{r}^2}\,|\mathbf{n}\times\mathbf{a}|^2\,\mathbf{n}.$$

Bibliography

[1] J. S. Ames and D. Murnaghan. *Theoretical Mechanics: An Introduction to Mathematical Physics*. Dover Publications, New York, NY, USA, 1958.

[2] V.I. Arnold. *Mathematical Methods of Classical Mechanics*. Springer, 2nd edition, 1989.

[3] A.P. Arya. *Introduction to Classical Mechanics*. Prentice Hall, 2nd edition, 1998.

[4] S. Banach. *Mechanics*. Elsevier, 1951.

[5] V. Barger and M. Olsson. *Classical Mechanics: A Modern Perspective*. McGraw-Hill, New York, NY, USA, 1995.

[6] H. Baruh. *Analytical Dynamics*. McGraw-Hill, New York, NY, USA, 1999.

[7] A. Bettini. *A Course in Classical Physics 1: Mechanics*. Springer, 2016.

[8] A. Brizard. *An Introduction to Lagrangian Mechanics*. World Scientific, Singapore, 2008.

[9] S. B. Cahn and B. E. Nadgorny. *A Guide to Physics Problems: Part 1 – Mechanics, Relativity, and Electrodynamics*. Springer, 1994.

[10] M. G. Calkin. *Lagrangian and Hamiltonian Mechanics*. World Scientific, Singapore, 1996.

[11] E. Corinaldesi. *Classical Mechanics for Physics Graduate Students*. World Scientific, Singapore, 2015.

[12] O. L. de Lange and J. Pierrus. *Solved Problems in Classical Mechanics*. Oxford University Press, Oxford, UK, 2010.

[13] A. Deriglazov. *Classical Mechanics: Hamiltonian and Lagrangian Formalism*. Springer, 2010.

[14] M. H. Emam. *Covariant Physics: From Classical Mechanics to General Relativity and Beyond*. Springer, 2011.

[15] Richard P. Feynman, Robert B. Leighton, and Matthew Sands. *The Feynman Lectures on Physics, Volume 1*. Addison-Wesley, Reading, MA, USA, 1964.

[16] J. M. Finn. *Classical Mechanics.* Springer, 2008.

[17] B. Finzi and P. Udeschini. *Esercizi di Meccanica Razionale.* Elsevier, 4th edition, 1986.

[18] R. Fitzpatrick. *Classical Mechanics.* Taylor & Francis, 2011.

[19] G. R. Fowles and G. L. Cassiday. *Analytical Mechanics.* Thomson Brooks/Cole, 7th edition, 2005.

[20] A. P. French. *Special Relativity.* W. W. Norton & Company, New York, NY, USA, 1968.

[21] G. Gallavotti. *The Elements of Mechanics.* Springer, 1983.

[22] F. Gantmacher. *Lectures in Analytical Mechanics.* Mir Publishers, Moscow, Russia, 1975.

[23] C. Gignoux and B. Silvestre-Brac. *Solved Problems in Lagrangian and Hamiltonian Mechanics.* Oxford University Press, Oxford, UK, 2009.

[24] H. Goldstein. *Classical Mechanics, 2nd Edition.* Addison-Wesley, Reading, MA, USA, 1980.

[25] H. Goldstein, C. Poole, and J. Safko. *Classical Mechanics, 3rd Edition.* Addison-Wesley, San Francisco, CA, USA, 2002.

[26] D. T. Greenwood. *Classical Dynamics.* Dover Publications, Mineola, NY, USA, 1977.

[27] R. D. Gregory. *Classical Mechanics.* Cambridge University Press, Cambridge, UK, 2006.

[28] W. Greiner. *Classical Mechanics: Point Particles and Relativity.* Springer, 2003.

[29] W. Greiner. *Classical Mechanics: Systems of Particles and Hamiltonian Dynamics.* Springer, 2003.

[30] L. N. Hand and J. Finch. *Analytical Mechanics.* Cambridge University Press, Cambridge, UK, 1998.

[31] T. M. Helliwell and V. V. Sahakian. *Modern Classical Mechanics.* Cambridge University Press, Cambridge, UK, 2019.

[32] O. D. Johns. *Analytical Mechanics for Relativity and Quantum Mechanics.* Oxford University Press, Oxford, UK, 2nd edition, 2011.

[33] J. V. José and E. J. Saletan. *Classical Dynamics: A Contemporary Approach.* Cambridge University Press, Cambridge, UK, 1998.

[34] T. W. B. Kibble and F. H. Berkshire. *Classical Mechanics.* Imperial College Press, London, UK, 5th edition, 2004.

[35] G. L. Kotkin and V. G. Serbo. *Exploring Classical Mechanics: A Collection of 350+ Solved Problems for Students, Lecturers, and Researchers*. Springer, 2nd revised edition, 2020.

[36] Cornelius Lanczos. *The Variational Principles of Mechanics*. Dover Publications, Mineola, NY, USA, 1986.

[37] L. D. Landau and E. M. Lifshitz. *Course of Theoretical Physics, Volume 1: Mechanics*. Butterworth-Heinemann, Oxford, UK, 3rd edition, 1976.

[38] R. B. Lindsay and H. Margenau. *Foundations of Physics*. Dover Publications, New York, NY, USA, 1936.

[39] J. H. Lowenstein. *Essentials of Hamiltonian Dynamics*. Cambridge University Press, Cambridge, UK, 2012.

[40] A. Malthe-Sorenssen. *Elementary Mechanics Using Python: A Modern Course Combining Analytical and Numerical Techniques*. Springer, 2015.

[41] L. Meirovitch. *Methods of Analytical Dynamics*. Dover Publications, Mineola, NY, USA, 2003.

[42] D. Morin. *Introduction to Classical Mechanics*. Cambridge University Press, Cambridge, UK, 2008.

[43] W. F. Osgood. *Mechanics*. Dover Publications, New York, NY, USA, 1937.

[44] A. Rañada. *Dinámica Clásica*. Editorial Reverté, Barcelona, Spain, 1994.

[45] W. Rindler. *Introduction to Special Relativity*. Clarendon Press, Oxford, UK, 2nd edition, 1991.

[46] W. Rindler. *Relativity: Special, General, and Cosmological*. Oxford University Press, Oxford, UK, 2nd edition, 2006.

[47] F. Scheck. *Mechanics: From Newton's Laws to Deterministic Chaos*. Springer, 2010.

[48] A. Sommerfeld. *Mechanics*. Academic Press, New York, NY, USA, 1964.

[49] M. Spivak. *Elementary Mechanics from a Mathematician's Viewpoint*. Publish or Perish, Houston, TX, USA, 1990.

[50] D. Strauch. *Classical Mechanics: An Introduction*. Springer, 2009.

[51] G. J. Sussman, J. Wisdom, and M. E. Mayer. *Structure and Interpretation of Classical Mechanics*. MIT Press, Cambridge, MA, USA, 2001.

[52] K. R. Symon. *Mechanics*. Addison-Wesley, Reading, MA, USA, 3rd edition, 1971.

[53] J. L. Synge and B. A. Griffith. *Principles of Mechanics*. McGraw-Hill, New York, NY, USA, 2nd edition, 1959.

[54] E. F. Taylor and J. A. Wheeler. *Spacetime Physics*. W. H. Freeman and Company, New York, NY, USA, 2nd edition, 1992.

[55] J. R. Taylor. *Classical Mechanics*. University Science Books, Sausalito, CA, USA, 2005.

[56] W. Thirring. *A Course in Mathematical Physics: Classical Dynamical Systems and Classical Field Theory*. Springer, 1981.

[57] S. T. Thornton and J. B. Marion. *Classical Dynamics of Particles and Systems*. Cengage Learning, Belmont, CA, USA, 5th edition, 2003.

[58] J. D. Walecka. *Introduction to Classical Mechanics*. World Scientific, Singapore, 2020.

[59] A. G. Webster. *The Dynamics of Particles and of Rigid, Elastic and Fluid Bodies*. Macmillan, London, UK, 1930.

[60] E. T. Whittaker. *A Treatise on the Analytical Dynamics of Particles and Rigid Bodies*. Cambridge University Press, Cambridge, UK, 4th edition, 1937.

Index

For Product Safety Concerns and Information please contact our EU
representative GPSR@taylorandfrancis.com
Taylor & Francis Verlag GmbH, Kaufingerstraße 24, 80331 München, Germany

www.ingramcontent.com/pod-product-compliance
Lightning Source LLC
Chambersburg PA
CBHW052117230326
41598CB00080B/3789
* 9 7 8 1 0 3 2 9 8 7 9 7 2 *